Sustainable Chemistry and Pioneering Green Engineering Solutions

Shrikaant Kulkarni
Sanjivani University, India & Victorian Institute of Technology, Australia

Hemantkumar Akolkar
Rayat Shikshan Sanstha's Abasaheb Marathe Arts and New Commerce Science College, India

IGI Global
Scientific Publishing
Publishing Tomorrow's Research Today

Vice President of Editorial Melissa Wagner
Director of Acquisitions Mikaela Felty
Director of Book Development Jocelynn Hessler
Production Manager Mike Brehm
Cover Design Jose Rosado

Published in the United States of America by
IGI Global Scientific Publishing
701 East Chocolate Avenue
Hershey, PA, 17033, USA
Tel: 717-533-8845
Fax: 717-533-7115
Website: https://www.igi-global.com E-mail: cust@igi-global.com

Library of Congress Cataloging-in-Publication Data

Names: Kulkarni, Shrikaant editor | Akolkar, Hemantkumar, 1986- editor
Title: Sustainable chemistry and pioneering green engineering solutions / edited by Shrikaant Kulkarni, Hemantkumar Akolkar.
Description: Hershey, PA : IGI Global Scientific Publishing, [2026] | Includes bibliographical references and index. | Summary: "This book delves into the critical role of chemistry in addressing global sustainability challenges within engineering disciplines. The book bridges fundamental chemical principles with engineering practices to develop sustainable solutions for energy, materials, and environmental systems"-- Provided by publisher.
Identifiers: LCCN 2025003940 (print) | LCCN 2025003941 (ebook) | ISBN 9798337314099 h/c | ISBN 9798337314105 s/c | ISBN 9798337314112 ebook
Subjects: LCSH: Sustainable engineering | Green chemistry--Industrial applications | Renewable energy sources--Industrial applications | Waste products--Environmental aspects
Classification: LCC TA163 .S855 2025 (print) | LCC TA163 (ebook)
LC record available at https://lccn.loc.gov/2025003940
LC ebook record available at https://lccn.loc.gov/2025003941

British Cataloguing in Publication Data
A Cataloguing in Publication record for this book is available from the British Library.

Table of Contents

Detailed Table of Contents

This chapter focused on the development of nano-scaled carbon-based materials based on simple, green approach using natural resources such as extracts of various plants or fruit/vegetable peel waste, tea, coffee, alovera, as well as citrus fruits- grape, lemons, oranges etc. for antioxidant, anti-bacterial and anti-cancer applications. The surface functionalization with the high level of polyphenolic compounds, that resist the body from free-radical damage, have been confirmed through FTIR, EDX analysis which can enhance their stability, impart bio-compatibility, anti-bacterial, anti-microbial and anti-cancer activity. The antioxidant activities have been investigated using DPPH free-radical scavenging assay. Furthermore, the bacterial activity of the synthesized nano-materials against various pathogenic organism (gram-positive and gram-negative bacterial strains) has been evaluated. Finally, dose-dependent anti-cancer activity in both in-vitro and in-vivo setting has established the anti-proliferative properties of the nano-materials against different cancer cell-lines.

Green nanomaterials refer to those materials synthesized using environmentally benign processes, renewable resources, and minimal energy consumption, thereby aligning with principles of green their individual components. When these materials are produced using eco-friendly approaches, they are termed green nanocomposites. Together, green nanomaterials and nanocomposites form a critical domain that bridges the gap between advanced material science and environmental responsibility.The driving force behind the rise of green nanotechnology is the pressing need to reduce environmental degradation caused by conventional synthesis and manufacturing methods. Traditional techniques often involve toxic chemicals, high energy consumption, and unsustainable resource utilization. In contrast, green nanotechnology emphasizes the use of non-toxic precursors, renewable feedstocks, and energy-efficient synthesis methods. The eco-friendly synthesis of these materials, such as using biomass-derived carbon sources, further reduces their environmental footprint.

The spiked environmental concerns associated with conventional plastics have rushed the evolution and adoption of eco-friendly substitutes like green polymers and biodegradable plastics. These materials arise from green energy resources such as plants, animals, and microbes, offering a viable option to minimize plastic pollution and carbon footprints. This chapter delves into the classification, properties, synthesis, and applications of green polymers and biodegradable plastics, emphasizing their benefits over traditional plastics. It also addresses challenges such as high production costs, mechanical properties, and degradation under different environmental conditions. Emphasis is placed on innovative technologies and future trends that aim at improving the efficiency, production, and scalability of these biopolymers. This chapter delivers a thorough exploration of the crucial role of green polymers and biodegradable plastics in promoting a sustainable environment and advancing a more inclusive circular economy.

Abstract: Hydrogen is widely regarded as a promising clean energy carrier, yet its effective storage presents significant challenges that hinder widespread adoption. This chapter provides an in-depth exploration of various hydrogen storage technologies, examining their advantages, limitations, and potential solutions. We focus on physical, chemical, and material-based storage methods, evaluating key factors such as energy efficiency, safety, and economic feasibility. Additionally, we highlight emerging advancements in novel materials, hybrid storage systems, and policy frameworks that could play a pivotal role in addressing current storage challenges. The chapter

Preface

The world is currently confronted with unprecedented environmental challenges, including pollution, resource depletion, and climate change. It is impossible to overestimate the contribution that engineering and chemistry provide to innovation and solutions as we work to build a more sustainable future. In order to tackle some of the most important problems of our day, this book, "Sustainable Chemistry and Pioneering Green Engineering Solutions," brings together state-of-the-art research and knowledge from all around the world.

From the creation of biodegradable plastics and green nanomaterials to the problems and solutions in hydrogen energy and storage, the book covers a lot of ground. It also looks at how waste management, water treatment, and sustainable chemistry and agriculture interact, showing how creative solutions have the power to change communities and industries.

This book's chapters, which are authored by top authorities in respective domains, provide an overview of the fascinating innovations and advancements influencing the direction of green engineering and sustainable chemistry. This book offers a thorough summary of the most recent findings and developments in the sector, covering everything from precision farming and the circular economy to the green synthesis of carbon-based nanomaterials.

CHAPTER OVERVIEW

Chapter 1: A Study on Challenges and Barriers in the Adoption of Hydrogen Energy

Suchismita Satapathy explores the obstacles hindering hydrogen energy deployment, highlighting technological and economic challenges. Using the SWARA method, this chapter prioritizes key issues that must be addressed to unlock hydrogen's full potential in a clean energy future.

Chapter 2: Air Pollution Control and Green Chemistry Solutions

Selvakumar P. and colleagues assess the sources and health impacts of air pollution, emphasizing particulate matter (PM2.5) and its implications. The chapter underscores the role of green chemistry in mitigating air pollutants and calls for integrated strategies to improve air quality and public health.

Chapter 3: Battery Chemistry Li-Ion Solid-State and Beyond

Himadri Das and co-authors provide a comparative analysis of lithium-ion and emerging battery chemistries. The discussion spans solid-state batteries, lithium-sulfur, and sodium-ion systems, with a focus on sustainability, safety, and scalability for future energy demands.

Chapter 4: Exploring the Chemistry of Energy Storage Systems

Krishna Kumar and Gulshan Kumar delve into the chemical foundations of advanced energy storage, including batteries, supercapacitors, and hydrogen systems. The chapter highlights innovations in materials and electrochemical design, aiming to meet the demands of renewable energy integration.

Chapter 5: Green Biosynthesis of Carbon-Based Nanomaterials

Amit Dutta discusses eco-friendly synthesis of carbon nanomaterials using plant and food waste. These materials exhibit promising antioxidant, antibacterial, and anticancer properties, presenting a sustainable approach to biomedical and environmental applications.

Chapter 6: Green Nanomaterials and Nanocomposites

Selvakumar P. and collaborators present the synthesis and applications of green nanomaterials and nanocomposites. Focusing on eco-friendly production methods, the chapter emphasizes renewable resources and reduced toxicity in nanotechnology.

Chapter 7: Green Polymers and Biodegradable Plastics

Prachi Saxena and C. Dogra examine the development, properties, and applications of green polymers and biodegradable plastics. The chapter addresses both the potential and the challenges of replacing conventional plastics with sustainable alternatives.

Chapter 8: Hydrogen Storage Challenges and Solutions

Himadri Das and team analyze different hydrogen storage technologies, discussing their efficiencies, limitations, and recent advancements. The chapter also explores policy frameworks and material innovations that could facilitate wider hydrogen adoption.

Chapter 9: Precision Agriculture Meets Sustainable Chemistry

Azmirul Hoque and co-authors explore how precision agriculture and sustainable chemistry intersect to enhance farming efficiency. Techniques like AI, blockchain, and biodegradable agrochemicals are examined for their role in reducing environmental impact and boosting productivity.

Chapter 10: Sustainable Chemistry and Pioneering Green Engineering Solutions

Shrikant Ghodse and colleagues showcase sustainable industrial chemical processes, especially for pharmaceuticals and agrochemicals. Case studies illustrate the shift toward solvent-free, catalytic, and biocatalytic methods aligned with green chemistry principles.

Chapter 11: Waste Management and Circular Economy

Ahass Sreenath and collaborators present a GIS-based approach to urban waste management, using Kozhikode as a case study. The chapter illustrates how spatial analysis can support circular economy practices and identify optimal waste disposal sites.

Chapter 12: Water Treatment and Sustainable Chemistry: A Review

Dattatraya Pansare and co-authors offer a comprehensive review of sustainable water treatment technologies. Emphasizing green adsorbents, nanomaterials, and circular economy practices, the chapter provides solutions for addressing water scarcity and pollution.

In order to build a more sustainable future, we hope that this book will encourage scholars, decision-makers, and business executives to collaborate. We can create novel answers to the most important problems facing the globe and create a better world for future generations by utilising the power of chemistry and engineering.

Dr. Shrikaant Kulkarni
Dr. Hemantkumar N. Akolkar

Chapter 1
A Study on Challenges and Barriers in the Adoption of Hydrogen Energy

Suchismita Satapathy
https://orcid.org/0000-0002-4805-1793
KIIT University, India

ABSTRACT

The Future of Hydrogen, provides a clear and fair look at hydrogen. It covers current conditions, possible uses, and how hydrogen could help build a safe, affordable, and sustainable energy system. Several technological and economic factors influence the cost of making hydrogen from natural gas. Overcoming more barriers and lowering costs will need bold, immediate steps. To do this, a detailed review of existing studies has been done. This review identifies the main challenges and ranks them using the MARCOS(Measurement of Alternatives and Ranking According to the Compromise Solution) method.

INTRODUCTION

Energy use worldwide rises with population growth and development. Almost everyone depends on fossil fuels to meet energy needs. While fossil fuels are essential for power, they are also highly controversial because of their environmental harm. Burning coal, oil, and natural gas releases large amounts of carbon dioxide and other greenhouse gases. These gases are major causes of global warming and climate change. Extracting, processing, and transporting fossil fuels can lead to

DOI: 10.4018/979-8-3373-1409-9.ch001

deforestation, habitat destruction, and pollution of soil and water. Hydrogen energy is gaining popularity over fossil fuels because it is better for the environment, more efficient, and more sustainable. It also benefits from being more developed in technology. Still, many people are not familiar with hydrogen fuel. To address this, rules and laws need to be made. These regulations should aim to stop pollution caused by fossil fuels and protect the environment.

BACKGROUND

Nanda et al. (2021) explain that hydrogen energy has gained a lot of attention recently. It is a renewable solution that can replace fossil fuels (Singh et al. 2022). Using fossil fuels like oil, coal, and natural gas harms the environment and intensifies global warming (Ali et al. 2024). Hydrogen has a high energy density; it stores about seven times more energy per weight than fossil fuels (Tarhan & Çil 2021). Yang et al. (2009) argue that hydrogen will be a key energy source in the future, but some problems still need fixing. Once hydrogen is made, transporting and storing it can be difficult, as Murray (2018) points out. Baykara (2018) claims hydrogen energy is friendly to the environment and climate. In the future, hydrogen could replace oil in cars, planes, trains, and ships (Farrell et al. 2003). The European Union predicts this will happen. The High-Level Group for Hydrogen (2003) forecasted that by 2040, 35% of vehicles in Europe will run on hydrogen.

RESEARCH METHODOLOGY

A thorough review of existing studies on hydrogen energy was carried out to identify its challenges. A questionnaire with 17 questions was created and shared with experts from academia and industry. Respondents were asked to rate each statement on a scale of 1 to 3, where 1 means disagree, 2 indicates no opinion, and 3 stands for agreement. Based on their responses, the main barriers to adopting hydrogen fuel were identified. The MARCOS method was used to rank these challenges in order of importance. Table 1 lists the key obstacles faced when adopting hydrogen energy.

Table 1. Obstacles of Adoption Challenges of Hydrogen energy

Sl no	Question	1	2	3
1.	ProdProducing energy through hydrogen is expensive.. ucing energy through hydrogen is expensive			
2.	It needs special facilities for making, storing, moving, and dispensing.			
3.	Renewable energy sources are essential but require large amounts of them.			
4.	Storing and transporting hydrogen presents many problems			
5.	Safety is a concern during production, storage, and transit			
6.	Technology must improve to make hydrogen production more efficient.			
7.	Meeting big energy needs calls for a lot of investment and scaling up.			
8.	Without strong policies and clear rules, funding and spreading hydrogen projects become harder.			
9.	Hydrogen has to compete with established fuels like coal and oil.			
10.	Public understanding and acceptance are still low, and more awareness campaigns are needed			
11.	Clear policies are critical for building hydrogen infrastructure, funding research, and encouraging adoption.			
12.	People need to see how hydrogen can benefit the environment			
13.	Better ways to store and move hydrogen are needed to lower costs.			
14.	Safety worries and doubts about regulations impact public support			
15.	Current methods like compression and liquefaction use a lot of energy and cost a lot			
16.	Building the right infrastructure takes time and money, which makes hydrogen expensive			
17.	Costs need to drop a lot to make hydrogen compete with traditional fuels.			

RESULTS & DISCUSSION

New applications for hydrogen, as well as efforts to provide clean hydrogen and build related infrastructure, are currently at their most risky stage in the adoption process. Small, targeted loans, guarantees, and other financial tools can support private companies in making investments, gaining experience, and sharing both risks and potential benefits. The MARCOS multi-criteria decision-making method has been used to analyze these challenges and prioritize factors affecting hydrogen energy adoption. The analysis results, shown in Table 2, highlight the key factors based on a factor analysis. Items with a Cronbach's alpha value above 0.5 were grouped into four main categories, including production costs, infrastructure growth, energy use during production, storage and transport, and safety issues. Afterward, the TISM method was applied to examine how these factors relate to each other.

Table 2. Factor analysis

	Item	F1	F2	F3	F4	F5	F6	F7	F8	F9	F10
Cost of production	1	0.512									
	15	0.563									
	17	0.602									
Infrastructure development	2		0.516								
	16		0.572								
Storage and transportation	4				0.523						
	13				0.533						
Safety concerns	5					0.526					
	14					0.547					

Marcos is a new method that can be used in many different ways. It was developed by looking at both the best possible solutions and the worst. Next, the usefulness of each option is measured, and several utility functions are created based on the results from other tools. This helps to determine the different priorities and their ranking. Stević and Brković (2020) talked about how an international transport company manages its transport system. To use the Marcos method for decision making, follow these steps:

The steps for implementation of multi-criteria decision making according to the Marcos method are as follows:

Step 1: Develop decision matrix

$$X = \begin{bmatrix} x_{11} & x_{1j} & x_{1n} \\ x_{21} & \cdots & x_{2n} \\ x_{i1} & x_{ij} & \vdots \\ x_{mn} & \cdots & x_{mn} \end{bmatrix} \tag{1}$$

Where: m is the number of alternatives, n is the number of criteria x_{ij} is the value of criterion j at alternative i, with $1 \leq i \leq m$; $1 \leq j \leq n$.

Step 2: Develope extended decision matrix by considering an ideal alternative (AI) and an anti-ideal alternative (AAI).

$$X = \begin{matrix} AAI \\ A_1 \\ A_2 \\ \vdots \\ A_m \\ AI \end{matrix} \begin{bmatrix} x_{aa1} & \cdots & x_{aan} \\ x_{11} & \cdots & x_{1n} \\ x_{21} & \cdots & x_{2n} \\ \vdots & \vdots & \vdots \\ x_{m1} & \cdots & x_{mn} \\ x_{ai1} & \cdots & x_{ain} \end{bmatrix} \quad (2)$$

For max criteria:

AAI = min (x_{ij}); $i = 1, 2, \ldots, m$; $j = 1, 2, \ldots, n$

AI = max (x_{ij}); $i = 1, 2, \ldots, m$; $j = 1, 2, \ldots, n$

For min criteria:

AAI = max (x_{ij}); $i = 1, 2, \ldots, m$; $j = 1, 2, \ldots, n$

AI = min (x_{ij}); $i = 1, 2, \ldots, m$; $j = 1, 2, \ldots, n$

Step 3: Find Normalized matrix by both max and min criteria

$$n_{ij} = \frac{x_{ij}}{x_{AI}} \quad (3)$$

For min criteria:

$$n_{ij} = \frac{x_{AI}}{x_{ij}} \quad (4)$$

Step 4: Find the the weight of the criteria.

$$v_{ij} = n_{ij} \cdot w_j \quad (5)$$

Where, w_j is the weight of the criterion j.

Step 5: Calculate the coefficients K_i^+ and K_j^- are calculated according to the equation (6) and (7)

$$K_i^- = \frac{S_i}{S_{AAI}} \quad (6)$$

$$K_i^+ = \frac{S_i}{S_{AI}} \quad (7)$$

Step 6: calculate

$$f(K_i^-) = \frac{K_i^+}{K_i^+ + K_i^i} \quad (8)$$

$$f(K_i^+) = \frac{K_i^-}{K_i^+ + K_i^i} \quad (9)$$

Step 7: Find the MARCOS score

$$f(K_i) = \frac{K_i^+ + K_i^-}{1 + \frac{1 - f(K_i^+)}{f(K_i^+)} + \frac{1 - f(K_i^-)}{f(K_i^-)}} \tag{10}$$

Table. 3 to 7 explains the steps of Marcos method used to find Marcos score.

Table 3. Decision Matrix

	Max	Min	Min	Max
	0.3	0.2	0.2	0.3
Anti ideal	3	5	5	2
	Residential	Commercial	Industrial	Research & Development
Anti ideal	4	3	3	4
Cost of prod	4	5	3	4
Infrastructure	3	5	3	5
Storage transportation	5	5	5	2
Safety concern	3	1	3	5
Ideal	5	1	3	5

Table 4. Extended Decision Matrix

	Residential	**Commercial**	**Industrial**	**Research & Development**
Anti ideal	0.6	1.666666667	1.666666667	0.4
A	0.8	3	1	0.8
Cost of prod	0.8	5	1	0.8
Infrastructure	0.6	5	1	1
Storage transportation	1	5	1.666666667	0.4
Safety concern	0.6	1	1	1
Ideal	1	1	1	1

Table 5. Normalized Matrix

	Residential	Commercial	Industrial	Research & Development	
Anti ideal	0.18	0.333333333	0.333333333	0.12	Min
A	0.24	0.6	0.2	0.24	
Cost of prod	0.24	1	0.2	0.24	
Infrastructure	0.18	1	0.2	0.3	
Storage transportation	0.3	1	0.333333333	0.12	
Safety concern	0.18	0.2	0.2	0.3	
Ideal	0.3	0.2	0.2	0.3	Max

Table 6. Marcos score Calculation

	0.966666667	K-	K+	f(K-)	f(K+)	f(K)	Rank
Cost of prod	1.68	2.163090128	1.68	3.163090128	2.68	3.843090128	2nd
Infrastructure	1.68	2.135593219	1.68	3.135593219	2.68	3.815593219	3rd
Storage transportation	1.753333333	2.228813558	1.753333333	3.228813558	2.753333333	3.982146892	ist
Safety concern	0.88	1.118644067	0.88	2.118644067	1.88	1.998644067	4th
Ideal	1						

The MARCOS Analysis shows Storage and transportation is the main barrier, ranked first followed by cost of production 2nd and infrastructure third and safety concern 4th barrier. Hence steps must be taken to mitigate these barriers.

Then TISM(total interpretive structural modelling) is used to find interrelationship between the nine barriers.

Interpretive Structural Modelling (ISM) builds on Interpretive Structural Modelling (ISM). It not only finds and arranges relationships between variables but also explains what those links mean. It's a popular method for making decisions and understanding complex systems in research.

The process of ISM involves several steps:

First, identify the main variables or factors affecting the system. Use source reviews, expert interviews, or brainstorming sessions to do this.

Next, define how these variables relate to each other. For example, ask if one element influences or supports another.When i is the base to achieve the j factor, relation is V, Similarly, if j is the basis for factor I, the relation is A. For both ways, , relationship between I and the j factor is X and for no relation, O is written

Then, create a Structural Self-Interaction Matrix (SSIM). This matrix captures expert opinions on the direction of influence between pairs of variables. Symbols are used to show whether one influences the other, they influence each other, or if no influence exists.

The SSIM is converted into an initial reachability matrix by applying simple rules. For example, if the relationship is "V," the matrix becomes 1 for influence and 0 for no influence.The next step involves adding transitive links. If variable A influences B, and B influences C, then A likely influences C as well. These links are added to complete the reachability matrix.Afterward, the levels of variables are identified through level partitioning. This process groups variables based on the sets they can reach and those that can reach them.Using these levels, a hierarchical diagram (digraph) is created. Links go from lower to higher levels, showing influence flow.The digraph is then converted into an interpretive model. Each node is labeled with the variable's name, and expert opinions explain why one factor impacts another. This interpretive step makes TISM different from traditional ISM.Optionally, a MICMAC analysis can classify variables. They fall into groups: autonomous, dependent, linkage, or independent drivers, based on their influence and reliance on others.Finally, the model is checked with experts. They review the structure and explanations to ensure clarity and accuracy.

Table 7. SSIM Matrix

	0.966666667	**K-**	**K+**	**f(K-)**	**f(K+)**	**f(K)**	**Rank**
Cost of prod	1.68	2.163090128	1.68	3.163090128	2.68	3.843090128	2nd
Infrastructure	1.68	2.135593219	1.68	3.135593219	2.68	3.815593219	3rd
Storage transportation	1.753333333	2.228813558	1.753333333	3.228813558	2.753333333	3.982146892	ist
Safety concern	0.88	1.118644067	0.88	2.118644067	1.88	1.998644067	4th
Ideal	1						

Table 8. Initial reach-ability Matrix

	C9	**C8**	**C7**	**C6**	**C5**	**C4**	**C3**	**C2**	**C1**
C1	V	V	O	O	V	V	A	V	A
C2	V	V	A	O	V	V	V	V	
C3	V	O	O	V	O	V	V		
C4	O	V	O	A	O	V			
C5	O	O	O	V	O				
C6	V	V	V	O					

continued on following page

Table 8. Continued

	C9	C8	C7	C6	C5	C4	C3	C2	C1
C7	O	O	V						
C8	V	V							
C9	A								
C10									

Table 9. Final reach-ability matrix

	C1	C2	C3	C4	C5	C6	C7	C8	C9
C1	0	0	0	1	1	1	1	1	1
C2	0	1	1	1	1	1	1	1	1
C3	1	1	1	1	1	1	1	1	1
C4	1	0	0	1	1	1	1	1	1
C5	1	0	0	1	1	1	0	1	0
C6	0	1	0	0	1	0	1	1	1
C7	1	1	0	1	1	1	1	1	1
C8	0	1	0	1	0	1	0	1	1
C9	0	1	0	1	0	1	1	0	1

Table 10. Level Partition

	C1	C2	C3	C4	C5	C6	C7	C8	C9	Drive power
C1	0	0	0	1	1	1	1	1	1	6
C2	0	1	1	1	1	1	1	1	1	8
C3	1	1	1	1	1	1	1	1	1	10
C4	1	0	0	1	1	1	1	1	1	7
C5	1	0	0	1	1	1	0	1	0	5
C6	0	1	0	1	1	0	1	1	1	5
C7	1	1	0	1	1	1	1	1	1	8
C8	0	1	0	1	0	1	0	1	1	5
C9	0	1	0	1	0	1	1	0	1	5
Dependence	4	6	2	9	7	8	7	8	8	

Table 11. Iteration-1

Challenges	Reach ability set	Antecedent Set	Intersection set	Level
C1	4,5,6,7,8,9	3,4,5,7	4,5,7	
C2	2,3,6,7,8,9	2,3,4,5,6,7,8,9	2,4,6,7,8,9	I
C3	1,2,3,4,5,6,7, 8, 9, 10	2,3	2,3	
C4	1,4,5,6,7,8,9	1,2,3,4,5,6,7,8,9	1,4,5,6,7,8,9	I
C5	1,4,5,6,8	1,2,3,4,5,6,7	1,4,5,6	
C6	2,4,5,7,8,9	1,2,3,4,5,7,8,9	2,4,5,7,8,9	I
C7	1,2,4,5,6,7,8,9	1,2,3,4,6,7,9	1,2,4,6,7,9	
C8	2,4,6,8,9	1,2,3,4,5,6,7,8	2,4,6,8	
C9	2,4,6,7,9	1,2,3,4,6,7,8,9	2,4,6,7,9	I

Table 12. Iteration 2

Challenges	Reach ability set	Antecedent Set	Intersection set	Level
C1	5,7,8	3,5,7	5,7	
C3	1,3,5,7, 8	3	3	
C5	1,5,8	1,3,5,7	1,5	
C7	1,5,7,8,	1,3,7	1,7	
C8	8	1,3,5,7,8	8	II

Table 13. Iteration 3

Challenges	Reach ability set	Antecedent Set	Intersection set	Level
C1	5,7	3,5,7	5,7	III
C3	1,3,5,7	3	3	
C5	1,5	1,3,5,7	1,5	III
C7	1,5,7	1,3,7	1,7	

Table 14. Iteration 4

Challenges	Reach ability set	Antecedent Set	Intersection set	Level
C3	3,7	3	3	
C7	7	3,7	7	IV

Figure 1. ISM Model

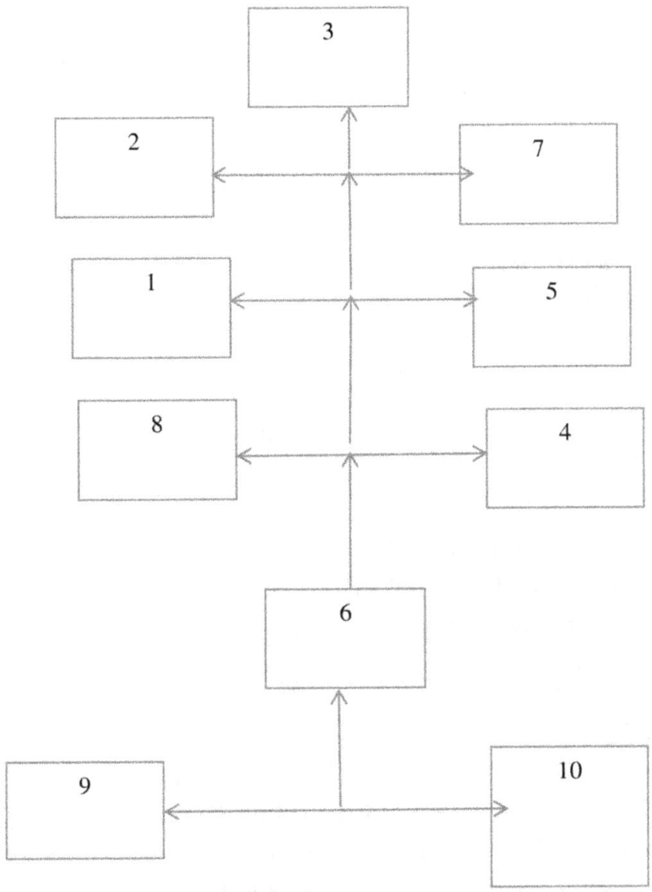

Figure 2. MicMAC Analysis

										Drive power
9							c7			6
8		c3						c8		8
7								c6		9
6									c9	7
5				c1		c2		c5	c4	5
4										5
3										8
2										5
1										5
Dependence	4	6	2	9	7	8	7	8	8	
	1	2	3	4	5	6	7	8	9	

From figure.1 and 2 it is clear that C1 is the key variable, c2, c4, c5,c6,c7,c8, c9 are having high dependency and low drive power. Where as c3 is having high drive-power and low dependency.

CONCLUSION

Hydrogen is seen as the energy of the future because it is clean and produces almost no carbon emissions. Right now, its uses are limited because the infrastructure is lacking and costs are high. As more areas begin to use hydrogen and production and storage methods improve, prices will fall. From making hydrogen to storing it, every step is important. Since most hydrogen is produced from fossil fuels, which release carbon dioxide, this method should be phased out. Instead, hydrogen should be made using electricity from renewable sources like solar, wind, or hydroelectric power. Producing hydrogen this way will be cheaper in the coming years, making it a sustainable and clean energy source. Its role in transportation and power generation shows how important hydrogen can become. It could be used widely to fuel vehicles in the future. Since cars currently mostly run on oil, demand for oil may decrease while demand for hydrogen increases. This shows how big the hydrogen economy could grow. Hydrogen isn't found freely in nature and must be produced from molecules such as water or methane. It's not a source of energy on its own but needs energy to be created. About half of all hydrogen made goes into making nitrogen fertilizer through the Haber process. Around a quarter is used to convert low-grade oil, especially from sands. It's already an important chemical in making liquid fuels for transportation. While it's not as simple as using natural gas, we do have a lot of experience working with hydrogen on a large scale. Right now, most hydrogen is made by breaking down natural gas or coal, both producing carbon dioxide. Future growth depends on producing hydrogen without emissions. Many plans focus on using electricity from renewable energy sources for electrolysis. Sometimes, excess power from nuclear plants can be used too. One promising method is using high-temperature nuclear reactors to split water with heat in thermochemical processes. The growing need for hydrogen from refineries and factories favors cheaper ways to produce it. Because hydrogen pipelines are limited, production sites may be far from where it is needed. Renewable energy sources like solar and wind are unpredictable. Power output can vary, causing dips and surges in the grid. To balance this, large energy storage systems are used. Today, most store excess renewable power in big batteries for short times. But batteries become very expensive over longer periods. This increases the cost of electricity. Hydrogen offers a way to store large amounts of energy for longer times. Overcoming the hurdles in hydrogen energy is key to its growth.

DECLARATIONS

Author contributions

Dr. S. Satapathy made a major contribution in writing, analyzing the manuscript.

Funding

There was no funding for carrying out this research.

COMPLIANCE WITH ETHICAL STANDARDS

Conflict of interest

The authors declare of having no conflict of interest.

Ethical approval

All studies were conducted in accordance with principles for human experimentation as defined in the declaration of Helsinki and its later amendments or comparable ethical standards.

Informed consent

Informed consent after explaining the nature of investigation was obtained from each participant in this study.

Animal Studies

No animal studies were carried out by the authors for this article.

Data and Code Availability

Data and code are available and can be sent as per the request

Consent for publication

We, the authors of this paper,provide our consent for the publication of identifiable details, which can include photograph(s) and/or videos and/or case history and/or details within the text ("Material") to be published in the above Journal and Article.

REFERENCES

Baykara, S. Z. (2018). Hydrogen: A brief overview on its sources, production and environmental impact. *International Journal of Hydrogen Energy*, *43*(23), 10605–10614. https://sdg.iisd.org/news/world-population-to-reach-9-9-billion-by-2050/. DOI: 10.1016/j.ijhydene.2018.02.022

Ewan, B. C. R., & Allen, R. W. K. (2005). A figure of merit assessment of the routes to hydrogen. *International Journal of Hydrogen Energy*, *30*(8), 809–819. DOI: 10.1016/j.ijhydene.2005.02.003

Farrell, A. E., Keith, D. W., & Corbett, J. J. (2003). A strategy for introducing hydrogen into transportation. *Energy Policy*, *31*(13), 1357–1367. DOI: 10.1016/S0301-4215(02)00195-7

Murray, L. J., Dincă, M., & Long, J. R. (2009). Hydrogen storage in metal–organic frameworks. *Chemical Society Reviews*, *38*(5), 1294–1314. DOI: 10.1039/b802256a PMID: 19384439

Singh, R., Altaee, A., & Gautam, S. (2020). Nanomaterials in the advancement of hydrogen energy storage. *Heliyon*, *6*(7), e04487. DOI: 10.1016/j.heliyon.2020.e04487 PMID: 32743097

World Health Organization. (2018). COP24 special report: health and climate change. https://www.energy.gov/eere/articles/hydrogen-clean-flexible-energy-carrier [Access 05/11/2020].

Yang, J., Sudik, A., Wolverton, C., & Siegel, D. J. (2010). High capacity hydrogen storage materials: Attributes for automotive applications and techniques for materials discovery. *Chemical Society Reviews*, *39*(2), 656–675. Retrieved May 11, 2020, from https://www.energy.gov/eere/articles/hydrogen-clean-flexible-energy-carrier. DOI: 10.1039/B802882F PMID: 20111786

Chapter 2
Air Pollution Control and Green Chemistry Solutions

P. Selvakumar
https://orcid.org/0000-0002-3650-4548
Nehru Institute of Technology, Coimbatore, India

Nikitha Sreekanthaswamy
https://orcid.org/0009-0009-2627-9365
Dayananda Sagar Business Academy, India

Anubhav
https://orcid.org/0009-0009-2707-8223
Chandigarh Group of Colleges, Jhanjeri, India

Santosh Kumar Nathsharma
https://orcid.org/0000-0003-3935-2938
Stewart Science College, India

Satyajit M. Deshmukh
https://orcid.org/0009-0000-8551-8113
Datta Meghe College of Engineering, Navi Mumbai, India

A. Revathi
https://orcid.org/0000-0003-4387-2442
Kongu Engineering College, India

T. C. Manjunath
https://orcid.org/0000-0003-2545-9160
Rajarajeswari College of Engineering, India

ABSTRACT

These substances, which include particulate matter (PM), nitrogen oxides (NOx), sulfur dioxide (SO2), carbon monoxide (CO), volatile organic compounds (VOCs), and ozone (O3), are primarily produced by human activities such as industrial processes,especially in urban areas, poses a growing concern for public health, ecosystem stability, and the overall well-being of the planet. is well-documented, with a substantial body of scientific evidence linking exposure to various air pollutants to a wide range of chronic and acute health conditions. Particulate matter, particularly

DOI: 10.4018/979-8-3373-1409-9.ch002

fine particles with a diameter of 2.5 micrometers or smaller (PM2.5), is one of the most dangerous forms of air pollution. These tiny particles can penetrate deep into the lungs and enter the bloodstream, leading to respiratory diseases such as asthma, chronic obstructive pulmonary disease (COPD), and lung cancer. governments and health organizations worldwide. Beyond respiratory and cardiovascular health, air pollution has a profound effect on the environment.,

INTRODUCTION TO AIR POLLUTION AND ITS IMPACT ON HUMAN HEALTH AND THE ENVIRONMENT

These substances, which include particulate matter (PM), nitrogen oxides (NOx), sulfur dioxide (SO2), carbon monoxide (CO), volatile organic compounds (VOCs), and ozone (O3), are primarily produced by human activities such as industrial processes,especially in urban areas, poses a growing concern for public health, ecosystem stability, and the overall well-being of the planet. is well-documented, with a substantial body of scientific evidence linking exposure to various air pollutants to a wide range of chronic and acute health conditions. Particulate matter, particularly fine particles with a diameter of 2.5 micrometers or smaller (PM2.5), is one of the most dangerous forms of air pollution. These tiny particles can penetrate deep into the lungs and enter the bloodstream, leading to respiratory diseases such as asthma, chronic obstructive pulmonary disease (COPD), and lung cancer. governments and health organizations worldwide.

Beyond respiratory and cardiovascular health, air pollution has a profound effect on the environment., droughts, and heatwaves, which in turn have far-reaching consequences for ecosystems, agriculture, and human settlements.Another significant environmental Additionally, air pollution can have a detrimental effect on soil quality, contaminating it with heavy metals and chemicals that can harm plant life and reduce agricultural productivity. The electrification of transportation, along with the promotion of public transport, cycling, and walking, can also help reduce emissions from vehicles, one of the primary sources of air pollution in urban areas. In addition,, and encouraging sustainable agricultural practices can significantly lower emissions and improve air quality.On the regulatory side, governments must enact and enforce strict air quality standards, monitor pollution levels, and International cooperation is also essential in addressing transboundary pollution and sharing knowledge and technology to mitigate air pollution on a global scale. programs can help individuals understand the sources and impacts of air pollution and encourage more sustainable lifestyles.In conclusion, highlights the need for immediate and sustained action to address its causes and mitigate its effects. The reduction of air pollution requires concerted efforts from governments, the challenge of air pollution will only intensify, making it imperative for society to prioritize efforts that ensure cleaner air, healthier populations, and a more sustainable planet for future generations.

FUNDAMENTALS OF AIR POLLUTION CONTROL: TECHNOLOGIES AND STRATEGIES

Understanding the fundamental principles of air pollution control, the technologies employed, and the strategies for their implementation is key to addressing the growing challenges of begins with the identification and classification of pollutants. Pollutants can be broadly categorized into two types: primary and secondary pollutants. Primary pollutants are those directly emitted into the atmosphere from sources such as factories, Secondary pollutants, on the other hand,. Ozone (O3),(Al-Thani et al., 2018) which forms through the photochemical reaction of VOCs and NOx in the presence of sunlight, is one of the most common secondary pollutants. The basic harmful pollutants in the atmosphere to levels that are safe for human health and the emission control technologies, and regulatory frameworks. Source control, also known as pollution prevention, aims to reduce or eliminate the generation of pollutants at their source. This approach is often more cost-effective and environmentally friendly than attempting to remove pollutants after they have been emitted. and the adoption of cleaner transportation options.Emission control technologies, which are designed to reduce the concentration of pollutants in exhaust gases before they are released control. These technologies are applied at various points in the pollution source chain, from industrial facilities to mobile sources like vehicles. There are several categories of air pollution control technologies, each suited to different types of pollutants and sources., including:Electrostatic Precipitators (ESPs): These devices use electric fields exhaust gas, causing them to be attracted to oppositely charged collection plates. ESPs are highly effective at removing fine particulate matter and are commonly used in industries like power generation and cement production.Fabric Filters (Baghouses): Baghouses are large filtration systems that use fabric bags to trap particles as exhaust gases pass through them., including steel manufacturing and chemical processing.Cyclones: . While less efficient than ESPs and fabric filters for fine particulate matter, they are relatively inexpensive and are used in applications with lower particulate concentrations.(Anderson et al., 2012). Strategies for controlling ozone include:Emission Reductions of NOx and VOCs: Regulations aimed at reducing emissions from mobile and stationary in controlling ozone formation. Technologies like catalytic converters in vehicles and low-NOx burners in industrial furnaces are commonly used to reduce NOx emissions.

Photocatalysis: In some cases, photocatalysis using titanium dioxide (TiO2) can help break down VOCs and other ozone precursors. This process is used in air purifiers and certain building materials, which can help reduce indoor air pollution.. Mobile Source Control: Mobile sources, such as automobiles, trucks, and airplanes, are major contributors to air pollution, particularly in urban areas. Control strategies for mobile sources include:Engine Emission Controls: from internal combustion engines. These devices help remove pollutants such as NOx, CO, and particulate matter from exhaust gases.Alternative Fuels: vehicles, hydrogen fuel cells, and bio-fuels, can significantly reduce the environmental impact of transportation. Electric vehicles, in particular, are seen Public Transportation and Active Mobility: Promoting related air pollution.In addition to these technological approaches, air pollution control strategies must also be supported by regulatory frameworks and policies that set air quality standards, monitor pollutant levels, and enforce compliance. the actions taken by industries and governments to reduce emissions. Monitoring networks that track air pollution levels provide essential data for policymakers, Public awareness and community engagement are also help reduce behaviors that contribute to pollution, such as excessive energy consumption, waste generation, and reliance on fossil-fuel-powered vehicles. Furthermore, citizens can play an active role in advocating for stronger air quality regulations and participating in community efforts to reduce pollution.In conclusion, air pollution control(Arunachalam et al., 2015) health, requiring a combination of advanced technologies, regulatory measures, and societal engagement. From the implementation of pollution prevention strategies to the deployment of state-of-the-art emission control systems, a variety of solutions are available to mitigate the harmful effects of air pollution. However, effective air pollution

GREEN CHEMISTRY PRINCIPLES AND APPLICATIONS IN AIR POLLUTION CONTROL

The 12 principles of green chemistry, introduced by Paul Anastas and John Warner, provide a framework to guide the development of sustainable solutions across industries. These principles emphasize waste prevention, atom economy, safer solvents and auxiliaries, renewable feedstocks, energy efficiency, and the reduction of derivatives, among other guidelines. acid rain, respiratory illnesses, and global climate change. Traditional approaches to air pollution control often involve end-of-pipe treatments that remove or mitigate pollutants after their generation. While these methods are effective to some extent, they are reactive and may involve significant energy consumption,(Babar & Shareefdeen, 2014).The role of catalysts in reducing air pollution is another prominent example of green chemistry applications. Catalysts enable chemical reactions to proceed under milder conditions with reduced energy consumption and waste generation(Bailey & Solomon, 2004). MOFs, in particular, exhibit high surface area and tunable properties, making them effective for adsorption-based pollution control technologies. Moreover, research into biodegradable and bio-based materials for air filtration systems has reduced the environmental impact of traditional pollution control technologies.In urban environments, where, eliminate tailpipe emissions, contributing to improved air quality. Advances in battery technologies, such as the development of lithium-ion and solid-state batteries, align with green chemistry principles by using less toxic materials and enhancing energy efficiency.In holistic and sustainable control by focusing on pollution prevention, cleaner production methods, and innovative materials and technologies., reduce environmental impact, and improve air quality.(Bandyopadhyay et al., 2014) From cleaner fuels and advanced catalysts to sustainable materials and energy-efficient processes, green chemistry plays a pivotal role in addressing the challenges of air pollution while promoting a greener and healthier future. As society continues to prioritize sustainability, the integration of green chemistry into air pollution control strategies will remain essential for achieving environmental and economic goals.

PARTICULATE MATTER CONTROL: TECHNOLOGIES AND STRATEGIES FOR REDUCING PM EMISSIONS

Fine and ultrafine problems such as reduced visibility (haze), climate change through its role as a climate-forcing agent, and damage to ecosystems(Cariolet et al., 2018). Given the severity of its impacts, the control of PM emissions has become a critical component of air pollution management strategies worldwide. A combination of advanced technologies, innovative materials, and effective regulatory policies has been Sources of Particulate MatterPM emissions can originate volcanic eruptions, dust storms, sea spray, and biological particles such as pollen and spores. However, human activities are the dominant contributors to PM emissions, particularly in urban and industrial areas. Major anthropogenic sources include:Industrial Processes: Cement manufacturing, mining, metal processing, and chemical production release large amounts of particulate matter.. Residential Heating and Cooking: The burning of biomass, wood, and coal in households is a significant source of PM, especially in developing regions.Agricultural Activities: Tilling, harvesting, and the burning of agricultural waste generate PM emissions. Construction and Demolition: Dust generated during construction, road paving, and demolition contributes to PM pollution.Given the diverse nature of PM sources, an integrated and multi-faceted approach is required to effectively control and reduce PM emissions Technologies for Particulate Matter Control Various technologies have been developed to reduce PM emissions at their sources or capture(Chae & Park, 2011) . ESPs use an electric field to charge PM in a gas stream, causing the charged particles to be attracted to and deposited on collector plates. Once collected, the particles are periodically removed. ESPs are highly efficient (with efficiencies exceeding 99%) for removing fine PM (PM2.5) and can handle large gas volumes. However, they require significant energy input and periodic maintenance to ensure optimal performance.

Recent advancements in ESP technology include the use of advanced materials for electrodes and improved electrical controls to enhance efficiency and reduce energy consumption. highly effective at capturing particulate matter, including ultrafine particles. In baghouses, polluted gas streams are passed through filter bags made of woven or non-woven fabrics, where PM is trapped on the fabric surface while clean air passes through. Fabric filters can achieve removal efficiencies exceeding 99.9% and are widely used in cement plants, chemical manufacturing,(Chao, 2008) and metal processing industries. Modern developments in filter materials, such as PTFE (polytetrafluoroethylene) membranes and nanofiber coatings, have improved filter durability, efficiency, and resistance to high temperatures and corrosive gases.Wet scrubbers use liquid (typically water) to capture and remove PM from gas streams through impaction, interception, and absorption. Scrubbers are particularly effective for capturing PM and soluble gases simultaneously, making them suitable for industrial processes with mixed emissions. However, wet scrubbers generate wastewater that requires treatment, and they may not be as effective for ultrafine PM compared to ESPs or baghouses. Advanced scrubber designs, such as Venturi scrubbers, have been developed to improve particle capture efficiency, especially for smaller PM.Hybrid systems combine two or more control technologies to enhance PM removal efficiency. For example, hybrid ESP-baghouse systems combine the advantages of ESPs (pre-collection of larger particles) with baghouses (fine PM capture), resulting in overall improved performance. Hybrid systems are increasingly being adopted in power plants and heavy industries to meet stringent emission standards.Control Strategies for Reducing PM Emissions(Dhupia et al., 2011)In addition to end-of-pipe technologies, a range of strategies can be employed to minimize PM emissions at the source. These strategies include process optimization, the adoption of cleaner fuels, and the implementation of regulatory policies. Transitioning from high-emission fuels, such as coal and biomass, to cleaner alternatives like natural gas, biofuels, and hydrogen can significantly reduce PM emissions. Combustion optimization techniques, including the use of advanced burners and oxygen-enriched combustion, improve fuel efficiency and minimize incomplete combustion.Industries can implement cleaner production processes, such as substituting raw materials that generate fewer particulates and optimizing equipment operation. For example, the use of pre-calcination technology in cement kilns reduces PM emissions by improving combustion efficiency.Governments and international organizations have established emission standards and regulations to control PM pollution.

VOLATILE ORGANIC COMPOUND (VOC) CONTROL: TECHNOLOGIES AND STRATEGIES FOR REDUCING VOC EMISSIONS

These pollutants pose serious risks to human health, causing respiratory problems, eye irritation, neurological disorders, and even cancer with prolonged exposure. Moreover, VOCs impact the environment by degrading air quality and contributing to climate change as greenhouse gases or precursors to aerosols. Common sources of VOC emissions include industrial processes (e.g., chemical manufacturing, petrochemical refineries), vehicular emissions, solvent usage (paints, coatings, adhesives), fuel storage, and biomass burning. Given their environmental and health hazards, controlling VOC emissions has become a priority worldwide, leading to the development of advanced technologies and strategies for VOC abatement. Biogenic VOCs, such as isoprene and monoterpenes, are emitted from vegetation, accounting for a significant portion of global VOC emissions. However, anthropogenic sources, particularly in urban and industrialized regions, are the major contributors to harmful VOC concentrations. Key anthropogenic sources include:(Fajersztajn et al., 2013). Industrial Processes: Chemical manufacturing, petroleum refining, pharmaceutical production, and printing processes release VOCs as fugitive emissions or byproducts.. Solvent Use: VOCs are released during the use of paints, coatings, adhesives, and cleaning agents, which contain solvents that readily evaporate.. Fuel Storage and Handling: Gasoline storage, distribution, and refueling processes contribute to significant VOC emissions through evaporation and leaks. Agricultural Practices: Pesticides, herbicides, and fertilizers often contain VOCs that evaporate during application(Erisman & Schaap, 2004).. Residential Activities: Household products, such as air fresheners, aerosols, and cleaning chemicals, are also sources of indoor VOC emissions.The control of VOC emissions requires targeted approaches that address both large-scale industrial emissions and diffuse sources, such as transportation and consumer products.Technologies for VOC Control echnological advancements have enabled the development of effective VOC control systems that either destroy VOCs or capture them for recovery. These technologies are categorized into destruction-based methods (e.g., thermal and catalytic oxidation) and recovery-based methods (e.g., adsorption, absorption, and condensation).Thermal oxidation involves combusting VOC-laden air streams at high temperatures (typically 700-1,200°C) to convert VOCs into carbon dioxide (CO_2) and water vapor. This method is highly effective for destroying VOCs with efficiencies exceeding 99%. Thermal oxidizers are classified into three types:Direct Flame Thermal Oxidizers: VOCs are combusted in a simple combustion chamber using supplemental fuel. Regenerative Thermal Oxidizers (RTOs): RTOs use heat exchangers with ceramic beds to recover and reuse heat, making them more energy-efficient for treating large airflows with low VOC concentrations.(Fann et al., 2011)Recuperative Thermal Oxidizers: These systems recover heat using metallic heat exchangers, improving efficiency but with lower heat recovery than RTOs.Catalytic oxidation reduces the temperature required for VOC destruction (typically 250-500°C) by using catalysts such as platinum, palladium, or other metal oxides. LDAR systems use infrared cameras, ultrasonic detectors, and gas analyzers to detect leaks for timely repair. Cleaner Fuels and Transportation Governments worldwide have implemented stringent regulations to control VOC emissions. Examples include the Clean Air Act in the United States, which sets limits on VOC emissions, and the European Union's VOC Solvents Directive. Regulations incentivize industries to adopt low-VOC technologies and cleaner practices., such as paints, cleaning agents, and personal care items, helps reduce VOC emissions at the household level. Public awareness campaigns play a critical role in encouraging behavioral changes and sustainable choices.Conclusion The control and reduction of VOC emissions are essential for . A combination of advanced technologies, such as thermal and catalytic oxidation, adsorption, absorption, and biological treatments, enables effective VOC abatement across industrial and transportation sectors. Preventive strategies, including process optimization, cleaner fuels, and regulatory frameworks, further complement these technologies. (Friel et al., 2009)By integrating innovative solutions and promoting sustainable practices, significant progress can be made in reducing VOC emissions and fostering a cleaner, healthier environment.

NITROGEN OXIDES (NOX) AND SULFUR DIOXIDE (SO_2) CONTROL: TECHNOLOGIES AND STRATEGIES FOR REDUCING NOX AND SO_2 EMISSIONS

Diesel engines, in particular, are major contributors to NOx emissions.NOx emissions are generated during high-temperature processes such as cement manufacturing, glass production, and chemical synthesis.SO_2 (Giuliano et al., 2018a). Secondary temperatures, reducing NOx emissions.minimizing NOx generation.. By recirculating a portion of exhaust gas into the intake air, the combustion temperature is lowered, suppressing NOx formation.Technologies for Controlling SO_2 Emissions SO_2 control technologies primarily focus on removing sulfur from fuels before combustion or treating exhaust gases to eliminate SO_2. Pre-combustion desulfurization involves removing sulfur from fuels, such as coal and oil, before combustion. Coal can be washed to reduce its sulfur content, while oil undergoes hydrodesulfurization (HDS) to produce low-sulfur fuels.Flue Gas Desulfurization (FGD): FGD systems, also known as scrubbers, are the most effective and widely used post-combustion technology for SO_2 control. They can achieve removal efficiencies exceeding 95%. FGD technologies include:Wet FGD: This method uses an alkaline scrubbing solution (e.g., limestone or lime slurry) to react with SO_2, forming gypsum (calcium sulfate), which can be used in construction materials. Wet FGD is highly efficient but generates wastewater that requires treatment.Dry FGD: In dry scrubbing, a dry sorbent such as reacting with SO_2 to form solid byproducts that can be captured in a baghouse. Dry FGD systems are suitable for smaller installations with lower SO_2 concentrations.Spray Dry Scrubbers: These systems combine wet and dry methods, .In coastal power plants, seawater is used as a natural scrubbing medium to remove SO_2. The SO_2 reacts with seawater to form sulfate, which can be safely discharged back into the ocean after neutralization.Sorbent injection involves injecting powdered sorbents such as lime, sodium bicarbonate, or trona directly . The byproducts are collected in particulate control devices like baghouses. Advanced Oxidation Technologies:Emerging technologies, such as oxidation and catalytic processes, can convert SO_2 into sulfuric acid or sulfur, enabling its recovery and reuse in industrial applications.Strategies for Reducing NOx and SO_2 Emissions

GREEN CATALYSTS FOR AIR POLLUTION CONTROL: DESIGN, SYNTHESIS, AND APPLICATIONS

To mitigate these pollutants, catalytic technologies have emerged as a highly effective and widely adopted solution, particularly due to their ability to accelerate chemical reactions under mild conditions while minimizing energy consumption and harmful byproducts. In recent years, the development and application of green catalysts—catalysts that are environmentally benign, energy-efficient, and sustainable—have gained immense attention. Green catalysts offer an eco-friendly approach to air pollution control by improving process efficiency, reducing the use of hazardous chemicals, and enabling the conversion of pollutants into harmless or useful products. This includes their role in NOx reduction(Hou et al., 2015), VOC elimination, CO oxidation, and CO_2 capture or transformation. The design, synthesis, and application of green catalysts have evolved through innovative advancements in materials science, nanotechnology, and sustainable chemistry, paving the way for a cleaner and greener environment.Green Catalysts and Their CharacteristicsGreen catalysts are designed to adhere to emphasize the reduction of hazardous materials, energy efficiency, and waste minimization. Unlike traditional catalysts, green catalysts focus on the use of non-toxic, earth-abundant materials and renewable resources. Characteristics that define green catalysts include: High Activity and Selectivity: Green catalysts facilitate targeted chemical reactions, ensuring the effective conversion of pollutants with minimal side products. Stability and Reusability: Long-term stability and the ability to be reused without significant deactivation reduce costs and environmental impact. Eco-friendly Materials: Utilizing non-toxic, non-precious metals, and biocompatible supports minimizes environmental hazards.(Ibáñez-Forés et al., 2013) Energy Efficiency: Catalysts that operate at low temperatures and pressures reduce energy consumption, lowering the carbon footprint of pollution control processes.Scalability: Green catalysts are designed for easy synthesis and application in industrial-scale operations. innovative approaches to incorporate eco-friendly materials, enhance catalytic performance, and minimize resource consumption. Recent advancements include:Heterogeneous Catalysts: Heterogeneous green catalysts are solid materials that enable catalytic be easily separated and reused. Common designs include metal oxides, supported noble metals, and nanostructured catalysts.Metal Oxide Catalysts: are widely used for air pollution control due to their low toxicity, high stability, and ability to catalyze oxidation-reduction reactions.Non-Precious Metal Catalysts: To replace expensive noble metals like platinum (Pt) and palladium (Pd), researchers are developing catalysts using earth-abundant metals such as iron (Fe), nickel (Ni), and copper (Cu). These catalysts are cost-effective and sustainable.Support Materials: Green catalysts often use environmentally benign support materials such as activated (Invernizzi et al., 2019)carbon, silica, and bio-based materials like cellulose and chitosan to enhance catalytic performance.. Nanostructured Catalysts: Nanotechnology has revolutionized catalyst design by enabling the synthesis of catalysts with high surface area, improved reactivity, and controlled particle size. Nanocatalysts, including nanoparticles, nanowires, and nanoclusters, provide enhanced activity for air pollution control. For instance, titanium dioxide (TiO_2) nanoparticles are widely used in photocatalysis for VOC and NOx removal. Biocatalysts: Enzymes and microorganisms are being explored as green catalysts for air pollution control due to their natural ability to degrade pollutants under mild conditions. For example, laccase enzymes can oxidize VOCs, while biofilters utilize microbial activity to remove NOx and SO_2.

Photocatalysts: Photocatalysts harness solar energy to drive redox reactions for pollutant degradation. Semiconductor photocatalysts such as TiO_2, ZnO, and $BiVO_4$ have been extensively studied for air purification.

Photocatalysis offers a sustainable, energy-efficient method for removing VOCs, NOx, and SO_2 without generating secondary pollutants. Recent developments focus on doping, nanostructuring, and hybridizing photocatalysts to improve visible light absorption and efficiency. Electrocatalysts: Green electrocatalysts facilitate electrochemical processes for pollutant removal, such as NOx and CO_2 reduction.

Electrocatalytic systems are energy-efficient and can be powered by and transition metal phosphides are being developed for advanced electrocatalytic applications. Bioinspired and Biomimetic Catalysts: Inspired by natural catalytic processes, bioinspired catalysts mimic the activity of enzymes or natural materials for pollutant degradation. These include hybrid catalysts combining natural enzymes with synthetic materials, offering high efficiency under mild conditions.Applications of Green Catalysts in Green catalysts play a pivotal role in the abatement of major air pollutants through various catalytic processes, including oxidation, reduction, and conversion.Selective Catalytic Reduction (SCR):(Karka et al., 2017) SCR systems use green catalysts such as metal-doped zeolites and transition metal oxides to convert NOx into harmless nitrogen (N_2). Iron- and copper-based catalysts are promising alternatives to conventional vanadium-based SCR catalysts due to their lower toxicity and environmental compatibility.Photocatalytic NOx Removal: TiO_2-based photocatalysts degrade NOx under UV or visible light, converting it into nitrates, which can be washed away or collected. Modified photocatalysts with doped elements improve efficiency under visible light.. SO_2 Removal:Green catalysts such as metal oxides and alkaline earth materials are used for flue gas desulfurization (FGD) processes, where SO_2 is converted into gypsum or other reusable byproducts. Photocatalytic oxidation of SO_2 into sulfate ions offers a sustainable method for SO_2 abatement. Catalytic Oxidation: Green catalysts, including metal oxides and non-precious metals, facilitate the complete oxidation of VOCs into CO_2 and water. Catalysts such as MnO_2-CeO_2 composites and perovskite oxides exhibit high efficiency and stability. Photocatalytic Degradation: Semiconductor photocatalysts break down VOCs into harmless products using solar energy. TiO_2-based nanocatalysts and hybrid photocatalysts are widely applied in air purification systems.Non-precious metal catalysts, such as copper oxides and cerium oxides, are being developed for CO oxidation in vehicular and industrial emissions, replacing expensive noble metal catalysts. Green catalysts facilitate the conversion of CO_2 into value-added chemicals or fuels through electrocatalytic and photocatalytic processes. For instance, copper-based electrocatalysts enable CO_2 reduction to methanol or hydrocarbons.Green catalysts are integrated into air purifiers, coatings, and building materials to remove indoor pollutants such as formaldehyde, benzene, and NOx. Photocatalytic coatings on walls and glass surfaces ensure continuous air purification using ambient light.Conclusion

BIO-BASED ADSORBENTS FOR AIR POLLUTION CONTROL: DEVELOPMENT, CHARACTERIZATION, AND APPLICATIONS

These materials have shown great promise in removing airborne pollutants through adsorption processes, contributing to cleaner air and reduced environmental footprints.Advantages of Bio-Based Adsorbents Bio-based adsorbents offer several advantages over conventional adsorbents, such as activated carbon or synthetic resins: Renewability: Derived from renewable biomass sources, bio-based adsorbents contribute to resource sustainability and circular economies. Low Cost: Agricultural residues,(Kelly et al., 2003) forestry waste, and industrial byproducts are abundant, inexpensive, and often considered waste materials, making them economically viable.Biodegradability: Bio-based adsorbents can degrade naturally, reducing long-term environmental pollution.. High Porosity and Surface Area: Biomass-based adsorbents can be modified to exhibit enhanced porosity and functional groups, enabling high adsorption capacity.(Lai et al., 2011) Versatility: These adsorbents can remove a wide range of air pollutants, including VOCs, heavy metals, acidic gases (SO_2, NOx), and PM. Development and Synthesis of Bio-Based AdsorbentsBio-based adsorbents are developed through various physical, chemical, and biological modifications of raw biomass. The development process involves enhancing the surface area, porosity, and adsorption efficiency of natural biomass to ensure effective pollutant removal.Raw Materials Agricultural Residues: Common agricultural wastes, such as rice husks, wheat straw, corn stalks, coconut shells, and sugarcane bagasse, serve as ideal precursors for bio-based adsorbents due to their availability and lignocellulosic composition.Forestry Waste: Sawdust, bark, wood chips, and forest residues are rich in cellulose and lignin, making them suitable for adsorbent production.Algae and Aquatic Biomass: Microalgae and seaweed contain functional groups (e.g., carboxyl, hydroxyl) that enhance adsorption capacity. Biochar: Biochar is a carbon-rich material produced through pyrolysis (thermal decomposition) of biomass in an oxygen-limited environment. It is characterized by its high porosity, surface area, and chemical stability.Physical Activation: Biomass is carbonized at high temperatures (400–800°C) under an inert atmosphere to produce porous biochar with a high surface area. Physical activation using steam or CO_2 further enhances porosity and adsorption efficiency.Chemical Activation: Biomass or biochar is treated with activating agents such as potassium hydroxide (KOH), phosphoric acid (H_3PO_4), or zinc chloride ($ZnCl_2$) to create micropores and mesopores. This improves pollutant adsorption due to increased surface area and functional groups.Surface Functionalization: Chemical treatment with acids, bases, or oxidizing agents introduces functional groups (e.g., hydroxyl, carboxyl, amino) that enhance the affinity of bio-based adsorbents for specific pollutants. Developing cost-effective regeneration techniques to ensure long-term use and minimize waste generation. Scaling Up: Scaling up bio-based adsorbent production for industrial applications while maintaining sustainability and economic feasibility. Pollutant-Specific Adsorbents: Designing adsorbents tailored to specific pollutants for improved selectivity and efficiency.Conclusion Bio-based adsorbents represent a promising, sustainable, and cost-effective solution for air pollution control. Derived from renewable biomass sources, these adsorbents offer high porosity, functional versatility,(Leurent & Windisch, 2011).

GREEN SOLVENTS FOR AIR POLLUTION CONTROL: PROPERTIES, APPLICATIONS, AND ENVIRONMENTAL IMPACT

Green solvents are environmentally friendly solvents that offer reduced toxicity, lower volatility, biodegradability, and renewable origin, aligning with the principles of green chemistry. By replacing conventional solvents, green solvents can improve the efficiency and sustainability of pollution control processes, reduce secondary waste generation, and minimize the overall environmental impact. Their applications range from VOC capture and CO_2 absorption to pollutant scrubbing and emission mitigation, providing an eco-friendly pathway to cleaner air.Properties of Green Solvents Green solvents are characterized by properties that make them safer for the environment and human health while maintaining high performance in industrial processes. Key properties include: Low Toxicity: Green solvents have minimal health hazards and reduced ecotoxicity, making them safer to handle and use.Biodegradability: These solvents break down naturally in the environment, reducing the risk of long-term pollution.Low Volatility: Unlike traditional volatile organic solvents, green solvents have low vapor pressures, minimizing emissions into the atmosphere. Renewable Origin: Green solvents are often derived from biomass, natural resources, or waste materials, contributing to sustainability and circular economies. Non-Flammability: Many green solvents are inherently non-flammable, enhancing safety in industrial applications. Energy Efficiency: Green solvents enable pollution control processes to operate under mild temperature and pressure conditions, reducing energy consumption and greenhouse gas emissions.Types of Green Solvents Several types of green solvents have been developed and applied for air pollution control. These include: non-volatile, thermally stable, and tunable for specific applications. Their unique properties make them ideal for absorbing gaseous pollutants such as CO_2, SO_2, and VOCs. Functionalized ionic liquids can further enhance absorption capacity and selectivity., which form eutectic liquids at low temperatures. They are biodegradable, low-cost(Mainka et al., 2015), and customizable, making them suitable for pollutant absorption and scrubbing applications. diffusivity and liquid-like density. It is non-toxic, non-flammable, and widely used for VOC extraction and pollutant remediation in industrial processes., starches, and plant oils. Examples include glycerol, ethanol, and lactate esters. These solvents are biodegradable, non-toxic, and effective in pollutant scrubbing systems. Water: Water, as a universal solvent, is inherently green and widely used for gas scrubbing and particulate control in industries. Functionalized water-based solvents, such as aqueous amine solutions, have shown significant promise in CO_2 capture Hydrofluoroethers (HFEs): HFEs are low-toxicity, non-ozone-depleting solvents with low global warming potential (GWP). They are used for gas scrubbing and air purification in clean industrial processes.Applications of Green Solvents in Air Pollution Control Green solvents have been successfully applied in various air pollution control processes to remove or mitigate hazardous pollutants. Key applications include:(Maizlish et al., 2013). Acidic gases like SO_2 and NOx contribute to acid rain, respiratory illnesses, and environmental degradation.

SUSTAINABLE MATERIALS FOR AIR POLLUTION CONTROL: DEVELOPMENT, CHARACTERIZATION, AND APPLICATIONS

Air pollution remains one of the most pressing global environmental challenges, with adverse impacts on public The need for effective and sustainable air pollution control technologies has grown exponentially as conventional methods often rely on non-renewable, energy-intensive materials that generate secondary pollutants or are difficult to recycle. In response to these challenges, sustainable materials have emerged as promising alternatives for air pollution mitigation. These materials, derived from renewable sources, industrial waste, natural minerals, or innovative synthetic approaches, prioritize environmental sustainability, energy efficiency, and high performance. Through careful design, development, and characterization, sustainable materials can effectively capture, degrade, or transform air pollutants, offering eco-friendly solutions for cleaner air and a healthier environment. Materials such as titanium dioxide (TiO_2), zinc oxide (ZnO), and doped semiconductors harness solar energy to degrade VOCs, NOx, and other gaseous pollutants via photocatalytic reactions.. Hydrogels and Aerogels: Sustainable hydrogels and bio-based aerogels exhibit high porosity and adsorption capacity, enabling them to capture airborne particles and gaseous pollutants effectively.Development of Sustainable Materials The development of sustainable materials for air pollution control focuses on designing materials that are environmentally friendly, cost-effective, and efficient in removing a wide range of pollutants. Key strategies include Utilization of Renewable Resources: Sustainable materials are developed using agricultural residues, forestry byproducts, and bio-based polymers to reduce dependence on non-renewable resources(Perez et al., 2015). For example, functionalized MOFs and biochar exhibit enhanced pollutant removal efficiency. Hybrid and Composite Materials: Hybrid materials combine the advantages of multiple sustainable components, such as biochar-metal composites or clay-nanoparticle hybrids, to achieve superior performance in air pollution control applications.Characterization of Sustainable Materials(Porritt et al., 2011) Comprehensive characterization is essential to evaluate the properties and performance of sustainable materials for air pollution control. Key characterization techniques include:Gas Adsorption and Desorption Tests: These tests quantify the adsorption capacity and kinetics of materials for specific pollutants, such as CO_2, SO_2, and VOCs. Applications of Sustainable Materials in Air Pollution Control Sustainable materials have been successfully applied in various air pollution control technologies, including adsorption, catalysis, filtration, and photocatalysis.

Lower Carbon Footprint: By reducing energy consumption and emissions, sustainable materials contribute to climate change mitigation.. Biodegradability and Reusability: Bio-based and natural materials degrade naturally or can be regenerated for repeated use, minimizing environmental persistence. hallenges and Future Directions Despite their potential, several challenges remain in the widespread implementation of sustainable materials:. Scalability: Scaling up the production of sustainable materials while maintaining performance and cost-effectiveness is a challenge.. Regeneration and Reuse: Ensuring the long-term reusability and regeneration of adsorbents without performance loss requires further research.. Selectivity: (Ren et al., 2012)Developing materials with high selectivity for specific pollutants is critical for targeted pollution control.conclusion Sustainable materials represent an innovative and eco-friendly approach to air pollution control, offering solutions that align with environmental sustainability and green chemistry principles. Bio-based materials, natural minerals, MOFs, and advanced nanomaterials have demonstrated remarkable potential in pollutant capture, degradation, and mitigation. Through careful design, functionalization, and characterization,

CASE STUDIES: SUCCESSFUL APPLICATIONS OF GREEN CHEMISTRY SOLUTIONS FOR AIR POLLUTION CONTROL

This document highlights key case studies that demonstrate the successful application of green chemistry in addressing air pollution, showcasing innovations, challenges, and measurable impacts The automotive industry is a major contributor to air pollution due to vehicle emissions,(Ribeiro et al., 2015) particularly nitrogen oxides (NOx), carbon monoxide (CO), hydrocarbons (HCs), and particulate matter (PM). The adoption of green chemistry solutions through catalytic converters has significantly reduced harmful emissions. Catalytic converters, widely implemented in vehicles, utilize transition metals such as platinum, palladium, and rhodium to promote the conversion of toxic gases into less harmful products, including carbon dioxide (CO_2), nitrogen (N_2), and water (H_2O).For instance, Toyota introduced advanced three-way catalytic converter technology to comply with stringent emission regulations. Through optimization of catalyst design and material efficiency, this technology enhances the conversion of NOx, CO, and HCs simultaneously. Additionally, advancements in nanotechnology have allowed the reduction of catalyst loading, ensuring cost-effectiveness and resource conservation while maintaining high efficiency.A notable example includes a collaboration between the automotive sector and research institutions to develop sustainable catalyst systems with reduced precious metal content. This innovation not only lowers emissions but also addresses concerns regarding the depletion of critical materials. The successful integration of catalytic converters in vehicles worldwide has led to measurable reductions in urban air pollutants, improving air quality and public health.. Traditional solvent-based products release high levels of VOCs, contributing to ground-level ozone and smog formation. Green chemistry has enabled the development of low-VOC and VOC-free coatings, offering a sustainable alternative to conventional solvent-based solutions.One successful case study is Sherwin-Williams' development of waterborne coatings for automotive and architectural applications. By replacing hazardous organic solvents with water as the primary medium, VOC emissions were reduced by up to 90%. These coatings utilize innovative formulations that ensure performance comparable to traditional paints, including durability, adhesion, and aesthetic quality.Similarly, AkzoNobel introduced powder coatings, which eliminate the need for solvents altogether. Powder coatings are applied as dry powders and cured using heat, leading to zero VOC emissions(Rodriguez de San Miguel, 2019). The adoption of these green chemistry solutions in the construction, automotive, and consumer goods sectors has significantly reduced air pollution while meeting regulatory requirements.Additionally, companies have implemented bio-based solvents derived from renewable resources, such as soy and citrus extracts, further enhancing the sustainability of coatings and minimizing environmental impact. Research conducted in the Netherlands demonstrate reducing methane (CH_4) and ammonia emissions in agriculture. When applied to soils, biochar enhances nutrient retention, mitigates greenhouse gas emissions, and improves soil health.The implementation of these green chemistry solutions in agriculture highlights their potential to address air pollution while promoting sustainable food production systems conclusion(Sarigiannis et al., 2017) The case studies presented demonstrate the transformative potential of green chemistry solutions in controlling and reducing air pollution across various sectors. Innovations such as catalytic converters, low-VOC coatings, cleaner combustion technologies, industrial emission control systems, and bio-based agricultural solutions illustrate the practical applications of green chemistry principles. By minimizing hazardous substances, promoting resource efficiency, and fostering sustainable practices, green chemistry provides a pathway toward cleaner air and a healthier environment.

INDUSTRIAL APPLICATIONS OF GREEN CHEMISTRY SOLUTIONS FOR AIR POLLUTION CONTROL: CHALLENGES AND OPPORTUNITIES

Green chemistry,, has become a cornerstone in addressing industrial air pollution. Industrial processes, which include energy production, manufacturing, transportation, and chemical processing, are significant contributors to air pollution. to public health and the environment. Green chemistry offers sustainable alternatives that reduce emissions, enhance process efficiency, and minimize waste generation. However, the application of green chemistry in industry also faces numerous challenges, including technological,(Scheepers et al., 2014) economic, and regulatory barriers. This document explores industrial applications of green chemistry solutions for air pollution control, highlighting key opportunities, challenges, and examples of successful implementation.The principles of green chemistry—such as waste prevention, energy efficiency, the use of renewable feedstocks, and safer chemical design—align with the goals of reducing air pollution. Several opportunities exist for industries to adopt green chemistry solutions that address emissions and improve sustainability:Fossil fuel combustion in energy generation remains a primary source of air pollution. Green chemistry solutions have enabled the adoption of cleaner fuels and alternative energy sources, such as biodiesel, bioethanol, and hydrogen fuel. Biofuels, derived from renewable biomass, offer lower emissions of SOx, NOx, and particulate matter compared to conventional fuels.Case Study: In Brazil, biodiesel produced from waste cooking oil and soybean oil has been integrated into transportation fuels, reducing emissions of NOx and PM by up to 40%. This success highlights biofuel production as a viable industrial opportunity to mitigate air pollution. This approach addresses the dual challenges of reducing GHG emissions and promoting resource circularity.(Sellitto et al., 2015)Case Study: Carbon Clean Solutions successfully implemented CCU technologies in cement and steel plants, capturing CO_2 emissions and converting them into chemicals like methanol and calcium carbonate.Process intensification strategies, such as integrating multiple operations into single units and reducing energy consumption, align with green chemistry principles. Dow Chemical's development of low-NOx burner technology for industrial furnaces reduced NOx emissions by 40%. By optimizing combustion efficiency, Dow improved both environmental performance and operational costs. Example: LafargeHolcim implemented SCR technology and alternative fuels (e.g., waste-derived fuels) to reduce SOx and NOx emissions. These solutions allowed the company to comply with emission standards while reducing reliance on fossil fuels. Integrating green chemistry solutions with circular economy principles, such as resource recovery and waste valorization, can create economic and environmental benefits.Training programs Conclusion The industrial applications of green chemistry solutions for air pollution control demonstrate significant opportunities to reduce emissions, enhance efficiency, and promote sustainability. By addressing challenges such as high costs, technological limitations, and policy barriers, industries can accelerate the adoption of green chemistry innovations. Successful case studies across sectors, including energy, chemicals, cement, and steel, highlight the transformative potential of green chemistry to minimize environmental impacts while ensuring economic growth.

MUNICIPAL AND COMMUNITY APPLICATIONS OF GREEN CHEMISTRY SOLUTIONS FOR AIR POLLUTION CONTROL: CHALLENGES AND OPPORTUNITIES

Air pollution remains one of the most critical environmental and public health challenges facing municipalities and communities worldwide. Urban areas, characterized by dense populations, extensive vehicular traffic, industrial hubs, and waste management activities, are often epicenters of air pollution. The pollutants of concern include particulate matter (PM), nitrogen oxides (NOx), sulfur oxides (SOx), volatile organic compounds (VOCs), carbon monoxide (CO), and greenhouse gases (GHGs) These solutions aim to minimize pollution sources, enhance efficiency, and promote the use of non-toxic alternatives. This document examines successful municipal and community applications of green chemistry, while addressing the challenges and highlighting opportunities for broader implementation.The transportation sector is a dominant contributor to air pollution in urban communities due to emissions from internal combustion engines, including CO, NOx, VOCs, and PM. Municipalities have increasingly adopted green chemistry solutions, such as biofuels, electric vehicles, and catalytic technologies, to mitigate emissions.derived from renewable biomass, such as waste cooking oil, sugarcane, and algae. Biofuels significantly reduce SOx, NOx, and PM emissions compared to conventional diesel. For instance, Stockholm's buses operate on bioethanol and biogas, reducing CO(Tong et al., 2016)$_2$ emissions by up to 70%.Although electric vehicles (EVs) are not strictly green chemistry innovations, their batteries rely on advanced chemistry principles to store energy efficiently. Cities such as Shenzhen, China, have transitioned entire bus fleets to electric vehicles, eliminating tailpipe emissions and improving urban air quality. Innovations in battery chemistries, such as lithium iron phosphate ($LiFePO_4$), have enhanced safety and sustainability in municipal EV programs.(Venkataraman et al., 2010). Composting programs, such as those in Seattle and Milan, reduce the volume of organic waste sent to landfills, thereby minimizing methane emissions. Municipalities are increasingly promoting biodegradable and compostable materials derived from green chemistry innovations to address waste pollution.Cities like Amsterdam have adopted advanced WTE technologies that use cleaner combustion processes and catalytic filters to convert waste into energy while reducing NOx, SOx, and particulate emissions. These technologies optimize combustion efficiency and incorporate emissions control systems aligned with green chemistry principles. For example, transitioning to biofuels or electric vehicles requires significant investments in fueling stations and charging infrastructure.(Vlachokostas et al., 2011)Inconsistent or weak air quality policies in some regions limit the adoption of green chemistry solutions. Stronger municipal regulations, incentives, and public-private partnerships are needed to drive change. Lack of awareness or resistance to change within communities can slow the adoption of green chemistry initiatives. Effective communication and education programs are essential to build public support.To overcome these challenges, municipalities can harness the following opportunities:1. Public-Private Partnerships: Collaborating with industry leaders, academic institutions, and technology providers can accelerate the development and implementation of green chemistry solutions.2. Incentive Programs: Municipalities can offer tax incentives, subsidies, and grants to businesses and residents adopting green chemistry technologies.. Regulatory Frameworks: Strong policies mandating the use of low-emission technologies, sustainable materials, and waste-to-energy solutions can drive widespread adoptio. Community Engagement: Building awareness through education campaigns and pilot projects can demonstrate the benefits of green chemistry solutions, fostering public support. Integration with Circular Economy Models: Municipalities can promote circular economy practices, such as resource recovery, waste valorization, and renewable energy generation, to achieve sustainable pollution control Conclusion Green chemistry offers municipalities and communities a transformative opportunity to address air pollution through sustainable, efficient, and innovative solutions. Applications in transportation, building materials, waste management, and industrial pollution prevention demonstrate the practical benefits of green chemistry. However, challenges such as funding constraints, technological barriers, and policy gaps must be addressed to realize its full potential.

FUTURE DIRECTIONS FOR GREEN CHEMISTRY SOLUTIONS IN AIR POLLUTION CONTROL: EMERGING TRENDS AND TECHNOLOGIES

Air pollution remains one of the most pressing global challenges, contributing to climate change, environmental degradation, and public health crises. As industrialization, urbanization, and energy consumption continue to rise, innovative solutions are needed to mitigate emissions and control air quality. Green chemistry—a science-driven approach to designing processes and products that minimize the generation of hazardous substances—presents promising opportunities for future advancements in air pollution control. The integration of cutting-edge technologies, renewable materials, and sustainable practices is opening new pathways to address emissions across multiple sectors. Nano-sized particles offer greater surface area, enhancing their reactivity.Low Precious-Metal Catalysts: Efforts are underway to reduce dependence on scarce and expensive materials like platinum and palladium in emission control catalysts. Researchers are exploring metal-free and earth-abundant catalysts, such as graphene-based or transition-metal alternatives, to lower costs and improve sustainability. Advanced sorbents, such as metal-organic frameworks (MOFs), are being designed to maximize CO_2 adsorption while minimizing energy consumption.Green hydrogen, produced through electrolysis powered by renewable energy, (Woodcock et al., 2009)is a cleanalternative to fossil fuels in industrial processes and transportation. Hydrogen combustion emits only water vapor, making it a zero-emission energy carrier. Green chemistry innovations are focused on developing sustainable and cost-effective electrolysis processes using non-toxic materials and efficient catalysts. Advanced Air Filtration and Purification Technologies:Green chemistry is driving the development of next-generation filtration systems that capture and degrade pollutants.Photocatalytic Filters: Photocatalytic oxidation (PCO) technology uses light-activated catalysts to break down air pollutants like VOCs, NOx, and PM into harmless byproducts. Materials such as titanium dioxide (TiO_2) are being modified to enhance their performance under visible light.Biomimetic Filtration Systems: Inspired by natural processes, biomimetic filters integrate bio-based materials and enzymes to degrade pollutants efficiently. For example, enzyme-coated membranes can neutralize(Wu et al., 2017) VOCs and CO_2.The integration of AI and machine learning (ML) with green chemistry is accelerating the discovery and optimization of sustainable solutions. AI algorithms can model chemical processes, identify optimal catalysts, and predict pollutant behavior, enabling the development of targeted emission control strategies.Example: AI-driven simulations are being used to design catalytic materials with enhanced efficiency and durability for industrial applications.Green chemistry innovations are being applied across sectors to target specific air pollution challenges.The transition to renewable energy sources, such as solar, wind, and bioenergy, is central to reducing emissions. Green chemistry enables the development of efficient storage solutions, such as green batteries and fuel cells, to complement renewable energy.Solid-State Batteries: Advances in solid-state battery chemistries using non-toxic materials are replacing traditional lithium-ion batteries, reducing environmental risks.Green chemistry solutions for transportation include cleaner fuels, lightweight materials, and advanced emission control technologies.Hydrogen Fuel Cells: Hydrogen-powered vehicles are emerging as zero-emission alternatives to internal combustion engines. Green chemistry innovations are improving the efficiency and safety of hydrogen storage systems. Bio-Based Lubricants and Additives: Sustainable, biodegradable lubricants reduce friction and emissions in vehicle engines, enhancing fuel efficiency.Industries such as cement, steel, and chemical manufacturing are significant emitters of CO_2 and other pollutants. Green chemistry solutions include process optimization, carbon capture, and substitution of hazardous inputs.Electrification of Industrial Processes: Green hydrogen and renewable energy are replacing fossil fuels in high-temperature processes, reducing emissions.Carbon-Neutral Feedstocks: Industries are shifting toward bio-based and recycled feedstocks to lower their carbon footprint.In urban areas, emerging green chemistry technologies are addressing indoor and outdoor air pollution.Smog-Eating Coatings: Photocatalytic coatings applied to building surfaces and roads can degrade pollutants like NOx and VOCs, improving urban air quality.(Xia et al., 2015)Air-Purifying Plants and Biofilters: Engineered plants and biofilters, enhanced with green chemistry-based additives, are being deployed to capture and neutralize pollutants in cities.

Resource Availability: The dependence on rare materials for catalysts and batteries poses sustainability concerns, necessitating the development of earth-abundant alternatives.Policy and Regulation Gaps: Inconsistent global regulations on air quality and emissions control slow the adoption of advanced solutions. Stronger policy frameworks are needed to incentivize green chemistry innovations.Public Awareness and Acceptance: Successful implementation requires community engagement and awareness about the benefits of green chemistry solutions. Governments, industries, and academic institutions must collaborate to accelerate innovation in green chemistry solutions(Yang et al., 2017)Circular Economy Integration: Emphasizing circular economy principles, such as resource recovery and waste valorization, will enhance the sustainability of green chemistry technologies.Global Policy Frameworks: International agreements and local policies should incentivize the adoption of clean technologies and penalize polluting practices.Education and Workforce Development: Building a skilled workforce trained in green chemistry principles will drive innovation and implementation.Public-Private Partnerships: Partnerships between governments, industries, and research institutions can provide funding and technical expertise for scaling up green chemistry solutions.The future of air pollution control is deeply connected to the advancements in green chemistry, which offer transformative, sustainable, and scalable solutions to reduce emissions and protect environmental and public health. Emerging technologies, such as bio-based materials, advanced catalysts, carbon capture and utilization, hydrogen energy, and AI-driven process optimization, represent the next frontier in air quality management.

THE ROLE OF GREEN CHEMISTRY SOLUTIONS IN ADDRESSING AIR POLLUTION CHALLENGES

This document outlines the role of green chemistry in addressing air pollution, focusing on its principles, key applications, challenges, and opportunities across various sectors.mong these principles, the key strategies relevant to air pollution control include preventing waste generation, designing less hazardous chemical syntheses, maximizing energy efficiency, and using renewable feedstocks. Unlike traditional methods of pollution control, which focus on managing pollutants after they have been released into the atmosphere, green chemistry adopts a proactive approach by preventing emissions at the source. This paradigm shift enables industries, urban communities, and governments to minimize environmental harm while improving human health outcomes.The role of green chemistry in air pollution control involves three critical aspects: Pollution Prevention: Developing processes that reduce or eliminate emissions during production and transportation.Cleaner Substitutes: Replacing hazardous substances with environmentally benign alternatives.. Emission Control Innovations: Enhancing technologies that capture, degrade, or utilize pollutants for beneficial purposes. By integrating these principles into various industries and community initiatives, green chemistry enables a transition to cleaner air and a healthier environment.

Catalysis, a cornerstone of green chemistry, enhances the efficiency and sustainability of industrial processes while minimizing pollutant formation.Selective Catalytic Reduction (SCR): SCR systems, driven by advanced catalysts, convert harmful NOx emissions into harmless nitrogen and water vapor in industries and power plants.Green Catalysts: Research is focused on developing earth-abundant, non-toxic catalysts as alternatives to rare or precious metals, making emission control more sustainable. Carbon Capture and Utilization (CCU): Green chemistry innovations have enabled industries to capture carbon emissions and transform them into value-added products. CO_2(Younger et al., 2008) Conversion Technologies: Captured CO_2 can be converted into chemicals, fuels, or polymers using energy-efficient processes. For example, CO_2-derived methanol is used as a cleaner fuel alternative. Industrial Symbiosis: Industries are adopting circular economy models where waste emissions from one facility become feedstocks for another, reducing overall pollution.The transportation sector is a leading contributor to urban air pollution, emitting NOx, VOCs, CO, and particulate matter from vehicular exhaust. Green chemistry offers viable solutions for cleaner transportation systems Sustainable Fuels and Biofuels:iofuels, derived from renewable biomass such as agricultural residues, algae, and waste oils, are replacing fossil fuels in vehicles, reducing emissions.Biodiesel and Bioethanol: Cities such as Stockholm and Curitiba have adopted biodiesel and ethanol in public transit, significantly reducing SOx,(Zhao et al., 2017) CO, and PM emissions.Green Hydrogen: Hydrogen fuel cells, powered by green hydrogen (produced using renewable energy), emit only water vapor, offering a zero-emission alternative for heavy transport. Lightweight Materials and Engine Efficiency:Green chemistry supports the development of advanced materials, such as lightweight composites and bio-based polymers, to improve vehicle fuel efficiency and reduce emissions. Emission Control Technologies:green chemistry-driven innovations in catalytic converters and filters effectively reduce vehicular emissions. For example, three-way catalytic converters transform NOx, CO, and hydrocarbons into nitrogen, CO_2, and water.Improper waste management generates methane, VOCs, and other air pollutants, especially in landfills and incineration processes. Although challenges remain, green chemistry offers a clear path toward cleaner air and a more sustainable future. By fostering innovation, strengthening policies, and engaging stakeholders,(Zanobetti & Schwartz, 2005) green chemistry will play a central role in solving air pollution challenges and driving global progress toward environmental and societal well-being.

CONCLUSION

Air pollution remains one of the most critical environmental challenges, threatening ecosystems, human health, and global economic stability. With industrialization, urbanization, and energy demands continuing to rise, traditional approaches to managing air quality are no longer sufficient to address the magnitude and complexity of the problem. Unlike conventional pollution management methods that emphasize end-of-pipe technologies, green chemistry offers a proactive pathway by designing safer chemical processes, materials, and products that inherently minimize environmental harm. This approach reduces the generation of harmful emissions such as particulate matter (PM), volatile organic compounds (VOCs), nitrogen oxides (NOx), sulfur oxides (SOx), and greenhouse gases (GHGs), among others, at their source. Through innovations such as bio-based materials, energy-efficient catalytic processes, carbon capture and utilization (CCU) technologies, and green hydrogen solutions, industries can significantly reduce their air pollution footprint while maintaining economic viability. The adoption of these technologies not only enhances air quality but also aligns with broader climate change mitigation goals, supporting global efforts to limit warming and achieve net-zero emissions., has benefited significantly from green chemistry advancements. Sustainable fuels, such as biofuels and green hydrogen, offer cleaner alternatives to fossil fuels, while lightweight, bio-based materials improve vehicle efficiency and reduce emissions. Coupled with advanced emission control systems, catalytic converters, and electric mobility technologies, these innovations represent a paradigm shift toward cleaner transportation systems that minimize environmental impact. However, while green chemistry solutions present transformative opportunities for air pollution control, their widespread adoption is not without challenges. High implementation costs, technological scalability, and limited infrastructure often pose barriers for industries and governments, particularly in developing economies. Research and development of green alternatives require significant investment, and the transition from traditional technologies to green chemistry-driven solutions demands a collaborative effort across sectors. Moreover, inconsistent policy frameworks, lack of incentives, and limited awareness among stakeholders continue to slow down progress. Addressing these challenges will require a multi-faceted approach, combining strong regulatory frameworks, financial incentives, public-private partnerships, and educational initiatives to drive the adoption and scaling of green chemistry innovations globally. The future of air pollution control depends on the commitment of governments, industries, and communities to embrace green chemistry as a central component of sustainable development strategies. By prioritizing investments in research, fostering technological innovation, and implementing supportive policies, societies can accelerate the adoption of cleaner and safer alternatives. Collaboration between scientists, policymakers, businesses, and local communities is essential to creating a s to tackle air pollution challenges efficiently.In conclusion, green chemistry offers a holistic, innovative, and sustainable approach to air pollution control, addressing . By designing processes and materials that are safer, cleaner, and resource-efficient, green chemistry empowers industries, urban planners, and governments to transition to sustainable models of development. The principles of green chemistry emphasize prevention over remediation, offering solutions that minimize emissions, protect public health, and preserve the environment for future generations.

REFERENCES

Al-thani, H., Koç, M., & Isaifan, R. J. (2018). A review on the direct effect of particulate atmospheric pollution on materials and its mitigation for sustainable cities and societies. *Environmental Science and Pollution Research International, 25*(28), 27839–27857. DOI: 10.1007/s11356-018-2952-8 PMID: 30128969

Anderson, J. O., Thundiyil, J. G., & Stolbach, A. (2012). Clearing the air: A review of the effects of particulate matter air pollution on human health. *Journal of Medical Toxicology; Official Journal of the American College of Medical Toxicology, 8*(2), 166–175. DOI: 10.1007/s13181-011-0203-1 PMID: 22194192

Arunachalam, S., Brantley, H., Barzyk, T. M., Hagler, G., Isakov, V., Kimbrough, E., Naess, B., Rice, N., Snyder, M. G., Talgo, K., & Venkatram, A. (2015). Assessment of port-related air quality impacts: Geographic analysis of population. *International Journal of Environment and Pollution, 58*(4), 231–250. DOI: 10.1504/IJEP.2015.077455

Babar, Z. B., & Shareefdeen, Z. (2014). Management and control of air emissions from electronic industries. *Clean Technologies and Environmental Policy, 16*(1), 69–77. DOI: 10.1007/s10098-013-0594-6

Bailey, D., & Solomon, G. (2004). Pollution prevention at ports: Clearing the air. *Environmental Impact Assessment Review, 24*(7), 749–774. DOI: 10.1016/j.eiar.2004.06.005

Bandyopadhyay, A., Bandyopadhyay, S., & Biswas, M. N. (2014). Fabrication and operation of an online photoelectric particulate matter monitoring device. *Clean Technologies and Environmental Policy, 17*(1), 265–271. DOI: 10.1007/s10098-014-0782-z

Cariolet, J. M., Colombert, M., Vuillet, M., & Diab, Y. (2018). Assessing the resilience of urban areas to traffic-related air pollution: Application in greater Paris. *The Science of the Total Environment, 615*(615), 588–596. DOI: 10.1016/j.scitotenv.2017.09.334 PMID: 28988095

Chae, Y., & Park, J. (2011). Quantifying costs and benefits of integrated environmental strategies of air quality management and greenhouse gas reduction in the Seoul metropolitan area. *Energy Policy, 39*(9), 5296–5308. DOI: 10.1016/j.enpol.2011.05.034

Chao, K. (2008). A new look at the cross-impact matrix and its application in futures studies. *Journal of Futures Studies, 12*(4), 45–52.

Dhupia, J., Adnanes, A. K., Lee, K. M., & Kennedy, L. (2011) Electrification of port and port operations. *Proceedings of MTEC2011*, pp. 1–6

Erisman, J. W., & Schaap, M. (2004). The need for ammonia abatement with respect to secondary PM reductions in Europe. *Environmental Pollution, 129*(1), 159–163. DOI: 10.1016/j.envpol.2003.08.042 PMID: 14749079

Fajersztajn, L., Veras, M., Barrozo, L. V., & Saldiva, P. (2013). Air pollution: A potentially modifiable risk factor for lung cancer. *Nature Reviews. Cancer, 13*(9), 674–678. DOI: 10.1038/nrc3572 PMID: 23924644

Fann, N., Roman, H. A., Fulcher, C. M., Gentile, M. A., Hubbell, B. J., Wesson, K., & Levy, J. I. (2011). Maximizing health benefits and minimizing inequality: Incorporating local-scale data in the design and evaluation of air quality policies. *Risk Analysis, 31*(6), 908–922. DOI: 10.1111/j.1539-6924.2011.01629.x PMID: 21615761

Friel, S., Dangour, A. D., Garnett, T., Lock, K., Chalabi, Z., Roberts, I., Butler, A., Butler, C. D., Waage, J., McMichael, A. J., & Haines, A. (2009). Public health benefits of strategies to reduce greenhouse-gas emissions: Food and agriculture. *Lancet, 374*(9706), 2016–2025. DOI: 10.1016/S0140-6736(09)61753-0 PMID: 19942280

Giuliano, A., Gioiella, F., Sofia, D., & Lotrecchiano, N. (2018a). A novel methodology and technology to promote the social acceptance of biomass Power plants avoiding Nimby syndrome. *Chemical Engineering Transactions, 67*, 307–312.

Hou, Y., Velthof, G. L., & Oenema, O. (2015). Mitigation of Ammonia, nitrous oxide and methane emissions from manure management chains: A meta-analysis and integrated assessment. *Global Change Biology, 21*(3), 1293–1312. DOI: 10.1111/gcb.12767 PMID: 25330119

Ibáñez-Forés, V., Bovea, M. D., & Azapagic, A. (2013). Assessing the sustainability of best available techniques (BAT): Methodology and application in the ceramic tiles industry. *Journal of Cleaner Production, 51*, 162–176. DOI: 10.1016/j.jclepro.2013.01.020

Invernizzi, S., Montagnoli, F., & Carpinteri, A. (2019). Fatigue assessment of the collapsed XXth century cable-stayed Polcevera bridge in Genoa. *Procedia Structural Integrity, 18*, 237–244. DOI: 10.1016/j.prostr.2019.08.159

Karka, P., Papadokonstantakis, S., & Kokossis, A. (2017). Cradle-to-gate assessment of environmental impacts for a broad set of biomass-to-product process chains. *The International Journal of Life Cycle Assessment, 22*(9), 1418–1440. DOI: 10.1007/s11367-017-1262-6

Kelly, F. J., Dunster, C., & Mudway, I. (2003). Air pollution and the elderly: Oxidant/antioxidant issues worth consideration. *The European Respiratory Journal, 40*(40 suppl), 70s–75s. DOI: 10.1183/09031936.03.00402903 PMID: 12762578

Lai, K. H., Lun, V. Y. H., Wong, C. W. Y., & Cheng, T. C. E. (2011). Green shipping practices in the shipping industry: Conceptualization, adoption, and implications. *Resources, Conservation and Recycling, 55*(6), 631–638. DOI: 10.1016/j.resconrec.2010.12.004

Leurent, F., & Windisch, E. (2011). Triggering the development of electric mobility: A review of public policies. *European Transport Research Review, 3*(4), 221–235. DOI: 10.1007/s12544-011-0064-3

Mainka, A., Zajusz-Zubek, E., & Kaczmarek, K. (2015). PM10 composition in urban and rural nursery schools in upper Silesia, Poland: Trace elements analysis. *International Journal of Environmental Research and Public Health, 12*(7), 7990–8008. DOI: 10.3390/ijerph120707990 PMID: 26184269

Maizlish, N., Woodcock, J., Co, S., Ostro, B., Fanai, A., & Fairley, D. (2013). Health cobenefits and transportation-related reductions in greenhouse gas emissions in the San Francisco Bay Area. *American Journal of Public Health, 103*(4), 703–709. DOI: 10.2105/AJPH.2012.300939 PMID: 23409903

Perez, L., Trüeb, S., Cowie, H., Keuken, M. P., Mudu, P., Ragettli, M. S., Sarigiannis, D. A., Tobollik, M., Tuomisto, J., Vienneau, D., Sabel, C., & Künzli, N. (2015). Transport-related measures to mitigate climate change in Basel, Switzerland: A health-effectiveness comparison study. *Environment International, 85*, 111–119. DOI: 10.1016/j.envint.2015.08.002 PMID: 26386464

Porritt, S., Shao, L., Cropper, P., & Goodier, C. (2011). Adapting dwellings for heat waves. *Sustainable Cities and Society, 1*(2), 81–90. DOI: 10.1016/j.scs.2011.02.004

Ren, W. X., Geng, Y., Xue, B., Fujita, T., Ma, Z., & Jiang, P. (2012). Pursuing co-benefits in China's old industrial base: A case of Shenyang. *Urban Climate, 1*(72), 55–64. DOI: 10.1016/j.uclim.2012.07.001

Ribeiro, I., Monteiro, A., Martins, H., Freitas, S., Borrego, C., & Lopes, M. (2015). How does the use of biodiesel affect urban air quality? *International Journal of Environment and Pollution, 58*(1/2), 79–88. DOI: 10.1504/IJEP.2015.076585

Rodriguez de San Miguel, J. A. (2019). Management of air pollution in Mexico. *Management of Environmental Quality, 30*(3), 578–592. DOI: 10.1108/MEQ-05-2018-0099

Sarigiannis, D. A., Kontoroupis, P., Nikolaki, S., Gotti, A., Chapizanis, D., & Karakitsios, S. (2017). Benefits on public health from transport-related greenhouse gas mitigation policies in southeastern European cities. *The Science of the Total Environment, 579*, 1427–1438. DOI: 10.1016/j.scitotenv.2016.11.142 PMID: 27919555

Scheepers, C. E., Wendel-Vos, G. C. W., den Broeder, J. M., van Kempen, E. E. M. M., van Wesemael, P. J. V., & Schuit, A. J. (2014). Shifting from car to active transport: A systematic review of the effectiveness of interventions. Transp Res. *Transportation Research Part A, Policy and Practice, 70*, 264–280. DOI: 10.1016/j.tra.2014.10.015

Sellitto, M. A., Borchardt, M., Pereira, G. M., & Bubicz, M. E. (2015). Tool for environmental performance assessment of city bus transit operations: Case studies. *Clean Technologies and Environmental Policy, 17*(4), 1053–1064. DOI: 10.1007/s10098-014-0892-7

Tong, Z., Baldauf, R. W., Isakov, V., Deshmukh, P., & Max Zhang, K. (2016). Roadside vegetation barrier designs to mitigate near-road air pollution impacts. *The Science of the Total Environment, 541*, 920–927. DOI: 10.1016/j.scitotenv.2015.09.067 PMID: 26457737

Venkataraman, C., Sagar, A. D., Habib, G., Lam, N., & Smith, K. R. (2010). The Indian National Initiative for advanced biomass cookstoves: The benefits of clean combustion. *Energy for Sustainable Development, 14*(2), 63–72. DOI: 10.1016/j.esd.2010.04.005

Vlachokostas, C., Achillas, C., Moussiopoulos, N., & Banias, G. (2011). Multicriteria methodological approach to manage urban air pollution. *Atmospheric Environment, 45*(25), 4160–4169. DOI: 10.1016/j.atmosenv.2011.05.020

Woodcock, J., Edwards, P., Tonne, C., Armstrong, B. G., Ashiru, O., Banister, D., Beevers, S., Chalabi, Z., Chowdhury, Z., Cohen, A., Franco, O. H., Haines, A., Hickman, R., Lindsay, G., Mittal, I., Mohan, D., Tiwari, G., Woodward, A., & Roberts, I. (2009). Public health benefits of strategies to reduce greenhouse-gas emissions: Urban land transport. *Lancet, 374*(9705), 1930–1943. DOI: 10.1016/S0140-6736(09)61714-1 PMID: 19942277

Wu, Y., Zhang, S., Hao, J., Liu, H., Wu, X., Hu, J., Walsh, M. P., Wallington, T. J., Zhang, K. M., & Stevanovic, S. (2017). On-road vehicle emissions and their control in China: A review and outlook. *The Science of the Total Environment, 574*, 332–349. DOI: 10.1016/j.scitotenv.2016.09.040 PMID: 27639470

Xia, T., Nitschke, M., Zhang, Y., Shah, P., Crabb, S., & Hansen, A. (2015). Traffic-related air pollution and health co-benefits of alternative transport in Adelaide, South Australia. *Environment International*, *74*, 281–290. DOI: 10.1016/j.envint.2014.10.004 PMID: 25454245

Yang, C., Zhao, W., Deng, K., Zhou, V., Zhou, X., & Hou, Y. (2017). The association between air pollutants and autism spectrum disorders. *Environmental Science and Pollution Research International*, *24*(19), 15949–15958. DOI: 10.1007/s11356-017-8928-2 PMID: 28540549

Younger, M., Morrow-Almeida, H. R., Vindigni, S. M., & Dannenberg, A. L. (2008). The built environment, climate change, and health. Opportunities for co-benefits. *American Journal of Preventive Medicine*, *35*(5), 517–526. DOI: 10.1016/j.amepre.2008.08.017 PMID: 18929978

Zanobetti, A., & Schwartz, J. (2005). The effect of particulate air pollution on emergency admissions for myocardial infarction: A multicity Case-crossover analysis. *Environmental Health Perspectives*, *113*(8), 978–982. DOI: 10.1289/ehp.7550 PMID: 16079066

Zhao, Z. Q., Bai, Z. H., Winiwarter, W., Kiesewetter, G., Heyes, C., & Ma, L. (2017). Mitigating ammonia emission from agriculture reduces PM2.5 pollution in the Hai River basin in China. *The Science of the Total Environment*, *609*, 1152–1160. DOI: 10.1016/j.scitotenv.2017.07.240 PMID: 28787789

Chapter 3
Battery Chemistry Li-ion Solid-State and Beyond

Himadri Sekhar Das
https://orcid.org/0000-0002-3509-3388
Haldia Institute of Technology, India

Sudipta Banerjee
https://orcid.org/0000-0003-0150-6794
Symbiosis Institute of Technology, Symbiosis International University, Pune, India

Heranmoy Maity
Ideal Institute of Engineering Kalyani, Nadia, India

Gourisankar Roymahapatra
https://orcid.org/0000-0002-8018-5206
Haldia Institute of Technology, India

ABSTRACT

Battery chemistry has undergone a transformative journey, propelling advancements in energy storage technologies that are pivotal to modern applications. This chapter delves into the evolution and comparative analysis of lithium-ion (Li-ion) batteries, solid-state batteries, and emerging alternatives. Li-ion batteries, known for their high energy density and longevity, have dominated the energy storage market, yet face challenges such as safety concerns and resource dependency. Solid-state batteries, featuring non-flammable solid electrolytes, promise enhanced safety, higher energy densities, and longer lifespans but are hindered by manufacturing complexities and cost. Beyond these, next-generation chemistries, including lithium-sulfur, sodium-ion, and multivalent systems, offer potential breakthroughs in affordability, sustainability, and performance. The chapter examines key materials, electrochemical mechanisms,

DOI: 10.4018/979-8-3373-1409-9.ch003

and technological hurdles, alongside discussing future directions such as recycling, sustainability, and integration into renewable energy systems.

INTRODUCTION

Battery technology is essential for contemporary energy storage systems, with uses ranging from portable devices and electric cars to the integration of renewable energy sources (Himadri Sekhar Das, (2025). This chapter examines the chemistry of lithium-ion (Li-ion) batteries, reviews progress in solid-state batteries, and looks into new technologies that hold the potential to go beyond current limitations (Parvizi et al., 2025).

Due to the swift progress in technology and the worldwide shift towards renewable energy, energy storage solutions have become a key focus of contemporary innovation. Batteries, which are crucial elements of energy systems, have undergone considerable development over the years. Lithium-ion (Li-ion) batteries have become a leading technology owing to their high energy density, efficiency, and versatility. Li-ion batteries are now essential to modern life, from powering portable electronics to facilitating electric vehicles and grid storage.

Despite their success, challenges such as safety concerns, limited resource availability, and environmental impacts have spurred research into alternative battery chemistries. Solid-state batteries, for instance, have garnered attention for their use of non-flammable solid electrolytes, offering superior safety and performance metrics. Similarly, emerging technologies like lithium-sulfur, sodium-ion, and multivalent systems hold the promise of overcoming current limitations, such as cost and resource constraints, while meeting the demands of next-generation applications (A. Machín et al, (2024).

This chapter explores the intricate chemistry behind these battery technologies, analyzing their materials, mechanisms, and advancements. Additionally, it explores the obstacles to commercialization and future outlooks, emphasizing how innovative battery systems could transform energy storage and aid in achieving sustainability. We seek to offer a thorough comprehension of how battery chemistry has developed and where it is headed, from Li-ion systems to innovative alternatives, through this investigation.

HISTORY OF DIFFERENT BATTERY TECHNOLOGIES

The development of batteries spans over two centuries, with each innovation representing a significant leap in energy storage technology:

Voltaic Pile (1800): Alessandro Volta invented the first true battery, the voltaic pile, which used alternating layers of zinc and copper separated by brine-soaked cloth. It provided a steady electric current and laid the foundation for modern batteries.

Daniell Cell (1836): John Frederic Daniell improved the voltaic pile by introducing the Daniell cell, which used copper and zinc electrodes in separate solutions to reduce hydrogen buildup and improve stability. This design became widely used in telegraph systems.

Lead-Acid Battery (1859): The first rechargeable battery, the lead-acid battery, was invented by Gaston Planté. It employed lead and lead dioxide plates immersed in sulfuric acid and continues to be an essential technology for automotive and backup power applications (Kurzweil, 2010).

Nickel-Cadmium (NiCd) Battery (1899): Waldemar Jungner launched the nickel-cadmium battery, which provided a higher energy density and was rechargeable, in contrast to lead-acid batteries. Its performance notwithstanding, modern applications of cadmium have been restricted due to environmental concerns (L. Wang,(2022).

Alkaline Battery (1950s): The alkaline battery, utilizing zinc and manganese dioxide with an alkaline electrolyte, was created by Lewis Urry. Due to its reliability and long shelf life, it became a favored option for consumer electronics.

Nickel-Metal Hydride (NiMH) Battery (1980s): NiMH batteries, developed from NiCd technology, substituted cadmium with a metal hydride alloy, resulting in higher capacity and lower toxicity. Prior to the dominance of lithium-ion technology, they were favored for portable devices (X. Yi et al., 2024).

Lithium-Ion (Li-Ion) Battery (1990s): The launch of Li-ion batteries by Sony in 1991 brought about a transformation in portable energy storage. With their high energy density, long cycle life, and efficiency, they became indispensable for smartphones, electric vehicles, and renewable energy systems.

Solid-State Battery (21st Century): Solid-state batteries were developed as a next-generation solution, using solid materials in place of liquid electrolytes. They promise enhanced safety, greater energy density, and extended lifespans, although challenges in manufacturing persist.

Emerging Chemistries: Modern advancements focus on lithium-sulfur, sodium-ion, and multivalent batteries, aiming to reduce costs, improve performance, and address resource sustainability challenges. Innovations in flow batteries and bio-batteries are also being explored for niche applications.

Each stage of battery development reflects a response to evolving energy demands and technological constraints, shaping the trajectory toward sustainable and high-performance energy storage solutions.

LITHIUM-ION BATTERIES

Historical Development

Li-ion batteries emerged as a revolutionary energy storage solution in the early 1990s, building on foundational research by pioneers like Whittingham, Goodenough, and Yoshino. These batteries quickly dominated the market due to their high energy density, relatively long cycle life, and decreasing costs (M.M. Hasan et al, (2025).

Historical Development

Lithium-ion (Li-ion) batteries marked a pivotal advancement in energy storage technology, emerging as a revolutionary solution in the early 1990s. Their development was underpinned by decades of foundational research by visionaries such as Stanley Whittingham, John Goodenough, and Akira Yoshino, whose collective efforts laid the groundwork for modern battery technology (J. Xie et al, (2020).

Whittingham's early work in the 1970s introduced the concept of intercalation electrodes, while Goodenough's development of lithium cobalt oxide ($LiCoO_2$) as a cathode material significantly increased energy density. Yoshino's contributions in the 1980s, including the incorporation of a carbon-based anode and critical safety improvements, culminated in the commercialization of the first Li-ion battery by Sony in 1991.

These batteries swiftly took over the market, transforming portable electronics and later influencing electric vehicles and grid storage. Their success is due to a mix of high energy density, a relatively long cycle life, and rapidly decreasing manufacturing costs resulting from extensive innovation and scaling.

Evolution of Lithium-Ion Battery

The lithium-ion (Li-ion) battery has become a cornerstone of modern energy storage, evolving through decades of scientific discovery and technological innovation:

Early Foundations (1970s): Stanley Whittingham pioneered the concept of rechargeable lithium batteries by demonstrating the intercalation of lithium ions into titanium disulfide (TiS_2) cathodes. Paired with a lithium metal anode, these early batteries offered high energy density but faced challenges such as dendrite formation, which caused short circuits and safety risks (V. M. Reddy, (2020). Fig.1 shows the history and improvement of Lithium–ion battery.

Figure 1. The history and improvement of Lithium–ion battery.

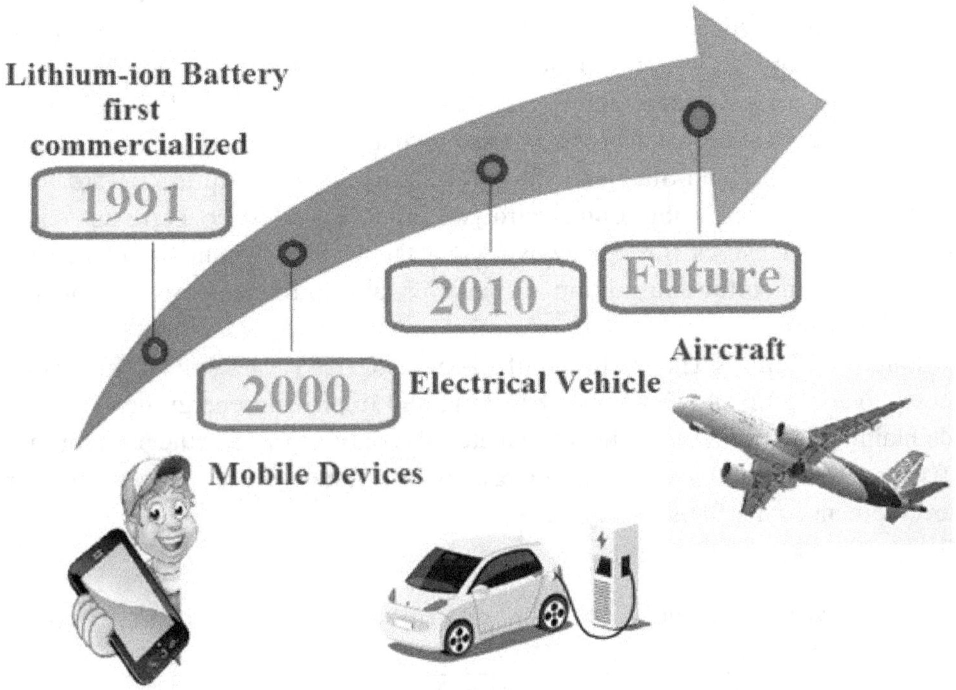

Key Materials Discovery (1980): John Goodenough identified lithium cobalt oxide ($LiCoO_2$) as a cathode material, significantly increasing energy density and stability. This breakthrough established the groundwork for commercial Li-ion batteries.

First Commercial Prototype (1985): Akira Yoshino advanced Li-ion technology by introducing a carbon-based anode, such as petroleum coke or graphite, to replace metallic lithium. This innovation enhanced safety by mitigating dendrite formation and enabled practical commercialization.

Commercialization (1991): Sony, in collaboration with Yoshino, launched the first commercial Li-ion battery. It featured a graphite anode, a $LiCoO_2$ cathode, and a liquid electrolyte, achieving a balance between energy density, safety, and rechargeability.

Advancements in Materials (1990s–2000s): Alternative cathode materials, such as lithium iron phosphate ($LiFePO_4$) and nickel-manganese-cobalt (NMC) oxides, were developed to improve safety, cycle life, and energy density. Improvements in electrolytes and separators enhanced thermal stability and reduced the risk of thermal runaway (Z. Tang et al, (2023).

Mass Adoption and Scaling (2000s): Li-ion batteries began dominating portable electronics, such as smartphones and laptops, due to their superior energy density and compact size. Large-scale production led to significant cost reductions.

Application Expansion (2010s):The demand for electric vehicles (EVs) and renewable energy storage drove innovations in high-capacity batteries. Chemistries like NMC and nickel-cobalt-aluminum (NCA) became prevalent in EV applications due to their high energy density and power output.

Emergence of Solid-State and Beyond (2020s):Research shifted toward solid-state Li-ion batteries, using solid electrolytes to address safety concerns and enable higher energy densities. Simultaneously, efforts to develop sustainable alternatives, such as lithium-sulfur and lithium-air batteries, aim to reduce costs and environmental impact.

Future Directions:The evolution of Li-ion batteries is ongoing, with advancements focusing on sustainability, recycling, and improving energy density while maintaining safety. Novel designs and materials continue to drive the push for more efficient, durable, and eco-friendly energy storage solutions.Table.1 shows the entire development from 1980s to till date.

Table 1. Year-wise evolution chart summarizing key developments in lithium-ion battery Technology:

Year	Key Development
1980s	Initial Development: John B. Goodenough, Rachid Yazami, and Akira Yoshino made significant contributions to lithium-ion battery chemistry, identifying key materials like lithium cobalt oxide ($LiCoO_2$) for the cathode.
1991	Commercialization: Sony and Asahi Kasei released the first commercially viable lithium-ion battery. This marked the first use of $LiCoO_2$ in commercial products.
1996	Battery Safety Improvements: Introduction of safety mechanisms to prevent overheating and thermal runaway, addressing concerns of the early battery designs.
2000s	Advancements in Anode Materials: Research shifted toward improving anode materials, with graphite being the dominant choice. Lithium iron phosphate ($LiFePO_4$) and lithium titanate ($Li_4Ti_5O_{12}$) also gained attention for their safety and cycle life.
2003	Lithium Iron Phosphate ($LiFePO_4$): Improved performance for safety and longevity; $LiFePO_4$ became a significant development for applications requiring long life cycles, such as electric vehicles.
2006	Battery Capacity and Energy Density: Research into silicon-based anodes began to gain traction, promising significant increases in capacity compared to traditional graphite.
2009	Electric Vehicle Integration: Lithium-ion batteries began to be widely adopted in electric vehicles, most notably by Tesla, which used lithium-ion batteries in its Roadster.
2012	Solid-State Battery Research: Initial progress toward solid-state lithium-ion batteries, with the aim of improving safety, energy density, and longevity.

continued on following page

Table 1. Continued

Year	Key Development
2015	Nickel Manganese Cobalt (NMC) and NCA Chemistries: These cathode materials emerged as prominent contenders in automotive applications due to their balance of safety, cost, and performance.
2017	Battery Recycling Efforts: Increased focus on recycling technologies to recover valuable materials (e.g., lithium, cobalt, nickel) to reduce environmental impacts and address supply chain challenges.
2020	Advanced Manufacturing and Cost Reduction: Significant breakthroughs in reducing the cost of lithium-ion batteries through advancements in large-scale production, driving down prices for electric vehicles and consumer electronics.
2023	Solid-State and Lithium-Sulfur Batteries: Solid-state batteries, which promise higher energy densities and better safety, continue to progress, while lithium-sulfur batteries begin to show potential for next-generation energy storage solutions.
2024	Sustainability Focus: Ongoing research focuses on improving the sustainability of lithium-ion batteries through better recycling methods, less reliance on rare materials, and advancements in solid-state and hybrid battery technologies.

Basics of Battery Mechanism

A battery is a device that stores electrical energy and releases it through electrochemical reactions. A battery comprises one or more electrochemical cells that transform chemical energy into electrical energy through redox (reduction-oxidation) reactions. The fundamental working mechanism of a battery can be divided into several essential components and processes. The mechanism of the Li-ion Battery is illustrated in Fig. 2.

Figure 2. shows the mechanism of Li-ion Battery.

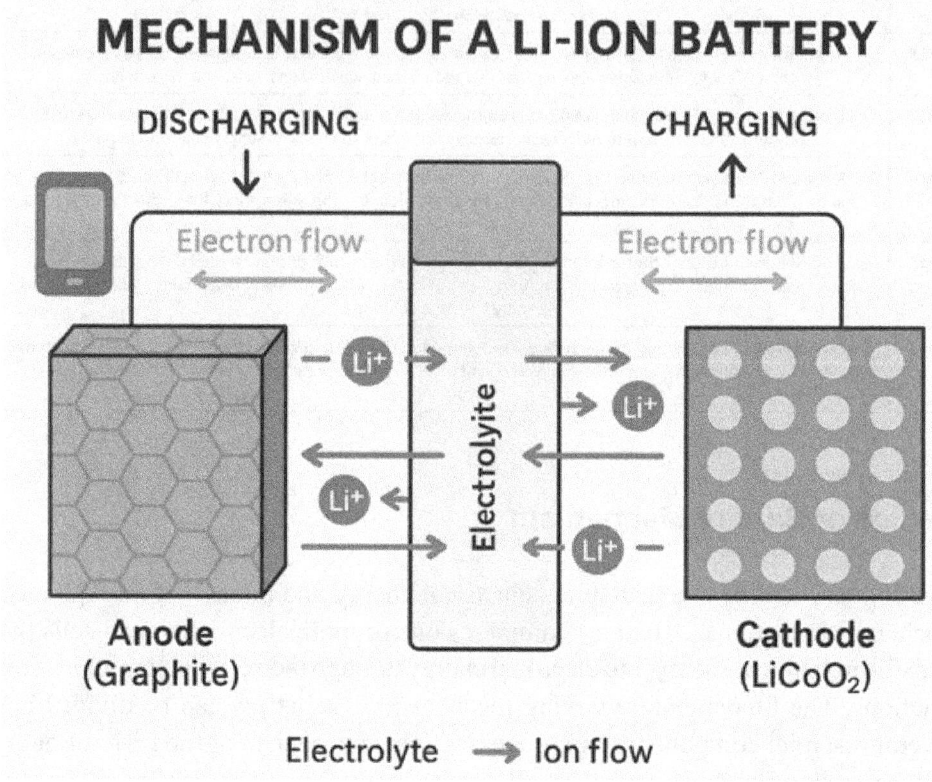

Key Components of a Battery

Anode (Negative Electrode): Oxidation takes place at the anode within the battery. While discharging (when the battery is powering a device), the anode releases electrons and undergoes oxidation. In a lithium-ion battery, the anode is usually composed of a substance that readily loses electrons, like lithium.

Cathode (Positive Electrode): The site of reduction is the cathode. It acquires the electrons released from the anode and undergoes a chemical reaction that results in energy storage. The cathode of lithium-ion batteries is usually composed of materials such as lithium cobalt oxide ($LiCoO_2$) or lithium iron phosphate ($LiFePO_4$).

Electrolyte: The electrolyte facilitates the movement of ions between the anode and cathode. It can take the form of a liquid (as seen in conventional Li-ion batteries), a gel, or a solid (as found in solid-state batteries). The battery relies on the electrolyte for its functioning, as it enables ion movement during charging and discharging.

Separator: The separator is a permeable membrane that keeps the anode and cathode apart while permitting the flow of ions through the electrolyte. It stops the two electrodes from short-circuiting.

External Circuit: The external circuit facilitates the movement of electrons from the anode to the cathode through the device being powered (like a smartphone or electric vehicle).

It is this flow of electrons that supplies the device with electrical energy.

Battery Operation: The Charge/Discharge Cycle

Discharge Cycle (Powering the Device):

When a battery is in use (discharging), the following process occurs:

Anode: At the anode, the material (e.g., lithium in Li-ion batteries) undergoes oxidation, losing electrons. These electrons flow through the external circuit to the cathode.

Cathode: The electrons arriving at the cathode are accepted by the cathode material, where they participate in a reduction reaction, storing energy.

Ions Flow: Meanwhile, positive ions (e.g., lithium ions in a Li-ion battery) move from the anode through the electrolyte and separator to the cathode to balance the electron flow.

Charge Cycle (Recharging the Battery):

During charging, the battery is connected to an external power source (like a charger). The following happens:

Anode: The external voltage forces electrons to flow back into the anode. This reduces the cathode material, and the ions from the cathode are driven back to the anode, restoring the chemical potential energy.

Cathode: At the cathode, the reduction reaction reverses, and the material releases the ions it previously stored.

Ion Movement: Positive ions flow from the cathode to the anode via the electrolyte, restoring the charge balance.

Energy Storage and Release: Energy Storage: As a result of chemical reactions at the anode and cathode, energy is stored in a battery. During the charging process, energy is accumulated by transferring ions from the cathode to the anode and by developing an electrical charge within the battery.

Energy Release: During the battery's discharge phase, the energy it has stored is converted into electrical energy. Devices are powered by the current of electrons moving from anode to cathode via an external circuit.

Factors Affecting Battery Performance

Several factors can influence a battery's performance:

Capacity: The overall quantity of charge a battery can retain, usually quantified in ampere-hours (Ah) or milliampere-hours (mAh).

Voltage: The potential difference that exists between the anode and cathode, which dictates how much electrical energy a battery can supply.

Efficiency: The battery's capability to sustain a high energy output while minimizing energy loss during charging and discharging cycles.

Cycle Life: The total number of complete charge and discharge cycles a battery can endure before its capacity deteriorates significantly.

Self-discharge: The natural loss of charge that occurs even when a battery is not in use.

The core operation of a battery relies on the movement of ions and electrons during charging and discharging cycles, propelled by oxidation and reduction reactions occurring at the anode and cathode. Through this mechanism, chemical energy can be transformed into electrical energy, which serves as power for devices ranging from small electronics to electric vehicles (Alex K. Koech et al., 2024). With the evolution of battery technologies, grasping these fundamental principles is crucial for enhancing performance, safety, and sustainability in a range of applications.

CHEMISTRY AND COMPONENTS

Li-ion batteries rely on the reversible intercalation of lithium ions between the anode and cathode through an electrolyte. Key components include:

Lithium-ion (Li-ion) batteries operate based on the reversible intercalation and deintercalation of lithium ions between the anode and cathode during charging and discharging cycles. This mechanism enables efficient energy storage and release, making Li-ion batteries a preferred choice for diverse applications (T. Rahman et al, (2024). The primary components and their roles are:

Anode: Typically composed of graphite, the anode provides a stable and conductive structure for lithium-ion intercalation during charging. Its stability and high capacity make it an ideal material for reversible lithium storage.

Cathode: The cathode is usually made from lithium metal oxides such as lithium cobalt oxide ($LiCoO_2$), lithium iron phosphate ($LiFePO_4$), and nickel-manganese-cobalt (NMC) variants. These materials determine the battery’s energy density, safety, and overall performance.

Electrolyte: A liquid or gel electrolyte containing lithium salts (e.g., lithium hexafluorophosphate, $LiPF_6$) dissolved in organic solvents facilitates the transport of lithium ions between the electrodes. The electrolyte must balance conductivity, stability, and safety.

Separator: A porous polymer membrane physically separates the anode and cathode, preventing electrical short circuits while allowing lithium-ion flow. This component is critical for maintaining safety and efficiency.

The synergistic interaction of these components ensures the high energy density, cycle life, and reliability of Li-ion batteries. However, ongoing research aims to improve these materials to address safety concerns, enhance performance, and reduce costs. Table.2 shows the various materials used in Lithium ion batteries.

Table 2. Summarizing various materials used in different components of lithium-ion batteries:

Component	Materials	Advantages	Challenges
Anode	- Graphite (Most common)	High conductivity, good cycle life	Limited capacity, potential dendrite formation
	- Lithium Titanate (LTO)	High safety, excellent cycling stability	Lower energy density
	- Silicon-based (e.g., Silicon-Graphite)	High capacity	Volume expansion, cycle instability
Cathode	- Lithium Cobalt Oxide ($LiCoO_2$)	High energy density	Limited lifespan, thermal instability, cobalt toxicity
	- Lithium Iron Phosphate ($LiFePO_4$)	High safety, long cycle life, lower cost	Lower energy density
	- Nickel Manganese Cobalt Oxide (NMC)	Good balance of energy density, lifespan, and safety	Cost of cobalt, thermal management
	- Nickel Cobalt Aluminum Oxide (NCA)	High energy density	Thermal instability, cost of cobalt
	- Lithium Nickel Manganese Oxide (LNMO)	High voltage, cobalt-free	Limited commercial availability
Electrolyte	- Liquid Electrolytes (e.g., Lithium Hexafluorophosphate in carbonate)	High ionic conductivity	Flammability, electrolyte leakage
	- Solid-State Electrolytes (e.g., Sulfides, Oxides, Polymers)	Non-flammable, higher energy density potential	Manufacturing complexity, cost
	- Gel Polymer Electrolytes	Improved safety compared to liquid electrolytes	Lower ionic conductivity than liquid electrolytes
Separator	- Polyethylene (PE)	Good thermal stability	Poor mechanical properties at high temperatures
	- Polypropylene (PP)	Good chemical resistance	Shrinkage at high temperatures

continued on following page

Table 2. Continued

Component	Materials	Advantages	Challenges
	- Ceramic-Coated Separators	Enhanced safety and thermal stability	Higher cost
	- Solid-State Separators	Integrated with solid electrolytes for advanced designs	Limited flexibility, manufacturing complexity

Discharge Process (Battery in use):

During discharge, the lithium ions move from the anode (typically graphite) to the cathode (typically lithium cobalt oxide $LiCoO_2$). The electrons flow through the external circuit, providing electric current.

At the anode (Graphite):

$$LiC_6=Li^++e^-+C_6$$

(Lithium ions move from the anode to the electrolyte, while electrons flow through the external circuit.)

At the cathode (Lithium Cobalt Oxide, LiCoO2)

$$LiCoO_2+Li^++e-=LiCoO_2$$

(Lithium ions are intercalated into the cathode materials)

Overall Discharge Reaction is: $LiC_6+LiCoO_2=C_6+Li_2CoO_2$

Carging Process (Battery Charging):

During charging the process is reversed and lithium ions move back from the cathode to the ande, storing energy.

At the anode (Graphite):

$$Li_6+Li^++e^-=LiC_6$$

At cathode (Lithium Cobalt Oxide,LiCoO2)

$$LiCoO_2=LiCoO_2+ Li^++e^-$$

Overall, the charge Reaction is:

$$Li_2CoO_2=LiCoO_2+Li^++e^-$$

Discharge: Lithium ions move from the anode to the cathode, release energy.

Charge: Lithium ions move from the cathode back to the anode, storing energy.

The efficiency of a battery is influenced by several factors, which can affect its overall performance, including its capacity, power output, cycle life, and energy efficiency. These factors can be broadly categorized into material-related, design-related, operational, and environmental factors. Below are the key factors that impact battery efficiency:

Battery Chemistry

Electrode Type: The efficiency of the battery is greatly influenced by the materials used for the anode and cathode (e.g., graphite, lithium cobalt oxide, lithium iron phosphate). Materials exhibiting greater conductivity and more stable electrochemical properties tend to deliver superior performance.

Electrolyte Composition: The electrolyte is essential for enabling the transfer of ions between the anode and cathode. The selection of electrolyte (liquid, gel, or solid) can influence the battery's internal resistance, ion transfer efficiency, and operational temperature range.

Internal Resistance

Ohmic Losses: Energy losses manifested as heat can occur due to the resistance of the battery components (electrodes, electrolyte, and separator), which diminishes overall efficiency. When internal resistance is high, heat generation during charging and discharging increases, potentially shortening the battery's lifespan.

Ionic Resistance: Efficiency is also influenced by how easily ions can move through the electrolyte. When ionic resistance is high, the battery's efficiency during charging and discharging cycles decreases.

Temperature

Effect on Chemical Reactions: The efficiency of batteries is greatly influenced by temperature. Elevated temperatures can lead to quicker deterioration of materials, particularly the electrolyte; conversely, reduced temperatures can hamper battery performance by decelerating ion movement and electrochemical processes.

Optimal Operating Range: Each battery chemistry has an optimal temperature range (e.g., Li-ion batteries typically perform best at around 20-25°C). Operating outside this range can reduce efficiency and cause irreversible damage.

Charge/Discharge Rates

Fast Charging and Discharging: Fast charging or discharging can lead to reduced efficiency due to higher internal resistance and heat generation. High charge/discharge rates can also cause the buildup of dendrites (in the case of lithium-ion) or stress the electrode materials, leading to reduced capacity and shorter cycle life.

C-rate: The C-rate indicates how fast a battery is charged or discharged compared to its maximum capacity. Efficiency and cycle life are usually diminished by higher C-rates (rapid charging/discharging).

State of Charge (SOC)

Voltage Window: The efficiency of a battery is affected by how deeply it is discharged and charged. Charging to full capacity or discharging too low can cause efficiency losses due to chemical degradation and increased wear on materials (I. N. Shchurov et al, (2021).

Depth of Discharge (DoD): Deeper discharge cycles typically result in more significant wear and lower efficiency, shortening the battery's lifespan.

Cycle Life and Aging

Degradation of Materials: With repeated charging and discharging cycles, the materials (like the anode and cathode) in batteries degrade over time, diminishing their ability to efficiently store and release energy. This results in a reduction of capacity and an escalation of internal resistance (Li et al., 2024).

Capacity Fade: With battery aging, efficiency declines due to the reduced ability of the electrodes to store ions. This is especially the case for lithium-ion and lithium-sulfur batteries, which encounter difficulties like dendrite formation and the polysulfide shuttle effect.

Battery Design and Construction

Electrode Surface Area: The surface area of the electrodes can influence the efficiency of ion exchange and energy storage. Larger surface areas typically result in better ion movement and reduced resistance, increasing efficiency (Lobato-Peralta et al, (2024).

Porosity of Electrodes: Higher porosity allows for better ion flow within the electrode, reducing resistance and improving efficiency.

Separator Properties: The separator between the anode and cathode must allow ions to pass freely while preventing short circuits. The material and design of the separator can influence the battery's efficiency by reducing or enhancing ionic flow.

Charging Algorithm

Charge Management: The efficiency of battery charging depends on the charging algorithm. Overcharging or charging too quickly can cause heating, reduce battery life, and decrease efficiency. Smart charging systems that control voltage, current, and temperature improve overall charging efficiency (Tomaszewska et al, (2019).

Constant Current vs. Constant Voltage: The charging strategy used can also affect efficiency. Constant current (CC) charging followed by constant voltage (CV) charging is a common method for lithium-ion batteries, as it optimizes the charging process to avoid overheating and degradation.

Environmental Factors

Humidity: Excess moisture can affect battery materials, especially the electrolyte, leading to reduced efficiency and potential short circuits.

Vibration and Shock: Mechanical stresses like vibration and shock can damage battery components, leading to internal resistance increases, reduced capacity, and lower efficiency.

External Load

Load Variability: The nature of the external load (e.g., devices or vehicles) affects battery efficiency. A constant, stable load is more efficient compared to a fluctuating load that causes the battery to repeatedly cycle through charge and discharge states.

Battery Management System (BMS)

Monitoring and Control: The BMS is essential for managing battery performance by tracking voltage, temperature, and state of charge, and ensuring the battery functions within safe parameters. An inefficiently designed or faulty BMS can result in ineffective charging and discharging cycles (Krishna et al, (2024).

Battery Packaging

Heat Dissipation: The physical design and packaging of the battery influence its ability to dissipate heat. Inefficient heat management can lead to thermal buildup and reduced efficiency.

Energy Density: High energy-density batteries are generally more compact but can generate more heat, so optimizing the energy density while ensuring efficient heat dissipation is a key factor in improving overall efficiency.

Energy Conversion Losses

Power Conversion Efficiency: In battery-operated devices, the efficiency of converting stored energy to usable power (through converters and inverters) can affect the total efficiency of the system. Energy lost during conversion reduces the net energy available from the battery (Orth et al, (2023).

The efficiency of a battery is influenced by various internal and external factors, such as the selection of materials, design features, charging and discharging methods, temperature, and usage conditions. To achieve the best battery performance, it is crucial to balance these factors so that the battery functions under conditions that reduce internal resistance, avoid degradation, and optimize energy storage and discharge. Continuous progress in materials science, battery design, and management systems is crucial for enhancing battery efficiency and prolonging battery life.

Lithium-ion (Li-ion) batteries are one of the most widely used energy storage technologies due to their high energy density, long cycle life, and relatively low self-discharge rate. They have found applications across various sectors. Below are some of the key applications of Li-ion batteries:

Consumer Electronics

Smartphones: Smartphones predominantly utilize Li-ion batteries as their power source, owing to the batteries' compact form factor and high energy density.

Laptops: Portable computing devices like laptops rely on Li-ion batteries for long-lasting power and lightweight design.

Tablets: Tablets use Li-ion batteries for portability and energy efficiency, providing hours of use on a single charge.

Smart watches and Wearables: Small and lightweight Li-ion batteries are ideal for wearable devices, providing long battery life without adding bulk.

Cameras: Digital cameras, including high-performance professional cameras, use Li-ion batteries for extended use and quick recharge times.

Electric Vehicles (EVs)

Passenger Cars: In electric vehicles (EVs), Li-ion batteries serve as the main energy storage solution, providing a high energy density and extended driving range. Electric vehicles from manufacturers such as Tesla, Nissan, and Chevrolet are powered by Li-ion batteries.

Electric Buses: EV buses utilize Li-ion batteries to cut greenhouse gas emissions and provide a more sustainable urban transport option (Das, 2025).

Electric Scooters, Motorcycles, and Bicycles: Li-ion batteries power various types of electric two-wheelers, making them efficient, lightweight, and practical for personal transport.

Hybrid Electric Vehicles (HEVs): HEVs combine internal combustion engines with Li-ion battery-powered electric motors to enhance fuel efficiency and reduce emissions.

Energy Storage Systems

Grid Storage: Lithium-ion batteries serve for energy storage at the grid level, making it possible to store electricity from renewable sources such as solar and wind. This contributes to grid stabilization and the handling of supply-demand variations (Tasneem et al., 2025).

Integration of Renewable Energy: They accumulate surplus energy generated by wind turbines or solar panels during periods of low demand and discharge it when demand for energy is elevated.

Home Energy Storage: Residential energy storage solutions like the Tesla Powerwall utilize Li-ion batteries to capture solar energy for future use, thereby enhancing energy autonomy and reliability.

Power Tools

Cordless Power Tools: Lithium-ion batteries are widely used in power tools such as drills, saws, and sanders. Compared to older battery technologies such as NiCd, they provide advantages of longer run time, quicker charging, and reduced weight (Waseem et al., 2025).

Electric Lawn Equipment: Mowers, trimmers, and leaf blowers powered by Li-ion batteries offer a lightweight and dependable energy source for landscaping.

Medical Devices

Portable Medical Equipment: Li-ion batteries power portable medical devices such as insulin pumps, hearing aids, pacemakers, and defibrillators. Their compact size and reliability are ideal for critical medical applications.

Wearable Health Monitoring Devices: Li-ion batteries are used in wearable health devices like fitness trackers, ECG monitors, and continuous glucose monitors, which need lightweight and long-lasting batteries.

Electric Wheelchairs: Powered wheelchairs and mobility scooters often use Li-ion batteries for efficient, long-lasting power to assist with mobility.

Aerospace and Aviation

Electric Aircraft: Li-ion batteries are being explored for use in electric aircraft, where their high energy density and light weight make them suitable for short-range flights.

Unmanned Aerial Vehicles (UAVs): Drones use Li-ion batteries to power their flight systems. The batteries' energy density allows drones to achieve longer flight times, making them ideal for applications such as surveillance, delivery, and inspection.

Satellite Power Systems: Some satellites use Li-ion batteries to store energy for operations when they are not in direct sunlight. The lightweight and high-efficiency properties of Li-ion batteries are well-suited for space applications.

Robotics

Robotic Systems: Li-ion batteries power various types of robots, including autonomous mobile robots (AMRs), drones, and robotic arms. Their long-lasting charge and compact size enable robots to operate for extended periods without recharging (T. Mikołajczyk et al, (2023).

Industrial Robots: Li-ion batteries are used in warehouse robots, material handling robots, and delivery robots, enhancing their productivity by providing continuous, reliable power.

Military and Defense

Portable Power Systems: Li-ion batteries provide portable power solutions for military applications, including communication devices, night vision equipment, and portable energy storage for remote operations.

Electric Military Vehicles: Military vehicles, including electric and hybrid-electric armored vehicles, utilize Li-ion batteries due to the importance of reduced emissions and enhanced operational stealth.

Unmanned Systems: Li-ion batteries are used in military drones, robots, and unmanned ground vehicles (UGVs) for reconnaissance, surveillance, and tactical operations.

Backup Power Systems

Uninterruptible Power Supplies (UPS): Li-ion batteries are used in UPS systems to provide backup power during electrical outages, offering a compact and efficient solution for homes, offices, and data centers.

Emergency Lighting Systems: Li-ion batteries provide reliable backup power for emergency lighting in commercial, industrial, and residential buildings.

Telecommunications: Telecommunication infrastructure, including cell towers and data centers, uses Li-ion batteries for backup power to ensure continuous operation during power failures.

Transportation Infrastructure

Electric Trains: Some electric trains use Li-ion batteries to provide power for specific segments of their journey, particularly in areas where overhead power lines are not feasible.

Trolleybuses: Li-ion batteries are used in trolleybuses and other electric public transportation systems, where they help power the vehicle between charging stations or reduce reliance on overhead cables.

Smart Grids

Energy Management: Li-ion batteries are essential components of smart grids, aiding in more efficient energy storage, the incorporation of renewable energy sources, and improvements to the grid's overall stability and reliability.

Demand Response: Li-ion battery systems are used in demand response applications, where they help balance supply and demand by providing backup power during peak times and storing energy when demand is low.

Due to their efficiency, longevity, and versatility, lithium-ion batteries have emerged as the leading energy storage solution. Their uses extend to consumer electronics, electric vehicles, energy storage solutions, medical devices, aerospace applications, military uses, backup power systems, and beyond. As technology continues to evolve, new applications for Li-ion batteries are emerging, especially in renewable energy integration, smart grids and electric transportation, making them a critical component of modern energy systems. Fig.3 shows few of the applications of Li-ion Battery.

Figure 3. Shows few of the applications of Li-ion Battery.

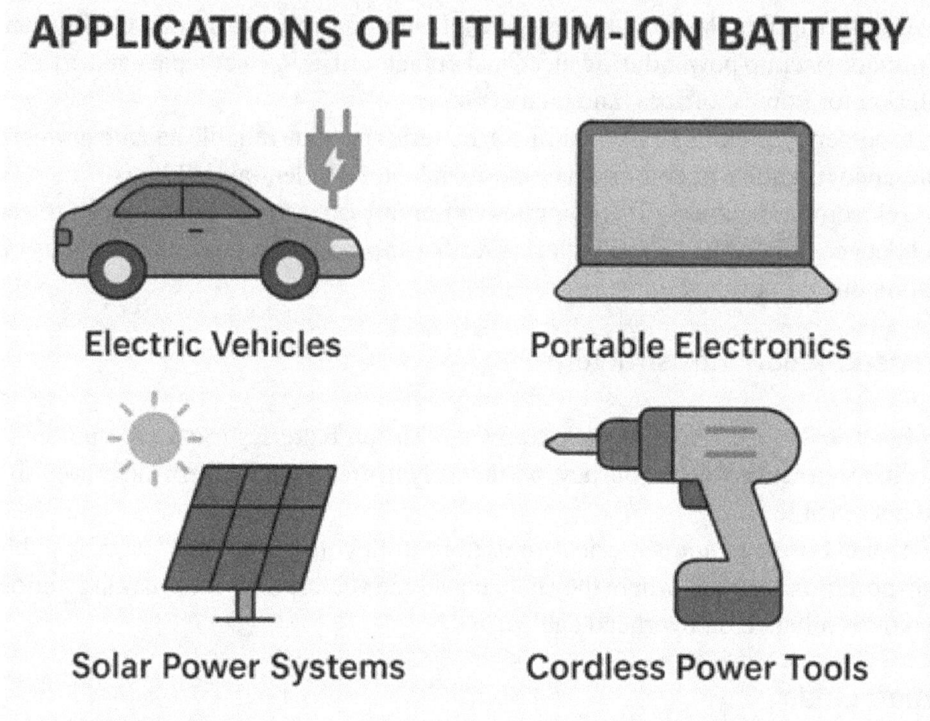

Challenges of Lithium-Ion Batteries

Despite their widespread success and adoption in a variety of applications, lithium-ion (Li-ion) batteries face several significant challenges that hinder their performance, scalability, and long-term sustainability. These challenges are crucial to address as demand for high-performance energy storage solutions continues to grow.

Limited Energy Density (200-300 Wh/kg):

Issue: With an energy density of 200-300 Wh/kg, lithium-ion batteries are limited in their use for applications where weight and size are crucial factors, including portable electronics, electric vehicles (EVs), and aerospace. Although enhancements have been implemented, more progress is required to satisfy the growing energy needs of these sectors (Koech et al, (2024).

Impact: The battery size and weight must be balanced due to the low energy density, which decreases the range of electric vehicles and restricts the energy storage capacity of portable devices. Li-ion batteries must enhance their energy density to stay competitive with alternative energy storage technologies (such as solid-state and lithium-sulfur batteries).

Safety Concerns, Including Thermal Runaway:

Issue: Li-ion batteries are susceptible to safety risks such as thermal runaway, which can occur when the battery overheats, leading to a fire or explosion. Overcharging, short-circuiting, physical damage, and manufacturing defects can all trigger thermal runaway. Due to the flammability of the liquid electrolyte found in standard Li-ion batteries, fires are more likely (Chen et al., 2021).

Impact: Due to safety concerns, the use of Li-ion batteries in certain crucial applications—like electric vehicles and large-scale energy storage systems—is restricted. Manufacturers and researchers continue to develop improved battery management systems (BMS) and incorporate safety mechanisms (e.g., thermal protection layers, non-flammable electrolytes) to mitigate these risks.

Dependence on Critical Materials Like Cobalt:

Issue: Lithium-ion batteries rely on several critical materials, including cobalt, nickel, and lithium. Cobalt, in particular, has raised concerns due to its limited availability, high cost, and ethical sourcing issues. Cobalt extraction mainly occurs in areas such as the Democratic Republic of Congo (DRC), where mining practices frequently include child labor and hazardous working environments. Additionally, the supply of cobalt is vulnerable to geopolitical risks and market fluctuations.

Impact: Due to the reliance on cobalt and other essential resources, Li-ion battery manufacturing is expensive and cannot be sustained over time. To guarantee that Li-ion batteries remain sustainable and affordable, the pressure is mounting to find substitutes for these materials, lessen dependence on them, or improve their recycling. To address these challenges, work is being done to create battery chemistries that do not contain cobalt (such as lithium iron phosphate ($LiFePO_4$) and nickel-manganese-cobalt (NMC) alternatives).

Limited Cycle Life and Degradation:

Issue: Over time, lithium-ion batteries experience degradation, leading to a decrease in capacity and cycle life. This is primarily due to factors such as the growth of lithium dendrites, loss of active material in the electrodes, and breakdown of the electrolyte. In typical consumer electronics, this degradation manifests as shorter battery life, while in electric vehicles, it can lead to reduced driving range and higher replacement costs (S. S. Madani et al, (2025).

Impact: The limited lifespan of Li-ion batteries affects their long-term cost-effectiveness and sustainability, particularly in industries that require frequent cycling, such as EVs and grid storage. Research is focused on developing better materials, battery management systems, and charging strategies to improve battery longevity and mitigate degradation.

Environmental Impact and Recycling:

Issue: The environmental impact of lithium-ion batteries is a growing concern, particularly with regard to their end-of-life disposal and recycling. Batteries contain hazardous materials (e.g., heavy metals, solvents) that can pose environmental risks if not properly managed. While lithium and cobalt can be recycled, the recycling rates remain relatively low, and the process is energy-intensive and expensive.

Impact: The increasing demand for Li-ion batteries raises concerns about the sustainability of the supply chain and the environmental footprint of battery production and disposal. Recycling efforts need to be scaled up, and new technologies must be developed to improve the efficiency and economic viability of recycling processes.

High Manufacturing Costs:

Issue: Although the price of lithium-ion batteries has seen a substantial decrease in the last ten years, thanks to improved manufacturing processes and economies of scale, they are still comparatively costly when measured against other energy storage alternatives. This holds particularly for high-performance variants employed in electric vehicles and massive storage systems.

Impact: The high costs associated with manufacturing restrict the broad use of electric vehicles and renewable energy storage systems. In order to lower costs even more, it is crucial to continue researching materials that are less expensive and more abundant, as well as to enhance manufacturing processes.

Addressing These Challenges

Efforts are underway to address these challenges in several ways:

Material Substitution: Researchers are investigating alternative materials to lessen or eradicate the reliance on cobalt and other rare elements, including lithium iron phosphate ($LiFePO_4$), sodium-ion, and solid-state batteries.

Battery Design Innovations: New designs like solid-state batteries and lithium-sulfur batteries are promising to overcome the limitations of energy density, safety, and cycle life.

Battery Recycling and Reuse: Improved recycling technologies and battery reuse initiatives are being developed to reduce environmental impacts and recover valuable materials from spent batteries.

Advanced Manufacturing Techniques: Automation, improved quality control, and novel materials are being leveraged to reduce the cost and improve the scalability of Li-ion battery production.

These advancements are critical for addressing the current limitations of Li-ion batteries and enabling the transition to more sustainable, high-performance energy storage solutions.

Solid-State Batteries:

Solid-state batteries (SSBs) represent a sophisticated category of energy storage device, in which the liquid electrolyte found in conventional lithium-ion batteries is substituted with a solid electrolyte. This transition presents various unique benefits, positioning solid-state batteries as a compelling option for numerous applications, including consumer electronics, electric vehicles (EVs), and grid energy storage. Solid-state batteries offer key characteristics and advantages, including:

Advantages of Solid-State Batteries

Enhanced Safety: Conventional lithium-ion batteries utilize liquid electrolytes that are combustible and can cause thermal runaway or fires if they are harmed or charged incorrectly. With their non-flammable solid electrolytes, solid-state batteries provide a significantly improved level of safety, minimizing the risks of leakage, combustion, and short-circuiting.

Greater Energy Density: Solid-state batteries could reach energy densities that exceed those of standard lithium-ion batteries. This is because it can utilize materials like lithium metal for the anode, which can hold more lithium ions than graphite, as well as solid electrolytes that can handle greater voltages.

Longer Cycle Life: The solid-state design reduces issues like dendrite formation (a key cause of failure in conventional lithium-ion batteries) and degradation of the electrolyte, thereby offering longer cycle life and better stability over time.

Accelerated Charging: The greater ionic conductivity of the solid electrolyte and the option of using lithium metal for the anode may allow for reduced charging durations, which is essential for uses such as electric cars.

Lighter Weight: By substituting the liquid electrolyte and some metal components, solid-state batteries can decrease the overall weight of the battery, aiding in a more compact and efficient design.

Challenges and Current Limitations

Manufacturing Complexity: It is more difficult to produce solid-state batteries than to produce conventional lithium-ion batteries. To guarantee efficiency and reliability, it is necessary to produce high-quality solid electrolytes and to optimize the interface between these electrolytes and the electrodes.

Cost: At present, the cost of solid-state batteries is much greater than that of traditional lithium-ion batteries, mainly because of the complex materials and manufacturing processes involved. Nevertheless, current investigations are aimed at lowering costs via new materials and production techniques that can be scaled up.

Material Selection: A range of materials can be used to create solid electrolytes, including ceramics (like lithium phosphorus oxynitride (LiPON)), sulfides, or polymers. Each of these options comes with its own set of benefits and challenges, including factors such as ionic conductivity, chemical stability, and ease of fabrication.

Low Temperature Performance: Some solid electrolytes exhibit poor ionic conductivity at low temperatures, which could limit their performance in colder environments unless further advancements are made in material science.

Types of Solid Electrolytes Used in Solid-State Batteries

Ceramic Solid Electrolytes: Notable examples are lithium garnet ($Li_7La_3Zr_2O_{12}$) and lithium lanthanum zirconium oxide (LLZO). These electrolytes exhibit excellent stability and high ionic conductivity, but they are usually brittle and demand precise interfaces with electrodes.

Polymer Electrolytes: Polymers such as polyethylene oxide (PEO) or poly(ethylene glycol) (PEG) are used in solid-state batteries for their flexibility and ease of processing. However, they tend to have lower ionic conductivity compared to ceramics.

Sulfide Electrolytes: Sulfide-based electrolytes, like lithium thiophosphate (Li_3PS_4), offer high ionic conductivity and better mechanical properties than ceramics. However, they can be sensitive to moisture and may have lower stability under certain conditions.

Future of Solid-State Batteries: The potential of solid-state batteries to revolutionize energy storage is immense, especially in sectors where safety, energy density, and longevity are critical, such as in electric vehicles and portable electronics. Ongoing research aims to address the technical and manufacturing challenges to make solid-state batteries commercially viable and cost-effective. As new materials and fabrication techniques emerge, solid-state batteries are expected to become more widely available, marking a significant leap in battery technology.

Solid Electrolyte Materials

Solid electrolytes fall into three main categories:

- **Ceramic:** Materials like garnet-type $Li_7La_3Zr_2O_{12}$ (LLZO) exhibit high ionic conductivity and stability.
- **Polymeric:** Organic polymers such as PEO (Polyethylene oxide) are flexible but exhibit lower ionic conductivities.
- **Composite:** Hybrid materials combining ceramic and polymeric properties.

Advantages

- Enhanced safety due to reduced flammability.
- Higher energy densities facilitated by the use of lithium metal anodes.
- Potential for extended cycle life.

Current Limitations

- Manufacturing challenges, including cost and scalability.
- Interface stability between solid electrolyte and electrodes.
- Limited ionic conductivity at ambient temperatures for some materials.

Beyond Li-Ion: Emerging Technologies

With the demand for energy storage growing, the limitations of Lithium-Ion (Li-ion) batteries—such as their energy density, dependence on critical materials, and safety issues—have prompted investigations into new battery technologies. The goal of these alternative technologies is to address the inherent limitations of Li-ion batteries, with the potential for higher energy densities, reduced costs, and enhanced sustainability. Here’s an overview of several promising battery technologies that go beyond lithium (Hasan et al, (2025):

Sodium-Ion Batteries::

Sodium-ion batteries (Na-ion) offer a sustainable alternative to Li-ion technology by utilizing sodium (Na) in place of lithium (Li). Due to sodium's greater abundance, lower cost, and widespread availability, Na-ion batteries present a promising option for large-scale energy storage solutions.

Advantages:

Abundant Material: Sodium is much more abundant than lithium, and its cost is significantly lower.

Safety: Sodium-ion batteries are considered safer than Li-ion batteries due to their lower reactivity and reduced risk of thermal runaway.

Challenges:

Lower Energy Density: Na-ion batteries have a lower energy density (around 100-150 Wh/kg) compared to Li-ion batteries (150-250 Wh/kg). This is due to the larger size of sodium ions, which makes it more difficult to achieve the same packing density.Cycle Life: Na-ion batteries generally have a shorter cycle life, as sodium ions tend to cause more stress on the electrodes during cycling.

Material Development: The development of suitable cathode and anode materials that can effectively accommodate sodium ions is still a challenge.

Applications:

Large-scale energy storage, grid storage, and potentially in applications where weight is less of a concern.

Lithium-Sulfur (Li-S) Batteries::

With one of the highest theoretical energy densities of all battery chemistries, lithium-sulfur (Li-S) batteries are highly appealing for use in electric vehicles (EVs) and aerospace applications that require long-lasting power.

Advantages:

High Theoretical Energy Density: The theoretical energy density of Li-S batteries is much higher (around 2,500 Wh/kg) than that of Li-ion batteries (150-250 Wh/kg), which makes them promising for lightweight, high-capacity storage solutions.

Cost-Effective Materials: Sulfur is abundant, inexpensive, and environmentally friendly compared to materials like cobalt used in Li-ion batteries.

Challenges:

Low Conductivity of Sulfur: The electrical conductivity of sulfur and lithium sulfide (Li_2S) is low, which restricts the performance of Li-S batteries.

Polysulfide Shuttle Effect: When the battery cycles, polysulfides generated at the cathode may dissolve in the electrolyte and travel back to the anode. This process results in a loss of active material and a considerable reduction in battery capacity over time.

Cycle Life: The cycle life of Li-S batteries is relatively low (usually around 300-500 cycles) compared to that of Li-ion batteries, which restricts their practical applications.

Applications:

- Potential for use in electric vehicles, drones, and energy storage systems, but commercial viability remains a challenge.

Solid-State Lithium-Metal Batteries::

Solid-state lithium-metal batteries, which utilize lithium metal anodes and solid electrolytes, have the potential to reach energy densities far beyond those of conventional Li-ion batteries, thereby advancing battery technology (Nicola Boaretto et al, (2021).

Advantages:

Increased Energy Density: By employing lithium metal as an anode, rather than conventional graphite anodes, which has greater energy density, it is possible to achieve energy densities that are significantly higher and may surpass 500 Wh/kg.

Safety: Solid electrolytes (like lithium phosphorus oxynitride or garnet-based materials) do not catch fire and remove the risk of dendrite formation linked to liquid electrolytes.

Enhanced Efficiency: The solid-state configuration can diminish energy loss and enhance performance throughout the battery's lifecycle.

Challenges:

Dendrite Formation: Despite the theoretical safety of solid-state batteries, dendrites (spiky lithium metal structures) can form on lithium metal anodes, penetrating the solid electrolyte and causing short circuits and failures.

Challenges in Manufacturing: The production of solid-state batteries involves complexities and high costs, stemming from the necessity for processes with great precision and the challenges posed by establishing stable interfaces between the electrodes and the solid electrolyte.

Temperature Sensitivity: The ionic conductivity of solid-state electrolytes is often low at room temperature, which requires either an increase in operating temperature or the creation of new materials that can function effectively at lower temperatures.

Applications:

- High-performance applications, such as electric vehicles, aerospace, and military use, where energy density and safety are critical.

Beyond Lithium: Multivalent and Flow Batteries::

Magnesium-Ion and Calcium-Ion Batteries

Magnesium (Mg) and calcium (Ca) batteries use multivalent ions (Mg^{2+} and Ca^{2+}) rather than monovalent lithium ions. These ions can intercalate into electrode materials more efficiently, offering the potential for higher energy densities (Mousa Almousa et al, (2025).

Advantages:

Higher Capacity: Multivalent ions (Mg^{2+} and Ca^{2+}) have a higher charge than lithium, which theoretically offers greater energy storage capacity.

Abundant and Cheap Materials: Magnesium and calcium are abundant and non-toxic, which could lower costs and environmental impact compared to lithium.

Challenges:

Electrolyte Design: The development of efficient and stable electrolytes for magnesium and calcium batteries is ongoing, as these ions interact with materials in ways that differ from lithium ions.

Limited Cycle Life: Because multivalent ions are generally less mobile than lithium ions, this can lead to slower charging and discharging cycles and a reduced cycle life.

Applications:

Large-scale energy storage, grid-scale applications, and potentially for electric vehicles if technological challenges are overcome.

Redox Flow Batteries::

Redox flow batteries store energy in two electrolyte solutions, one positive and one negative, which are stored in external tanks and flow through the system during charge and discharge cycles.

Advantages:

Scalability: Redox flow batteries can be easily scaled for large energy storage systems, making them ideal for grid-scale applications and renewable energy storage.

Longevity: These batteries typically have a long cycle life, as their performance is not affected by the same issues as traditional batteries, such as dendrite formation or electrode degradation.

Challenges:

Low Energy Density: The energy density of redox flow batteries is lower than that of Li-ion or solid-state batteries, which limits their use in applications where high energy density is essential, such as in electric vehicles.

Complexity: The design of redox flow batteries is more complex and requires large storage tanks, which can make them less practical for certain consumer applications.

Applications:

Energy storage on a large scale for grid applications, including the balancing of renewable energy sources (like wind and solar) and backup power systems.The search for battery technologies that are more efficient, cost-effective, and sustainable is propelling innovations in various directions, transcending the boundaries of Li-ion batteries. Although Sodium-Ion batteries present a promising low-cost and plentiful alternative, they encounter difficulties regarding energy density and cycle life. Although Lithium–Sulfur batteries have the potential for very high energy densities, their development is impeded by problems with material conductivity and cycle life limitations. Solid-State Lithium-Metal batteries hold the potential to transform the energy storage industry due to their extremely high energy densities and enhanced safety. However, challenges such as dendrite formation and manufacturing complexity persist.

Meanwhile, multivalent batteries, like magnesium and calcium-ion, and redox flow batteries show potential for large-scale, sustainable energy storage, especially for grid applications, but face hurdles in energy density and efficiency.

In conclusion, while these emerging technologies have the potential to outpace Li-ion batteries in the future, significant research, development, and commercial scaling are required before they can compete on a large scale. Their success will depend on overcoming technical challenges, improving performance metrics, and reducing costs to make them viable alternatives to Li-ion in various applications.

FUTURE DIRECTIONS

Sustainability and Recycling

Recycling initiatives aim to recover critical materials like lithium, cobalt, and nickel. Research into sustainable alternatives and second-life applications for batteries is accelerating.

Advanced Manufacturing Techniques

Techniques such as 3D printing, roll-to-roll processing, and AI-driven material discovery are poised to revolutionize battery production.

Integration with Renewable Energy

Next-generation batteries will play a key role in grid-level energy storage, enabling higher penetration of renewable energy sources and reducing reliance on fossil fuels.

Comparative Study of Different Battery Technologies

Various battery technologies have been developed over the years, each offering distinct advantages and drawbacks. Here, we provide a comparative study of the most prominent battery technologies, including **Lithium-Ion (Li-Ion)**, **Solid-State Batteries**, **Lithium-Sulfur (Li-S)**, **Sodium-Ion**, **Lead-Acid**, and **Nickel-Metal Hydride (NiMH)** batteries. Table.3 Different types of battery and their properties.

Table 3. Comparative study of different types batteries.

Property/Technology	Lithium-Ion (Li-Ion)	Solid-State Batteries	Lithium-Sulfur (Li-S)	Sodium-Ion (Na-Ion)	Lead-Acid	Nickel-Metal Hydride (NiMH)
Energy Density	150-250 Wh/kg	300-500 Wh/kg (theoretical)	500-600 Wh/kg (theoretical)	100-150 Wh/kg	30-50 Wh/kg	60-120 Wh/kg
Voltage	3.6-4.2 V	3.0-4.5 V (depending on material)	2.1-2.5 V	3.0-3.6 V	2.0 V	1.2 V
Cycle Life	500-1500 cycles	1000-5000 cycles (theoretical)	300-500 cycles	2000-4000 cycles	300-500 cycles	500-1000 cycles
Charge Time	1-4 hours	1 hour or less (potential)	1-3 hours	2-5 hours	4-8 hours	2-4 hours
Safety	Risk of thermal runaway	Higher safety (non-flammable electrolyte)	Less stable, dendrite formation	Better than Li-Ion but lower than Solid-State	Low (acid spills, corrosion)	Moderate (hydrogen release)
Cost	\$100-\$300/kWh	High (\$300-\$500/kWh or more)	Moderate (\$100-\$200/kWh)	\$80-\$150/kWh	Low (\$50-\$100/kWh)	Moderate (\$100-\$200/kWh)
Environmental Impact	High (mining of lithium, cobalt, etc.)	Potentially lower (solid electrolytes may be more sustainable)	High (difficult recycling of sulfur)	Moderate (less environmentally harmful than Li-ion)	High (lead contamination)	Moderate (less toxic than Lead-Acid)
Temperature Range	-20°C to 60°C	-20°C to 60°C (subject to materials)	-20°C to 60°C	-20°C to 60°C	-18°C to 50°C	-20°C to 60°C

continued on following page

Table 3. Continued

Property/Technology	Lithium-Ion (Li-Ion)	Solid-State Batteries	Lithium-Sulfur (Li-S)	Sodium-Ion (Na-Ion)	Lead-Acid	Nickel-Metal Hydride (NiMH)
Applications	Consumer electronics, EVs, energy storage	Electric vehicles, portable electronics, energy storage	Electric vehicles, grid storage	Electric vehicles, grid storage	Backup power, automotive	Hybrid vehicles, consumer electronics
Technological Readiness	Mature and widely used	Still under development, promising but expensive	Under development, limited commercial use	Emerging, growing research focus	Mature but declining	Mature, primarily in hybrid vehicles

Lithium-Ion (Li-Ion) Batteries

- **Advantages**:
  - High energy density and efficiency.
  - Fast charge times and long cycle life.
  - Widely used in consumer electronics and electric vehicles.
  - Well-established infrastructure for manufacturing and recycling.
- **Drawbacks**:
  - Safety risks due to the flammability of the liquid electrolyte and potential for thermal runaway.
  - Dependency on critical raw materials like cobalt, which have environmental and ethical concerns.
  - High manufacturing costs compared to some older technologies.

Solid-State Batteries

- **Advantages**:
  - Higher safety profile due to non-flammable solid electrolytes.
  - Potential for higher energy densities (up to 500 Wh/kg) and longer cycle life.
  - Could lead to lighter, more compact batteries with faster charging times.
- **Drawbacks**:
  - High manufacturing cost and challenges in scaling production.
  - Solid-state electrolytes may suffer from low ionic conductivity at lower temperatures.

- Still in the developmental stage, with commercial mass production not yet realized.

Lithium-Sulfur (Li-S) Batteries

- **Advantages**:
  - High theoretical energy density (500-600 Wh/kg).
  - Use of sulfur, a cheaper and more abundant material, compared to cobalt and nickel.
  - Potential for significant cost reduction in battery production.
- **Drawbacks**:
  - Lower cycle life (300-500 cycles).
  - Difficulty in maintaining stability and preventing the dissolution of sulfur during cycling.
  - Less developed than Li-ion and solid-state technologies.

Sodium-Ion (Na-Ion) Batteries

- **Advantages**:
  - Sodium is abundant and inexpensive compared to lithium.
  - Safer than Li-ion as it is less reactive and has a lower tendency for thermal runaway.
  - Can perform similarly to Li-ion in terms of energy density and voltage.
- **Drawbacks**:
  - Lower energy density (100-150 Wh/kg) than lithium-ion batteries.
  - Lower cycle life and relatively slower charge/discharge rates.
  - Commercial production and large-scale adoption still in early stages.

Lead-Acid Batteries

- **Advantages**:
  - Low cost and long-established technology.
  - Well-suited for high-power applications like automotive starting and backup power.
  - Highly recyclable, reducing environmental impact.
- **Drawbacks**:
  - Low energy density (30-50 Wh/kg), making them unsuitable for applications requiring compact, high-energy storage.
  - Shorter cycle life and degradation over time.
  - Environmental impact due to lead contamination during disposal.

Nickel-Metal Hydride (NiMH) Batteries

- **Advantages**:
  - o Moderate energy density (60-120 Wh/kg) and safer than Li-ion.
  - o Widely used in hybrid electric vehicles (HEVs) and other mid-range power applications.
  - o Relatively safe, with no flammable liquid electrolytes.
- **Drawbacks**:
  - o Lower energy density compared to Li-ion batteries.
  - o Higher self-discharge rate.
  - o Heavier than Li-ion batteries, making them less suitable for lightweight applications.

Battery technology's future involves enhancing the performance and sustainability of current systems, as well as investigating new and emerging chemistries. Due to their energy density, safety improvements, and cost-effectiveness, Lithium-Ion batteries dominate the current market. However, their dependence on critical materials and safety concerns remain significant challenges. Although Solid-State Batteries promise improved safety and energy density, they remain under development (Satyam Rawat et al., 2025). **Lithium-Sulfur** and **Sodium-Ion** batteries provide alternative solutions with potential for lower costs and higher energy densities but face issues related to cycle life and commercial viability. **Lead-Acid** and **NiMH** batteries, while cost-effective and widely used, are being gradually phased out in favor of more efficient and higher-performance technologies.

CONCLUSION

The development of battery chemistry from Li-ion to solid-state and new technologies highlights the ongoing quest for greater energy densities, enhanced safety, and sustainability. Continued innovation will be critical to addressing global energy challenges and driving the transition to a cleaner, more efficient energy future.

There have been major developments in battery chemistry, with lithium-ion (Li-ion) batteries at the forefront of energy storage for a range of uses, including portable devices and electric vehicles (EVs). Though Li-ion technology has transformed the industry, it encounters multiple intrinsic challenges that impede its sustainability, efficiency, and safety over the long term. Research into alternative battery chemistries and designs has been driven by the constraints of energy density, safety concerns, reliance on essential materials such as cobalt, and environmental effects.

Lithium-ion batteries have demonstrated their worth across various applications owing to their relatively high energy density, rapid charge/discharge rates, and established technology. Nonetheless, in order to meet the rising demand for electric vehicles, renewable energy storage systems, and consumer electronics, considerable advancements in Li-ion batteries are required. Some of the current solutions aimed at enhancing their performance and sustainability include innovations in anode/cathode materials, improved management systems, and advancements in recycling technologies.

Solid-State Batteries, which show promise as an alternative, tackle numerous issues associated with Li-ion batteries, including safety, energy density, and lifespan. These batteries are less susceptible to thermal runaway and can achieve higher energy density due to the substitution of a solid for the liquid electrolyte, making them suitable for applications that demand compact, safe, and high-capacity energy storage, such as electric vehicles. Nevertheless, there are still difficulties regarding the choice of materials, manufacturing intricacies, and expenses. Advancements are continuing with the aim of enhancing solid electrolyte materials, scalability, and performance across various temperatures in order to make solid-state batteries more commercially viable.

Beyond Lithium-Ion: The future of battery chemistry is exploring even more innovative solutions that could potentially address the fundamental limitations of current technologies. Dies umfasst: Lithium-Schwefel-Batterien: Mit einer hohen Energiedichte und kostengünstigen Materialien stellt die Lithium-Schwefel-Technologie eine vielversprechende Alternative dar. Nonetheless, challenges like restricted cycle life and inadequate conductivity still require attention. Sodium-Ion Batteries: Sodium-ion batteries present a promising option because sodium is plentiful and less expensive, but their energy density and cycle life do not yet match those of lithium-ion batteries.

Next-Generation Chemistries: Innovative technologies like solid-state lithium-metal batteries, magnesium-ion, and aluminum-ion batteries are surfacing as potential contenders for the future, each offering distinct benefits related to energy density, cost, or availability of materials.

REFERENCES

Chen, Y., Kang, Y., Zhao, Y., Wang, L., Liu, J., Li, Y., & Li, B. (2021). A review of lithium-ion battery safety concerns: The issues, strategies, and testing standards. *Journal of Energy Chemistry*, *59*, 83–99.

Das, H. S. (2025). Life Cycle Assessment of Emerging Battery Technologies for Electric Vehicles. In Optimized Energy Management Strategies for Electric Vehicles (pp. 87-118). IGI Global Scientific Publishing.

Das, H. S., Mishra, S., & Roymahapatra, G. (2025). Advanced Nano-Structured Materials for Energy Storage Devices. *Advances in Chemical and Materials Engineering*, *2025*, 34. DOI: 10.4018/979-8-3693-5320-2.ch001

Hasan, M. M., Haque, R., Jahirul, M. I., Rasul, M. G., Fattah, I. M. R., Hassan, N. M. S., & Mofijur, M. (2025). Advancing energy storage: The future trajectory of lithium-ion battery technologies. *Journal of Energy Storage*, *120*, 116511.

Koech, A. K., Mwandila, G., Mulolani, F., & Mwaanga, P. (2024). Lithium-ion battery fundamentals and exploration of cathode materials: A review. *South African Journal of Chemical Engineering*.

Krishna, G., Singh, R., Gehlot, A., Shaik, V. A., Twala, B., & Priyadarshi, N. (2024). IoT-based real-time analysis of battery management system with long range communication and FLoRa. *Results in Engineering*, *23*, 102770.

Kurzweil, P. (2010). Gaston Planté and his invention of the lead–acid battery—The genesis of the first practical rechargeable battery. *Journal of Power Sources*, *195*(14), 4424–4434.

Li, Y., Guo, W., Stroe, D. I., Zhao, H., Kristensen, P. K., Jensen, L. R., & Gurevich, L. (2024). Evolution of aging mechanisms and performance degradation of lithium-ion battery from moderate to severe capacity loss scenarios. *Chemical Engineering Journal*, *498*, 155588.

Lobato-Peralta, D. R., Okoye, P. U., & Alegre, C. (2024). A review on carbon materials for electrochemical energy storage applications: State of the art, implementation, and synergy with metallic compounds for supercapacitor and battery electrodes. *Journal of Power Sources*, *617*, 235140.

Machín, A., Cotto, M. C., Díaz, F., Duconge, J., Morant, C., & Márquez, F. (2024). Environmental Aspects and Recycling of Solid-State Batteries: A Comprehensive Review. *Batteries*, *2024*(10), 255. DOI: 10.3390/batteries10070255

Madani, S. S., Shabeer, Y., Allard, F., Fowler, M., Ziebert, C., Wang, Z., Panchal, S., Chaoui, H., Mekhilef, S., Dou, S. X., See, K., & Khalilpour, K. (2025). A Comprehensive Review on Lithium-on Battery Lifetime Prediction and Aging Mechanism Analysis. *Batteries*, *2025*(11), 127. DOI: 10.3390/batteries11040127

Mikołajczyk, T., Mikołajewski, D., Kłodowski, A., Łukaszewicz, A., Mikołajewska, E., Paczkowski, T., Macko, M., & Skornia, M. (2023). Energy Sources of Mobile Robot Power Systems: A Systematic Review and Comparison of Efficiency. *Applied Sciences (Basel, Switzerland)*, *2023*(13), 7547. DOI: 10.3390/app13137547

Orth, N., Munzke, N., Weniger, J., Messner, C., Schreier, R., Mast, M., & Quaschning, V. (2023). Efficiency characterization of 26 residential photovoltaic battery storage systems. *Journal of Energy Storage*, *65*, 107299.

Parvizi, P., Jalilian, M., Amidi, A. M., Zangeneh, M. R., & Riba, J. R. (2025). From Present Innovations to Future Potential: The Promising Journey of Lithium-Ion Batteries. *Micromachines*, *16*(2), 194.

Rahman, T., & Alharbi, T. (2024). Data-Driven Degradation Estimation Techniques, and Sustainable Directions for Energy Storage Systems. *Batteries*, *2024*(10), 220. DOI: 10.3390/batteries10070220

Reddy, V. M., Mauger, A., Julien, C. M., Paolella, A., & Zaghib, K. (2020). Brief History of Early Lithium-Battery Development. *Materials (Basel)*, *2020*(13), 1884. DOI: 10.3390/ma13081884 PMID: 32316390

Shchurov, I. N., Dedov, S. I., Malozyomov, B. V., Shtang, A. A., Martyushev, N. V., Klyuev, R. V., & Andriashin, S. N. (2021). Degradation of Lithium-Ion Batteries in an Electric Transport Complex. *Energies*, *2021*(14), 8072. DOI: 10.3390/en14238072

Tang, Z., Feng, D., Xu, Y., Chen, L., Zhang, X., & Ma, Q. (2023). Safety Issues of Layered Nickel-Based Cathode Materials for Lithium-Ion Batteries: Origin, Strategies and Prospects. *Batteries*, *2023*(9), 156. DOI: 10.3390/batteries9030156

Tasneem, O., Tasneem, H., & Xian, X. (2025). Lithium-ion Battery Technologies for Grid-Scale Renewable Energy Storage. Next Research, 100297.

Tomaszewska, A., Chu, Z., Feng, X., O'kane, S., Liu, X., Chen, J., & Wu, B. (2019). Lithium-ion battery fast charging: A review. *ETransportation*, *1*, 100011.

Wang, L., Wang, J., Wang, L., Zhang, M., Wang, R., & Zhan, C. (2022). A critical review on nickel-based cathodes in rechargeable batteries. *International Journal of Minerals Metallurgy and Materials*, *29*(5), 925–941. DOI: 10.1007/s12613-022-2446-z

Waseem, M., Lakshmi, G. S., Ahmad, M., & Suhaib, M. (2025). Energy storage technology and its impact in electric vehicle: Current progress and future outlook. *Next Energy*, *6*, 100202.

Xie, J., & Lu, Y.-C. (2020, May 19). A retrospective on lithium-ion batteries. *Nature Communications*, *11*(1), 2499. DOI: 10.1038/s41467-020-16259-9 PMID: 32427837

Yi, X., Kirk, C., & Robertson, N. (2024). Nickel hydroxide-based energy storage devices: Nickel-metal hydride batteries vs. nickel hydroxide hybrid supercapacitors. *Carb Neutrality*, *3*(1), 39. DOI: 10.1007/s43979-024-00114-7

Chapter 4
Exploring the Chemistry of Energy Storage Systems Batteries, Supercapacitors, and Energy Storage

Krishna Kumar
https://orcid.org/0009-0001-9707-3155
Guru Gobind Singh Indraprastha University, India

Gulshan Kumar
Guru Gobind Singh Indraprastha University, India

ABSTRACT

The rising global energy demand combined with the rising trend on renewable energy systems underlines the necessity of having efficient, sustainable, and scalable storage technology for energy. This chapter looks into the foundational chemistry of advanced energy storage systems, which lies in next-generation batteries, supercapacitors, and hydrogen storage technologies. The chapter highlighted challenges associated with energy storage devices along with the role of innovative materials and electrochemical mechanisms. It then discusses some recent breakthroughs in material science, with an emphasis on the development of high-performance electrodes, electrolytes, and solid-state materials. The chapter concludes by outlining future trends which push the design and optimization to accelerate energy storage systems, so that the next generation of technologies can set the stage for a more sustainable and energy-efficient future.

DOI: 10.4018/979-8-3373-1409-9.ch004

1. INTRODUCTION

The global energy landscape is undergoing profound change, pushed by the growing urgency to diminish reliance on fossil fuels and mitigate adverse impacts of climate change. Among these alternatives are solar and wind as renewable sources, which have made tremendous gains but still offer several challenges because they are inherently intermittent and variable. It will require the design and deployment of energy storage systems that are critical enablers for the global energy transition by ensuring efficient utilization and management of energy. Such energy storage systems play a very fundamental role as they help to integrate energy supply and demand at various scales and provide much-needed flexibility, reliability, and resilience in power grids, thus enabling the widespread integration of renewable sources of energy (Raut et al., 2024). Beyond grid-scale applications, energy storage technologies are indispensable in modern advancements such as portable electronics, electric vehicles, and smart grids, where efficient, durable, and sustainable energy management is paramount. At its heart, chemistry underlies any energy storage technology that governs the capture, storage, and release of energy. Consequently, chemical principles dictate the efficiency, capacity, charge-discharge dynamics, safety, and lifespan of the systems. From this myriad of energy storage technologies, the two most significant ones to date are batteries and supercapacitors, largely because of their complementary advantages. Batteries work through electrochemical reactions, where the energy is stored in the form of chemical potential and released through redox reactions (Oyedotun & Mamba, 2024). The development of battery technology has been quite phenomenal, from the traditional lead-acid and nickel-cadmium batteries to the advanced lithium-ion batteries that dominate the market today. The high energy density and impressive cycle life associated with the ability to serve applications varying from consumer electronics to large-scale energy storage are among the reasons for the high penetration of lithium-ion batteries. These, however, come with problems such as resource availability, thermal instability, and environmental issues, and researchers have been directed into new generations of battery technologies: solid-state batteries, lithium-sulfur-based batteries, and others including sodium-ion or magnesium-ion systems (Lim et al., 2023). Instead, supercapacitors store their energy by charging electrostatic charge build-up and, by contrast to battery storage, there is no consumption of chemicals by chemical reactions involved; this provides fundamentally different supercapacitor designs. Supercapacitors therefore demonstrate exceptionally high-power densities, ultrafast charge/discharge rates, long cycle lives. Supercapacitors have strong potential for real-time energy distribution applications; besides, it represents a lot in the utilization, such as systems of regenerative braking, supplementary power supply backup, hybrid solution for energy sources (Venkateswarlu et al., 2023). However, they have a major limitation in that their energy density is limited compared to batteries; therefore, their possible application would be limited if used in long-duration storage. New breakthroughs in carbon-based materials including graphene and activated carbon, and transition metal oxides, indicate the improvement of their energy storage capacity closer to traditional batteries. Batteries and supercapacitors in hybrid energy storage systems appear as promising candidates in optimizing the energy storage performance with the exploitation of both these technologies (K. Kumar, Tyagi, Maity, Singh, et al., 2024). Battery will offer higher energy density and suitability for long duration storage; meanwhile, the use of a supercapacitor guarantees fast delivery of energy as well as a charge cycle capability which increases efficiency in the whole system and lengthens its lifetime. The introduction of hybrid energy storage systems has been proposed for numerous applications, such as electric vehicles, grid stabilization, and industrial energy management (Rashid Khan & Latif Ahmad, 2025). However, the vast implementation and scalability of such systems are hindered by significant challenges regarding material sustainability, cost-effectiveness, thermal management, and environmental impact. The overcoming of such challenges calls for innovation in material designs, system integration, and energy management strategies to realize practical yet economically viable solutions. A detailed comprehension of the chemistry involved in energy storage systems is a key prerequisite for the development of future technologies that will provide greater performance, safety, and sustainability. Electrode materials, electrolytes, and electrochemical reactions are central elements in defining the efficiency and durability of energy storage devices. A number of state-of-the-art research activities focused on nanostructured materials, solid electrolytes, and green chemistry have recently been concentrated on improving the current level of technology. Nanomaterials found to have higher surface areas with tunable properties have a huge potential for charge storage capacity and conductivity (Lobato-Peralta et al., 2024). Solid electrolytes are developed in particular within solid-state battery systems to remove problems such as leakage or thermal instability associated with liquid electrolytes. Green chemistry principles are increasingly prominent in energy storage research, where materials used are benign, synthesis routes are sustainable, and components are recyclable to minimize the ecological footprint of energy storage technologies. A comprehensive understanding of the underlying chemical mechanisms ruling energy storage is essential for pushing innovation and solving some of the current global energy challenges (Smdani et al., 2023). As the world moves toward a carbon-neutral balance and sustainable energy systems, advancements in the field of energy storage will change the way in which energy can be generated, stored, and distributed. Development of efficient, cost-effective, and environmentally friendly storage technologies will determine the shift toward a resilient and sustainable global energy economy. The integration of interdisciplinary research in materials science, electrochemistry, and engineering can offer the future promise of making a cleaner, more efficient, and more sustainable energy landscape possible.

2. BATTERY CHEMISTRY: LI-ION, SOLID-STATE, AND BEYOND

Modern technology has taken the application of batteries as integral parts for powering everything from simple consumer electronics, through electric vehicles and grid-scale infrastructure. In this application, a battery stores energy and expends it according to electrochemical principles by using redox reactions. A range of key thermodynamic and kinetic parameters defines a battery system's efficiency and viability-including energy density, power density, cycle life, and charging rate. The amount of energy that can be stored per unit weight or volume defines the energy density in Wh/kg or Wh/L, which plays a significant role in applications that demand compact and lightweight storage (Han et al., 2023). The rate at which energy can be delivered is described by power density, expressed in W/kg or W/L, which is essential in high-power applications, such as the acceleration of electric vehicles and grid stabilization. Similarly, important for reliability assurance in battery systems are other factors such as safety and lifespan. Concerns ranging from electrode corrosion, electrolyte instability, to thermal management cause great impact and should therefore prompt consistent innovation with the battery chemical configuration and design. LIBs represent the most used battery technology among existing ones because of their high energy density, long cycle life, and efficiency. LIBs work based on reversible intercalation of lithium ions between the anode and cathode that allows repeated cycles of charging and discharging. Cathode materials play a paramount role in selecting appropriate energy capacity, safety, and stability, with layered oxides ($LiCoO_2$, $LiNiMnCoO_2$) and phosphates ($LiFePO_4$) being a few examples. Anode materials are typically composed of graphite; however, recent silicon-based anodes offer much higher capacities. Still, these materials are prone to volume expansion upon lithiation, which affects the cycle life and mechanical stability (as shown in Fig. 1). Electrolytes, organic solvents with lithium salts dissolved in them, help transport ions but also pose a safety risk since they are flammable and might decompose. Despite these advantages, LIBs suffer from capacity degradation due to electrode instability, side reactions, and electrolyte decomposition. Another critical safety concern is thermal runaway, where excessive heat generation leads to catastrophic failure. Overcoming these challenges will require the development of next-generation materials and battery architectures that enhance performance and safety. An alternative to traditional LIBs, solid-state batteries (SSBs) replace liquid electrolytes with solid ionic conductors. This brings safety advantages, in that there is no risk of leakage and thermal runaway, in addition to a possibility to replace traditional graphite anodes by lithium-metal anodes with theoretical capacities (Yu et al., 2023). Ionic conductivity along with electrochemical stability is delivered by such solid electrolytes as sulphides, oxides, and polymers. Some problems persist, including those at interfaces between solid electrolytes and electrodes, where very high interfacial resistance is known to inhibit the transport of ions and diminish battery efficiency in general. Finally, the mechanical integrity of a solid electrolyte is an issue because repeated cycling can create microcracks within it, which also degrade performance. Current research focuses on composition optimization of electrolytes, interface improvement between electrodes and electrolytes, as well as development of scalable manufacturing techniques to make SSBs commercially viable. The rising demand for sustainable and cost-effective energy storage solutions has brought into focus the alternative battery chemistries beyond lithium (Kilic et al., 2023). As sodium is a more abundant and inexpensive raw material, its ionic battery, SIBs, has been considered for large-scale applications. Even though SIBs have an inferior energy density compared with LIBs, they have advantages with respect to the availability of raw materials and environmental sustainability. Magnesium-ion batteries also represent another possibility towards better volumetric capacity and safety by their non-dendritic behaviour upon plating. Nevertheless, their practicality is badly restricted by poor diffusion kinetics and limited possible electrolytes. Zinc-ion batteries offer improved safety, cost-effectiveness, and stability and are promising for stationary applications of energy storage. High-capacity battery systems include lithium-sulfur and metal-air batteries, offering great advances in energy storage capabilities. Lithium-sulfur batteries have the advantage of achieving high theoretical energy densities by using sulfur cathodes; however, the polysulfide dissolution and low cycle stability pose a practical challenge. The most prominent type of metal-air battery is based on the reaction of oxygen from the air with a metal cathode (Srivastava et al., 2024). Metal-air batteries achieve theoretical capacities that are exceptionally high, but issues include electrolyte stability, oxygen reduction kinetics, and design of the air-electrode interface. Redox-flow batteries, specifically those based on vanadium, represent a scalable technology for grid-level energy storage. Energy and power capacity can be tuned independently. These flow-based systems have a long cycle life and stability but are limited commercially due to the high cost and availability of the vanadium electrolyte. Future development in battery chemistry is gradually shifted toward sustainability, recyclability, and reduction in environmental footprint. The most significant challenges facing large-scale production include the use of eco-friendly electrode and electrolyte materials and developing effective recycling processes for batteries (Y. Chen et al., 2023). Advancements are made in nanostructured electrodes, hybrid material systems, and novel compositions of electrolytes for improved energy density, durability, and economy. The hybrid storage systems integrated with the battery, either in parallel or as supercapacitors, open up the prospect for optimization through diverse applications. In addition to this, solid-state technologies and emerging architectures for batteries will have a significant role to play in determining the next generation of energy storage solutions. A future, where the global appetite for renewable energy sources keeps growing, will ensure continued improvement in battery technologies to meet all the increasing demands for efficient, safe, and sustainable energy storage (K. Kumar et al., 2025). This wave will be triggered by breakthroughs in materials science and electrochemical engineering, including manufacturing processes, for better performance with high-performance battery systems that support a low-carbon future. Novel battery chemistries are being explored along with improvements in existing technologies. Energy storage solutions that are more efficient and durable and environmentally responsible and economically viable will emerge.

Figure 1. (A) Digital twin-driven 3D structures and lithium-ion dynamics in graphite-silicon electrodes with varying silicon particle sizes (B) cycle discharge, charge capacities, and coulombic efficiency of synthesized electrodes (C) Representation of electrode composites: Cases 1 and 2 show active material-solid electrolyte electrodes, while Case 3 depicts a Si/carbon nanofiber-solid electrolyte composite (D) Composition and electrochemical properties of synthesized electrode. Reproduce from (Boorboor Ajdari et al., 2024) under an open access Creative Commons CC BY 4.0 license

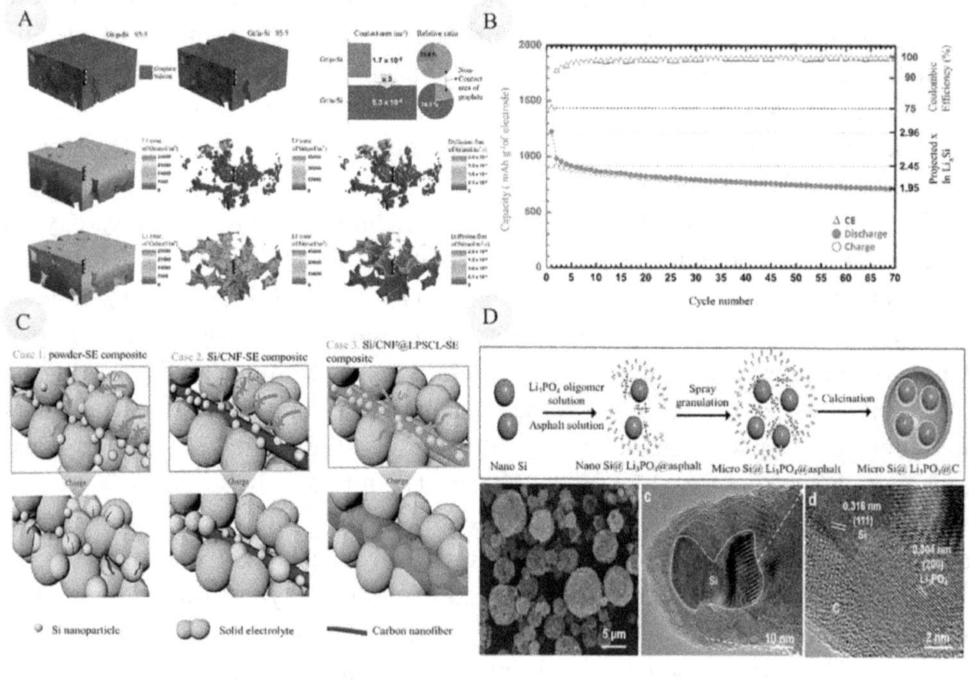

3. SUPERCAPACITORS: MATERIALS AND MECHANISM

Supercapacitors are electrochemical energy storage devices that act at a cross-stage between conventional capacitors and batteries, hence they have higher power density, faster rates of charge/discharge, and longer lifespan. SCs work differently from a battery because its power source is not based on chemical reactions but rather on electrostatic and faradaic mechanisms, allowing for the generation of power instantly. Their unique properties make them highly suitable for applications requiring frequent, fast energy bursts. Indeed, high regenerative braking forces in the electric vehicle, grid stabilization, and portable electronic devices are a few examples. Although their energy density remains lower than that of batteries, their energy densities are highly promising, and continuous advancements in materials and hybrid storage systems can make them more suitable (K. Kumar, Tyagi, Maity, & Kumar, 2024). Supercapacitor performance is intimately dependent on electrode materials, as these electrodes control the mechanism of energy storage, conductivity, and overall electrochemical stability. Carbon-based materials such as activated carbon, graphene, and CNTs are widely used in supercapacitors due to their high surface area, excellent electrical conductivity, and tunable porosity, which facilitate efficient charge storage via the EDLC mechanism. Of the ones developed, activated carbon remains the most commercially viable material due to its efficiency and well-developed microporous structure with high electrochemically active surface area. Graphene and CNTs present better conductivity and improved charge transport characteristics, but synthesis at large scale and restacking of materials make them challengeable when it comes to effective surface area (Alam et al., 2023). Besides EDLC-based materials, great attention has been directed toward pseudocapacitive materials, mainly conducting polymers and transition metal oxides (TMOs), to expand the energy storage capabilities of supercapacitors. Charge storage of conducting polymers, such as polyaniline (PANI) and polypyrrole (PPy), occurs due to reversible faradaic redox reactions at the electrode-solution interface that increase capacitance significantly beyond what EDLC materials might achieve. However, the long-term cycling stability is still an issue due to structural degradation on repeated charge-discharge cycles. Pseudocapacitive TMOs include RuO_2, MnO_2, and NiO, showing high theoretical capacitance values due to rapid redox reactions (Dissanayake & Kularatna-Abeywardana, 2024). The potential of TMOs has been compromised due to low electrical conductivity and instability in the structure; hence, hybrid materials combining carbon matrices with TMOs are needed for better charge transport and mechanical stability. The two primary mechanisms of energy storage in supercapacitors are electric double-layer capacitance and pseudocapacitance. The EDLC mechanism is highly dependent on the specific surface area and conductivity of the electrode material, as it involves electrostatic accumulation of charges at the interface between the electrode surface and electrolyte ions. In contrast, pseudocapacitance is due to faradaic processes, which include redox reactions, ion intercalation, and electrosorption, and yield a higher energy density than the purely electrostatic mechanisms (as shown in Fig. 2). TMOs and conducting polymers are dominated by pseudocapacitive behavior, while carbon-based materials are dominated by the EDLC mechanism. Recently, researchers have turned to hybrid supercapacitors, which operate using both mechanisms simultaneously, so that the energy density gap between supercapacitors and batteries can be overcome at the cost of high-power output (Ma et al., 2023). Despite their many benefits, supercapacitors have limitations that prevent them from being widely applied, especially when higher energy density is required. One major limitation is the relatively low storage capacity of a supercapacitor compared with a battery; this limits the application of supercapacitors to long-duration energy supply applications. Another area requiring improvement is the retention of capacitance over many cycles and the stability of operation under extreme operating conditions. To overcome these issues, there have been significant efforts toward the development of advanced electrode materials, such as graphene-based composites, heteroatom-doped carbon nanomaterials, and hybrid systems that integrate both pseudocapacitive and EDLC mechanisms. Sustainable alternatives like biomass-derived carbon materials have also gained attention because they are more cost-effective and environmentally friendly compared to other supercapacitor electrode fabrication methods (Qiu et al., 2023). Precursors like banana peel, potato peel, and tea residue have been explored to produce highly porous activated carbon, with promising electrochemical properties for high-performance energy storage. Future trends in the field of supercapacitor research and development would involve developing materials sustainably, innovating in hybrid storage systems, and scale and cost effectiveness. Hybrid energy storage has come to represent integration with a battery where it benefits from its high energy density but gets a boost of the charge-discharge properties offered by a supercapacitor. In addition, there is still more research and development in flexible and solid-state supercapacitors with the high demand for wearable electronics and safer energy storage solutions (Pandey et al., 2024). Thus, the supercapacitor further develops with the advancing field of material science and device engineering and plays an important role in the next-generation energy storage system. It will thus be the primary role of supercapacitors in enabling high-performance and environmentally friendly energy storage technologies as global energy demands increasingly shift toward more sustainable and efficient solutions.

*Figure 2. (**a**) Exfoliated graphene-mediated GO films for supercapacitor application with tunable interlayer spacing for volumetric capacitance (**b**) Multilayer-folded graphene ribbon films and their gravimetric capacitances (**c**) Electrolyte-mediated densification of graphene films and their volumetric capacitances (**d**) Carbon dots pillared porous graphene framework and its rate performance. Reproduce from (Z. Liu et al., 2020) under an open access Creative Commons CC BY 4.0 license.*

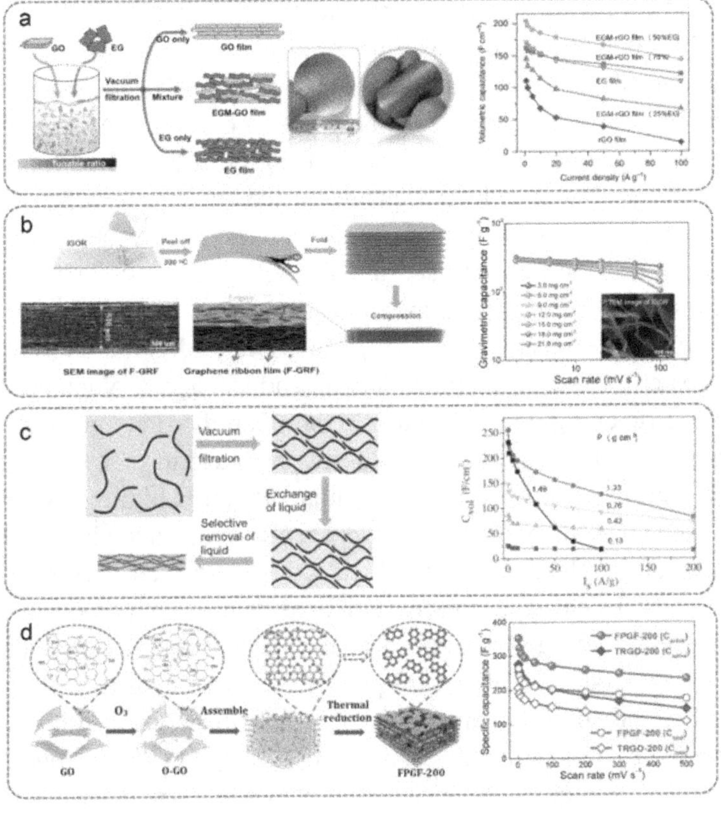

4. HYDROGEN STORAGE: CHALLENGES AND SOLUTIONS

Hydrogen represents a critical factor in the quest toward carbon-neutralization, for it is not only a clean energy carrier, but also a potential vehicle for transportation and heavy industries, along with grid-scale energy storage. With the fight against carbon intensifying, this element is already being considered a possible alternative for fossil fuels that have been conventional so far. Hydrogen storage cells offer significantly higher energy density than batteries and supercapacitors, making them ideal for long-term and large-scale energy storage (Table 1). However, storage challenges hinder the high uptake of hydrogen-based technologies. Hydrogen has a very low energy density when in gaseous form, thus forcing the need for proper storage solutions to allow for its considerable use (Mulky et al., 2024). The performance of hydrogen storage systems is mainly evaluated against two key indicators: gravimetric capacity that refers to how much hydrogen is stored by a unit of mass of material, and volumetric capacity or the storage density per unit of volume. They are the quantities that determine feasibility of different storage options for practical implementations. Methods for hydrogen storage are broadly categorized as physical and chemical. In physical techniques, there is either compression or liquefaction to increase the hydrogen density (as depicted in Fig. 3). Generally, in high-pressure tanks, the storage pressures of hydrogen are carried out between 350–700 bar. This is the technique that the majority of fuel cell vehicles and industries use (Ma et al., 2024). Although this method can facilitate fast refueling and rather simple system integration, it suffers from major drawbacks: the energy-intensive compression required, the use of materials resistant to high pressure yet light enough, and possible safety issues, such as hydrogen embrittlement and leakage. Cryogenic liquefaction offers a volumetric energy density much higher than the first approach by cooling hydrogen to near -253°C. This method is particularly useful for aerospace and long-distance transport applications. However, liquefaction is highly energy-intensive, consuming about 30-40% of the stored hydrogen's energy, and also associated with boil-off losses due to heat transfer, which reduces long-term storage efficiency (Navaid et al., 2023). Other alternative storage methods include chemical storage. In such a method, hydrogen is bonded to materials through chemical means for storage under moderate conditions. In particular, metal hydrides like magnesium hydride, MgH_2, and sodium alanate, $NaAlH_4$, have been used in high-volumetric storage densities by reversible absorption and desorption of hydrogen. However, these materials often require elevated temperatures, that are usually around 300°C for the release of hydrogen, and have slow kinetics, which limit their practical application. More complex hydrides such as borohydrides, e.g., $LiBH_4$ and $NaBH_4$, and amides, e.g., $LiNH_2$, offer higher storage capacities but have issues of poor reversibility, lack of thermal stability, and low hydrogen desorption rates (Massaro et al., 2023). Another interesting class of chemical storage material includes LOHCs such as methylcyclohexane and dibenzyltoluene. These chemically store hydrogen at ambient conditions and release it on catalytic dehydrogenation, which presents a scalable alternative for long-distance transport. However, the efficiency of LOHCs is often constrained by the need for high-performance catalysts and the significant energy input required for the dehydrogenation process. With the progression of storage technology, there are many issues that need to be conquered. The physical techniques such as compression and liquefaction are extremely energy-intensive, causing efficiency losses. High-pressure storage materials are prone to mechanical degradation and hydrogen embrittlement, which makes them dangerous and undermines their effectiveness for long-term storage. In chemical storage, the kinetics and thermodynamics of hydrogen absorption and desorption are suboptimal, requiring high temperatures and pressures that decrease the overall efficiency of the system. Storage materials also often have reversibility problems and slow reaction rates, making them impractical for real-world applications (Ghorbani et al., 2023). To overcome these challenges, interdisciplinary research is needed to improve material properties, optimize storage processes, and integrate hydrogen storage with existing energy infrastructures. Recent advances in characterization techniques have improved our understanding of hydrogen-material interactions. Therefore, there has been tremendous scope for developing new-generation storage materials. The insight into the process of hydrogen diffusion and bonding in the storage materials was obtained from in situ neutron scattering, whereas electronic structure and chemical environment at the atomic level were analyzed with X-ray absorption spectroscopy. Temperature-programmed desorption techniques would be used in monitoring hydrogen release kinetics and thermodynamics accurately in order to optimize the performance of the material (Hassan, Sameen, Salman, et al., 2023). Moreover, high-resolution electron microscopy proved to be the most insightful in revealing microstructural evolution under the cycling of hydrogen, making it a crucial information source for the design of future storage systems. Future R&D areas include lightweight composite tanks that have excellent mechanical properties, nanostructured hydrides with improved kinetics, and optimized LOHC systems that require less energy to release hydrogen. Incorporation of more advanced catalysts and innovative material compositions is likely to provide higher storage efficiency, and incipient fabrication techniques, such as additive manufacturing and nanostructuring, may also promote scalability of these future hydrogen storage technologies. Continuing research in material science and system design will advance further hydrogen storage technology so that the potential for sustainable energy landscape without dependency on fossil fuel inputs is feasible in the future.

Table 1. Comparative Analysis of Supercapacitors, Batteries, and Hydrogen Storage Systems Based on Key Performance Metrics

Property	Supercapacitors	Batteries	Hydrogen Storage Systems	Ref.
Energy Density (Wh/kg)	5-50 (activated carbon-based)	100-300 (Li-ion), 50-150 (Na-ion)	33,600 (compressed H_2 at 700 bar)	(Pelosi et al., 2023)
Power Density (W/kg)	1,000–10,000 (high surface area materials)	100-1,500	~1,500 (fuel cells)	(Tawalbeh et al., 2024)
Charge/Discharge Time	1-10 s (rapid response)	10 min to several hours	Hydrogen refueling (3-5 min)	(Sahoo & Timmann, 2023)
Cycle Life (cycles)	10^6-10^7 (high durability)	500–10,000 (Li-ion), ~3,000 (Na-ion)	~10,000 (metal hydrides)	(Lazaroiu et al., 2023)
Efficiency (%)	90-98	80-95	40–60 (fuel cell), 50-70 (electrolysis)	(D. Wang et al., 2024)
Self-Discharge Rate (%/day)	5-40%	0.1-2%	Negligible (chemical storage)	(Irham et al., 2024)
Operating Temperature (°C)	-40 to 70	-20 to 60	-253 (liquid H_2), -20 to 80 (metal hydrides)	(Wali et al., 2023)
Specific Energy (Wh/L)	1-10 (low volumetric energy density)	250-700 (LIBs)	2,600 (liquid H_2), 1,000 (metal hydrides)	(Y. Yang et al., 2024)
Applications	Short bursts of power, hybrid systems	Consumer electronics, EVs, grid	Fuel cells, transportation, grid storage	(Mohammed et al., 2023)

*Figure 3. (**a**) Fundamental approaches for modifying the kinetics of hydrogenation and dehydrogenation for hydrogen storage application (**b**) Traditional models of rate-limiting steps (**c**) Depiction of homogeneous and heterogeneous catalysts (**d**) Catalysis in hydrides for heterogeneous systems in hydrogen storage. Reproduce from (Salman et al., 2022) under an open access Creative Commons CC BY 4.0 license.*

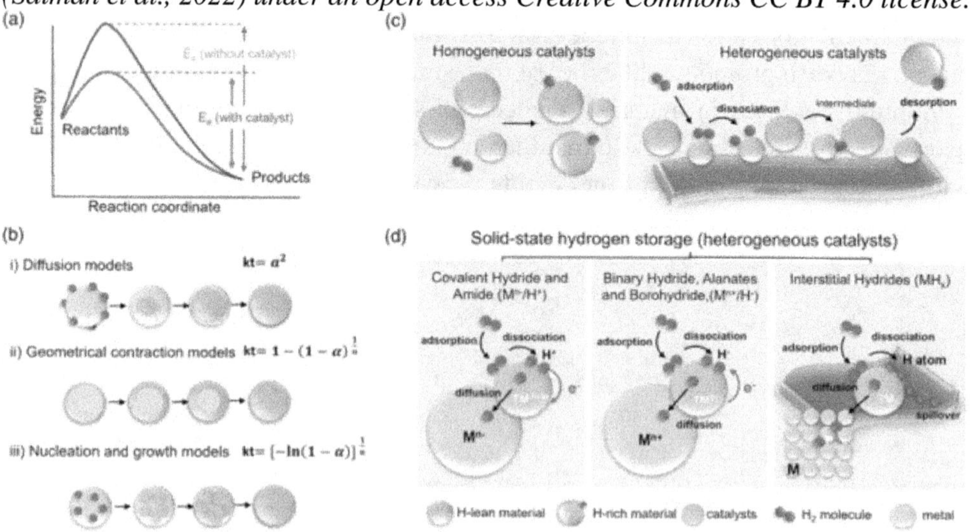

5. ELECTROCHEMICAL MODELING AND SIMULATIONS

Electrochemical modeling and simulations are the crucial technologies associated with the progress and advancement of energy storage systems. These carry critical information regarding the internal processes and behavior of the technologies in question, such as batteries, supercapacitors, and hydrogen storage. They can predict the system behavior in different conditions and allow for the design of newer systems that are much more efficient and durable. Modeling bridges the gulf between pure electrochemical principles and practical applications, representing a means through which material properties can be optimized, performance is predicted, and next-generation energy storage technologies come to market quickly (as shown in Fig. 4). The range of electrochemical modeling in battery systems varies from the simplest equivalent circuit models to complex physics-based models that consider precise electrochemical processes. Equivalent circuit models approximate battery behavior with electrical elements such as resistors, capacitors, and inductors, which are helpful in studying general system performance, such as charge/discharge rates, impedance, and efficiency under various loading conditions. Even though the above models prove insightful, these do not depict the actual reactions involved within the chemistry of the battery. Physic based models can detail more information with regard to electrochemical reactions involving both electrodes as well as an interface. It describes electrochemical reactions Butler-Volmer kinetics rate of which will be calculated through models of an ion being transferred throughout the electrolyte with Nernst-Planck equations (Zhou et al., 2023). The P2D model, a more sophisticated model, approximates the electrode as a porous material and considers processes such as diffusion of ions, electron conduction, and reaction kinetics. In addition to these models, several thermal models have been developed to predict heat generation during charging and discharging, which is an important aspect in battery management, prevents overheating, and ensures safety. Degradation models also prove to be significantly important in studying the battery and how internal factors like cycle life, temperature cycling, and even overcharging work to degrade and shorten the length of the battery lifespan. Since supercapacitors store energy by accumulation of electrostatic charges and faradaic processes, their simulations with electrochemical elaborations are also widely performed. Equivalent circuit models in the context of supercapacitors - similar to battery system simulations - often analyze performance metrics, including energy density and power density and charge/discharge rates - but are not very informative with regard to electrochemical mechanisms controlling the behavior of supercapacitors themselves (Zeng et al., 2023). To gain a deeper insight, electrochemical impedance spectroscopy models are mainly used, that is, applying alternate currents to the supercapacitor and finding the resulting voltage. These spectra give information related to the inner resistance, capacitance, and also the dynamics related to the ions moving within the material of the electrodes. In particular, pseudocapacitance is very important in materials such as transition metal oxides or conducting polymers, as energy is stored through faradaic processes at the electrode surface. Modeling pseudocapacitance involves simulating redox reactions that occur during charge/discharge cycles. Moreover, detailed models of electrode structure are necessary to accurately predict supercapacitor performance, especially when using advanced materials such as graphene or carbon nanotubes. The above simulations help in understanding the pore size distribution, surface area, and ion diffusion rates, which have a direct influence on capacitance and energy storage. Electrochemical modeling is essential for optimizing material selection and storage configurations in hydrogen storage systems, which are critical for the hydrogen economy (P. Kumar et al., 2024). High-pressure tanks can physically store hydrogen, while liquid hydrogen at cryogenic temperatures can also hold the gas. Hydrogen can also be stored chemically in the form of metal hydrides, complex hydrides, and liquid organic hydrogen carriers, or LOHCs. Thermodynamic models are applied in high-pressure storage to simulate the relationships of pressure and temperature. This helps ensure the storage tanks are designed to handle the stresses and thermal effects as a result of compressing hydrogen. For cryogenic storage, optimization is attained during liquefaction of hydrogen based on energy requirements and management of phase transitions as well as cooling processes. Models of chemical storage methods, such as metal hydrides, simulate hydrogen absorption and desorption rates to be important elements in calculating the efficiency and capacity of a storage system. The models identify the best thermodynamic and kinetic conditions for hydrogen release and improve the overall performance and safety of hydrogen storage technology. Electrochemical modeling further extends toward device-level and even system-level optimisation for optimal performance from combinations of the respective devices with some simulations having one or multiple energy storage systems integrating batteries, supercapacitors, or units of hydrogen-based storage for developing hybrid systems with advanced energy and power density alongside others (Miranda et al., 2023). These simulations are then critical to running hybrid systems. Their performance, one piece at a time, is critical and has to operate efficiently and reliably, especially during periods of grid storage or electric vehicle usage. Electrochemical simulations also play an important role in material discovery and screening. Material electrochemical performance is simulated based on theoretical models before synthesis for potential next-generation energy storage devices. Computational techniques, like high-throughput screening and machine learning, allow one to quickly find materials with the optimal properties to accelerate the development of new storage technologies. There are still important challenges even with the striking progress achieved in electrochemical modeling. Electrochemical systems, namely involving multiphysics phenomena, ion transport, electron conduction, and electrochemical reactions, are inherently hard to model with accuracy and computational efficiency. Material heterogeneity introduces further complexities because real materials are not uniform in composition, structure, or properties. Many of the models are based on assumptions idealized, which do not always correspond to the variability seen in practice. In addition, parameterization of the model remains a problem; getting the desired material parameters accurate, such as diffusion coefficients for ions, conductivity, and reaction rates, is difficult and varies with temperature, pressure, and other factors (Abeng & Anadebe, 2023). However, electrochemical modeling remains the most powerful way to advance new energy storage technologies because it provides complete insight into energy storage devices' internal mechanics, thus driving the optimization of materials, predictability of device behavior, and innovativeness with solutions to existing challenges. According to further increases in computational power and improvements in modeling techniques, electrochemical simulations will play a more important role in the design and optimization of efficient, sustainable, and scalable energy storage systems.

Table 2. Electrochemical Modeling and Simulation Techniques for Energy Storage Systems

Modeling Approach	Governing Principles	Mathematical Framework	Key Advantages	Limitations	Ref.
Equivalent Circuit Model (ECM)	Represents electrochemical behavior using lumped electrical components (resistors, capacitors, voltage sources).	Ohm's Law, Kirchhoff's Laws, RC circuit modeling.	Computationally efficient, suitable for real-time applications.	Lacks detailed representation of electrochemical and degradation processes.	(C. Liu et al., 2023)
Physics-Based Model (PBM)	Describes ion transport, reaction kinetics, and thermodynamic behavior at the electrode-electrolyte interface.	Nernst-Planck equation, Butler-Volmer equation, porous electrode theory, conservation laws.	Provides detailed insight into charge/ discharge processes and failure mechanisms.	High computational cost, requires extensive experimental parameterization.	(Ge et al., 2023)
Molecular Dynamics (MD) Simulations	Simulates atomic/ molecular interactions governing ion transport, diffusion, and interfacial phenomena.	Newton's equations of motion, interatomic potentials (e.g., Lennard-Jones, Coulombic).	Atomistic-level resolution, enables prediction of material properties before experimental synthesis.	Limited by short simulation timescales (nanoseconds) and system sizes.	(Pordanjani et al., 2023)
Density Functional Theory (DFT) Calculations	Quantum mechanical approach to predict electronic structures and electrochemical interactions.	Schrödinger equation, Kohn-Sham density functional formalism.	High precision in electronic structure calculations, enables discovery of novel materials.	Computationally expensive, restricted to small molecular systems and static conditions.	(B. Chen et al., 2023)
Finite Element Analysis (FEA)	Numerical method for solving coupled electrochemical, thermal, and mechanical phenomena in complex geometries.	Partial differential equations (PDEs), mesh-based numerical solutions (e.g., COMSOL Multiphysics, ANSYS).	Spatially resolved insights, integrates multiphysics simulations (electrochemical, thermal, mechanical).	Requires accurate input parameters, computationally intensive for large-scale systems.	(Guo et al., 2024)

Figure 4. Battery aging prediction framework using real vehicle data and transfer learning. Reproduce from (K. Yang et al., 2024) under an open access Creative Commons CC BY 4.0 license.

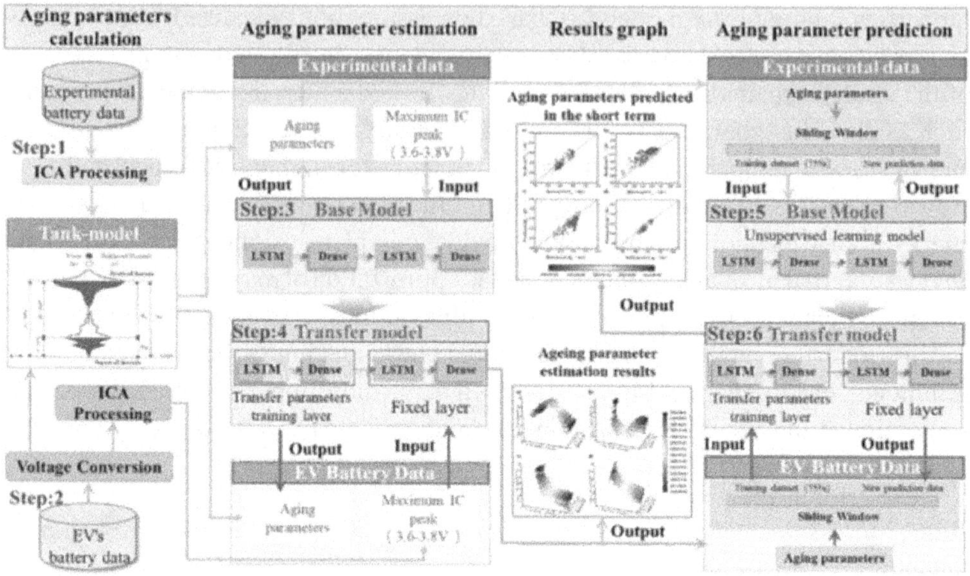

6. ROLE OF ARTIFICIAL INTELLIGENCE AND MACHINE LEARNING IN ENERGY STORAGE

With emergent AI and ML into the energy storage domain, numerous challenges can now be solved innovatively, starting from optimizing performance and enhancing the safety of components to discovering new materials. It can catalyze accelerated development toward efficient and sustainable energy storage systems through automated data analysis, optimum system designs, and prognoses of future behavior for storage components. The advent of AI and ML in the domain of material discovery is revolutionizing how we discover new materials for the battery, supercapacitors, and hydrogen storage. Usually, in material discovery, time and labor-intensive efforts are taken along with a procedure of trial and error for finding the correct composition of different material combinations (Rojek et al., 2023). AI, in contrast, can automatically do this process by performing high-throughput screening of materials, analyzing large datasets, and predicting the performance of novel materials before they are physically tested. This ML algorithm can predict crucial characteristics like ion conductivity, energy density, and stability of the energy storage systems through huge datasets of material properties (Table 3). In addition, the application of AI-driven techniques such as molecular simulations allow researchers to explore combinations of elements and structures that may never have crossed their minds by considering traditional methods. This is significant in the design of highly efficient and durable materials for energy storage. Optimizing Battery Management Systems (BMS) is another area of application where AI/ML is playing a significant role. BMS are crucial in ensuring the safe operation of batteries while maintaining efficiency by observing how many charge/discharges cycles they can perform before deteriorating in health in general. State-of-Charge and State of Health prediction: the overall performance of batteries can be easily predicted and managed by developing estimation techniques of ML, so they are quite improved compared to classical estimation methods; in BMS, they might predict more precisely and accurately for both SOC and SOH. This, in turn, improves the quality of charging and discharge cycles and maximizes the lifespan of the battery as much as possible to prevent issues such as overcharging and deep discharging. In addition, AI-based BMS could predict potential faults or degradation patterns before they appear, thus lessening the probability of failure and increasing the lifetime of the battery. Predictive maintenance, driven by AI, can identify early signs of malfunction, allowing for timely intervention and reducing downtime in energy storage systems. Supercapacitors are highly utilized in applications requiring rapid surges of power (Entezari et al., 2023). In that case, AI and ML similarly improve performance by optimizing performance in such devices. Supercapacitors are highly valued for their high-power density and very rapid charge/discharge rates but generally offer a lower energy density than batteries. ML will predict the capacity, cycle life, and overall efficiency of a supercapacitor based on its materials used and conditions for operation (as shown in Fig. 5). The AI models can also identify new materials that could be useful in supercapacitor applications by optimizing the structure of electrode materials and optimally improved the electrolytes. Hybrid systems that integrate both batteries and supercapacitors are gaining popularity due to the high energy and high-power densities required for various applications. Here, AI could optimize the interaction between the battery and the supercapacitor to make sure each operates within its optimum range. Thus, hybrid systems can produce superior performance than single batteries or supercapacitors. The third potential beneficiary is hydrogen storage, essential to widespread application of hydrogen as a clean energy resource. Hydrogen can be stored physically, using for example high-pressure tanks or cryogenic systems, or chemically in materials, like metal hydrides. AI and ML may further improve hydrogen storage systems performance by predicting kinetics for the release and absorption of hydrogen under a wide range of conditions. Machine learning models may be trained on the thermodynamic and kinetic properties of hydrogen storage materials to make predictions that enable the design of systems operating more efficiently at reduced pressures and temperatures (Olabi et al., 2023). Optimization through this process will be critical to scaling hydrogen storage technologies for practical use in transport, grid, or other applications. But aside from individual parts, AI and ML play important roles in combining different energy storage technologies in hybrid systems. Hybrid systems composed of batteries, supercapacitors, and hydrogen storage have the potential for higher energy efficiency and better performance compared to a single technology. Artificial intelligence will have an added utility by predicting power requirements and giving instantaneous decisions concerning the usage, be it on batteries, a supercapacitor, or even hydrogen fuel. Optimization done at this systems level means achieving a highly optimal use of power through storage equipment that produces minimized wastage or dissipation losses while in practical performance. AI-driven energy management systems can predict energy demand patterns, ensuring energy is stored and distributed in the most efficient manner possible. This is crucial for integrating intermittent renewable energy sources, such as solar and wind, which necessitate efficient energy storage systems to ensure grid stability. Moreover, safety is one major concern in energy storage systems, primarily in batteries, where the problems of thermal runaway, overcharging, and so can cause a catastrophic failure. AI and ML can make safety protocols better by providing real-time monitoring and early warnings of potential risks. For instance, AI models can analyze temperature data to predict overheating issues before they actually materialize, allowing for corrective actions like reduction in charging rates or activation of cooling systems. Continuous performance data analysis by AI systems can therefore detect anomalies that may indicate an early onset of failure, and thus measures can be taken to prevent that before actual damage is done. This predictive capability is important to ensure the reliability and safety of energy storage systems, especially in high-risk applications such as electric vehicles or large-scale grid storage (H. Yang et al., 2023). Therefore, AI and ML are emerging to be great catalysts for progress in the domain of energy storage technologies. They facilitate discoveries of new materials, optimization of system performance, safety improvements, and hybrid integration and thus accelerate the development of innovation that can make energy storage more efficient, sustainable, and reliable. Therefore, with ongoing developments in AI and ML, their contribution toward energy storage systems is bound to increase and expedite the transformation toward a cleaner and safer energy future.

Table 3. Role of Artificial Intelligence and Machine Learning in Energy Storage

AI/ML Application	Energy Storage Domain	Scientific Contributions	Representative Examples	Ref.
Computational Materials Discovery	Battery and Supercapacitor Electrodes & Electrolytes	Accelerates the identification and optimization of novel high-capacity electrode and electrolyte materials through predictive modeling and data-driven simulations	AI-driven density functional theory (DFT) calculations for Li-ion cathode materials screening	(Cordeiro-Costas et al., 2023)
Electrode Microstructure Optimization	Lithium-Ion and Hybrid Batteries	Enhances the electrochemical performance by optimizing electrode porosity, conductivity, and ion transport pathways	ML-assisted prediction of Li-ion diffusion rates in nanostructured anodes	(Zayed et al., 2023)
Battery Management Systems (BMS) Optimization	Lithium-Ion and Solid-State Batteries	Improves real-time monitoring, extending battery lifespan and enhancing safety through accurate predictions of State of Charge (SOC) and State of Health (SOH)	Deep learning models for real-time SOC estimation in high-energy-density batteries	(Giglio et al., 2023)
Anomaly Detection and Failure Prediction	Grid-Scale Energy Storage and Electric Vehicle (EV) Batteries	Enables early-stage fault detection, mitigating thermal runaway and degradation risks	AI-driven thermal imaging analysis for early detection of dendrite formation in Li-metal batteries	(Xu et al., 2023)
Cycle Life and Degradation Prediction	Supercapacitors and Battery Systems	Forecasts electrochemical aging trends, optimizing charge/discharge protocols to minimize capacity fade	Recurrent neural networks (RNNs) for cycle life prediction of Li-ion and sodium-ion batteries	(Balakumar et al., 2023)
Electrochemical Performance Modeling	Energy Storage Device Engineering	Facilitates the design of energy storage systems by correlating physicochemical parameters with electrochemical performance	AI-based modeling of charge transfer kinetics and diffusion dynamics in supercapacitor electrodes	(Mhlanga, 2023)
Manufacturing Process Optimization	Battery and Supercapacitor Fabrication	Enhances efficiency in synthesis routes, electrode fabrication, and electrolyte formulation for scalable production	AI-driven process parameter optimization in solid-state electrolyte synthesis	(Mehrenjani et al., 2023)
Energy Storage Grid Integration	Renewable Energy and Hybrid Storage Systems	Improves load balancing and predictive control of energy storage units in smart grids	AI-based time-series forecasting for renewable energy storage demand prediction	(Lemaoui et al., 2023)
Hydrogen Storage System Optimization	Fuel Cells and Hydrogen Storage Technologies	Enhances hydrogen adsorption/desorption efficiency, reducing storage losses and improving thermodynamic stability	ML-assisted computational modeling of metal-organic frameworks (MOFs) for hydrogen storage applications	(Ruan et al., 2023)
Next-Generation Energy Storage Innovations	Emerging Electrochemical and Solid-State Storage Technologies	Identifies novel energy storage materials and architectures by analyzing large-scale experimental and theoretical datasets	AI-driven discovery of high-energy-density solid-state electrolytes for post-Li-ion battery chemistries	(Yazici et al., 2023)

Figure 5. Machine learning applications in optimizing material constituents for EESC include: (A) predicting novel olivine-type LiMXO4 solid-state electrolytes (B) designing Li-Si-(Mn, Fe, Co)-O cathode materials (C) identifying single-atom catalyst descriptors linked to hydrogen evolution reaction (HER) activity(D & E) uncovering perovskite descriptors associated with oxygen evolution reaction (OER) activity. Reproduce from (H. Yang et al., 2023) under an open access Creative Commons CC BY 4.0 license.

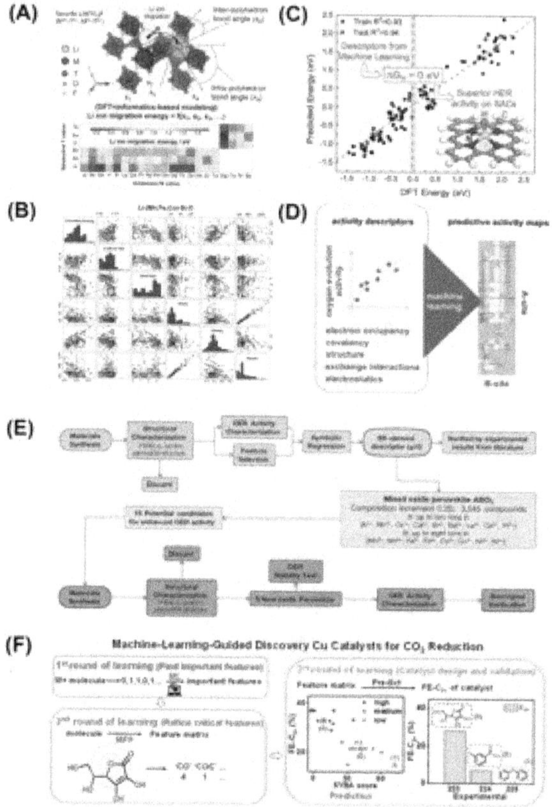

7. FUTURE TRENDS IN ENERGY STORAGE TECHNOLOGIES

Integration of multiple energy storage solutions, sustainable materials, and advancements of computational tools are shaping the future of energy storage technologies. Hybrid energy storage systems account for one of the most crucial innovations related to this field in that they strategically combine batteries, supercapacitors, and hydrogen technologies to optimize energy storage performance. Batteries, for example, lithium-ion and sodium-ion systems, have high energy density through redox reactions occurring between the anode and cathode, thus allowing efficient retention of energy. Supercapacitors, on the other hand, offer rapid charge-discharge cycles because of their electrostatic storage mechanism and surface redox reactions, making them highly suitable for applications requiring instantaneous power delivery. To complement such technologies, hydrogen-based storage systems serve as long-term carriers of energy (Table 4). These systems store excess energy for long periods through adsorption-based or chemical storage mechanisms in the form of hydrogen (Elalfy et al., 2024). The combination of these technologies within the system enables renewable energy to be stored more efficiently, flexibly, and with diminished intermittency of sources such as solar and wind energies. Machine learning and predictive analytics are becoming increasingly important for this integration, allowing for real-time monitoring and optimization of charge-discharge cycles, thermal management, and energy flow, thus extending system longevity and performance. One of the major trends in energy storage innovation is the shift toward sustainability, mainly driven by the need to minimize environmental impact and align with circular economy principles. Bio-derived and recyclable materials are an alternative to conventional resource-intensive materials. Carbon from agricultural residues, lignocellulosic biomass, and biochar is becoming an increasingly popular choice as an electrode material in batteries and supercapacitors owing to its large surface area, tunable porosity, and favorable electrochemical properties. At the same time, electrolyte design is rapidly evolving toward less toxic, renewable, and recyclable options like ionic liquids, deep eutectic solvents, and polymer-based gel electrolytes, which significantly improve stability and safety while diminishing environmental risks. These materials and processes will be guided by the principles of green chemistry to focus on techniques such as hydrothermal carbonization, sol-gel processing, and electrospinning that yield synthesis products minimizing hazardous byproducts. Lifecycle assessments are also conducted to quantify the environmental footprint of energy storage technologies so that it meets the material and process-related sustainability goals of the energy technology while maintaining the highest performance possible. Through advancements in nanotechnology and computational modeling, there have been great increases in the improvement of electrochemical properties and scaling of energy storage systems (Yi et al., 2023). Enhancement in nanostructure electrode materials from graphene derivatives and MXenes towards metal-organic frameworks (MOFs) enhanced conductivity, increased ion transport properties, and better surface area towards higher energy density and charge-storage capabilities. In addition, solid-state electrolytes are being explored as alternatives to the traditional liquid electrolytes, which are flammable and prone to leakage, posing safety risks. The development of ceramic- and polymer-based solid-state electrolytes is enhancing the safety and efficiency of next-generation batteries by offering higher ionic conductivity and improved thermal stability. Moreover, artificial intelligence and machine learning break the time and cost barriers in developing high-performance energy storage solutions, thereby accelerating material discovery by predicting ion transport mechanisms and optimizing electrode architectures and failure modes. These innovations encompass a paradigm shift in energy storage technology and serve as an integrated multifaceted use of mechanisms, environmentally sustainable materials, and advanced computational tools (Zhang et al., 2023). Together, these factors will lead to improvements in efficiency and density while improving sustainability factors greatly and bring about revolutionary changes in the capabilities of energy storage systems, which are liable to make them viable enough to support global transitions to renewable energy.

Table 4. Future approaches in trending energy storage technologies

Technology	Scientific Basis	Advantages	Challenges	Ref.
Solid-State Batteries (SSBs)	Use of solid electrolytes (ceramic, polymer, or sulfide-based) instead of liquid electrolytes to enhance ionic conductivity and stability.	Higher energy density, improved thermal stability, and enhanced safety.	High fabrication cost, dendrite formation, and interfacial resistance.	(Salkuti, 2023)
Sodium-Ion Batteries (SIBs)	Sodium-based cathodes (e.g., $Na_3V_2(PO_4)_3$, $Na_{0.66}[Mn_{0.5}Fe_{0.5}]O_2$) and hard carbon anodes as an alternative to lithium-based systems.	Abundant raw materials, cost-effectiveness, and promising for grid-scale storage.	Lower energy density, limited cycle life, and sluggish kinetics.	(Ramachandran et al., 2023)
Lithium-Sulfur Batteries (Li-S)	Sulfur-based cathodes and lithium metal anodes with polysulfide shuttle suppression strategies.	Theoretical energy density ~2600 Wh/kg, lightweight, and environmentally friendly.	Poor conductivity of sulfur, rapid capacity fading, and polysulfide dissolution.	(Satpathy et al., 2023)
Dual-Ion Batteries (DIBs)	Simultaneous storage of cations (Li^+, Na^+) and anions (PF_6^-, BF_4^-) in intercalation-type electrodes.	High voltage, improved energy efficiency, and reduced material costs.	Limited anion host materials and structural stability.	(Muzaffar et al., 2023)
Metal-Air Batteries (MABs)	Use of metal (e.g., Zn, Li, Al) anodes with air cathodes catalyzing oxygen reduction reactions (ORR).	High theoretical energy density, lightweight, and suitable for grid storage.	Cathode degradation, sluggish kinetics, and moisture/CO_2 sensitivity.	(Nahirniak et al., 2023)
Hybrid Supercapacitors (HSCs)	Integration of Faradaic (battery-like) and non-Faradaic (capacitor-like) electrodes.	Higher energy density than conventional supercapacitors and faster charge-discharge rates than batteries.	Trade-off between power and energy density, electrode degradation.	(Amir et al., 2023)

continued on following page

Table 4. Continued

Technology	Scientific Basis	Advantages	Challenges	Ref.
Redox Flow Batteries (RFBs)	Electrochemical energy storage using redox-active species (e.g., vanadium, organic, or metal-based redox couples) in liquid electrolytes.	Scalable energy storage, long cycle life, and decoupled power-energy capacity.	Low energy density, high electrolyte cost, and membrane degradation.	(Rana et al., 2023)
Hydrogen Storage Materials	Utilization of metal hydrides (e.g., MgH_2, TiFe), porous carbons, and MOFs for hydrogen adsorption and release.	High hydrogen storage capacity, enabling fuel cell applications.	Material degradation, slow kinetics, and high operating temperatures.	(Agajie et al., 2023)
Flexible & Wearable Batteries	Use of polymer-based electrolytes, ultrathin electrodes, and nanostructured materials to achieve mechanical flexibility.	Enables energy storage in wearable electronics and biomedical devices.	Limited mechanical durability, low energy density, and fabrication challenges.	(Hassan, Sameen, Olapade, et al., 2023)
Recyclable & Green Batteries	Development of biodegradable and non-toxic battery components, including aqueous electrolytes and cellulose-based separators.	Reduces environmental impact, improves sustainability, and enhances recyclability.	Lower energy density, poor long-term stability, and industrial scalability issues.	(H. Wang et al., 2023)
Quantum Dot-Based Batteries	Incorporation of quantum dots (e.g., graphene QDs, perovskite QDs) in electrodes and electrolytes for enhanced charge transport.	High surface area, improved conductivity, and tunable electronic properties.	High synthesis cost, agglomeration, and photochemical instability.	(Hussain et al., 2023)
AI-Optimized Energy Storage	Machine learning-driven battery management systems for predictive maintenance, charge optimization, and degradation modeling.	Enhances efficiency, extends lifespan, and enables real-time energy management.	Requires large datasets, complex algorithm training, and sensor integration.	(Li et al., 2023)

8. CONCLUSION

This chapter offered a detailed presentation of the chemical drivers of further progress in the energy storage of batteries, supercapacitors, and other hydrogen storage devices. In times of increased worldwide energy demand as well as intensified shifts toward renewables, the necessary solution lies within efficient, sustainable, and scaleable energy storages. We have presented a detailed review of lithium-ion, solid-state, and alternative battery technologies as well as next-generation supercapacitors and hydrogen storage mechanisms that have shown some of the major challenges and breakthroughs shaping the future of energy storage. From enhanced cathodes and anodes to the development of new electrolytes and hybrid systems,

REFERENCES

Abeng, F. E., & Anadebe, V. C. (2023). Combined electrochemical, DFT/MD-simulation and hybrid machine learning based on ANN-ANFIS models for prediction of doxorubicin drug as corrosion inhibitor for mild steel in 0.5 M H2SO4 solution. *Computational & Theoretical Chemistry, 1229*, 114334. DOI: 10.1016/j.comptc.2023.114334

Agajie, T. F., Ali, A., Fopah-Lele, A., Amoussou, I., Khan, B., Velasco, C. L. R., & Tanyi, E. (2023). A Comprehensive Review on Techno-Economic Analysis and Optimal Sizing of Hybrid Renewable Energy Sources with Energy Storage Systems. *Energies, 16*(2), 642. DOI: 10.3390/en16020642

Alam, S., Khan, M. I., Fiaz, F., Iqbal, M. Z., Alam, F., Ahmad, Z., & Hegazy, H. H. (2023). Advancements in asymmetric supercapacitors: Material selection, mechanisms, and breakthroughs with metallic oxides, sulfides, and phosphates. *Journal of Energy Storage, 72*, 108208. DOI: 10.1016/j.est.2023.108208

Amir, M., Deshmukh, R. G., Khalid, H. M., Said, Z., Raza, A., Muyeen, S. M., Nizami, A.-S., Elavarasan, R. M., Saidur, R., & Sopian, K. (2023). Energy storage technologies: An integrated survey of developments, global economical/environmental effects, optimal scheduling model, and sustainable adaption policies. *Journal of Energy Storage, 72*, 108694. DOI: 10.1016/j.est.2023.108694

Balakumar, P., Vinopraba, T., & Chandrasekaran, K. (2023). Deep learning based real time Demand Side Management controller for smart building integrated with renewable energy and Energy Storage System. *Journal of Energy Storage, 58*, 106412. DOI: 10.1016/j.est.2022.106412

Boorboor Ajdari, F., Asghari, P., Molaei Aghdam, A., Abbasi, F., Rao, R. P., Abbasi, A., Ghasemi, F., Ramakrishna, S., & Mikaeili Chahartagh, N. (2024). Silicon Solid State Battery: The Solid-State Compatibility, Particle Size, and Carbon Compositing for High Energy Density. *Advanced Functional Materials, 34*(30), 2314822. Advance online publication. DOI: 10.1002/adfm.202314822

Chen, B., Deng, Q., Yang, G., Zhou, Y., Chen, W., Cai, Y., & Tu, Z. (2023). Numerical study on heat transfer characteristics and performance evaluation of PEMFC based on multiphase electrochemical model coupled with cooling channel. *Energy, 285*, 128933. DOI: 10.1016/j.energy.2023.128933

Chen, Y., Lennartz, P., Liu, K. L., Hsieh, Y., Scharf, F., Guerdelli, R., Buchheit, A., Grünebaum, M., Kempe, F., Winter, M., & Brunklaus, G. (2023). Towards All-Solid-State Polymer Batteries: Going Beyond PEO with Hybrid Concepts. *Advanced Functional Materials, 33*(32), 2300501. Advance online publication. DOI: 10.1002/adfm.202300501

Cordeiro-Costas, M., Villanueva, D., Eguía-Oller, P., & Granada-Álvarez, E. (2023). Intelligent energy storage management trade-off system applied to Deep Learning predictions. *Journal of Energy Storage, 61*, 106784. DOI: 10.1016/j.est.2023.106784

Dissanayake, K., & Kularatna-Abeywardana, D. (2024). A review of supercapacitors: Materials, technology, challenges, and renewable energy applications. *Journal of Energy Storage, 96*, 112563. DOI: 10.1016/j.est.2024.112563

Elalfy, D. A., Gouda, E., Kotb, M. F., Bureš, V., & Sedhom, B. E. (2024). Comprehensive review of energy storage systems technologies, objectives, challenges, and future trends. *Energy Strategy Reviews, 54*, 101482. DOI: 10.1016/j.esr.2024.101482

Entezari, A., Aslani, A., Zahedi, R., & Noorollahi, Y. (2023). Artificial intelligence and machine learning in energy systems: A bibliographic perspective. *Energy Strategy Reviews, 45*, 101017. DOI: 10.1016/j.esr.2022.101017

Ge, R., Boyce, A. M., Shui Zhang, Y., Shearing, P. R., Cumming, D. J., & Smith, R. M. (2023). Discrete element method and electrochemical modelling of lithium ion cathode structures characterised by X-ray computed tomography. *Chemical Engineering Journal, 465*, 142749. DOI: 10.1016/j.cej.2023.142749

Ghorbani, B., Zendehboudi, S., Saady, N. M. C., & Dusseault, M. B. (2023). Hydrogen storage in North America: Status, prospects, and challenges. *Journal of Environmental Chemical Engineering, 11*(3), 109957. DOI: 10.1016/j.jece.2023.109957

Giglio, E., Luzzani, G., Terranova, V., Trivigno, G., Niccolai, A., & Grimaccia, F. (2023). An Efficient Artificial Intelligence Energy Management System for Urban Building Integrating Photovoltaic and Storage. *IEEE Access : Practical Innovations, Open Solutions, 11*, 18673–18688. DOI: 10.1109/ACCESS.2023.3247636

Guo, S., Li, J., Wang, Y., & Wang, Z. (2024). Electrochemical-thermal coupling model of lithium-ion battery at ultra-low temperatures. *Applied Thermal Engineering, 240*, 122205. DOI: 10.1016/j.applthermaleng.2023.122205

Han, Z., Fang, R., Chu, D., Wang, D.-W., & Ostrikov, K. (2023). Introduction to Supercapacitors. *Nanoscale Advances, 5*(16), 4015–4017. DOI: 10.1039/D3NA90074F PMID: 37560421

Hassan, Q., Sameen, A. Z., Olapade, O., Alghoul, M., Salman, H. M., & Jaszczur, M. (2023). Hydrogen fuel as an important element of the energy storage needs for future smart cities. *International Journal of Hydrogen Energy*, *48*(78), 30247–30262. DOI: 10.1016/j.ijhydene.2023.03.413

Hassan, Q., Sameen, A. Z., Salman, H. M., Jaszczur, M., & Al-Jiboory, A. K. (2023). RETRACTED: Hydrogen energy future: Advancements in storage technologies and implications for sustainability. *Journal of Energy Storage*, *72*, 108404. DOI: 10.1016/j.est.2023.108404

Hussain, I., Lamiel, C., Javed, M. S., Ahmad, M., Sahoo, S., Chen, X., Qin, N., Iqbal, S., Gu, S., Li, Y., Chatzichristodoulou, C., & Zhang, K. (2023). MXene-based heterostructures: Current trend and development in electrochemical energy storage devices. *Progress in Energy and Combustion Science*, *97*, 101097. DOI: 10.1016/j.pecs.2023.101097

Irham, A., Roslan, M. F., Jern, K. P., Hannan, M. A., & Mahlia, T. I. (2024). Hydrogen energy storage integrated grid: A bibliometric analysis for sustainable energy production. *International Journal of Hydrogen Energy*, *63*, 1044–1087. DOI: 10.1016/j.ijhydene.2024.03.235

Kilic, A., Oral, B., Eroglu, D., & Yildirim, R. (2023). Machine learning for beyond Li-ion batteries: Powering the research. *Journal of Energy Storage*, *73*, 109057. DOI: 10.1016/j.est.2023.109057

Kumar, K., Tyagi, U., Maity, S. K., & Kumar, G. (2024). Next-Gen Energy Storage Devices: Carbon-Based Materials for Flexible Supercapacitors. In *Innovative Materials for Industrial Applications: Synthesis, Characterization and Evaluation* (pp. 135–168). IGI Global Scientific Publishing. DOI: 10.4018/979-8-3693-7505-1.ch005

Kumar, K., Tyagi, U., Maity, S. K., Singh, S., Nikita, , & Kumar, G. (2025, April). A strategic approach towards the synthesis of jackfruit peel-based activated carbon: Assessment of different activating agents and operating conditions. *Biomass Conversion and Biorefinery*, *15*(8), 12387–12408. Advance online publication. DOI: 10.1007/s13399-024-06030-1

Kumar, K., Tyagi, U., Sirohi, S., Kumar, R., Kumar Maity, S., Nikita, , Singh, S., & Kumar, G. (2025). A critical review on nanostructure-doped carbonized biomass: A new Era in sustainable supercapacitor technology. *Fuel*, *381*, 133707. DOI: 10.1016/j.fuel.2024.133707

Kumar, P., Kumar Jain, A., Srivastava, J. P., Kumar, R., Saxena, K. K., Prakash, C., & Buddhi, D. (2024). Multiphysics Simulation of the Shape Prediction and Material Removal Rate in Electrochemical Machining Process. *Advances in Materials and Processing Technologies, 10*(2), 772–784. DOI: 10.1080/2374068X.2023.2192132

Lazaroiu, G., Gmal Osman, M., & Strejoiu, C.-V. (2023). Performance Evaluation of Renewable Energy Systems: Photovoltaic, Wind Turbine, Battery Bank, and Hydrogen Storage. *Batteries, 9*(9), 468. DOI: 10.3390/batteries9090468

Lemaoui, T., Darwish, A. S., Almustafa, G., Boublia, A., Sarika, P. R., Jabbar, N. A., Ibrahim, T., Nancarrow, P., Yadav, K. K., Fallatah, A. M., Abbas, M., Algethami, J. S., Benguerba, Y., Jeon, B.-H., Banat, F., & AlNashef, I. M. (2023). Machine learning approach to map the thermal conductivity of over 2,000 neoteric solvents for green energy storage applications. *Energy Storage Materials, 59*, 102795. DOI: 10.1016/j.ensm.2023.102795

Li, Y., Wei, Y., Zhu, F., Du, J., Zhao, Z., & Ouyang, M. (2023). The path enabling storage of renewable energy toward carbon neutralization in China. *ETransportation, 16*, 100226. DOI: 10.1016/j.etran.2023.100226

Lim, J. M., Jang, Y. S., & Van, T. Nguyen, H., Kim, J. S., Yoon, Y., Park, B. J., Seo, D. H., Lee, K.-K., Han, Z., Ostrikov, K. (Ken), & Doo, S. G. (. (2023). Advances in high-voltage supercapacitors for energy storage systems: Materials and electrolyte tailoring to implementation. *Nanoscale Advances, 5*(3), 615–626. DOI: 10.1039/D2NA00863G PMID: 36756532

Liu, C., Lombardo, T., Xu, J., Ngandjong, A. C., & Franco, A. A. (2023). An experimentally-validated 3D electrochemical model revealing electrode manufacturing parameters' effects on battery performance. *Energy Storage Materials, 54*, 156–163. DOI: 10.1016/j.ensm.2022.10.035

Liu, Z., Zhang, S., Wang, L., Wei, T., Qiu, Z., & Fan, Z. (2020). High-efficiency utilization of carbon materials for supercapacitors. *Nano Select, 1*(2), 244–262. DOI: 10.1002/nano.202000011

Lobato-Peralta, D. R., Okoye, P. U., & Alegre, C. (2024). A review on carbon materials for electrochemical energy storage applications: State of the art, implementation, and synergy with metallic compounds for supercapacitor and battery electrodes. *Journal of Power Sources, 617*, 235140. DOI: 10.1016/j.jpowsour.2024.235140

Ma, N., Yang, D., Riaz, S., Wang, L., & Wang, K. (2023). Aging Mechanism and Models of Supercapacitors: A Review. *Technologies, 11*(2), 38. DOI: 10.3390/technologies11020038

Ma, N., Zhao, W., Wang, W., Li, X., & Zhou, H. (2024). Large scale of green hydrogen storage: Opportunities and challenges. *International Journal of Hydrogen Energy*, *50*, 379–396. DOI: 10.1016/j.ijhydene.2023.09.021

Massaro, M. C., Biga, R., Kolisnichenko, A., Marocco, P., Monteverde, A. H. A., & Santarelli, M. (2023). Potential and technical challenges of on-board hydrogen storage technologies coupled with fuel cell systems for aircraft electrification. *Journal of Power Sources*, *555*, 232397. DOI: 10.1016/j.jpowsour.2022.232397

Mehrenjani, J. R., Gharehghani, A., Ahmadi, S., & Powell, K. M. (2023). Dynamic simulation of a triple-mode multi-generation system assisted by heat recovery and solar energy storage modules: Techno-economic optimization using machine learning approaches. *Applied Energy*, *348*, 121592. DOI: 10.1016/j.apenergy.2023.121592

Mhlanga, D. (2023). Artificial Intelligence and Machine Learning for Energy Consumption and Production in Emerging Markets: A Review. *Energies*, *16*(2), 745. DOI: 10.3390/en16020745

Miranda, D., Gonçalves, R., Wuttke, S., Costa, C. M., & Lanceros-Méndez, S. (2023). Overview on Theoretical Simulations of Lithium-Ion Batteries and Their Application to Battery Separators. *Advanced Energy Materials*, *13*(13), 2203874. Advance online publication. DOI: 10.1002/aenm.202203874

Mohammed, A. S., Atnaw, S. M., Salau, A. O., & Eneh, J. N. (2023). Review of optimal sizing and power management strategies for fuel cell/battery/super capacitor hybrid electric vehicles. *Energy Reports*, *9*, 2213–2228. DOI: 10.1016/j.egyr.2023.01.042

Mulky, L., Srivastava, S., Lakshmi, T., Sandadi, E. R., Gour, S., Thomas, N. A., Shanmuga Priya, S., & Sudhakar, K. (2024). An overview of hydrogen storage technologies – Key challenges and opportunities. *Materials Chemistry and Physics*, *325*, 129710. DOI: 10.1016/j.matchemphys.2024.129710

Muzaffar, N., Afzal, A. M., Hegazy, H. H., & Iqbal, M. W. (2023). Recent advances in two-dimensional metal-organic frameworks as an exotic candidate for the evaluation of redox-active sites in energy storage devices. *Journal of Energy Storage*, *64*, 107142. DOI: 10.1016/j.est.2023.107142

Nahirniak, S., Ray, A., & Saruhan, B. (2023). Challenges and Future Prospects of the MXene-Based Materials for Energy Storage Applications. *Batteries*, *9*(2), 126. DOI: 10.3390/batteries9020126

Navaid, H., Emadi, H., & Watson, M. (2023). A comprehensive literature review on the challenges associated with underground hydrogen storage. *International Journal of Hydrogen Energy*, *48*(28), 10603–10635. DOI: 10.1016/j.ijhydene.2022.11.225

Olabi, A. G., Abdelghafar, A. A., Maghrabie, H. M., Sayed, E. T., Rezk, H., Al Radi, M., Obaideen, K., & Abdelkareem, M. A. (2023). Application of artificial intelligence for prediction, optimization, and control of thermal energy storage systems. *Thermal Science and Engineering Progress*, *39*, 101730. DOI: 10.1016/j.tsep.2023.101730

Oyedotun, K. O., & Mamba, B. B. (2024). New trends in supercapacitors applications. *Inorganic Chemistry Communications*, *170*, 113154. DOI: 10.1016/j.inoche.2024.113154

Pandey, D., Kumar, K. S., & Thomas, J. (2024). Supercapacitor electrode energetics and mechanism of operation: Uncovering the voltage window. *Progress in Materials Science*, *141*, 101219. DOI: 10.1016/j.pmatsci.2023.101219

Pelosi, D., Baldinelli, A., Cinti, G., Ciupageanu, D.-A., Ottaviano, A., Santori, F., Carere, F., & Barelli, L. (2023). Battery-hydrogen vs. flywheel-battery hybrid storage systems for renewable energy integration in mini-grid: A techno-economic comparison. *Journal of Energy Storage*, *63*, 106968. DOI: 10.1016/j.est.2023.106968

Pordanjani, A. H., Aghakhani, S., Afrand, M., Zhang, P., Tang, R., Mahian, O., Wongwises, S., & Rashidi, M. M. (2023). Thermo-electrochemical simulation of the cooling process in a compact battery pack considering various configurations. *Journal of Power Sources*, *553*, 232112. DOI: 10.1016/j.jpowsour.2022.232112

Qiu, C., Jiang, L., Gao, Y., & Sheng, L. (2023). Effects of oxygen-containing functional groups on carbon materials in supercapacitors: A review. *Materials & Design*, *230*, 111952. DOI: 10.1016/j.matdes.2023.111952

Ramachandran, T., Sana, S. S., Kumar, K. D., Kumar, Y. A., Hegazy, H. H., & Kim, S. C. (2023). Asymmetric supercapacitors: Unlocking the energy storage revolution. *Journal of Energy Storage*, *73*, 109096. DOI: 10.1016/j.est.2023.109096

Rana, M. M., Uddin, M., Sarkar, M. R., Meraj, S. T., Shafiullah, G. M., Muyeen, S. M., Islam, M. A., & Jamal, T. (2023). Applications of energy storage systems in power grids with and without renewable energy integration — A comprehensive review. *Journal of Energy Storage*, *68*, 107811. DOI: 10.1016/j.est.2023.107811

Rashid Khan, H., & Latif Ahmad, A. (2025). Supercapacitors: Overcoming current limitations and charting the course for next-generation energy storage. *Journal of Industrial and Engineering Chemistry*, *141*, 46–66. DOI: 10.1016/j.jiec.2024.07.014

Raut, K., Shendge, A., Chaudhari, J., Lamba, R., & Alshammari, N. F. (2024). Modeling and simulation of photovoltaic powered battery-supercapacitor hybrid energy storage system for electric vehicles. *Journal of Energy Storage*, *82*, 110324. DOI: 10.1016/j.est.2023.110324

Rojek, I., Mikołajewski, D., Mroziński, A., & Macko, M. (2023). Machine Learning- and Artificial Intelligence-Derived Prediction for Home Smart Energy Systems with PV Installation and Battery Energy Storage. *Energies*, *16*(18), 6613. DOI: 10.3390/en16186613

Ruan, Y., Liang, Z., Qian, F., Meng, H., & Gao, Y. (2023). Operation strategy optimization of combined cooling, heating, and power systems with energy storage and renewable energy based on deep reinforcement learning. *Journal of Building Engineering*, *65*, 105682. DOI: 10.1016/j.jobe.2022.105682

Sahoo, S., & Timmann, P. (2023). Energy Storage Technologies for Modern Power Systems: A Detailed Analysis of Functionalities, Potentials, and Impacts. *IEEE Access : Practical Innovations, Open Solutions*, *11*, 49689–49729. DOI: 10.1109/ACCESS.2023.3274504

Salkuti, S. R. (2023). Advanced Technologies for Energy Storage and Electric Vehicles. *Energies*, *16*(5), 2312. DOI: 10.3390/en16052312

Salman, M. S., Pratthana, C., Lai, Q., Wang, T., Rambhujun, N., Srivastava, K., & Aguey-Zinsou, K.-F. (2022). Catalysis in Solid Hydrogen Storage: Recent Advances, Challenges, and Perspectives. *Energy Technology (Weinheim)*, *10*(9), 2200433. Advance online publication. DOI: 10.1002/ente.202200433

Satpathy, S., Misra, N. K., Shukla, D., Goyal, V., Bhattacharyya, B. K., & Yadav, C. S. (2023). An in-depth study of the electrical characterization of supercapacitors for recent trends in energy storage system. *Journal of Energy Storage*, *57*, 106198. DOI: 10.1016/j.est.2022.106198

Smdani, G., Islam, M. R., Ahmad Yahaya, A. N., & Bin Safie, S. I. (2023). PERFORMANCE EVALUATION OF ADVANCED ENERGY STORAGE SYSTEMS: A REVIEW. *Energy & Environment*, *34*(4), 1094–1141. DOI: 10.1177/0958305X221074729

Srivastava, M., MR, A. K., & Zaghib, K. (2024). Binders for Li-Ion Battery Technologies and Beyond: A Comprehensive Review. *Batteries*, *10*(8), 268.

Tawalbeh, M., Farooq, A., Martis, R., & Al-Othman, A. (2024). Optimization techniques for electrochemical devices for hydrogen production and energy storage applications. *International Journal of Hydrogen Energy*, *52*, 1058–1092. DOI: 10.1016/j.ijhydene.2023.06.264

Venkateswarlu, S., Vallem, S., Umer, M., Jyothi, N. V. V., Giridhar Babu, A., Govindaraju, S., Son, Y., Jong Kim, M., & Yoon, M. (2023). Recent progress on MOF/MXene nanoarchitectures: A new era in coordination chemistry for energy storage and conversion. *Journal of Energy Chemistry*, *86*, 409–436. DOI: 10.1016/j.jechem.2023.07.044

Wali, S. B., Hannan, M. A., Abd Rahman, M. S., Alghamdi, H. A., Mansor, M., Ker, P. J., Tiong, S. K., & Mahlia, T. M. I. (2023). Usage count of hydrogen-based hybrid energy storage systems: An analytical review, challenges and future research potentials. *International Journal of Hydrogen Energy*, *48*(89), 34836–34861. DOI: 10.1016/j.ijhydene.2023.05.298

Wang, D., Liu, N., Chen, F., Wang, Y., & Mao, J. (2024). Progress and prospects of energy storage technology research: Based on multidimensional comparison. *Journal of Energy Storage*, *75*, 109710. DOI: 10.1016/j.est.2023.109710

Wang, H., Fu, F., Huang, M., Feng, Y., Han, D., Xi, Y., Xiong, W., Yang, D., & Niu, L. (2023). Lignin-based materials for electrochemical energy storage devices. *Nano Materials Science*, *5*(2), 141–160. DOI: 10.1016/j.nanoms.2022.01.002

Xu, K., Guo, Y., Lei, G., & Zhu, J. (2023). A Review of Flywheel Energy Storage System Technologies. *Energies*, *16*(18), 6462. DOI: 10.3390/en16186462

Yang, H., He, Z., Zhang, M., Tan, X., Sun, K., Liu, H., Wang, N., Guan, L., Wang, C., Wan, Y., Wang, W., Hu, H., & Wu, M. (2023). Reshaping the material research paradigm of electrochemical energy storage and conversion by machine learning. *EcoMat*, *5*(5), e12330. Advance online publication. DOI: 10.1002/eom2.12330

Yang, K., Zhang, L., Wang, W., Long, C., Yang, S., Zhu, T., & Liu, X. (2024). Multiscale modeling for enhanced battery health analysis: Pathways to longevity. *Carbon Neutralization*, *3*(3), 348–385. DOI: 10.1002/cnl2.124

Yang, Y., Wu, Z., Yao, J., Guo, T., Yang, F., Zhang, Z., Ren, J., Jiang, L., & Li, B. (2024). An overview of application-oriented multifunctional large-scale stationary battery and hydrogen hybrid energy storage system. *Energy Reviews*, *3*(2), 100068. DOI: 10.1016/j.enrev.2024.100068

Yazici, İ., Shayea, I., & Din, J. (2023). A survey of applications of artificial intelligence and machine learning in future mobile networks-enabled systems. *Engineering Science and Technology, an International Journal, 44*, 101455. DOI: 10.1016/j.jestch.2023.101455

Yi, Z., Chen, Z., Yin, K., Wang, L., & Wang, K. (2023). Sensing as the key to the safety and sustainability of new energy storage devices. *Protection and Control of Modern Power Systems*, *8*(1), 27. DOI: 10.1186/s41601-023-00300-2

Yu, X., Chen, R., Gan, L., Li, H., & Chen, L. (2023). Battery Safety: From Lithium-Ion to Solid-State Batteries. *Engineering (Beijing), 21*, 9–14. DOI: 10.1016/j.eng.2022.06.022

Zayed, M. E., Kabeel, A. E., Shboul, B., Ashraf, W. M., Ghazy, M., Irshad, K., Rehman, S., & Zayed, A. A. A. (2023). Performance augmentation and machine learning-based modeling of wavy corrugated solar air collector embedded with thermal energy storage: Support vector machine combined with Monte Carlo simulation. *Journal of Energy Storage*, *74*, 109533. DOI: 10.1016/j.est.2023.109533

Zeng, L., Peng, J., Zhang, J., Tan, X., Ji, X., Li, S., & Feng, G. (2023). Molecular dynamics simulations of electrochemical interfaces. *The Journal of Chemical Physics*, *159*(9), 091001. Advance online publication. DOI: 10.1063/5.0160729 PMID: 37671960

Zhang, H., Gao, S., & Zhou, P. (2023). Role of digitalization in energy storage technological innovation: Evidence from China. *Renewable & Sustainable Energy Reviews*, *171*, 113014. DOI: 10.1016/j.rser.2022.113014

Zhou, Y., Ouyang, Y., Zhang, Y., Li, Q., & Wang, J. (2023). Machine Learning Assisted Simulations of Electrochemical Interfaces: Recent Progress and Challenges. *The Journal of Physical Chemistry Letters*, *14*(9), 2308–2316. DOI: 10.1021/acs.jpclett.2c03288 PMID: 36847421

Chapter 5
Green Biosynthesis of Carbon-Based Nanomaterials for Antioxidant, Anti-Bacterial, and Anti-Cancer Applications

Amit Kumar Dutta
https://orcid.org/0000-0002-1235-766X
Bangabasi Morning College, University of Calcutta, India

ABSTRACT

This chapter focused on the development of nano-scaled carbon-based materials based on simple, green approach using natural resources such as extracts of various plants or fruit/vegetable peel waste, tea, coffee, alovera, as well as citrus fruits- grape, lemons, oranges etc. for antioxidant, anti-bacterial and anti-cancer applications. The surface functionalization with the high level of polyphenolic compounds, that resist the body from free-radical damage, have been confirmed through FTIR, EDX analysis which can enhance their stability, impart bio-compatibility, anti-bacterial, anti-microbial and anti-cancer activity. The antioxidant activities have been investigated using DPPH free-radical scavenging assay. Furthermore, the bacterial activity of the synthesized nano-materials against various pathogenic organism (gram-positive and gram-negative bacterial strains) has been evaluated. Finally, dose-dependent anti-cancer activity in both in-vitro and in-vivo setting has established the anti-proliferative properties of the nano-materials against different cancer cell-lines.

DOI: 10.4018/979-8-3373-1409-9.ch005

INTRODUCTION

Carbon is one of the most abundant elements on the Earth and also most useful in the Periodic Table. Carbon brushes, Graphite, a more stable form of Carbon, are well known essential component of electric motors and have been extensively utilised as lubricant (Morstein & Dienwiebel, 2021). Again, activated charcoal or activated carbon granules are very popular and have been widely used in cosmetics, personal care products and water filtration process over the past three decades (W. A. Helbig, 1946). Now-a-days, so many research works have been concentrated to develop alternative cost-effective and next-generation multi-functional materials with the help of these virtually unlimited and free carbon-based sources, though so many metal-based multi-functional nanomaterials provide incredible benefits to the society and have already been used in everyday consumer products. Metal-based nanomaterials, while offering unique properties, present drawbacks like potential toxicity, environmental concerns, and limitations in sustainable synthesis and recycling (Dutta, 2023). Moreover, carbon-based materials usually possess unique structural dimension, outstanding electrical conductivity, excellent mechanical, thermal, optical properties, enabled them attracted significant interest in diverse area. Carbon also exists in a wide variety of allotropic forms with 0D to 3D structures. Again, in the field of nano-technology, when the dimension of a material be so small, in the nano-scaled range, new interesting properties have been generated such as interesting chemical and physical properties, large specific surface area, large specific surface to volume ratio, greater surface tension, greater surface activity, high catalyst-loading capacity and better light absorption capacity. As a whole, carbon-based nano-materials become 'Revolutionary Materials' which have been developed for enormous attainable applications and can be utilised in daily life and everyday consumer products. Various types of Carbon-nanomaterials such as fullerene, nano-diamond, different nano-sized carbon allotropes, more popular carbon nanotubes (CNT), graphene, graphene oxide (GO), reduced graphene oxide (rGO) nano-sheet, carbon nano-fibers, carbon quantum dots (CQDs), carbon dots (CDs), polymeric graphite carbon nitride (g-C_3N_4), carbonized polymer dots (CPDs) etc. have been thoroughly investigated and started to be commercialized for the sake of society. Graphene nano-materials is commonly known as wondering material owing to possessing of unique 2D flat-layered graphite-like honeycomb lattice structure (graphitic allotropes), usually possess Π-conjugated electrons in the sp^2 hybridised domains which can effectively accelerate the electron transfer, scavenge photo-generated electron from semiconductor surface, enhance photo-generated charge-carrier separation and electron mobility or migration efficiency. This delocalised Π-electron cloud also responsible for Π-plasmon resonances and extensively used in plasmonic (LSPR)-based sensing application in health or biomedical, environmental and food testing systems over the years (Gupta et al., 2019). Carbon-based nanomaterials also possess unique optical properties. Carbon quantum dots (CQDs), carbon dots (CDs), polymeric graphite carbon nitride (g-C_3N_4), carbonized polymer dots (CPDs) with size below 10 nm show excellent photoluminescence behaviour (quantum confinement effect) and have been widely applied in luminous bodies. Within very small dimension, abundant sp^2/sp^3 Carbon sources promote integrity of Π -conjugation/ conjugated system and improve photoluminescence capabilities. These nanomaterials also possess large numbers of surface functional groups such as hydroxyl (-OH), carboxyl (-COOH), epoxy (-O-), amino (-NH_2) which are generated during synthesis process from nitrogen and oxygen containing raw materials and incorporated into the typical Π -electron network. These core structure and surface functional groups are main responsible for unique optical properties. Carbon quantum dots (CQDs) with below 10 nm can directly absorb light in the UV-vis range as the electrons move from HOMO to LUMO of the sp2 C=C and C=O bonds corresponding to Π-Π* transition and n-Π* transition. Again, doping with heteroatom N, O, P, S, or Boron has been proven to be a practical approach to tuning the optical properties of the Carbon-nanomaterials. Owing to the similar atomic size to the Carbon atom, especially N and O can incorporate into the main structure of carbon and also Π -electron network leading to the enhancement of optical properties, effectively broaden the light absorption with high fluorescence quantum yield. Variation of N and O content in carbon core can affect the size of Π -conjugated domain, can improve electron delocalization, the electron transport properties, longer carrier lifetime, broaden absorption region. Yang's group had prepared red/ NIR emissive carbon quantum dots (CQDs), deep red emissive carbonized polymer dots (CPDs) where efficient conjugated Π -system have been generated with N-heterocycle and aromatic rings of polymer chain through delocalisation of electrons. As a result, they exhibit exceptional sunlight-harvesting capacities, and improve broad visible light absorption. Carbon dots (CDs) also are promising materials, have strong light absorption, unique electron donor, acceptor capabilities and can also do light-harvesting over very broad spectral range from UV to NIR. Sometimes, its high degree of crystallinity promotes Π-Π* and n-Π* transition and results exceptional solar-light harvesting capacities, tuneable photo-luminescence, photo-dynamic effect and effective photo-exited electron transfer properties. Moreover, CDs especially N-doped CDs (heteroatom such as P or S or N or Boron are doped with CDs) show outstanding photon up-conversion (UC properties) ability to convert long wavelength (550-850 nm) NIR-light to short wavelength (440-500 nm) UV-Vis light (Richards et al., 2021) for enhancing full spectral-range (320-800 nm) solar energy harvesting and full-spectrum-solar-light driven application. Owing to presence of highest percentage (49%) of NIR in the solar spectrum, NIR-light active nano-catalysis system is required which can possess lower energy gap and can harvest low photonic energy. long wave-length NIR photons show special ability to penetrate deeper into solution/reaction medium and provide more uniform activation of both nano-catalyst and substrate materials, which will be benefited for large-scale industrial application.

Now-a-days, with tremendous development of science and technology, so many catalysis-research-works have been devoted on innovation and development of interesting technologies to be utilized as potential antioxidant, anti-bacterial, anti-cancer agent for commercial applications (H. Kumar, 2020). Recently, in cancer research, nano-materials have been designed in such a way that it can be bind to specific sites on cancer cells or tumors to deliver drugs more effectively, after that the antioxidant functionalized nano-materials efficiently aggregate at the target location, release their surface-loaded anti-cancer agents to a particular targeted sites during cancer therapy. As an anti-cancer material, researcher have now paid considerable attention on virtually unlimited and free carbon-based sources (also non-toxic bio-compatible) which have become suitable candidates for full-spectrum-solar-light driven (320-800 nm) photo-thermal therapy (PTT), photo-dynamic therapy (PDT). Now-a-days, red/NIR emissive CDs have been widely used in cancer treatment which can enhance photo-catalytic response towards reactive oxygen species (ROS) generation and destroy cancerous cell's structure. Simultaneously, the deep penetration power and photo-thermal effect of NIR-light raises the temperature of tumors and encourage PTT, PDT, where light-to-heat conversion efficiency kill cancer cells (Han et al., 2024). It has been noted that the cancer cells are more heat sensitive than normal cells and the cancer cells die at the temperature above about 45°C. CDs can encourage photo-dynamic therapy (PDT) where, they act as photo-sensitizer and can generate ROS, that cause apoptotic and necrotic cell death and trigger oxidative damage of the amino acids of protein, fatty acids in lipids and the nucleo-bases in DNA of tumor cells. CDs can transfer the light energy to the surrounding oxygen molecules and generate Singlet oxygen that interacts with adjacent cancer tissues to induce apoptosis, necrosis-related cell death (Han et al., 2024). Similarly, metal free carbon-based quantum dots (CQDs), Graphene, GO, CNT can do light-harvesting over very broad spectral range from UV to NIR (Wang et al., 2017, Serge-Correales et al., 2022). Polymeric graphite carbon nitride (g-C_3N_4), as an organic semiconductor, can strongly prevent the re-combination of photo-generated electron–hole pairs, enhance the electron-transport behavior and improve their photo-catalysis performances (Cao & Yu, 2014). Owing to the presence of unique 2D flat-layered structure, graphene-based materials, particularly Graphene Oxide (GO) and Graphene Quantum Dots (GQDs), exhibit antioxidant activity by scavenging free radicals, inhibit the oxidative-damage of healthy cells and protecting all important biomolecules including DNA, proteins, membranes. Hence, Graphene's antibacterial activity, the ability to interact with bacterial cells arises through potentially damaging their membranes or inducing oxidative stress. Single-walled carbon nanotubes (SWCNTs) have been demonstrated as strong anti-microbial agent, can entire the cell wall and cytoplasmic membrane of micro-organisms and viruses, that survive on surfaces for a long time in most cases and can able to induce the oxidative interference of proteins, metabolic activities, or damage in bacterial DNA and enzyme structures (S. E. Abo-Neima, 2020). Very recent, it has extensively been utilized in ointment/ cream for burns, wounds to inhibit bacterial infections. (E. Urnukhsaikhan, 2021). Again, carbon nanotubes (CNTs), carbon dots (CDs), and fullerenes, as an emerging class of novel materials, can exhibit a considerable antimicrobial activity, have been incorporated in cosmetic products formulation and also serve as photo-sensitizer in photo-catalysis process (M. Azizi-Lalabadi, 2020). Within the past few years, activated charcoal nanomaterials have been actively used in cosmetic products. Due to its adsorbing properties, ability to adsorb impurities, oil, and toxins from the skin, it is being used in all sorts of beauty products from face masks to skin cleansers and even soaps. In some cosmetic products such as creams, lotions, gel, sprays, sticks etc., Carbon Dots (CDs) have been included to create UV-filters that block ultraviolet (UV) radiation and exhibit high sun protection factor (SPF).

Bio-synthesis of nanoparticles is a laboratory synthesis process, eco-friendly green-chemistry-based techniques, where different biological resources such as parts-of-plants (leaves, flowers, barks, seeds etc.), micro-organisms have been used to prepare different carbon-based metal-free nanomaterials (G. C.Vidana-Gamage, 2021). As traditional ayurvedic medicine, different parts of medicinal plants have already been used since very old days which are important herb in Ayurvedic system with great medicinal values. Owing to possessing wide range of chemical constituents (bioactive phytochemicals) such as flavonoids, polyphenols, terpenoids, alkaloids, quinines, tannins etc., the parts-of-plants exhibit high anti-oxidant, anti-microbial, anti-bacterial, anti-diabetic, anti-inflammatory, anti-obesity, anti-cancer activities and can decrease the risk of developing different chronic non-communicable diseases. In recent years', these the parts-of-plants have been extensively studied through bio-synthesis process for development of more eco-friendly, non-toxic, most sustainable, cost-effective environmentally friendly different nano-materials which can extensively be applicable in bio-medical field. The process also can minimize the deployment of hazardous/toxic agents and expensive and rare materials/equipment to avoid adverse environmental impacts. In natural plant extract the antioxidant-riched phytochemicals act as good chelating/reducing agent where electrons have been transferred from the phytochemicals to inorganic compound, enabling the production of stable bulk metallic nano-materials with controlled size and shapes as well as capping/stabilizing agents, preventing the nano-particles from agglomerate to each other (R. Chatterjee, 2022, A. Demirbas, 2019). In addition, the bio-active phytochemicals components have automatically been incorporated onto the surface of the nano-materials, ie. surface functionalization, during bio-synthesis process. As a results, the potent antioxidant activity of the plant extract has been transmitted into the nano-particles, make them more bio-compatible and have been widely employed as more effective, cheaper, lower toxic new therapeutic agent, nano-carriers for nano-delivery (H. Kumar, 2020). So many in-vitro and in-vivo experiments have been carried out in the treatment or management of chronic diseases (A.C.Gonçalves, 2022). It has been proved that antioxidant functionalized nano-materials exhibits higher free-radical scavenging activity (lower IC50 value) than bare antioxidant containing natural extract and have been widely used as potential antioxidant, anti-bacterial, anti-cancer agent for commercial applications (H. Kumar, 2020). Recently, in cancer research, nano-materials have been designed through bio-synthesis process in such a way that it can bind to specific sites on cancer cells or tumors to deliver drugs more effectively, after that the antioxidant functionalized nano-materials efficiently aggregate at the target location, release their surface-loadded anti-cancer agents to a particular targeted sites during cancer therapy. Different carbon-nanomaterials such as carbon nanotubes (CNT), graphene, carbon quantum dots (CQDs), carbon dots (CDs), polymeric graphite carbon nitride ($g-C_3N_4$), carbonized polymer dots (CPDs) are most popular and acceptable bio-synthesized and bio-compatible material in medical sector (I. Fatimah, 2020, R. S. George, 2021). Several natural resources like crab shell, prawn shell, green tea leaves, coffee, garlic, ginger, and bio-wastes have become prominent in developing fluorescent carbon dots (CDs) due to their eco-friendliness, low cost, therapeutic properties and high availability. The use of plant parts as green resources not only meets the pressing need for largescale carbon-based material production, but also encourages the development of sustainable applications.

With these evidences, this chapter focused on bio-synthesis of different carbon-based nano-materials using antioxidant-riched plant extract and assess their anti-oxidant, cytotoxic activities. The antioxidant activities of the proposed nano-materials have been investigated through 2,2-diphenyl-1-picrylhydrazyl (DPPH) free-radical scavenging assay and the corresponding IC50 values have been determined. Furthermore, the bacterial activity of the synthesized nano-materials against various pathogenic organism (gram-positive and gram-negative bacterial strains) has been compared and discussed. Finally, dose-dependent cytotoxicity analysis, anti-cancer activity in both in-vitro and in-vivo setting has been elaborated.

Green bio-synthesis of Carbon-based nanomaterials

At first, different parts of medicinal plants ie. roots, leaves, flowers seeds etc. have been collected from natural sources (commonly known as important herbs in Ayurvedic system with great medicinal values). The corresponding extracts have been prepared by grinding about 25 g of fresh flower, leaves etc. followed by maceration using 50 mL of water, followed by filtering by using whatman 41 filter paper. Sometimes the extract needed to be centrifuged at 600 rpm for 5min. The clear extracts have been collected into a sterilized bottle for pyrolysis or carbonization process' and preserved at 4°C until further use.

Large-scale carbon dots (CDs) have been very easily synthesized through eco-friendly way from almost any plant-based organic precursor compounds such as leaves, seeds, tree bark, vegetables, fruits, wheat flour, any bio-degradable waste or any chemical compounds that have C, H and O bonds present in them such as glucose, urea, ascorbic acids, citric acids etc. (Figure 1).

Figure 1. Schematic diagram for synthesis process of carbon dots (CDs)

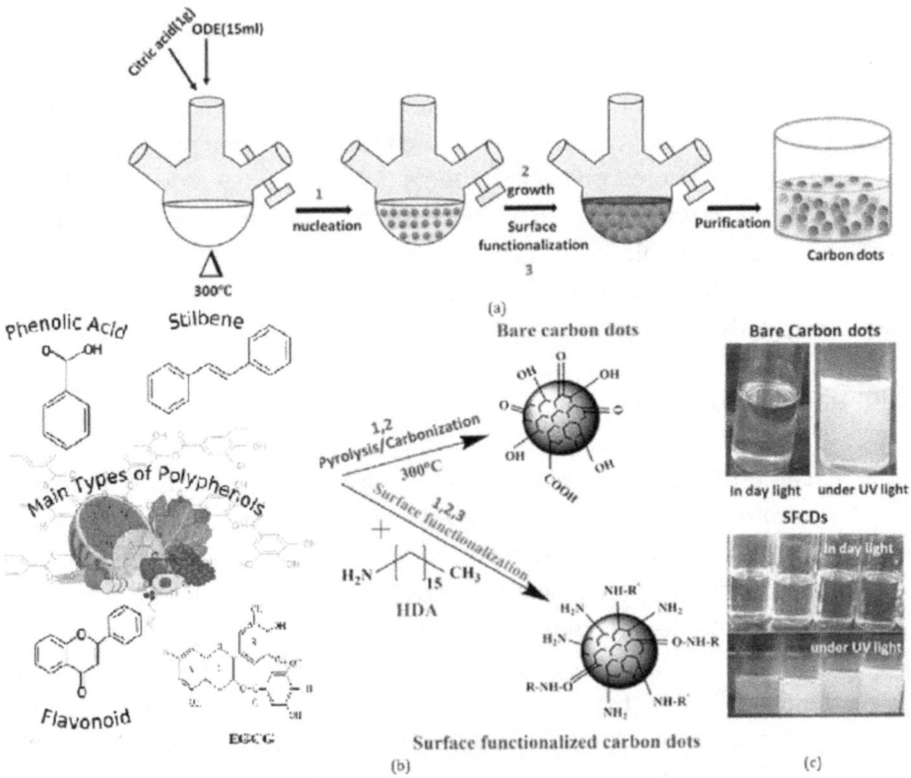

Gedda et al. 2023 reported a green-biosynthesis approach for synthesizing CDs from coriander leaves by one-step carbonization mediated synthesis, where coriander leaves solely served as the carbon source and passivation agent for CDs. According to the research group, at first the leaves have been sun-dried and pulverised into fine powder using agate mortar and pestle. 2.0 g of that powder has been added to 50.0 mL of demineralized water (DM water) under stirring for one hour. The supernatant has passed through a Whitman filter no.1 paper to separate large and undissolved particulate. The resulted filtrate has been carefully collected into a round bottomed flask and placed on a sand bath. The reflux reaction has been maintained at uniform temperature (180 °C) for 12 h under constant stirring. Formation of CDs has been confirmed with the development of reddish-brown coloured solution.

Barbero et al. 2025 have successfully synthesized Carbon nanoparticles with precise control of the diameter (15-150 nm) through bottom-up hydrothermal autoclave under comparable low reaction temperature (<300°C) and self-generated pressure in aqueous medium.

During green biosynthesis of carbon nanotubes (CNTs), Mostafavi et al.,2022 have reported the development of CNTs using various natural hydrocarbon precursors, including plant extracts (*e.g.*, tea-tree extract), essential oils (*e.g.*, eucalyptus and sunflower oil), biodiesel, milk, honey, and eggs, among others. Additionally, agricultural bio-wastes have been widely studied for synthesizing CNTs.

The water-soluble fluorescence carbon dots (CDs) have been developed using neem (*Azadirachta indica*) leaves in a sand bath without the usage of passivating agents. Again, water-soluble CDs, graphene quantum dots (GQDs), can also be synthesized from various domestic waste like tea leaves and food scraps through processes like pyrolysis carbonization.

Results and discussions

Different carbon-based nanomaterials have been generally synthesized using organic materials as a carbon source through two main approaches, one is "top-down" approach to creating carbon nanoparticles involves breaking down larger carbon structures (like graphite, soot, or carbon nanotubes) into smaller nanoparticles. The most useful "bottom-up" approach to creating carbon nanoparticles involves building up nanoparticles from smaller components, like atoms or molecules, rather than breaking down larger structures. This method focuses on assembling carbon units into desired structures through various chemical reactions, very simple carbonization, pyrolysis, hydrothermal, hydrothermal carbonization, solvothermal, microwave pyrolysis, ultrasonication and chemical vapor deposition processes. These methods allow for the formation of different carbon-based nanomaterials with desired properties, including water solubility and fluorescence, by carefully controlling the reaction conditions and precursor selection. Pyrolysis carbonization is a basic approach for preparing different carbon-based nano-materials, where complex organic molecules in the extract have been decomposed into carbon-rich fragments and different carbon-based nano-materials through one step process (without excessive equipment or complicated processes) under high temperature and pressure. Hydrothermal carbonization (HTC) is most cost-effective, with minimum equipment, less synthesis time containing and environmental-friendly one-step thermo-chemical process that converts organic materials, especially unlimited-free natural plant-based products and biomasses, into solid valuable carbonaceous materials. HTC of carbohydrates has been reported as a sustainable and green technique to develop different carbonaceous micro- and nano-materials having specific morphology and structure for several applications, including catalysis, separation science, metal ion adsorption and nanomedicine.

The plant leaf encompasses major active constituents' quercetin – a flavonoid, limonoids comprising nimbin derivatives, flavonoids, proteins, carbohydrates, sitosterol and carotene. These active biomolecules comprise carbon, oxygen, nitrogen as core elements. Coriander, Neem etc. leaves are edible and naturally contain carbohydrate and proteins which are abundant in carbon, nitrogen and oxygen elements. Coriander leaves contain natural antioxidants, mainly due to their high content of polyhenolic compounds like caffeic acid, protocatechuic acid, and glycitin. Consequently, the formation of CDs using Coriander leaves may involve functionalization of that bioactive compounds on its surface through dehydration and carbonization of coriander leaves followed by *in situ* surface passivation under high temperature and pressure conditions during hydrothermal treatment. Bottom-up approach has been adapted to transform small biomolecules to water soluble nontoxic CDs. The process initiates hydrolysis and condensation. Subsequently, self-polymerization/aggregation has been taken place. Finally, transformation to core and surface passivation occurs due to elevated temperature over a period of time. Sometimes, carbon nanomaterials have been developed from various domestic waste sources such as food waste, grass waste, rice husk, and other biomass materials. By utilizing these waste materials, researchers can simultaneously address waste-management challenges and generate valuable innovative materials for various applications, including sensing, wastewater treatment, and biomedical applications.

Formation of CDs has confirmed with the development of reddish-brown coloured solution. As the CDs exhibits unique fluorescence properties, UV-visible and fluorescence spectrophotometer has primarily been used to monitor the formation of the nano-materials. The water-soluble fluorescence CDs exhibited bright blue color fluorescence under exposure to UV light (365 nm) and pale-yellow color in daylight. The obtained UV–Vis absorbance spectra of CDs exhibited peaks at 270, and 350 nm with a tail spread over the visible region (Figure 2a). The absorption peak at 270 nm was observed due to π-π^* transition of the sp^2 conjugated system in an aromatic ring containing C=C bonds. A peak at 350 nm corresponds to a n-π^* transition due to C=O bonds. Excitation and emission CDs spectra in water are shown in (Figure 2b). With an excitation wavelength of 370 nm, the strongest emission peak was at 460 nm. The C=O bonds lie as a surface functional group on CDs and give rise to fluorescence by trapping energy on the surface states. In FTIR spectrum, free O-H stretching band-signals and in EDX analysis, some carbon (21.63 %) and oxygen (4.54 %) signals confirmed the presence of bio-active compounds on the NPs surface.

*Figure 2. (**a**) UV-Vis spectra absorption spectra of CDs (Inset: Images of CDs under daylight and UV irradiation). (**b**) Excitation and emission spectra of the CDs.*

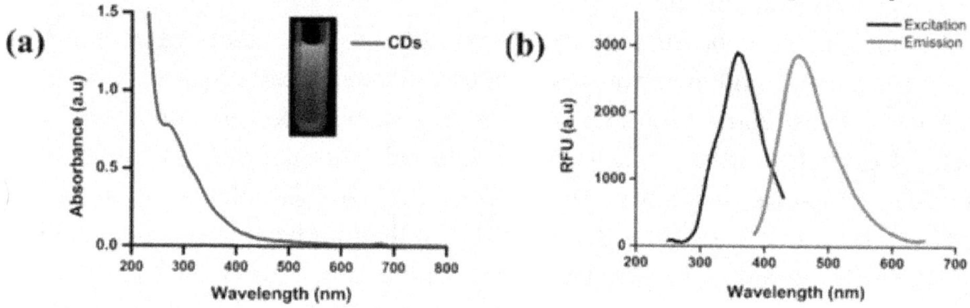

The morphological characteristics and size of CDs were visualised using High-resolution transmission electron microscopy (HR-TEM). The TEM images captured for single CDs at multiple magnifications showed no appearance of fringes, represented in Figure. 3 a, b, c. The HR-TEM image reveals a group of distinct and independent CDs with uniform black spherical spheres without aggregation, holding approximately 3.5 nm (average size). A broad hump is depicted in XRD analysis pattern of CDs at $2\theta=22°$ (approximately), which supports the amorphous nature of CDs. Absorption bands at 3657, 3100 and 2100 cm^{-1} are imputed to the vibratory stretching of O–H, C–H and Nitrile respectively. The weak peak at 1750 and 2850 cm^{-1} corresponds to the aldehyde group (stretching vibration). The absorption bands at 1625 cm^{-1} can be ascribed to the amide group. The large negative value (−11.0 mV) of zeta potential explains that CDs are possessed functional groups with negative charges on the surface of CDs.

Figure 3. (a, b, c) HRTEM images of CDs with 200,000X, 300,000X, and 400,000X magnification.

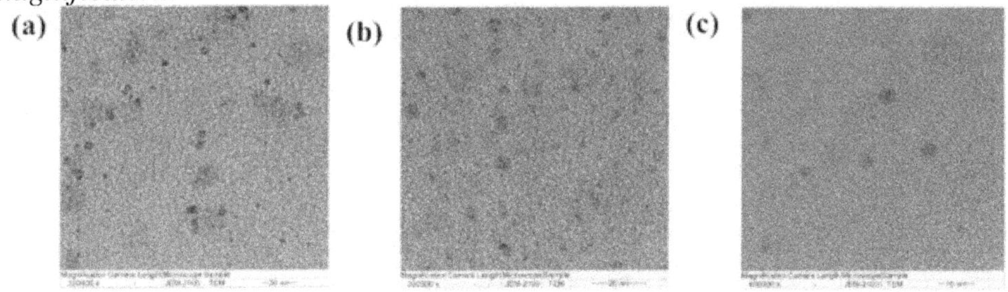

Determination of the Antioxidant capacity

Numerous studies showed that the plant extracts, rich in antioxidant components mainly possess the ability to act as free radical scavengers against harmful oxidants such as reactive oxygen (ROS) and nitrogen species (RNS) generated inside biological systems during electron transfer reactions by losing or accepting electrons. When, that antioxidant functionalized nano-materials has been synthesized through biosynthesis process, exhibits superior free-radical scavenging activity than bare antioxidant containing natural extract through interrupting ROS/RNS chain reactions, thereby preventing oxidation. Here, the proposed carbon-based nano-materials similarly can break down RNS/ROS and can effectively scavenge peroxyl (ROO•), hydroxyl (•OH), alkoxyl (RO•), $O2^{-}$•, NO• and other nitrogen or sulfur containing free radicals. The possible radical mechanism involves hydrogen atom transfer (HAT) or single electron transfer (SET) from carbon-nanomaterial's surfaces to DPPH•. The presence of so many conjugated double bond, phenolic hydroxyl groups, carboxyl (–COOH), hydroxyl (–OH), and amino (–NH_2, –NH–) groups presence in that functionalized carbon-nanomaterials allows the HAT or SET and reduction of DPPH• to DPPH-H (Figure 4). The unpaired electrons on carbon-nanomaterial's surfaces can be delocalized over the conjugated aromatic domain or through chemical bond rearrangement. According to Ngoc et. al, 2023, the blue fluorescence of Carbon dots (CDs) can be quenched by an electron transfer mechanism, in which nitroxide radicals act as electron acceptors. CDs can scavenge •OH radicals, one of primary sources of oxidative stress in biological systems, through conversion of that radical into a less reactive molecule thus preventing oxidative damage to biomolecules (e.g., lipids, proteins, and DNA). Furthermore, CDs possess unique photoluminescent properties owing to their quantum confinement and surface states. The excited states of the C-dots can interact with singlet oxygen molecules, which are ROS radicals that are highly responsible for oxidative damage, thus reducing the possibility of singlet oxygen formation. The quenching of the excited-state CDs plays an essential role in their antioxidant properties.

Figure 4. Mechanism of DPPH reduction by antioxidant CDs in aqueous media

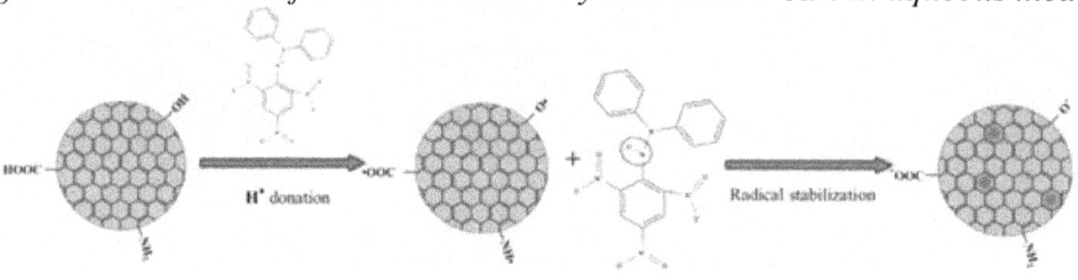

Recently, several research groups have investigated the antioxidant properties of carbon-based nanomaterials derived from different plant extracts. Gudimella et al. (2022), Sachdev & Gopinath, 2015 developed fluorescent CDs from Carica papaya leaves, coriendar leaves with significant antioxidant and anti-inflammatory properties. Antioxidant activity is a commonly used parameter which is defined in terms of ability of a material to scavenge or neutralize free radicals. The antioxidant activity is assessed via evaluating free radical content of a solution containing antioxidant nano-materials and comparing it to that in a solution without nano-materials (control).

The 2,2-diphenyl-1- picrylhydrazyl (DPPH) free radicals scavenging assay is one of the most commonly employed method used in antioxidant-research to determine the antioxidant properties of both plant-extract and synthesized nano-materials. DPPH-radical is commercially available a long-lived free radical, an organic nitrogen containing free radical, represent any kind of free radicals in our body, used to test antioxidant capabilities. That radical itself is deep purple in colour, which turns to yellow as soon as it interacts with an antioxidant. A compound is good antioxidant if it can easily donate an electron to the electron deficient DPPH radicals, stopping the chain of molecules stealing electrons from each other, inhibit the damage of healthy cells, all important biomolecules including DNA, proteins, membranes.

Figure 5. Proposed mechanism of antioxidant activities, Calibration curve for determination of IC50 value

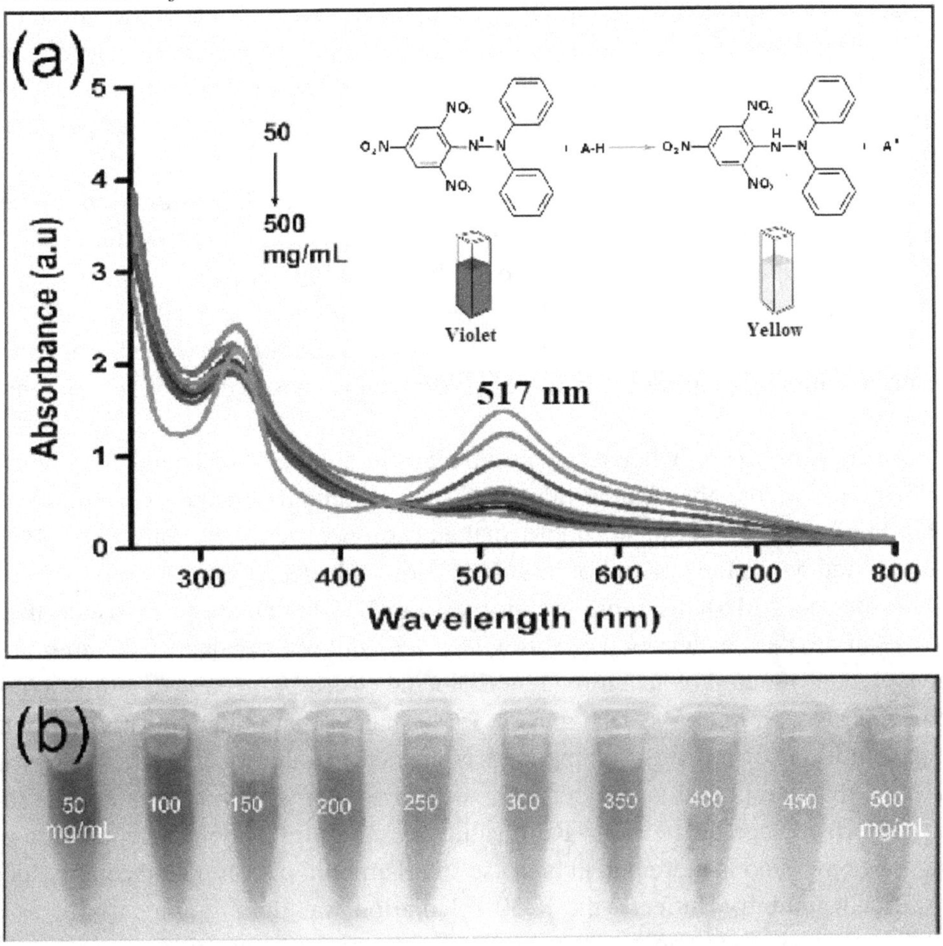

The experimental investigation has generally been carried out through spectrophotometrically because DPPH has an absorption band at 517 nm which disappears gradually upon reduction by an antioxidant compound (Figure 5a). When different concentration of proposed carbon-nanomaterials has been added to 50 μM methanolic DPPH solution, the color of the purple DPPH eventually transforms to yellow (Figure 5b) due to the addition of hydrogen/radical. The ability of carbon-nanomaterials to donate H^+ ions is based on the presence of COOH, OH/NH_2 surface functional groups. Corresponding, lower absorbance at 517 nm of the reaction mixture indicates higher DPPH free radical scavenging activity (S. Bhakya, 2015, T. Baygar, 2017). Ascorbic acid has been used as reference antioxidant because it can react with DPPH quickly reaching a steady state immediately. the radical scavenging activity of the nano-material has been found to increase in a dose-dependent manner.

Sachdev & Gopinath, 2015 has reported that when various concentrations of 05, 10, 20, 40, 60 and 70 μg/ml of biosynthesised Carbon dots (CDs) was added to 50 μM methanolic DPPH solution, there was a subsequent increase in the scavenging activity from 12 to 94%. Fascinatingly, the DPPH solution turned deep-purple to yellow with an increase in CDs concentrations.

Free radical scavenging activity was expressed as the % of inhibition, higher the percentage inhibition, the more antioxidant the particular compound was. The percentage of inhibition was calculated by the following equation

$$\% \text{ DPPH scavenging effect} = [(OD_{control} - OD_{sample}) / OD_{control}] \times 100,$$

where 'control' has been prepared without tested antioxidant tea sample or nano-material sample. A strong antioxidant, ascorbic acid, has been used as reference sample.

During experiment, when tested antioxidant sample has been added to Control DPPH solution, the absorbance value ie $OD_{control}$ gradually decreases owing to inhibition effect of antioxidant sample, make decompose some harmful free DPPH radicals and hence the amount of total DPPH radicals decreases.

Furthermore, Median inhibition concentration, IC 50 calculations are widely used to determine the concentration of antioxidant molecules that can do 50% inhibition ie. how much amount of an antioxidant molecule or drug is needed to inhibit a biological process by half (50%). High IC50 values means that a relatively high amount (concentration) of a drug or compound is needed to inhibit 50% of a biological process, that compound is less potent and requires a larger amount to achieve the desired effect whereas the lower IC50 value in microg/ml is more acceptable to us and the compound is more potent because small amount of compound can produce significant inhibitory effects. the IC50 calculation was done graphically using a calibration curve in the linear range by plotting the antioxidant concentration vs the corresponding scavenging effect.

From the experimental data, the more preferable low IC50 vale of 15 μg/ml for biosynthesised green Carbon dots (CDs) has been determined through constructing calibration curves which exhibits increasing % DPPH radical scavenging ability but concentration of 40 μg/ml of CDs exhibit highest antioxidant activity (~80%). Furthermore, Gudimella et al. (2022) developed fluorescent CDs from Carica papaya leaves with significant antioxidant properties with significantly lower IC50 value of 27.6 μg/mL in DPPH radical scavenging assay. Li et al. (2021) used biomass Salvia miltiorrhiza to generate high antioxidant capacity on fluorescent CDs through surface functionalization with S. miltiorrhiza, as a polymer molecule and exhibit excellent scavenging efficiencies of 71.4, 88.9 and 95.6%, respectively compared to plant extract alone (IC50 1920 μg/ml) indicating its strong anti-oxidant potential relevant to that of standard ascorbic acid.

Anti-bacterial Activity

Nano-materials especially carbon-based ones can inhibit bacterial growth through various mechanisms such as bacterial cell-membrane damage, production of reactive oxygen species (ROS) inside the bacterial-cell or interference with bacterial-metabolism, DNA-replication. When several natural resources like crab shell, prawn shell, green tea leaves, coffee, garlic, ginger, and bio wastes have been used as carbon sources in developing carbon-nanomaterials and high level of poly-phenolic bio-active compounds has been functionalized on the surface of the NPs, the anti-microbial, anti-inflammatory effect has been increasing much more. In this way, carbon-nanomaterials with high medicinal value exhibit free radical scavenging, total antioxidant, antibacterial, as well as antifungal activities. Different carbon nanomaterials such as carbon nanotubes and graphene, exhibit potent antibacterial and antifungal activities due to their unique physical and chemical properties. They can disrupt cell membranes, cause oxidative stress, and even penetrate biofilms, ultimately leading to bacterial-cell death. Carbon Dots, especially their ultra-small size, can penetrate bacterial cell-walls and membranes, leading to physical damage and membrane destabilization. Surface functionalization onto nao-materials can modulate surface charges and interaction with biological entities through electrostatic interaction. In carbon nanomaterials, carbon atoms are sp^2 hybridised state and can allow for easy attachment of functional hydroxyl, carboxyl, amino groups through halogenation or cycloaddition method. The functionalized NPs can interact very easily with bacterial cell-wall through attraction between the microbial cell wall's negative charge and NPs' positive charge (owing to possessing negative zeta potential). After interaction, the permeability function of the bacteria cell membrane changes and hence bacterial integrity has been disrupted and cause cell death (M. Yamanaka, 2005). Again, after interaction of the NPs with bacterial cell, the thiol group in the electron transport chain enzyme has been disrupted because the affinity of CDs to Sulphur and Phosphorus elements, which is abundantly found in bacterial cell wall, main responsible for anti-bacterial properties of the NPs. the capability of a NPs to enter and accumulated inside the wall of a bacterial-cell increases with decreasing the size of the NPs.

The synthesized NPs have been evaluated for anti-bacterial capabilities towards gram-positive and gram-negative bacterial stains through calculation of the inhibition zones and minimum inhibitory concentration (MIC) in μg/ml of the nano-antibiotics ie quantitative measurement of the lowest concentration of the NPs that inhibits the growth of a given strain of bacteria. Different bacterial stains show different diameter of inhibition-zone (DIZ) in mm unit (Figure 6) in experiment disc diffusion assay. But Minimum inhibitory concentration (MIC) and and minimal bactericidal concentration (MBC) has generally been determined with different bacteria in μg/ml unit during anti-bacterial activity testing (P. Wintachai, 2019). During anti-bacterial experiment, at first, the autoclaved petri dishes have been uniformly distributed with solidifying agar medium (Agar well diffusion method), followed by microbial inoculum onto the solidified agar medium surface. The each well was supplemented with CDs in different concentrations and subjected for 24 h' incubation under suitable conditions to observe bacterial growth. The zone of inhibition was calculated in diameter (mm) at each well carefully. Gedda et al. 2023 have investigated anti-bacterial activity of CDs using four different bacterial strains (two gram-positive and two gram-negative) and two fungal strains (Figure 6) in which the corresponding activity has been increased linearly with increased concentration of CDs ie. dose-dependent inhibition for antibacterial and antifungal activities.

*Figure 6. Antimicrobial activity of CDs (**a**, **b**) Antibacterial assay: Zone of inhibition against Staphylococcus aureus and Bacillus subtilis (gram-positive bacterial strains) and (**c**, **d**) Escherichia coli and pseudomonas aeruginosa (gram-negative bacterial strains), respectively (**e**, **f**) Zone of inhibition against candida and Aspergillus Niger (fungal strains) (scale bar is 10 mm for each). (S. Prabhu, 2021, I. Fatimah, 2020)*

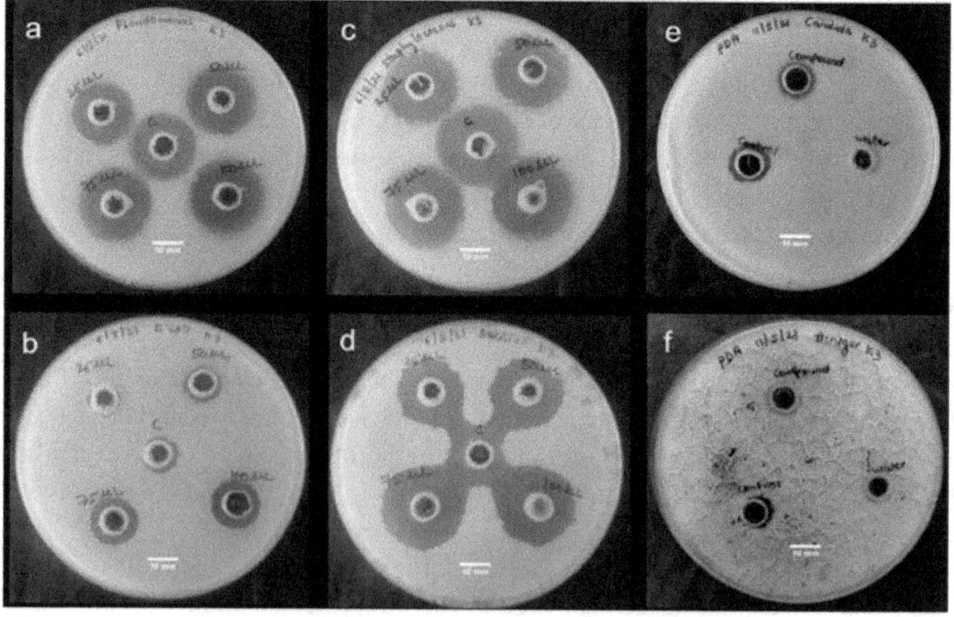

In case of CDs, the MIC has been evaluated as 0.5 mM (0.05 mg/mL) and the corresponding MBC, lowest concentration, as 4.5 mM that kills 99% of the initial bacterial population against E. Coli bacteria (P. Wintachai, 2019). Again, Antioxidant functionalized Carbon quantam dots (CQDs) has been found to be excellent anti-bacterial agent and using special ultra-sound assisted synthesis method, smaller and uniform CQDs particles illustrated enhanced anti-bacterial activity through providing more surface-contact with micro-organisms. S. aureus bacteria have been found to be most resistant to CQDs NPs with minimum inhibition-zone of ~20 mm. Moskvitina et al, 2023 have investigated the antimicrobial efficacy of multiwalled carbon nanotubes (MWNTs), catalytic filamentous carbon with different orientations of graphene blocks (coaxial–conical and stacked, CFC), ionic carbon (OLC), and ultrafine explosive nanodiamonds (NDs) through fluorescence microscopy (with staining of cells with SYTO 9 dye) and concluded that graphene blocks (coaxial–conical and stacked) completely suppressed the viability of E. coli and S. aureus leading to the blocking of basic cellular functions with the destruction of cell membranes. It has been clearly indicated in fluorescence microscopy image (a ×400 magnification) that all live E. coli bacterial-cells (green fluoresce image without exposure to carbon-based nanomaterials) have been converted to dead cells (fluoresce red) after 24 h of exposure to carbon nanomaterials (Figure 7).

Figure 7. Fluorescence microscopy images (SYTO 9 stain and the propidium iodide stain) live E. coli without exposure to graphene blocks (coaxial–conical and stacked) (fluoresce green) and Dead cells (fluoresce red) after 24 h of exposure to graphene blocks (coaxial–conical and stacked) (a ×400 magnification).

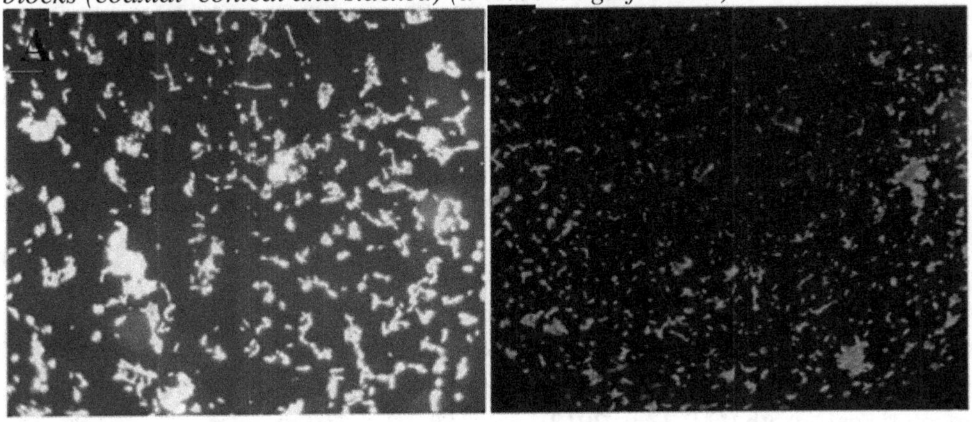

Anti-cancer activity

Variety of carbon-based nanomaterials, including carbon nanotubes, graphene, carbon quantum dots (CQDs), carbon dots (CDs), carbon nanotubes (CNTs), carbon nano onions, and carbon nano horns, emphasizing their unique properties become very promising to researchers to explore and create novel diagnostic and therapeutic methods. In case of cancer research, these carbon nanoparticles show promising anti-cancer activity with ability to induce apoptosis, suppress cell proliferation. They have been reported as suitable drug delivery carriers for anti-cancer drugs to the tumor sites. More interestingly, tiny carbon structure (1-10 nm) makes them unique class of nano-materials as fluorescence imaging agents suitable for early cancer detection and different bio-medical applications (bio-imaging, bio-sensing, photo-dynamic, photo-thermal etc.). CDs, CQDs-based bio-friendly nano-materials have been reported as novel approach for targeting and treating hepatocellular carcinoma (HCC) cells than in normal. Again, good antioxidant can oxidize itself and attacks cysteine residues of proteins in cancerous cells, inhibits their growth, induce apoptosis (programmed cell death). In this way, strong antioxidant containing materials can inhibits the development of various cancer cell, metabolic pathway that enhance apoptosis, suppress cell proliferation and inhibit angiogenesis, resulting carcinogenesis and cancer cell growth. With the help of nano-biotechnology, engineered NPs have been synthesized in such way that highly porous morphology, high surface-to-volume ratio of the NPs can enhance the anti-cancer effects. Again, when high level of polyphenolic compounds, which is abundantly presence in different plant-extract has been functionalized on the surface of the carbon-based NPs, the anti-tumor, anti-cancer effect has been increasing much more, which could help to improve the effectiveness of cancer therapy, cancer cell-imaging and prevention method. Actually, anthocyanin, catechin (EGCG) from tea extract, papaya leaves etc. possess high-level of cancer prevention action, have been widely used to manufacture of medicine-capsules and prescribe to cancer patients undergoing treatment with radiotherapy. Again, tea catechin (EGCG) show additional health-benefit as cancer preventive and can-do anti-tumor activity through inhibition of metallo-proteinase activity, induce the formation of ROS in the mitochondria of cancer cell-lines. According to Hsu et. al., anthocyanin in blue-tea plays cancer prevention through suppressing cell proliferation of COLO 320 DM (IC50 = 64.9 μg/ml) and HT-29 (IC50 = 55.2 μg/ml) by blocking cell cycle progression at the G0/G1 phase inducing apoptotic cell-death (C.P. Hsu, 2012).

Now-a-days, the inherent fluorescence properties of tiny carbon structure (2-10 nm) have been successfully utilized in both in vitro and in vivo bio-imaging application for the visualization of different cellular structures, processes, and even specific biomarkers. Carbon nanoparticles, particularly CQDs and CDs are more promising tools in bioimaging and fluorescence labelling due to their low toxicity, fairly good biocompatibility and permeability, weak interactions with proteins, resistance to swelling and photobleaching, easy clearance from the body, low cost, and immune system evasion. These features make them a solid base for diagnostics. Carbon dots can aid diagnosis by delivering information regarding the type, size, and position of tumors in humans. Their therapeutic functions can help them effectively deliver and controllable release of drugs to the specific part of the body. Yang et al., 2009 first time introduce non-toxic bio-synthesized CDs as contrast agents for optical imaging in vivo studies through direct injecting into live mice (Figure 8).

Figure 8. Mouse model as effective research tools for In vivo fluorescence imaging

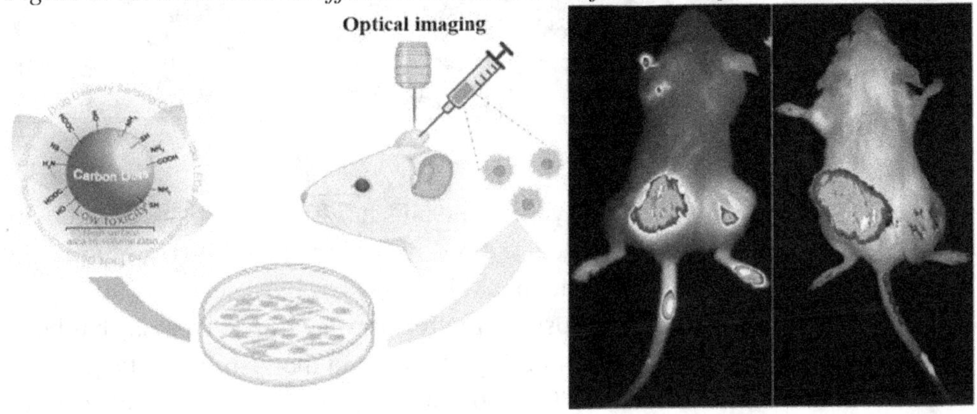

In case of cancer research, that bio-synthesized CDs becomes more promising in cancer cell imaging due to their ability to be functionalized with specific ligands and their intrinsic fluorescent properties. They can be used as fluorescent probes to visualize cancer cells, guide surgery, and potentially for targeted drug delivery. According to Gedda et al., (2023), when bio-synthesized CDs have been modified with Folic acid on their surface, the composite nanomaterials can be used as fluorescence target probe for specifically Folate-receptor which is generally predominantly found in cells of various tumors and cancers. In this way, Folic acid modified CDs interaction towards MCF7 human breast cancer cell-lines have been successfully enhanced corresponding bright fluorescence signal have been visualized under the confocal microscope upon 24h of treatment with CDs. Feng et al., (2015) have demonstrated N-doped CDs as effective fluorescence probes in bioimaging for hepG2 (liver hepatocellular carcinoma) cells (Figure 9).

Figure 9. Confocal fluorescence microscopic images of MCF7 cells, HeLa cells for 3–6 hours incubated with CDs.

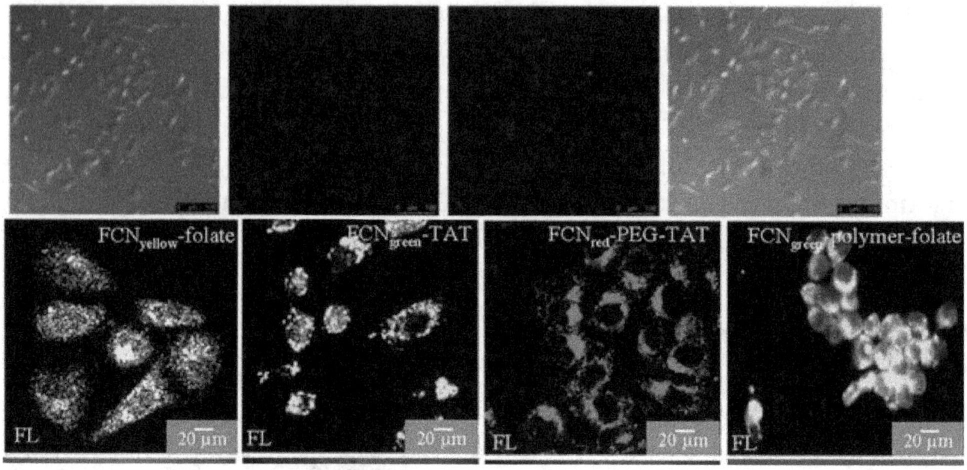

The anti-cancer activity of bio synthesized NPs has been investigated by S. Ansar et. al., 2020 through reduction of 3-(4,5-dimethyl thiazole-2-yl)2,5,-ditphenyl Tetrazolium bromide dye (MTT) and that MTT assay has been generally utilized to explore cytotoxic effects and apoptosis. A preliminary in vitro MTT assay has been performed to evaluate the biocompatibility of CDs and the viability of the cells in the presence of the nano-materials. The cells have treated with a varying concentration of CDs (20–1000 μg/mL) and subjected for incubation. The CDs showed good cytotoxicity even at higher concentrations after 24 h of incubation. There was not much decrease in the cell viability at lower concentrations, but a slight decrease up to 80% is observed at higher concentrations.

Similarly, in breast cancer cells (SUM159A), the viability has been investigated as 92%, 117%, 63%, 54%, 25%, and 9% at 50 100, 200, 300, 400, and 500 μg/mL concentrations respectively as shown in Figure 10 (Gandhi, 2024). As the concentration of CDs increases, the viability of the cells begins to decline. The observation of cancerous cells (SUM159A) demonstrates the anti-cancer characteristics of the nano-materials.

Figure 10. SUM159A cell viability at various concentrations of CDs ranging from 0-500

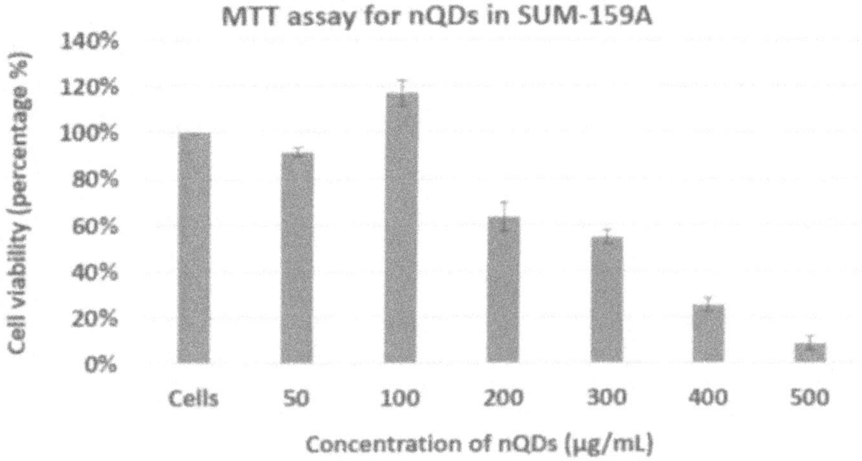

µg/mL (Gandhi, 2024)

Ansar et. al. has demonstrated CQDs as potent anti-cancer agent, which exhibits cytotoxic activity against MCF-7 breast cancer cell-lines with very low IC50 value of 17 µg/ml. Also exhibit cell death at higher concentration of nanoparticles ie. cytotoxicity increases proportionately with increasing concentration of CQDs with maximum effect at 100 µg/ml with IC 50 of 55 µg/ml (S. Ansar, 2020). In another study revealed that CQDs can able to kill almost all lung cancer cells (A549 and L-132) at 30 µg/ml concentration. Among the graphene based NPs, GO have been investigated for their anti-skin cancer, cytotoxic effect against cancer cell-lines, where dose-dependent cytotoxicity analysis reveals that higher concentration 500 or 250 µg/ml GO can significantly reduce the cell viability at 22% and 38%, respectively and IC50 of skin cancer has been reported as 409.7 µg/ml (B. Naiel,2022).

CONCLUSION

In summary, the proposed carbon-based nano-scaled materials have been successfully produced through green bio-synthesis route using antioxidant-riched green plant aqueous extract and natural renewable resources. Using extract as reducing and stabilizing agents for synthesis offers advantages, such as the fact that extract is readily available, sustainable, ecofriendly, cost effective, and non-hazardous. The synthesized carbon-based nano-materials has been widely explored as an efficient anti-bacterial and anti-cancer agent owing to the possessing superior anti-oxidant activity. They exhibit different anti-oxidant efficiencies based of different size, shape and morphology. The prepared carbon-based exhibited efficient antimicrobial activities against both Gram-positive and Gram-negative bacteria. Moreover, small number of carbon-based NPs can produce significant inhibitory effects against development of cancer cell. Therefore, the antioxidant-incorporated nano-materials can be used as potential anti-oxidant, anti-bacterial, anti-cancer agent for commercial applications.

ACKNOWLEDGMENT

The Author is grateful to Prof. Bibhutosh Adhikary, Department of Chemistry, Indian Institute of Engineering Science and Technology (IIEST), Shibpur, for helpful discussions. The Author is indebted to UGC, India for financial support through the MRP [PSW–066/15–16 (ERO)]. The author also acknowledges the MHRD (India) for providing instrumental facilities to the Department of Chemistry, IIEST Shibpur and RUSA Scheme of the Department of Higher Education, Government of West Bengal, for providing instrumental facilities to the Department of Chemistry, Bangabasi Morning College, Kolkata, India.

REFERENCES

Abo-Neima, S. E., Motaweh, H. A., & Elsehly, E. M. (2020). Antimicrobial activity of functionalised carbon nanotubes against pathogenic microorganisms. *IET Nanobiotechnology / IET*, *14*(6), 457–464. DOI: 10.1049/iet-nbt.2019.0342 PMID: 32755954

Ahmad, V., & Ansari, M. O. (2022). Antimicrobial Activity of Graphene-Based Nanocomposites: Synthesis, Characterization, and Their Applications for Human Welfare. *Nanomaterials (Basel, Switzerland)*, *12*(22), 4002. DOI: 10.3390/nano12224002 PMID: 36432288

Azizi-Lalabadi, M., Hashemi, H., Feng, J., & Jafari, S. M. (2020). Carbon nanomaterials against pathogens; the antimicrobial activity of carbon nanotubes, graphene/graphene oxide, fullerenes, and their nanocomposites. *Advances in Colloid and Interface Science*, *284*, 102250–102267. DOI: 10.1016/j.cis.2020.102250 PMID: 32966964

Cao, S., & Yu, J. (2014). g-C_3N_4-Based Photocatalysts for Hydrogen Generation. *The Journal of Physical Chemistry Letters*, *5*(12), 2101–2107. DOI: 10.1021/jz500546b PMID: 26270499

Dutta, A. K. (2023). Metal Chalcogenides-Based Multifunctional Nanomaterials for Solar-Light-Driven Photo-Catalysis and Electro-Chemical Sensing. In Kumar, A. (Ed.), *Innovative Multifunctional Nanomaterial for Photocatalysis, Sensing, and Imaging* (pp. 1–16). IGI Global Scientific Publishing., DOI: 10.4018/978-1-6684-8743-3.ch001

Feng, X., Jiang, Y., Zhao, J., Miao, M., Cao, S., Fang, J., & Shi, L. (2015). Easy synthesis of photoluminescent N-doped carbon dots from winter melon for bio-imaging. *RSC Advances*, *5*(40), 31250–31254. DOI: 10.1039/C5RA02271A

Gandhi, S., Jain, S., Bhatia, D., & Yadav, P. (2024). *Azadirachta indica:* derived, red-emitting, carbon nanoparticles for cellular bioimaging and potential therapeutics, *bioRxiv*, 02,26, 582094; DOI: 10.1101/2024.02.26.582094

Gedda, G., Sankaranarayanan, S. A., Putta, C. L., Gudimella, K. K., Rengan, A. K., & Girma, W. M. (2023). Green synthesis of multi-functional carbon dots from medicinal plant leaves for antimicrobial, antioxidant, and bioimaging applications. *Scientific Reports*, *13*(1), 6371. DOI: 10.1038/s41598-023-33652-8 PMID: 37076562

Gudimella, K., Gedda, G., Kumar, P. S., Babu, B. K., Yamajala, B., Rao, B. V., Singh, P. P., Kumar, D., & Sharma, A. (2022). Novel synthesis of fluorescent carbon dots from bio-based Carica Papaya Leaves: Optical and structural properties with antioxidant and anti-inflammatory activities. *Environmental Research*, *204*, 111854. DOI: 10.1016/j.envres.2021.111854 PMID: 34437850

Gupta, B. D., Pathak, A., & Semwal, V. (2019). Carbon-Based Nanomaterials for Plasmonic Sensors: A Review. *Sensors (Basel)*, *19*(16), 3536. DOI: 10.3390/s19163536 PMID: 31412590

Han, C., Kundu, B. K., Liang, Y., & Sun, Y. (2024). Near-Infrared Light-Driven Photocatalysis with an Emphasis on Two-Photon Excitation: Concepts, Materials, and Applications. *Advanced Materials*, *36*(5), 2307759–2307772. DOI: 10.1002/adma.202307759 PMID: 37703435

Helbig, W. A. (1946). Activated carbon. *Journal of Chemical Education*, *23*(2), 98–104. DOI: 10.1021/ed023p98

Li, Y., Li, W., Yang, X., Kang, Y., Zhang, H., Liu, Y., & Lei, B. (2021). Salvia Miltiorrhiza-Derived Carbon Dots as Scavengers of Reactive Oxygen Species for Reducing Oxidative Damage of Plants. *ACS Applied Nano Materials*, *4*(1), 113–120. DOI: 10.1021/acsanm.0c02419

Morstein, C. E., & Dienwiebel, M. (2021). Graphite lubrication mechanisms under high mechanical load. *Wear*, *477*, 203794. DOI: 10.1016/j.wear.2021.203794

Ngoc, L. T. N., Moon, J. Y., & Lee, Y. C. (2023, September 27). Plant Extract-Derived Carbon Dots as Cosmetic Ingredients. *Nanomaterials (Basel, Switzerland)*, *13*(19), 2654. DOI: 10.3390/nano13192654 PMID: 37836295

Richards, B. S., Hudry, D., Busko, D., Turshatov, A., & Howard, I. A. (2021). Photon Upconversion for Photovoltaics and Photocatalysis: A Critical Review. *Chemical Reviews, 121*(15), 9165–9195. DOI: 10.1021/acs.chemrev.1c00034 PMID: 34327987

Sachdev, A., & Gopinath, P. (2015). Green synthesis of multifunctional carbon dots from coriander leaves and their potential application as antioxidants, sensors and bioimaging agents. *Analyst, 140*(12), 4260–4269. DOI: 10.1039/C5AN00454C PMID: 25927267

Serge-Correales, Y. E., Ullah, S., Ferreira-Neto, E. P., Rojas-Mantilla, H. D., Hazra, C., & Ribeiro, S. J. L. A. (2022). UV-visible-NIR active smart photocatalytic system based on NaYbF4:Tm3+ upconverting particles and Ag_3PO_4/H_2O_2 for photocatalytic processes under light on/light off conditions. *Materials Advances, 3*(6), 2706–2715. DOI: 10.1039/D1MA01028J

Wang, X., Wang, F., Sang, Y., & Liu, H. (2017). Full-Spectrum Solar-Light-Activated Photocatalysts for Light–Chemical Energy Conversion. *Advanced Energy Materials, 7*(23), 1700473–1700489. DOI: 10.1002/aenm.201700473

Yang, S.-T., Cao, L., Luo, P. G., Lu, F., Wang, X., Wang, H., Meziani, M. J., Liu, Y., Qi, G., & Sun, Y.-P. (2009). Carbon dots for optical imaging in vivo. *Journal of the American Chemical Society, 131*(32), 11308–11309. DOI: 10.1021/ja904843x PMID: 19722643

Chapter 6
Green Nanomaterials and Nanocomposites

P. Selvakumar
https://orcid.org/0000-0002-3650-4548
Nehru Institute of Technology, Coimbatore, India

Ritu Dahiya
https://orcid.org/0000-0001-5788-006X
Chhotu Ram Arya College, India

P. Selvam
Karpaga Vinayaga College of Engineering and Technology, India

G. Shajitha
Providence College, Coonoor, India

T. C. Manjunath
https://orcid.org/0000-0003-2545-9160
Rajarajeswari College of Engineering, India

Santosh Kumar Nathsharma
https://orcid.org/0000-0003-3935-2938
Stewart Science College, India

Mohit Sharma
https://orcid.org/0009-0007-2280-8077
Maharshi Dayanand University, Rohtak, India

ABSTRACT

Green nanomaterials refer to those materials synthesized using environmentally benign processes, renewable resources, and minimal energy consumption, thereby aligning with principles of green their individual components. When these materials are produced using eco-friendly approaches, they are termed green nanocomposites. Together, green nanomaterials and nanocomposites form a critical domain that bridges the gap between advanced material science and environmental responsibility.The driving force behind the rise of green nanotechnology is the pressing need to reduce environmental degradation caused by conventional synthesis and manufacturing methods. Traditional techniques often involve toxic chemicals, high energy consumption, and unsustainable resource utilization. In contrast, green nanotechnology emphasizes the use of non-toxic precursors, renewable feedstocks,

DOI: 10.4018/979-8-3373-1409-9.ch006

and energy-efficient synthesis methods. The eco-friendly synthesis of these materials, such as using biomass-derived carbon sources, further reduces their environmental footprint.

INTRODUCTION TO GREEN NANOMATERIALS AND NANOCOMPOSITES

Green nanomaterials refer to those materials synthesized using environmentally benign processes, renewable resources, and minimal energy consumption, thereby aligning with principles of green their individual components. When these materials are produced using eco-friendly approaches, they are termed green nanocomposites. Together, green nanomaterials and nanocomposites form a critical domain that bridges the gap between advanced material science and environmental responsibility.The driving force behind the rise of green nanotechnology is the pressing need to reduce environmental degradation caused by conventional synthesis and manufacturing methods. Traditional techniques often involve toxic chemicals, high energy consumption, and unsustainable resource utilization. In contrast, green nanotechnology emphasizes the use of non-toxic precursors, renewable feedstocks, and energy-efficient synthesis methods. The eco-friendly synthesis of these materials, such as using biomass-derived carbon sources, further reduces their environmental footprint. In addition, natural and biodegradable polymers, including cellulose, chitosan, and starch, have gained significant attention as green nanomaterials. These polymers are renewable, biodegradable, and non-toxic, making them ideal for producing sustainable nanocomposites.Green nanocomposites are hybrid materials that combine a biodegradable or bio-based matrix with nanoscale fillers to achieve enhanced mechanical, thermal, and barrier properties. These composites often incorporate natural nanofillers, such as nanocellulose, nanosilica, clay nanoparticles, and biochar, to create materials with high strength, lightweight characteristics, and improved functionality.Nanocarriers made from biodegradable polymers, such as chitosan and polylactic acid, are used to deliver r enhances their therapeutic potential. Green nanocomposites are also finding applications in the food industry for developing sustainable packaging materials that provide enhanced barrier properties, antimicrobial activity, and reduced environmental impact. In conclusion, green nanomaterials and nanocomposites represent a sustainable and promising approach to address the challenges of environmental degradation, resource scarcity, and industrial pollution. By harnessing eco-friendly synthesis methods and renewable resources, these materials offer innovative solutions across future.

Fundamentals of Nanomaterials and Nanocomposites

Nanomaterials and nanocomposites represent a rapidly growing field within materials science, offering revolutionary properties and applications across various industries. At the heart of nanotechnology,, both of which dominate at the nanoscale. By leveraging these properties, nanomaterials are playing a transformative role in science, engineering, and technology, while nanocomposites, which integrate nanomaterials with a host matrix(Ali et al., 2017), further enhance functionality, performance, and application potential.Types and Classification of Nanomaterial-sNanomaterials are broadly categorized into four types based on their dimensionality and composition:Zero-dimensional nanomaterials (0D): These include nanoparticles where all dimensions are confined within the nanoscale range. Examples include quantum dots, metal nanoparticles, and nanoclusters. Quantum dots, for instance, are semiconductor optoelectronics and imaging. the nanoscale and include nanorods, nanotubes, and nanowires. Carbon nanotubes (CNTs), which are hollow cylindrical thermal stability. Similarly, metallic nanowires exhibit remarkable electrical properties for applications in sensors and electronics.two dimensions in the nanoscale and include nanosheets, graphene, and transition metal dichalcogenides (TMDs). Graphene, a one of the most researched nanomaterials. such as nanocomposites, nanoporous materials, and nanocrystals, where nanoscale features are distributed throughout the material volume.Additionally, nanomaterials are classified like fullerenes, carbon nanotubes, and graphene. Inorganic nanomaterials comprise metals (e.g., gold, silver, and copper nanoparticles) and metal oxides (e.g., titanium dioxide, zinc oxide, and iron oxide). Hybrid nanomaterials combine organic and inorganic components to exploit synergistic properties for diverse applications. bulk materials into nanoscale particles using mechanical, thermal, or chemical techniques. These composites offer lightweight characteristics, enhanced strength, and improved thermal stability, making them ideal for applications in aerospace, automotive, and packaging industries.

Metal-based nanocomposites: A metal matrix is reinforced with nanoparticles or nanowires to improve mechanical strength, conductivity, and wear resistance. Examples include aluminum or copper nanocomposites used in structural and electronic applications.Ceramic-based nanocomposites: These materials incorporate ceramic matrices with nanoscale fillers, offering excellent thermal stability, high strength, and resistance to corrosion. Such composites are extensively used in high-temperature and wear-resistant applications.Unique Properties of Nanomaterials and Nanocomposites Nanomaterials and nanocomposites exhibit several unique properties that make them superior to traditional materials:(Amooaghaie R et al., 2015) High surface area: Nanomaterials possess a larger surface Enhanced mechanical strength: Nanocomposites exhibit improved toughness, strength, and stiffness due to strong interfacial interactions between the matrix and nanofillers. Optical properties: Quantum size effects in nanoparticles result in size-dependent optical properties, such as fluorescence and plasmonic behavior, enabling applications in sensors and imaging. Applications of Nanomaterials and NanocompositesEnergy sector: Nanocomposites are employed in solar cells, batteries, fuel cells, and supercapacitors to improve energy efficiency and storage capabilities.Biomedical field: Nanomaterials are revolutionizing drug delivery,(Bankar A et al., 2010) diagnostics, and tissue engineering. Nanocarriers enable targeted drug delivery, while nanocomposites are used for bone scaffolds and wound healing.Electronics: Nanomaterials such as graphene and CNTs are used in flexible electronics, sensors, and conductive coatings.Aerospace and automotive. Principles of Green Chemistry in Nanomaterials, is built upon 12 key principles established by Paul Anastas and John Warner. These principles form the foundation for creating nanomaterials in a manner that reduces environmental impact while maintaining functionality and performance. Some of the most relevant principles in the context of nanomaterials are as follows:Prevention of Waste: Instead of managing or disposing of waste after its production, green chemistry emphasizes designing processes that prevent waste generation in the first place.

Designing Safer Chemicals and Materials: Nanomaterials must be designed to function effectively while minimizing toxicity to humans and the environment., which are widely used in medicine, diagnostics, and environmental applications. Energy Efficiency: Many nanomaterial synthesis(Belliveau BH et al., 1987) methods involve high temperatures, pressure, or energy inputs, leading to significant energy consumption.ultrasound-assisted synthesis, and room-temperature biosynthesis, to reduce energy demands while maintaining performance. Use of Renewable Feedstocks: Green chemistry prioritizes the use of renewable resources over non-renewable or depleting ones. Biomass, agricultural waste, and natural biopolymers serve as sustainable sources for producing nanomaterials. Techniques like solvent-free synthesis, room-temperature processing, and recycling of precursors ensure efficient utilization of resources and reduced energy demands., metal oxides, and biochar are used for water purification, soil remediation, and air filtration, supporting clean and sustainable environments. Additionally, nanocatalysts enable efficient energy generation through fuel cells, solar cells, and supercapacitors.Challenges and Future Prospects While green chemistry and sustainable development provide a roadmap for environmentally friendly nanomaterials, several challenges persist. CNFs can be functionalized for specific applications, such as absorbing contaminants from water or delivering drugs in controlled-release systems bacterial Nanocellulose (BNC): Unlike plant-derived cellulose, BNC is produced by bacteria, such as Acetobacter xylinum, under controlled fermentation conditions(Castro L et al., 2013). BNC exhibits high purity, biocompatibility, and a unique nanoscale network structure.

These properties make it ideal for biomedical applications, including wound dressings, artificial skin, and tissue-engineering scaffolds. Chitosan-based nanomaterials are also used in controlled-release fertilizers and pesticides, promoting sustainable agriculture.Protein-Based Nanomaterials Proteins are natural biopolymers composed of amino acids and are found in animal, plant, and microbial sources. Proteins such as silk fibroin, collagen, keratin, casein, zein, and whey protein are increasingly explored for the synthesis of nanomaterials due to their biodegradability, biocompatibility, and versatile functionality.Silk Fibroin Nanomaterials: Silk fibroin, derived from silkworms or spiders, is a protein with excellent mechanical strength, biocompatibility, and thermal stability. Silk fibroin nanoparticles and nanofibers are and as scaffolds for regenerative medicine due to their porous structure and ability to support cell growth.Collagen and Gelatin-Based Nanomaterials: Collagen, the most abundant protein in animal connective tissues, is widely used for synthesizing nanofibers, nanogels, and nanoparticles. Zein nanoparticles are used in food coatings, drug delivery, and packaging applications, providing barrier properties and controlled-release functionalities.Applications of Biopolymer-Based Nanomaterials Biopolymer-based nanomaterials have broad applications due to their sustainable and multifunctional properties:

chitosan nanoparticles, and protein-based scaffolds are ideal for supporting cell growth and controlled drug release.. Environmental Remediation: Biopolymeric nanomaterials such as chitosan and cellulose nanofibers adsorb pollutants, heavy metals, and dyes from water, providing sustainable solutions for water purification. Agriculture: Chitosan-based nanoparticles are employed for controlled-release fertilizers, pesticides, and antimicrobial coatings, promoting eco-friendly farming practices. Energy Storage and Sensors: Nanocellulose and protein-based materials are explored for fabricating flexible electrodes, supercapacitors, and biosensors. Conclusion Biopolymer-based nanomaterials derived from cellulose, chitin, proteins, and other natural polymers represent a sustainable and eco-friendly approach to nanotechnology. Their renewable nature, biocompatibility, and tunable properties enable diverse applications in medicine, packaging, environmental remediation, and energy sectors.

Metal-Organic Frameworks (MOFs) and Their Applications: MOFs as Green Nanomaterials for Energy, Environment, and Biomedical Applications

These materials have garnered immense interest due to their highly tunable structures, exceptional surface areas (often exceeding 7000 m^2/g), versatile chemical functionalities, and remarkable adsorption capacities. MOFs are often considered as green nanomaterials because they can be synthesized through environmentally friendly methods using mild conditions, water-based solvents, and sustainable precursors. Additionally, their structural flexibility and modularity allow researchers to tailor MOFs for specific applications while reducing energy and resource consumption. biomedical fields, aligning them with sustainable development and green chemistry principles MOFs in Energy Applications, make them excellent candidates for energy-related applications, including energy storage, gas storage, catalysis, and energy conversion.. Their high surface area and tunable pores provide efficient adsorption and storage of these gases under moderate conditions. MOFs such as MOF-177 and HKUST-1 have demonstrated exceptional hydrogen storage capacities, making them suitable for developing hydrogen-powered vehicles and portable energy devices.Gas Separation and Carbon Capture: MOFs are highly efficient for CO_2 capture and separation due to their selective adsorption properties. MOFs such as ZIF-8 and UiO-66 have been engineered to selectively adsorb CO_2 over other gases like N_2 and CH_4, enabling carbon capture from industrial emissions. This reduces greenhouse gas emissions and mitigates climate change, aligning with sustainable energy strategies.Batteries and Supercapacitors: MOFs and their supercapacitors. MOFs provide high ion diffusion pathways, excellent electrochemical stability, and large surface areas for storing energy, improving battery performance and longevity Photocatalytic(Chakravarty A et al., 2022) Energy Conversion: MOFs can function as photocatalysts for solar energy conversion and water splitting to produce hydrogen fuel. For instance, MOFs integrated with light-absorbing semiconductors enable efficient photocatalysis by harnessing solar energy. (Dahoumane SA et al., 2016)Their tunable properties allow for the design of MOFs that maximize photon absorption and catalytic activity, offering sustainable solutions for renewable energy generation.MOFs for Environmental Applications Environmental remediation has become one of the most prominent applications of MOFs, given their excellent adsorption capabilities, catalytic properties, water pollution, air contamination, and toxic waste management. Functionalized MOFs can also adsorb dyes and pharmaceuticals, offering sustainable solutions for wastewater treatment.MOFs with tailored pore sizes and functional groups selectively trap harmful pollutants, improving indoor and outdoor air quality while contributing to cleaner environments.breaking down toxic pollutants like phenols, pesticides, and pharmaceuticals into harmless byproducts. This photocatalytic activity is particularly beneficial for sustainable and energy-efficient environmental remediation. Desalination and Water Harvesting: MOFs such as ZIF-8 and MOF-801 have demonstrated the ability to capture and release water molecules from humid air, providing sustainable solutions for water harvesting in arid regions. MOFs are also used in membrane-based desalination processes to produce clean drinking water, addressing global water scarcity challenges.MOFs in Biomedis make them ideal carriers for biomedical applications. UiO-66, and ZIF-8 are used to encapsulate anticancer drugs, antibiotics, and anti-inflammatory agents, ensuring targeted delivery and minimizing side effects. MOFs' pH-responsive properties further enable site-specific drug release in tumor tissues or inflamed areas. nanoparticles, or radionuclides, for use in bioimaging and diagnostics. These hybrid MOF-based systems enable enhanced imaging through fluorescence, MRI, and PET scans, providing precise disease diagnosis and monitoring.(El-Kassas et al., 2014). Room-Temperature and Microwave-Assisted Synthesis: Energy-efficient methods, such as room-temperature reactions, ultrasound-assisted synthesis, and microwave heating, reduce energy consumption and production time.. Renewable Precursors: Organic linkers derived from natural sources or biomass are increasingly used in MOF construction, promoting sustainability.Conclusion Metal-Organic Frameworks (MOFs) have emerged as versatile green nanomaterials with transformative applications in energy, environment, and biomedical sectors. Their unique porous structures, tunable functionalities, and high surface areas make them invaluable for addressing critical challenges, including energy storage, environmental pollution, and advanced healthcare. and eco-friendly solutions that align with global efforts toward a cleaner, healthier, and more sustainable future. The continued development of MOFs promises to revolutionize industries while minimizing environmental impact and fostering sustainable development.

Green Synthesis and Applications of Graphene and Graphene-Based Nanomaterials

Biomass-Derived Precursors: Biomass, such as agricultural waste, plant extracts, and natural precursors (e.g., cellulose, glucose, starch), has been used to synthesize graphene via pyrolysis or hydrothermal methods. This process converts carbon-rich biomass into high-quality graphene without the need for toxic reagents. For example, glucose or starch can serve as carbon precursors to produce graphene nanosheets through controlled thermal decomposition.Green Reducing Agents: Conventional methods of producing reduced graphene oxide (rGO) involve hazardous reducing agents like hydrazine, which are toxic and environmentally harmful. Green synthesis replaces such agents with natural or non-toxic alternatives(Fernãndez JG et al., 2016), such as plant extracts (e.g., tea, coffee, aloe vera), ascorbic acid (vitamin C), and amino acids. These natural reducing agents facilitate the conversion of graphene oxide into rGO while reducing environmental impact. For instance, extracts from Magnolia leaves or green tea contain polyphenols that act as effective reducing and stabilizing agents.and pressure. This process avoids the use of toxic solvents, yielding high-quality graphene nanosheets or graphene composites. Similarly, microwave-assisted synthesis reduces energy consumption and shortens reaction times, making the process more sustainable. under mild conditions. This process involves the application of an electric potential in aqueous or green electrolyte solutions, avoiding the use of strong oxidizing agents. Electrochemical exfoliation is scalable, cost-effective, and environmentally friendly. f graphene-based materials. These biological entities produce reducing agents or assist in the reduction of graphene oxide to rGO under mild conditions. For example, Shewanella bacteria and fungal enzymes have been employed for sustainable graphene production.synthesized through green methods maintain their exceptional properties, enabling diverse applications across energy, environmental, biomedical, and industrial sectors.Energy Applications Supercapacitors and Batteries:(Fesharaki PJ et al., 2010) Graphene-based sensors are used for detecting environmental to provide rapid and accurate detection.

Biopolymer-Based Nanocomposites: Nanocomposites Derived from Biopolymers and Their Applications

Chitosan and Chitin Nanocomposites:(Gaumet M et al., 2008)Chitosan, derived from chitin (a structural polysaccharide found in crustacean shells), is widely used for its antimicrobial, film-forming, and biocompatible properties. Nanocomposites reinforced with nanoclays, metal nanoparticles, or graphene, drug delivery systems, and water purification membranes.Starch, a renewable polysaccharide derived from sources like corn, potato, and rice, is widely used in biopolymer-based nanocomposites. Natural proteins such as gelatin, silk fibroin, keratin, and soy protein are used as biopolymeric matrices in nanocomposites. These proteins are biocompatible and can interact with nanofillers to form strong hydrogen bonds,PLA is a biodegradable aliphatic polyester derived from renewable(Hameed S et al., 2019) sources like corn starch and sugarcane. PLA-based nanocomposites, reinforced with nanoclays, carbon nanotubes (CNTs), or graphene oxide (GO), exhibit enhanced tensile strength, thermal stability, and biodegradability. These materials are widely used in packaging, medical implants, and disposable products.HAs are microbial polyesters produced by bacteria under nutrient-limited conditions, whereas PCL is a synthetic biodegradable polyester. These materials offer excellent moisture resistance, antimicrobial activity, and prolonged shelf life, making them ideal for food packaging and pharmaceutical packaging applications.Drug Delivery Systems: Nanocomposites made from chitosan,(Ijaz F et al., 2017) PLA, or gelatin matrices can encapsulate and deliver drugs, proteins, or nanoparticles in a controlled and targeted manner.Tissue Engineering: Nanocomposites are used to fabricate scaffolds that mimic the extracellular matrix, promoting cell attachment, proliferation, and tissue regeneration. Materials such as cellulose-based or collagen-based nanocomposites are employed in bone, cartilage, and skin tissue engineering.Wound Dressings: Chitosan-based nanocomposites with antimicrobial nanoparticles (e.g., silver nanoparticles) accelerate wound healing and prevent bacterial infections.In agriculture, biopolymer-based nanocomposites are used for biodegradable mulch films, controlled-release fertilizers, and pesticide delivery systems. Starch or cellulose-based nanocomposites release nutrients or pesticides in a controlled manner, improving crop productivity while minimizing environmental impact.Biopolymer nanocomposites are employed in water and air purification due to their adsorption, filtration, and photocatalytic properties. For example:Water Purification: Chitosan-based nanocomposites loaded with nanoclays, carbon nanotubes, or metal oxides efficiently remove heavy metals, dyes, and pollutants from water.Air Filtration: Nanocomposites incorporating activated carbon or graphene are used in filters to remove particulate matter and gaseous contaminants. nature and renewable origins. For example, cellulose-based nanocomposites reinforced with graphene or carbon nanotubes exhibit high electrical conductivity and energy storage capacity.Biopolymer-based nanocomposites are used in flexible electronics, sensors, and conductive films. PLA or cellulose-based nanocomposites integrated with graphene or metallic nanoparticles provide excellent conductivity and mechanical flexibility, making them suitable for wearable electronics, biosensors, and environmental monitoring devices.Conclusion Biopolymer-based nanocomposites represent a promising class of sustainable materials that development. Continued advancements in biopolymer-based nanocomposites will environmental resources, a circular bioeconomy. These materials not only offer innovative solutions to global challenges but also pave the way for a greener and more sustainable future

Figure 1. Application of green nanotechnology

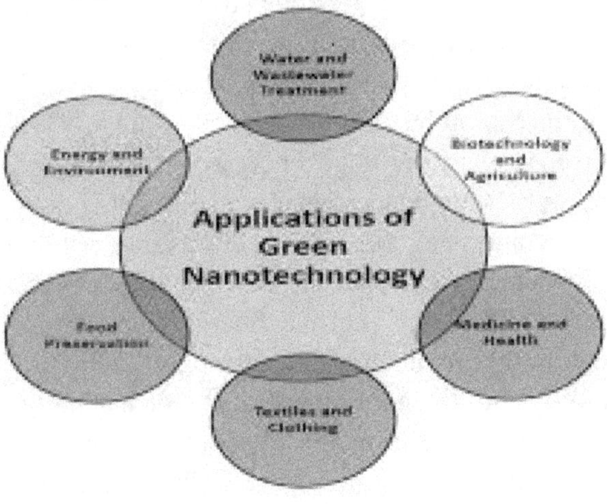

Carbon Nanotube-Based Nanocomposites: Applications in Energy, Environment, and Biomedical Sectors

CNTs Their high aspect ratio, large surface area, excellent electrical conductivity, exceptional mechanical strength, and thermal stability make CNTs ideal reinforcements for polymeric, ceramic, and metallic matrices in nanocomposites. By integrating CNTs with various host materials, carbon nanotube-based nanocomposites (CNT-NCs) exhibit enhanced physical, chemical, and functional properties, enabling applications across energy storage, environmental remediation, and biomedical fields. (Ilahi N et al., 2021)Key Properties of Carbon Nanotubes for NanocompositesMechanical Strength: CNTs have a tensile strength up to 100 times greater than steel while maintaining a lightweight nature(Jeyarani S et al., 2020). Their incorporation into nanocomposites enhances mechanical strength, elasticity, and toughness.Electrical Conductivity: CNTs are excellent conductors, which makes them suitable for conductive nanocomposites used in energy storage, electronics, and sensors.Thermal Stability: CNTs have exceptional thermal conductivity, allowing their application in thermally conductive and heat-resistant materials.High Aspect Ratio and Surface Area: The unique one-dimensional structure of CNTs enables efficient interaction with host materials, improving interfacial bonding and dispersion in the composite matrix.Synthesis and Fabrication of CNT-Based NanocompositesTo maximize the performance of CNT-based nanocomposites, surface functionalization of CNTs is often performed to improve their dispersion and compatibility with the host material. Functionalization involves introducing chemical groups or coatings to the CNT surface to enhance interaction with matrices, thereby preventing agglomeration. Applications of Carbon Nanotube-Based NanocompositesCNT-based nanocomposites play a pivotal role in energy storage, energy conversion, and renewable energy technologies superior, thermal properties,(Jevapatarakul D et al., 2020). CNTs are integrated into organic and perovskite solar cells as transparent electrodes, charge transport layers, or fillers in composites to enhance light absorption, conductivity, and energy conversion efficiency.Thermal Energy Harvesting: CNT-based nanocomposites are used in thermoelectric devices due to their ability to conduct heat and electricity simultaneously, enabling efficient thermal energy harvesting.CNT-based nanocomposites are highly effective in addressing environmental challenges, including water purification, air filtration, and pollutant remediation:Water Purification: CNT nanocomposites serve as advanced adsorbents and membranes for the removal of heavy metals, organic pollutants, dyes, and pathogens from water. CNTs functionalized with metal oxides, polymers, or biomaterials enhance adsorption efficiency, photocatalytic degradation, and filtration capacity. For instance, CNT-based membranes exhibit superior filtration performance due to their large surface area and unique nanoscale structure.Air Filtration: CNT-reinforced composites are utilized in air filters for the removal of particulate matter, toxic gases, and volatile organic compounds (VOCs).

Metal Oxide-Based Nanocomposites: Applications in Energy, Environment, and Biomedical Fields

The integration of metal oxides into composite structures enhances their mechanical strength, thermal stability, reactivity, and multifunctionality, enabling their widespread use in energy storage and conversion, environmental remediation(Kannan N et al., 2011), and biomedical fields. Many metal oxides (ZnO) possess strong photocatalytic properties, enabling pollutant degradation and self-cleaning surfaces under UV or visible light. Thermal and Mechanical Stability: Metal oxide nanocomposites improve the structural and thermal performance of host materials, providing stability for high-temperature applications.Synthesis and Fabrication The fabrication of metal oxide-based nanocomposites involves various synthesis techniques to achieve uniform dispersion and strong interfacial bonding between metal oxide nanoparticles and the matrix: resulting in metal oxide nanoparticles uniformly dispersed in the composite matrix.In-Situ Polymerization: Metal oxide nanoparticles are incorporated during the polymerization process, leading to well-dispersed nanocomposites. Electrospinning: Metal oxide nanoparticles are embedded into nanofibers, producing functional composites for filtration and biomedical scaffolds Physical Blending and Melt Processing: Mechanical mixing and extrusion are used to incorporate metal oxide nanoparticles into polymeric matrices for large-scale production.Functionalization of metal oxide nanoparticles with surfactants, polymers, or organic ligands is often performed to improve their dispersion, stability, and interaction with host materials. This step ensures the formation of high-performance nanocomposites with tailored properties and conversion technologies due to their optical and electronic properties Batteries and Supercapacitors:(Khalaj M et al., 2020). Fuel Cells: Metal oxide nanocomposites, such as cerium oxide (CeO_2) and zirconium oxide (ZrO_2), act as electrolytes and catalysts in solid oxide fuel cells (SOFCs), improving ionic conductivity and catalytic activity. Magnetic iron oxide nanocomposites enable site-specific drug delivery through magnetic targeting, minimizing side effects and enhancing therapeutic efficacy.Tissue Engineering: Metal oxide nanocomposites are used in scaffolds for bone and tissue regeneration due to their mechanical strength, biocompatibility, and osteogenic properties. Hydroxyapatite-cerium oxide composites, for example, enhance bone tissue growth and repair.Biosensors: Metal oxide nanocomposites are integrated into biosensors for detecting biomolecules, pathogens, and diseases with high sensitivity and selectivity. ZnO and TiO_2 nanocomposites are widely used in electrochemical and optical sensors.Biomedical Imaging: Iron oxide-based nanocomposites serve as contrast agents in magnetic resonance imaging

(MRI) for disease diagnosis and monitoring. Functionalized iron oxide nanoparticles provide enhanced imaging resolution.Antibacterial prevent bacterial infections, accelerate wound healing, and reduce biofilm formation.Conclusion applications spanning energy, environmental remediation, and biomedical fields. Their unique properties, including photocatalytic activity, electrical conductivity, antimicrobial efficacy, and mechanical stability, make them indispensable for addressing global challenges such as clean energy production, water purification, pollution control, and advanced healthcare solutions. By leveraging innovative fabrication techniques, functionalization strategies, and hybridization with other materials, metal oxide-based nanocomposites continue to demonstrate immense potential for driving technological advancements and promoting sustainability. As research progresses, these materials are expected to play a pivotal role in developing next-generation devices, smart materials, and eco-friendly solutions for a wide range of industrial and societal applications.

Hybrid Nanocomposites: Combining Different Nanomaterials for Enhanced Properties

The main goal is to exploit the combined strengths of individual components while mitigating their weaknesses. Hybridization can occur at the molecular, nanoscale, or microscale level, depending on the fabrication method and the intended application. For example, combining carbon nanotubes (CNTs) with metal oxides like TiO_2 results in improved photocatalytic and electrical properties, while adding polymers enhances processability, flexibility, and mechanical stability.Core-Shell Structures: One material acts as a core, while the other forms a shell, creating a hybrid structure with controlled interfaces (e.g., metal oxide-coated CNTs). Layered Structures: Nanomaterials like graphene or clay are combined with polymers or metal oxides in layered forms to enhance mechanical, thermal, and barrier properties.. Inorganic-Organic Hybrids: Combining inorganic nanomaterials (e.g., metal oxides) with organic polymers results in nanocomposites with enhanced optical, thermal, and electrical behavior. Hybrid Networks: Nanomaterials such as CNTs, graphene, or nanoparticles form a percolating network within a polymer matrix, improving conductivity and strength. Hierarchical Structures: Multi-phase nanocomposites integrate different nanomaterials in a controlled architecture to achieve tailored functionalities.Synthesis Techniques of Hybrid Nanocomposites The synthesis of hybrid nanocomposites involves various methods to ensure uniform dispersion, strong interfacial bonding, and controlled integration of components. Key techniques include Sol-Gel Processing: A versatile method for incorporating metal oxides or nanoparticles into a host matrix through hydrolysis and condensation of precursors. in-Situ Polymerization: Hybrid nanomaterials are incorporated into the polymer matrix during the polymerization process, ensuring uniform distribution.Electrospinning: Hybrid nanocomposites are fabricated as nanofibers for applications in filtration, tissue engineering, and energy devices.

Chemical Vapor Deposition (CVD): Used for growing carbon-based nanomaterials (e.g., graphene, CNTs) on substrates or other materials to create hybrid systems. Hydrothermal/Solvothermal Methods: These methods synthesize hybrid nanocomposites under controlled pressure and temperature to achieve specific morphologies and phase compositions.

Physical Blending and Mixing: Mechanical mixing and extrusion are employed for large-scale production of hybrid nanocomposites, especially for polymers reinforced with fillers.(Mahdavi M et al., 2013)To enhance interfacial interactions, surface functionalization of nanomaterials using chemical groups, surfactants, or biomolecules is often performed. This step prevents agglomeration and improves compatibility within the hybrid composite matrix Enhanced Properties of Hybrid Nanocomposites The synergistic effect of combining different nanomaterials results in the following enhanced properties Mechanical Strength and Toughness: Hybrid nanocomposites exhibit exceptional tensile strength, fracture toughness, and stiffness due to the reinforcement provided by nanomaterials like CNTs, graphene, or nanoclay. Electrical and Thermal Conductivity: The inclusion of conductive nanomaterials (e.g., graphene, CNTs) into polymers or ceramics results in materials with superior electrical and thermal conductivity, ideal for energy devices and electronics. Photocatalytic Activity: Combining metal oxides like TiO_2 or ZnO with carbon nanomaterials or noble metals enhances photocatalytic activity for environmental remediation and solar energy harvesting. Barrier and Protective Properties: Hybrid nanocomposites with layered structures exhibit excellent gas, moisture, and chemical barrier properties, useful for coatings and packaging applications. Optical and Magnetic Properties: Integrating metal nanoparticles (e.g., Au, Ag) or magnetic oxides (e.g., Fe_3O_4) with other materials leads to hybrid composites with unique optical, plasmonic, and magnetic behavior.(Majeed M et al., 2022). Photocatalytic hybrid materials degrade airborne pollutants and VOCs, improving air quality. Environmental Sensors: Hybrid nanocomposites are used in sensors to detect pollutants, gases, and toxic chemicals with high sensitivity and selectivity.. Biomedical Applications Drug Delivery Systems: Hybrid nanocomposites combine nanoparticles with biocompatible polymers to deliver drugs in a controlled and targeted manner. Magnetic iron oxide hybrids enable site-specific drug delivery under an external magnetic field.Tissue Engineering: Hybrid scaffolds integrating biopolymers, metal oxides, and carbon nanomaterials promote cell growth, differentiation, and tissue regeneration. Biosensors: Hybrid nanocomposites are used in biosensors for detecting biomolecules, pathogens, and diseases with enhanced signal sensitivity and response time.

Imaging and Cancer Therapy: Magnetic and plasmonic hybrid nanocomposites are used for imaging applications and cancer therapies, such as photothermal ablation of tumors. Structural and Industrial Applications Lightweight Materials: Hybrid nanocomposites reinforced with CNTs, graphene, or nanoclay are used in aerospace, automotive, and construction industries for lightweight, high-strength materials Corrosion-Resistant Coatins: Hybrid coatings offer superior corrosion protection, chemical resis tance, and durability for industrial and marine applications.

Smart Materials and Sensors: Hybrid nanocomposites are integrated into sensors, actuators, and smart devices for real-time monitoring and adaptive systems.Conclusion Hybrid nanocomposites, by combining different nanomaterials with tailored properties, represent a new frontier in material science, enabling multifunctional and high-performance materials. The synergistic interactions between constituent components result in enhanced mechanical, thermal, electrical, and optical properties, making hybrid nanocomposites suitable for applications in energy, environmental, biomedical, and structural fields. Their ability to address challenges in clean energy production, pollution (Nag M et al., 2021)control, advanced healthcare, and next-generation electronics highlights their immense potential in promoting sustainable technologies. Continued advancements in synthesis techniques, interfacial engineering, and large-scale production will drive the development of hybrid nanocomposites for transformative applications in science and industry.

Energy Applications of Green Nanomaterials and Nanocomposites: Energy Storage, Conversion, and Harvesting

Green nanomaterials and nanocomposites have gained significant friendly energy solutions has led to a focus on developing nanomaterials that are both efficient in energy storage, conversion, and harvesting and are derived from non-toxic, renewable, and eco-friendly sources. The green synthesis of nanomaterials, utilizing natural resources, plant extracts, and low-energy processes, has enabled the design of nanocomposites that not only demonstrate enhanced performance but also promote sustainability. These materials offer promising advancements in energy technologies, including batteries, supercapacitors, essential for storing electrical energy, whether it be for mobile devices, electric vehicles, or renewable energy systems. Green nanomaterials such as activated carbon derived from plant sources, carbon nanotubes, and conductive polymers (e.g., polyaniline, polypyrrole) are used to develop supercapacitor electrodes with high surface area and excellent electrical conductivity. Energy Conversion Applications: Fuel Cells and PhotovoltaicsThe conversion of energy from one form to another is another area where green nanomaterials and nanocomposites have nanomaterials play a central role in enhancing the efficiency, stability, and overall performance of these energy conversion technologies. and heat as by-products. They are a clean alternative to combustion-based power generation, with applications ranging from portable electronics to automotive power systems. Green nanomaterials, such as bio-derived carbon-based materials, metal oxide nanoparticles, and conducting polymers, are used as catalysts and electrodes in fuel cells. For instance, graphene oxide and carbon nanotubes have been shown to enhance the electrochemical reaction in proton exchange membrane fuel cells (PEMFCs) due to their high conductivity, large surface area, and stability. The use of green nanomaterials for thermoelectrics helps minimize the ecological footprint of these devices, contributing to a sustainable energy recovery system. are being investigated for their piezoelectric properties.(Nagalingam M et al., 2018) For instance, ZnO nanowires and nanorods exhibit piezoelectric effects when mechanically deformed, and when incorporated into hybrid nanocomposites with biopolymers or other carbon-based materials, they can significantly enhance the material's ability to harvest energy from small mechanical motions, such as human body movement or environmental vibrations. These piezoelectric devices are promising for powering low-energy devices in wearable electronics or remote sensing applications, particularly in remote or off-grid locations where conventional power sources are not available.While the potential of green nanomaterials in energy applications is vast, several challenges must be addressed to fully realize their capabilities.

These challenges include issues related to the scalability of green synthesis methods, the durability and stability of nanocomposites under operational conditions, (Omar et al., 2017)and the need for cost-effective production processes(Patel V et al., 2015). Additionally, the environmental impact of the disposal or recycling of nanomaterials, particularly metal-based nanomaterials, needs to be carefully evaluated.. The incorporation of renewable resources, such as plant-derived materials and natural fibers, into nanocomposites will further promote the sustainability of energy technologies.Green nanomaterials and nanocomposites are poised to revolutionize energy storage, conversion, and harvesting technologies by providing high-performance, sustainable alternatives to conventional materials. energy devices like batteries, supercapacitors, fuel cells, solar cells, thermoelectric, and piezoelectric systems.

Environmental Applications of Green Nanomaterials and Nanocomposites: Water Treatment, Air Purification, and Soil Remediation

Environmental pollution, including contaminated water, air, and soil, poses often come with challenges, including high energy consumption, toxic by-products, and inefficiency in addressing specific pollutants. As a result, there is a growing interest in the use of green nanomaterials and nanocomposites for environmental remediation. These materials, which are often derived fromwhich facilitates the binding of contaminants. Furthermore, by functionalizing the surface of these nanoparticles with green agents such as tannins or polysaccharides, it is possible to enhance the interaction between the nanomaterials and specific pollutants, improving their adsorption capacity. For example, green-synthesized TiO_2 nanoparticles coated with plant-derived molecules have been used to adsorb heavy metals like lead and copper from water. Antimicrobial and Pathogen Removal Green nanomaterials Green synthesis of silver nanoparticles, using plant extracts or microorganisms, further enhances their eco-friendliness while maintaining high antimicrobial efficacy.

These green-synthesized nanoparticles can be incorporated into filters, membranes, or hybrid materials to purify water from pathogenic bacteria, viruses, and other microorganisms.including the incorporation of plant-derived compounds into metal oxide nanoparticles, has been shown to enhance the photocatalytic efficiency and environmental compatibility of these materials. These green composites can break down toxic organic pollutants, pesticides,Air Purification Applications Air pollution is another major environmental concern, particularly in urban and industrial areas. The presence of particulate matter, volatile organic compounds (VOCs), carbon monoxide, and nitrogen oxides in the air can have serious health implications and contribute to climate change. Green nanomaterials offer innovative solutions for air purification due to their high surface area, adsorption capacity, and catalytic properties.Removal of Gaseous Pollutants:One purification is the removal of gaseous pollutants. Metal oxide-based nanomaterials, including TiO_2, ZnO, and CuO, synthesized using green methods, are effective at adsorbing and degrading harmful gases such as CO, NOx, and VOCs. For instance, TiO_2 nanoparticles synthesized from plant extracts have shown superior catalytic activity for the degradation of VOCs, including benzene and toluene, under UV light exposure. Additionally, the use of carbon-based nanomaterials, including activated carbon and graphene oxide, in air filtration systems allows for efficient removal of gaseous pollutants due to their large surface area and high porosity. The integration of these materials into composite membranes or filters offers a sustainable, low-cost solution for air purification. Photocatalytic Air Purification:Photocatalytic processes are also employed for air purification, as certain nanomaterials can degrade harmful substances through redox reactions under light exposure. Green-synthesized photocatalysts, particularly those based on TiO_2, ZnO, and CuO, can break down harmful pollutants such as VOCs, formaldehyde, and toluene, converting them into less toxic by-products like carbon dioxide and water. The green synthesis of these photocatalysts not only ensures their environmental friendliness but also enhances their photocatalytic activity by incorporating natural biomolecules that can act as stabilizing agents and promote the formation of reactive species. Photocatalytic air purifiers utilizing these green nanomaterials are increasingly being integrated into indoor air quality management systems, particularly in homes and commercial buildings.Soil Remediation Applications Soil pollution,, soil washing, and phytoremediation, often face challenges in terms of cost, efficiency, and environmental impact. Green nanomaterials offer an alternative by providing highly efficient, cost-effective, and environmentally benign solutions for the removal of pollutants from soil.and modified biochar, have been shown to effectively immobilize and remove heavy metals like lead, cadmium, arsenic, and chromium from contaminated soil. Green synthesis methods, such as the use of plant extracts or microorganisms(Patil RS et al., 2012), can further enhance the efficiency and sustainability of these nanoparticles. Additionally, the integration of green nanomaterials into multi-functional composites for combined water, air, and soil purification systems may provide more holistic and effective solutions to environmental pollution.Conclusion efficient, and cost-effective solutions for cleaning contaminated environments. The use of green synthesis methods further enhances the eco-friendliness of these materials, aligning with the goals of sustainable development and environmental protection. As research continues, green nanomaterials hold the potential to play a key role in the transition toward a cleaner, healthier, and more sustainable world.

Biomedical Applications of Green Nanomaterials and Nanocomposites: Biomedical Imaging, Drug Delivery, and Tissue Engineering Applications

nanotechnology in biomedicine has ushered in an era of precision medicine, offering groundbreaking solutions for diagnosis, treatment, and tissue repair. Green nanomaterials and nanocomposites, synthesized using environmentally friendly and sustainable processes, have emerged as key components in a range of biomedical applications. These materials, derived from natural, renewable resources, imaging, improve therapeutic outcomes, and support the regeneration of damaged tissues. This article explores the growing role of green nanomaterials and nanocomposites in these vital biomedical fields.Biomedical Imaging Applications Biomedical imaging plays a crucial role in disease diagnosis, monitoring, and therapeutic guidance. Nanomaterials, due to their unique optical, magnetic, and chemical properties, are increasingly being used including reduced toxicity, improved biocompatibility, and lower environmental impact during production.Magnetic Resonance Imaging (MRI):(Rahman MM et al., 2021a) In MRI, contrast agents are used to enhance the contrast of images, making it easier to detect abnormalities such as tumors, lesions, nanoparticles, synthesized using environmentally friendly methods, are gaining traction as alternatives to traditional MRI contrast agents. These nanoparticles offer high magnetic susceptibility, which allows for improved image resolution and sensitivity. Additionally, the surface of these nanoparticles can be functionalized with biomolecules such as antibodies or peptides to target specific tissues or tumors, enhancing the precision of MRI imaging. Green synthesis methods, such as using plant extracts or microorganisms, are particularly attractive for reducing the toxic by-products typically associated with the synthesis of metal oxide nanoparticles Fluorescence Imaging Fluorescence imaging, which relies on fluorescent dyes or nanoparticles to visualize cellular structures, tumors, or disease markers, is another area where green nanomaterials show significant promise. Quantum dots (QDs) made from eco-friendly materials such as carbon, silica, or biopolymer-based composites have been developed for fluorescence imaging. These QDs offer high fluorescence intensity, tunable emission wavelengths, and long-term stability. For example, green-synthesized carbon quantum dots (CQDs) derived from plant sources like fruit peels or vegetable extracts provide a non-toxic alternative to traditional semiconductor-based QDs.(Zhong-Yu L et al., 2001) The biocompatibility and minimal toxicity of these green QDs make them ideal for in vivo imaging, where safety is a primary concern Green synthesis techniques can further enhance the release profile of these nanomaterials by incorporating natural compounds such as essential oils, antioxidants, or plant-based compounds that act as stabilizing agents. For example, PLGA nanoparticles loaded with anti-inflammatory drugs, synthesized via a green approach using plant extracts, can be engineered to release the drug over a prolonged period, minimizing the need for frequent dosing and improving patient compliance. in treating genetic disorders, cancer, and viral infections. Green nanomaterials are emerging as valuable carriers for gene delivery, as they can encapsulate nucleic acids such as DNA or RNA and protect them from degradation in the bloodstream. Polymeric nanocarriers, such as those made from chitosan or polyethylenimine (PEI), have been used in gene delivery applications. Green synthesis of these carriers can enhance their biocompatibility and reduce the cytotoxicity typically associated with synthetic carriers. Moreover, these green nanomaterials can be functionalized with targeting ligands to direct the genetic material to specific cells, ensuring precise and efficient delivery.such as organ regeneration, wound healing, and bone repair. Nanomaterials play a critical role in tissue engineering by providing structural support, enhancing cell adhesion, and promoting tissue growth. Green nanomaterials offer several advantages in this field, including enhanced biocompatibility, reduced toxicity, and the ability to mimic the properties of natural extracellular matrices.. For instance, nanocellulose scaffolds, derived from plant fibers, are highly porous and have excellent mechanical strength, making them ideal for supporting cell growth in applications such as wound healing and bone regeneration. These materials can be functionalized with bioactive molecules such as growth factors, peptides, or extracellular matrix proteins to enhance cell attachment, proliferation, and differentiation.a aturally occurring mineral form of calcium phosphate, are commonly used in bone tissue engineering due to their similarity to the mineral component of bone. HA can be synthesized from natural sources such as corals or eggshells, and when combined with biopolymers like chitosan, collagen, or alginate, it forms nanocomposites that support bone cell adhesion and osteogenic differentiation. Additionally, green-synthesized hydroxyapatite nanoparticles can be functionalized with growth factors or other bioactive molecules to further enhance bone tissue regeneration.Conclusion Green nanomaterials and nanocomposites are, the potential of green nanomaterials to improve diagnostic precision, enhance therapeutic efficacy, and support tissue regeneration continues to grow. With continued innovation, these eco-friendly materials(Yehia et al., 2014) will play an integral role in the development of safer, more effective biomedical treatments, contributing to the future of personalized medicine and regenerative therapies.

Toxicity and Sustainability of Green Nanomaterials and Nanocomposites: Assessing the Toxicity and Sustainability of Green Nanomaterials and Nanocomposites

The rapid development of nanotechnology and the expanding use of nanomaterials in various fields, including biomedicine, energy, and environmental remediation, have raised important concerns regarding friendly and sustainable methods, are designed to offer solutions to these challenges, their toxicity and sustainability must be thoroughly evaluated. Green nanomaterials, often derived from natural, renewable sources such as plants, microorganisms, and biopolymers, are generally considered to be more environmentally benign compared to conventional, chemically synthesized nanomaterials. However, the potential risks associated with their use—ranging from toxicity to ecosystems and human health to their life cycle impact—require comprehensive assessment and regulatory oversight. This article examines the key factors involved in evaluating the toxicity and sustainability of green nanomaterials and nanocomposites, with a focus on their design, applications, environmental behavior, and the current challenges in toxicity assessment.Toxicity of Green Nanomaterials and NanocompositesThe toxicity of nanomaterials refers to their potential to cause harmful effects on living organisms, including humans, animals, and plants, as well as on microbial communities. Unlike bulk materials, nanomaterials exhibit unique properties due to their small size, large surface area, and high reactivity. These properties can lead to different biological interactions and toxicological effects, which are not always predictable based on traditional toxicity assessments of larger particles or chemicals. shape, surface charge, chemical composition, and surface functionalization of the materials. Smaller nanoparticles, for instance, tend to be more reactive and may interact more readily with biological systems compared to larger particles. This increased reactivity can lead to the generation of reactive oxygen species (ROS), which can cause oxidative stress and damage to cells, tissues, and organs. These tests help assess the potential for nanomaterials to induce cell death, DNA damage, or other harmful effects in cultured cells. In vivo studies, on the other hand,(Wang D et al., 2021) provide a more comprehensive understanding of the systemic effects of nanomaterials, including their distribution, bioaccumulation, and long-term impacts on organs and tissues. Animal studies using rodents or other model organisms are often employed to assess the potential for toxicity after exposure to green nanomaterials through different routes, such as inhalation, ingestion, or dermal contact.consideration, especially given the increasing demand for environmentally friendly materials in industrial and commercial applications. Unlike traditional nanomaterials, which are often synthesized using toxic chemicals, high-energy processes, or non-renewable resources, green nanomaterials are typically produced using renewable, non-toxic, and bio-based materials. These materials offer the potential for reducing the environmental footprint of nanotechnology and promoting sustainable development. However, assessing their overall sustainability requires a comprehensive understanding of their life cycle, from raw material extraction and synthesis to end-of-life disposal and potential recycling. For green nanomaterials, LCA helps assess the environmental benefits and trade-offs associated with their production, use, and disposal. An LCA typically considers several stages: raw material acquisition, synthesis, use, and end-of-life disposal or recycling(Vivek M et al., 2011). In the case of green nanomaterials, the use of renewable feedstocks, such as plant biomass or natural polymers, can significantly reduce the environmental impact during the raw material acquisition and synthesis stages. Moreover, the green synthesis of nanomaterials often employs lower-energy processes that result in fewer greenhouse gas emissions and less chemical waste compared to traditional methods.However, the end-of-life phase is an area of concern for the sustainability of green nanomaterials. Nanomaterials that accumulate in the environment, such as in soil or water, may have unintended ecological consequences. For example, the release of nanomaterials from consumer products, such as cosmetics or medical devices, could lead to the contamination of soil or water bodies. Once released, green nanomaterials may interact with microorganisms, plants, and animals, potentially disrupting ecosystems and food chains. The degradation or leaching of nanomaterials into the environment is another factor that must be carefully evaluated, as their long-term persistence could cause environmental harm.Eco-toxicological Assessment:In addition to assessing the direct toxicity of green nanomaterials to human health, it is essential to evaluate their potential eco-toxicity. Eco-toxicity refers to the harmful effects of materials on non-human organisms, including aquatic life, soil microbes, and plants. Green nanomaterials, despite their natural origin, can still pose risks to ecosystems if they accumulate in the environment or interfere with ecological processes. For example, biopolymer-based nanomaterials such as chitosan, cellulose, and starch have been shown to degrade naturally in the environment without leaving behind toxic residues. This biodegradability feature makes them more environmentally friendly than synthetic nanomaterials, which may persist in the environment for extended periods.In addition to biodegradability, the recyclability of green nanomaterials also contributes to their sustainability. Recycling processes for the recycling of green nanomaterials, which will further enhance their environmental sustainability.Challenges and Future Directions Despite the promising advantages of green nanomaterials, several challenges remain in fully assessing their toxicity and sustainability. The diversity of green nanomaterials, coupled with the variability in synthesis methods and raw materials, makes it difficult to develop universal testing standards. Furthermore, the long-term effects of green nanomaterials on both human health and ecosystems remain poorly understood, and more extensive studies are needed to evaluate their safety profiles under real-world conditions.methods and eco-toxicological assays will help provide more reliable data for regulatory agencies. Additionally, advances in the design of biodegradable and recyclable green nanomaterials will play a key role in mitigating environmental risks. Collaborations between material scientists, toxicologists, environmental engineers, and regulatory bodies will be essential to developing comprehensive guidelines for the safe and sustainable use of green nanomaterials. Conclusion Green nanomaterials and nanocomposites offer significant potential for improving sustainability in various industries, from biomedicine to environmental remediation. By utilizing renewable, non-toxic, and environmentally friendly raw materials, these materials offer eco-friendly alternatives to conventional synthetic nanomaterials. However, their toxicity and sustainability must be carefully assessed to ensure that their widespread use does not result in unintended environmental or health risks. A holistic approach that includes rigorous toxicity

Future Directions and Challenges in Green Nanomaterials and Nanocomposites: Emerging Trends, Challenges, and Future Research Directions

Green nanomaterials and nanocomposites, which are derived from renewable resources and synthesized through eco-friendly processes, represent a promising frontier in material science. The growing emphasis on sustainability and environmental responsibility has catalyzed the development of these materials, which offer potential solutions to global challenges in energy, environment, and healthcare. However, despite the significant progress made in green nanotechnology, numerous challenges remain in fully realizing the potential of green nanomaterials and nanocomposites. This article explores the emerging trends, challenges, and future research directions in the field of green nanomaterials, focusing on the continued development of sustainable synthesis methods, improved performance characteristics, safety and toxicity assessment, and integration into real-world applications.Emerging Trends in Green Nanomaterials and Nanocomposites non-toxic, and often biologically derived materials through more energy-efficient processes. The green synthesis of nanomaterials typically involves the use of natural plant extracts, microorganisms, or biopolymers, which act as reducing agents, stabilizers, or templates for nanoparticle formation. Recent developments in this area include the use of algae, fungi, and bacteria to produce nanoparticles(Samadi N et al., 2009), as well as new strategies that utilize waste products from agricultural or industrial processes as precursors for nanomaterial synthesis. The increasing focus on low-cost, scalable, and energy-efficient methods to produce green nanomaterials is expected to drive the widespread adoption of these materials in various industries.Nanocomposites, which combine nanomaterials with other matrix materials (e.g., polymers, metals, ceramics), offer unique properties that are superior to those of their individual components. The development of green nanocomposites that combine renewable or biodegradable polymers with nanomaterials derived from natural sources is gaining attention. These nanocomposites —such as carbon-based materials, biopolymers, and metal oxides—leading to materials with highly tailored properties for specific applications.As part of the sustainability agenda, there is an increasing focus on developing biodegradable and recyclable green nanomaterials. Biodegradable nanomaterials made from natural polymers such as cellulose, chitin, and, and recycled, thereby minimizing waste and conserving resources. The incorporation of environmentally friendly disposal methods into the lifecycle of green nanomaterials will play a crucial role in making them truly sustainable alternatives to conventional nanomaterials. another emerging trend is the functionalization of green nanomaterials to enhance their performance for specialized applications. Surface modification techniques, such as coating or grafting functional groups onto nanoparticles, can improve their dispersion, stability, and interaction with other materials. including their small size and large surface area, can lead to unexpected interactions with biological systems, potentially causing toxicity. While green nanomaterials are often perceived as being less toxic than synthetic nanomaterials, it is essential to conduct rigorous safety assessments to evaluate their effects on human health, ecosystems, and the environment. The development of standardized testing protocols for the toxicity of green nanomaterials, as well as guidelines for assessing their lifecycle impacts, is needed to ensure their safe and responsible use. Additionally, understanding the potential for bioaccumulation and environmental persistence is critical to avoid unintended long-term consequences. green nanomaterials creates uncertainty for manufacturers and researchers, which could slow the adoption of these materials. Regulatory agencies must develop clear policies and standards that address the unique properties and risks associated with green nanomaterials while promoting innovation and sustainability. International cooperation and harmonization of regulatory approaches will be essential to ensure the safe and responsible use of green nanomaterials worldwide. as well as addressing concerns about their potential risks, will be crucial for gaining widespread acceptance. Transparency in research, open communication between scientists and the public, and the establishment of safety protocols will help build trust in green nanomaterials and foster their adoption.

The future of green nanomaterials and nanocomposites lies in addressing the challenges outlined above and advancing our understanding of these materials through innovative research. Several key research directions are expected to shape the future of green nanotechnology.

Developing new, more efficient, and scalable green synthesis methods will be critical for advancing the field. Research into alternative renewable resources, such as waste biomass, agricultural byproducts, and food industry residues, offers opportunities for sustainable synthesis routes. Additionally, novel approaches such as 3D printing and microfluidics are being explored for the controlled synthesis of nanomaterials with precise shapes and properties. These techniques could enable the production of high-performance green nanomaterials at scale, opening up new possibilities for their commercial uses Future research will focus on the complete life cycle of green nanomaterials, from production to disposal. This includes developing methods to enhance the biodegradability and recyclability of nanomaterials, as well as designing products that can be easily repurposed or recycled at the end of their life. The integration of sustainability metrics, such as carbon footprint,(Rahman MM et al., 2021b) water usage, and energy consumption, into the design process will be a key research direction in ensuring that green nanomaterials contribute to a circular economy.Conclusion conditional nanocomposites, and the focus on biodegradable materials, point to exciting new directions for research and development. However, significant challenges remain, particularly in scaling up production, ensuring performance and durability, and addressing safety and regulatory concerns. By overcoming these challenges and focusing on sustainability, green nanomaterials have the potential to revolutionize multiple industries and contribute to a more sustainable future.

CONCLUSION

The rapid evolution of green nanomaterials and nanocomposites stands as a transformative milestone in the pursuit of sustainability and environmental stewardship within materials science and technology. These innovative materials, derived from renewable resources and synthesized via eco-friendly processes, offer unparalleled opportunities to address many of the pressing challenges faced by industries ranging from energy and environment to healthcare and electronics. Green nanomaterials, with their unique properties—ranging from enhanced mechanical and electrical characteristics to biodegradability and non-toxicity—promise to revolutionize the way we produce, consume, and dispose of materials in the modern world. However, the journey toward fully harnessing the potential of these materials is fraught with both challenges and opportunities, requiring further research, innovation, and collaboration across disciplines.One of the key factors driving the development of green nanomaterials is the urgent need to replace conventional, often environmentally harmful materials with alternatives that are not only high-performance but also sustainable. Traditional materials and their manufacturing processes often come with significant ecological footprints, including toxic waste, high energy consumption, and non-renewable resources. In contrast, green nanomaterials aim to address these issues by incorporating renewable feedstocks, such as biopolymers and natural extracts, and adopting environmentally friendly synthesis techniques, such as biosynthesis and low-energy methods. . Whether used in energy storage devices, water purification systems, biomedical applications, or packaging materials, green nanocomposites represent an important shift toward more sustainable manufacturing practices. By leveraging the synergy between nanomaterials and matrix materials, these composites offer opportunities for innovations that optimize resource usage while minimizing waste and pollution.Another challenge lies in regulatory and standardization issues. While green nanomaterials are often promoted as safer and more sustainable alternatives to conventional nanomaterials, the regulatory landscape for these materials is still under development. In many cases, the existing nanomaterial regulations do not differentiate between green and conventional nanomaterials, which can impede innovation and slow down the integration of green alternatives into commercial markets. Furthermore, while the promise of green nanomaterials is immense, their widespread adoption will depend on public perception and societal acceptance. Concerns about the potential health and environmental risks associated with nanotechnology persist, and without proper education and outreach, these concerns could hinder the full realization of green nanomaterials' potential. Addressing public skepticism through transparent communication, rigorous safety testing, and demonstrable real-world benefits will be crucial in fostering trust and confidence in these materials.. The key to unlocking their full potential lies in continued innovation, particularly in the development of novel synthesis techniques, the optimization of material properties, and the enhancement of material performance through functionalization. Researchers are exploring new biobased feedstocks, including waste biomass, agricultural byproducts, and organic residues, to create green nanomaterials that are not only sustainable but also cost-effective. In addition, hybrid nanocomposites, which combine different types of nanomaterials with distinct properties, promise to offer even more advanced materials with tailored characteristics for specific applications, further expanding their applicability in diverse industries.In conclusion, the development of green nanomaterials and nanocomposites represents a paradigm shift toward sustainable materials science, aligning technological progress with environmental responsibility. These materials . However, to fully realize the potential of green nanomaterials, a concerted effort is needed to overcome scalability issues, enhance material performance, ensure safety and sustainability, and foster public trust. With ongoing research, interdisciplinary collaboration, and advancements in synthesis and functionalization techniques, green nanomaterials and nanocomposites are poised to play and real-world application, ensuring that green nanotechnology contributes to a more sustainable, equitable, and healthier world.

REFERENCES

Ali, M. Y. S., Anuradha, V., Abishek, R., Yogananth, N., & Sheeba, H. (2017). In Vitro Anticancer Activity of Green Synthesis Ruthenium Nanoparticle from Dictyota Dichotoma Marine Algae. *NanoWorld Journal, 03*(4), 1. DOI: 10.17756/nwj.2017-049

Amooaghaie, R., Saeri, M. R., & Azizi, M. (2015). Synthesis, Characterization and Biocompatibility of Silver Nanoparticles Synthesized from Nigella Sativa Leaf Extract in Comparison with Chemical Silver Nanoparticles. *Ecotoxicology and Environmental Safety, 120*, 400–408. DOI: 10.1016/j.ecoenv.2015.06.025 PMID: 26122733

Bankar, A., Joshi, B., Kumar, A. R., & Zinjarde, S. (2010). Banana Peel Extract Mediated Novel Route for the Synthesis of Silver Nanoparticles. *Colloids and Surfaces. A, Physicochemical and Engineering Aspects, 368*(1-3), 58–63. DOI: 10.1016/j.colsurfa.2010.07.024

Belliveau, B. H., Starodub, M. E., Cotter, C., & Trevors, J. T. (1987). Metal Resistance and Accumulation in Bacteria. *Biotechnology Advances, 5*(1), 101–127. DOI: 10.1016/0734-9750(87)90006-1 PMID: 14543146

Castro, L., Blázquez, M. L., Muñoz, J. A., González, F., & Ballester, A. (2013). Biological Synthesis of Metallic Nanoparticles Using Algae. *IET Nanobiotechnology / IET, 7*(3), 109–116. DOI: 10.1049/iet-nbt.2012.0041 PMID: 24028809

Chakravarty, A., Ahmad, I., Singh, P., Ud Din Sheikh, M., Aalam, G., Sagadevan, S., & Ikram, S. (2022). Green Synthesis of Silver Nanoparticles Using Fruits Extracts of Syzygium Cumini and Their Bioactivity. *Chemical Physics Letters, 1*, 139493. DOI: 10.1016/j.cplett.2022.139493

Dahoumane, S. A., Yéprémian, C., Djédiat, C., Couté, A., Fiévet, F., Coradin, T., & Brayner, R. (2016). Improvement of Kinetics, Yield, and Colloidal Stability of Biogenic Gold Nanoparticles Using Living Cells of Euglena Gracilis Microalga. *Journal of Nanoparticle Research, 18*(3), 1. DOI: 10.1007/s11051-016-3378-1

El-Kassas, H. Y., & El-Sheekh, M. M. (2014). Cytotoxic Activity of Biosynthesized Gold Nanoparticles with an Extract of the Red Seaweed Corallina Officinalis on the MCF-7 Human Breast Cancer Cell Line. *Asian Pacific Journal of Cancer Prevention : APJCP, 15*(10), 4311–4317. DOI: 10.7314/APJCP.2014.15.10.4311 PMID: 24935390

Fernández, J. G., Fernández-Baldo, M. A., Berni, E., Camí, G., Durán, N., Raba, J., & Sanz, M. I. (2016). Production of Silver Nanoparticles Using Yeasts and Evaluation of Their Antifungal Activity against Phytopathogenic Fungi. *Process Biochemistry (Barking, London, England)*, *51*(9), 1306–1313. DOI: 10.1016/j.procbio.2016.05.021

Fesharaki, P. J., Nazari, P., Shakibaie, M., Rezaie, S., Banoee, M., Abdollahi, M., & Shahverdi, A. R. (2010). Biosynthesis of Selenium Nanoparticles Using Klebsiella pneumoniae and Their Recovery by a Simple Sterilization Process. *Brazilian Journal of Microbiology : Publication of the Brazilian Society for Microbiology*, *41*(2), 461–466. DOI: 10.1590/S1517-83822010000200028 PMID: 24031517

Gaumet, M., Vargas, A., Gurny, R., & Delie, F. (2008). Nanoparticles for Drug Delivery: The Need for Precision in Reporting Particle Size Parameters. *European Journal of Pharmaceutics and Biopharmaceutics*, *69*(1), 1–9. DOI: 10.1016/j.ejpb.2007.08.001 PMID: 17826969

Hameed, S., Khalil, A. T., Ali, M., Numan, M., Khamlich, S., Shinwari, Z. K., & Maaza, M. (2019). Greener Synthesis of ZnO and Ag-ZnO Nanoparticles Using Silybum marianum for Diverse Biomedical Applications. *Nanomedicine (London)*, *14*(6), 655–673. DOI: 10.2217/nnm-2018-0279 PMID: 30714480

Ijaz, F., Shahid, S., Khan, S. A., Ahmad, W., & Zaman, S. (2017). Green Synthesis of Copper Oxide Nanoparticles Using Abutilon Indicum Leaf Extract: Antimicrobial, Antioxidant and Photocatalytic Dye Degradation Activitie. *Tropical Journal of Pharmaceutical Research*, *16*(4), 743. DOI: 10.4314/tjpr.v16i4.2

Ilahi, N., Haleem, A., Iqbal, S., Fatima, N., Sajjad, W., & Sideeq, A.. (2021). Biosynthesis of Silver Nanoparticles Using Endophytic Fusarium Oxysporum Strain NFW16 and Their In Vitro Antibacterial Potential. *Microscopy Research and Technique*, *1*, 1. DOI: 10.1002/jemt.24018 PMID: 34888986

Jevapatarakul, D., T-Thienprasert, J., Payungporn, S., Chavalit, T., Khamwut, A., & T-Thienprasert, N. P. (2020). Utilization of Cratoxylum Formosum Crude Extract for Synthesis of ZnO Nanosheets: Characterization, Biological Activities and Effects on Gene Expression of Nonmelanoma Skin Cancer Cell. *Biomedicine and Pharmacotherapy*, *130*, 110552. DOI: 10.1016/j.biopha.2020.110552 PMID: 32739741

Jeyarani, S., Vinita, N. M., Puja, P., Senthamilselvi, S., Devan, U., Velangani, A. J., Biruntha, M., Pugazhendhi, A., & Kumar, P. (2020). Biomimetic Gold Nanoparticles for its Cytotoxicity and Biocompatibility Evidenced by Fluorescence-Based Assays in Cancer (MDA-MB-231) and Non-cancerous (HEK-293) Cells. *Journal of Photochemistry and Photobiology. B, Biology*, *202*, 111715. DOI: 10.1016/j.jphotobiol.2019.111715 PMID: 31790882

Kannan, N., Mukunthan, K. S., & Balaji, S. (2011). A Comparative Study of Morphology, Reactivity and Stability of Synthesized Silver Nanoparticles Using Bacillus Subtilis and Catharanthus Roseus (L.) G. *Colloids and Surfaces. B, Biointerfaces, 86*(2), 378–383. DOI: 10.1016/j.colsurfb.2011.04.024 PMID: 21592749

Khalaj, M., Kamali, M., Costa, M. E. V., & Capela, I. (2020). Green Synthesis of Nanomaterials - A Scientometric Assessment. *Journal of Cleaner Production, 267*, 122036. DOI: 10.1016/j.jclepro.2020.122036

Mahdavi, M., Namvar, F., Ahmad, M., & Mohamad, R. (2013). Green Biosynthesis and Characterization of Magnetic Iron Oxide (Fe3O4) Nanoparticles Using Seaweed (Sargassum Muticum) Aqueous Extract. *Molecules (Basel, Switzerland), 18*(5), 5954–5964. DOI: 10.3390/molecules18055954 PMID: 23698048

Majeed, M., Hakeem, K. R., & Rehman, R. U. (2022). Synergistic Effect of Plant Extract Coupled Silver Nanoparticles in Various Therapeutic Applications- Present Insights and Bottlenecks. *Chemosphere, 288*, 132527. DOI: 10.1016/j.chemosphere.2021.132527 PMID: 34637861

Nag, M., Lahiri, D., Sarkar, T., Ghosh, S., Dey, A., Edinur, H. A., Pati, S., & Ray, R. R. (2021). Microbial Fabrication of Nanomaterial and its Role in Disintegration of Exopolymeric Matrices of Biofilm. *Frontiers in Chemistry, 9*, 1. DOI: 10.3389/fchem.2021.690590 PMID: 34109159

Nagalingam, M., Kalpana, V. N., Rajeswari, V. D., & Panneerselvam, A. (2018). Biosynthesis, Characterization, and Evaluation of Bioactivities of Leaf Extract-Mediated Biocompatible Gold Nanoparticles from Alternanthera Bettzickiana. *Biotechnology Reports (Amsterdam, Netherlands), 19*, e00268. DOI: 10.1016/j.btre.2018.e00268 PMID: 29992102

Omar, H. H., & Bahabri, S., FM. El-Gend, A. (2017). Biopotential Application of Synthesis Nanoparticles as Antimicrobial Agents by Using Laurencia Papillosa. *International Journal of Pharmacology, 13*, 303–312. DOI: 10.3923/ijp.2017.303.312

Patel, V., Berthold, D., Puranik, P., & Gantar, M. (2015). Screening of Cyanobacteria and Microalgae for Their Ability to Synthesize Silver Nanoparticles with Antibacterial Activity. *Biotechnology Reports (Amsterdam, Netherlands), 5*, 112–119. DOI: 10.1016/j.btre.2014.12.001 PMID: 28626689

Patil, R. S., Kokate, M. R., & Kolekar, S. S. (2012). Bioinspired Synthesis of Highly Stabilized Silver Nanoparticles Using Ocimum Tenuiflorum Leaf Extract and Their Antibacterial Activity. *Spectrochimica Acta. Part A: Molecular and Biomolecular Spectroscopy, 91*, 234–238. DOI: 10.1016/j.saa.2012.02.009 PMID: 22381796

Rahman, M. M., Ferdous, K. S., Ahmed, M., Islam, M. T., Khan, M. R., Perveen, A., Ashraf, G. M., & Uddin, M. S. (2021a). Hutchinson-Gilford Progeria Syndrome: An Overview of the Molecular Mechanism, Pathophysiology and Therapeutic Approach. *Current Gene Therapy*, *21*(3), 216–229. DOI: 10.2174/1566523221666210303100805 PMID: 33655857

Rahman, M. M., Rahaman, M. S., Islam, M. R., Hossain, M. E., Mannan Mithi, F., Ahmed, M., Saldías, M., Akkol, E. K., & Sobarzo-Sánchez, E. (2021b). Multifunctional Therapeutic Potential of Phytocomplexes and Natural Extracts for Antimicrobial Properties. *Antibiotics (Basel, Switzerland)*, *10*(9), 1076. DOI: 10.3390/antibiotics10091076 PMID: 34572660

Samadi, N., Golkaran, D., Eslamifar, A., Jamalifar, H., Fazeli, M. R., & Mohseni, F. A. (2009). Intra/Extracellular Biosynthesis of Silver Nanoparticles by an Autochthonous Strain of Proteus mirabilis Isolated fromPhotographic Waste. *Journal of Biomedical Nanotechnology*, *5*(3), 247–253. DOI: 10.1166/jbn.2009.1029 PMID: 20055006

Vivek, M., Kumar, P. S., Steffi, S., & Sudha, S. (2011). Biogenic Silver Nanoparticles by Gelidiella Acerosa Extract and Their Antifungal Effects. *Avicenna Journal of Medical Biotechnology*, *3*, 143–148. PMID: 23408653

Wang, D., Xue, B., Wang, L., Zhang, Y., Liu, L., & Zhou, Y. (2021). Fungus-mediated green Synthesis of Nano-Silver Using Aspergillus sydowii and its Antifungal/antiproliferative Activities. *Scientific Reports*, *11*(1), 1. DOI: 10.1038/s41598-021-89854-5 PMID: 33990673

Yehia, R. S., & Al-Sheikh, H. (2014). Biosynthesis and Characterization of Silver Nanoparticles Produced by Pleurotus Ostreatus and Their Anticandidal and Anticancer Activities. *World Journal of Microbiology & Biotechnology*, *30*(11), 2797–2803. DOI: 10.1007/s11274-014-1703-3 PMID: 25053172

Zhong-Yu, L., Fu, J. K., Jin-Kun, F., Jian-Ming, W., Yue-Ying, L., & Hu, C. (2001). Preliminary Study on the Mechanism of Nonenzymatic Bioreduction of Precious Metal Ions. *Wuli Huaxue Xuebao*, *17*(5), 477–480. DOI: 10.3866/PKU.WHXB20010520

Chapter 7
Green Polymers and Biodegradable Plastics

Prachi Saxena
https://orcid.org/0000-0002-5161-9829
Lovely Professional University, India

C. Dogra
Lovely Professional University, India

ABSTRACT

The spiked environmental concerns associated with conventional plastics have rushed the evolution and adoption of eco-friendly substitutes like green polymers and biodegradable plastics. These materials arise from green energy resources such as plants, animals, and microbes, offering a viable option to minimize plastic pollution and carbon footprints. This chapter delves into the classification, properties, synthesis, and applications of green polymers and biodegradable plastics, emphasizing their benefits over traditional plastics. It also addresses challenges such as high production costs, mechanical properties, and degradation under different environmental conditions. Emphasis is placed on innovative technologies and future trends that aim at improving the efficiency, production, and scalability of these biopolymers. This chapter delivers a thorough exploration of the crucial role of green polymers and biodegradable plastics in promoting a sustainable environment and advancing a more inclusive circular economy.

DOI: 10.4018/979-8-3373-1409-9.ch007

1. INTRODUCTION

Plastics are essential in contemporary sphere because they are insubstantial, unreactive, highly versatile, and possess superior heat and electrical characteristics. These qualities make their prevalence in numerous sectors like home furnishings, circuit elements, equipment in the biomedical field, gadget making, and wrapping etc. The plastic output across the world is expected to increase by 450 megatons yearly by 2024 and anticipated to get twofold by the year 2040 (Sarah et al., 2024; ISW, 2024), and on the top, around 95%, being designed for single use and discarded after serving their purpose. The vast majority, i.e., 90% of the material used in packaging domain comes from plastics. A breakthrough has occurred during the 20th century (Bordes et al., 2009) in plastic mass production that transform manufacturing output. Derived from petroleum-based procedures, plastics being a byproduct is affordable to make and soon earned the title of "the wonder material". Some customary plastics employed in production domains are polyethylene terephthalate (PET), polypropylene (PP), polyvinyl chloride (PVC), high-density polyethylene (HDPE), and low-density polyethylene (LDPE). The utilization of plastics has been noted in nearly all sectors and transformed present-day life because of their adaptability, accessibility, and economic rates. Nevertheless, their extensive utilization has created notable threats to nature, such as inability of plastics to biodegrade, buildup in natural habitats, and excessive release of greenhouse gases (GHGs). These challenges have driven a global push towards eco-friendly substitutes, with green polymers and biodegradable plastics emerging as promising solutions. Made from alternative sources like vegetation, agricultural remains, and microbes, these materials not just intend to minimize dependence on carbon-based energy but also tackle the burning concern of plastic pollution. Within the habitat, these plastic debris breakdown into minute fragments known as "microplastics" due to sun exposure, weather-beaten, and flaming. Microplastics poses threats to both animals and the ecosystem as they interfere with food chains (Wabnitz & Nichols, 2010). Incinerating plastic trash liberates toxic gases like persistent organic pollutants (POPs), furan, and synthetic chemicals i.e., polychlorinated biphenyls (PCBs), into the surrounding. These noxious fumes can cause cancer, genetic mutations, nervous system failure, pulmonary disorders, cardiac infractions, shortness of breath, and affect genitals (Verma et al., 2016). Accordingly, there is an escalating interest in evolving sustainable alternatives like organic plastics. These materials can break down naturally, affecting the environment, delivering a viable alternative to the problems caused by non-decaying plastics. Biodegradable polymers (BDPs) have diverse characteristics, equipped them to compete with regular plastics. PLA and PHA, made from organic substances, are two frequently employed bioplastics (Averous, L. 2008).

Figure 1. Structure of some green polymers and biodegradable plastics

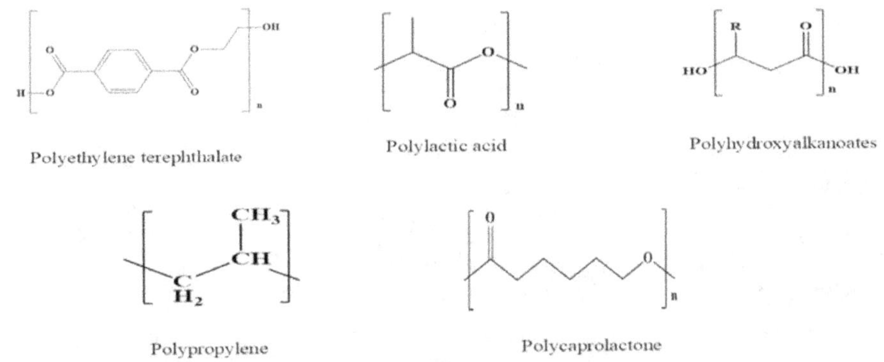

"Green" polymers are polymers made to some extent or entirely from sustainable organic resources excluding crude oil. Green polymers like polylactic acid (PLA), polyhydroxyalkanoates (PHA), polybutylene succinate (PBS), and starch-based plastics are eco-friendly materials sketched to minimize the ecological footprint associated with traditional polymers. These polymers are derived from renewable resources, such as plants, or are integrated with such procedures that minimize the use of energy and reduce waste generation. Examples include polylactic acid produced using sugar-brews (e.g., sugarcane), and polyhydroxyalkanoates (PHAs) produced by microorganisms, and cellulose-based polymers. Biodegradable plastics are a subset of polymers designed to disintegrate into organic substances. Examples include polyhydroxyalkanoates (PHAs), polylactic acid (PLA), and polycaprolactone (PCL) a synthetic biodegradable polymer. Figure 1. shows the structure of some green polymers and environmentally degradable materials. This chapter explores the journey of green polymers and biodegradable plastics by bridging the gap between revolution and environmental responsibility, these materials clear the way for a circular economy, where resources are reused, and waste is minimized.

2. CLASSIFICATION OF GREEN POLYMERS AND BIODEGRADABLE PLASTICS

Green polymers and biodegradable plastics are categorized based on their origin, composition, and degradability. The classification can be labeled into three essential factors: source, biodegradability, method of production, composition and applications. Figure 2. provides a concise classification of bio-based polymers (Gontard et al., 2018; Ghandarzadeh et al., 2013; Raza et al., 2020 Shah & Vasava, 2019; Karande et al., 2015; Hernandez et al., 2021).

Figure 2. Classification of bio-based polymers

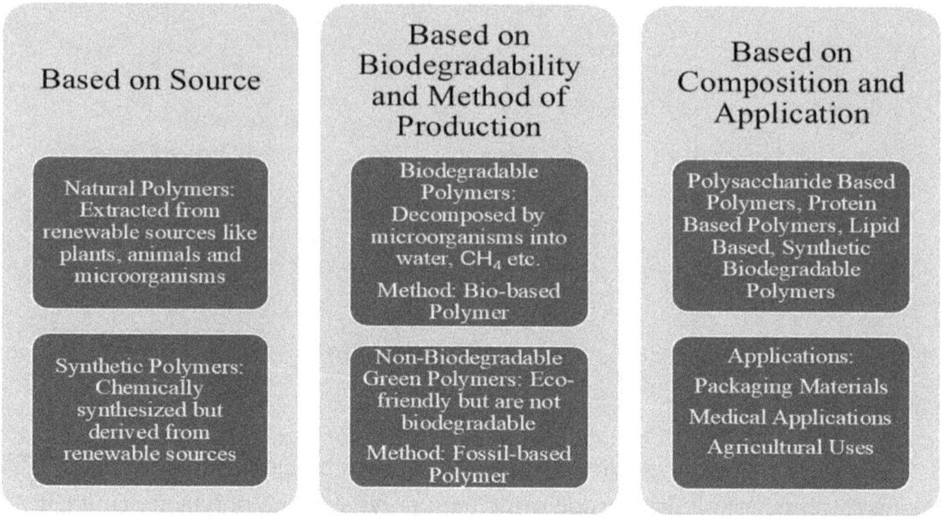

3. PROPERTIES OF GREEN POLYMERS AND BIODEGRADABLE PLASTICS

3.1. Biopolymers: Difference between Biodegradable and Bio-Based

If a polymer shows characteristics like biodegradability, extracted from raw materials that contain biomass, or one and the other then such polymer is considered as bioplastic (Ibrahim et al., 2021; Fico et al., 2022). Few researchers conclude bioplastics as those types of materials that can easily disintegrate into carbon dioxide, aqua, and compost driven through the activity of microscopic organisms and the enzymes they release. However, it's imperative to mention that not every eco-sourced polymer is compostable, and not every compostable polymer is a green source polymer. A nature-derived polymer comes from natural materials like organic matter that can be either partially or fully inexhaustible.

3.1.1. Biodegradability

Biopolymers are a prime feature of biodegradability, but they don't relate to the starting materials i.e., feedstocks. An environmentally degradable polymer can also be sourced from fossil fuels. Biodegradation is a transformative process where materials get converted into CO_2, H_2O, biofuel, and inorganic matter, with the help of eco-friendly decomposition. A biopolymer must breakdown at an equivalent pace to approved compostable materials to be considered as biodegradable and it is not supposed to leave any sort of toxic byproducts (Brizga et al., 2020). Although the petrochemical material can decay, if it doesn't meet the approved certifications, it ought to be recognized as non-decomposable despite being capable of biodegradation. Eventually, decomposable matter doesn't necessarily originate from natural routes. For instance, Oxygen-assisted biodegradation is a process where some polyolefins disintegrate into minute pieces by inducing oxygen, facilitating their breakdown into biodegradable forms. However, oxygen-induced degradation is not considered compostable for the time being as this process doesn't meet the desired approval (Green Dot Bioplastics., 2022).

3.1.2. Bio Composites

The shift from petrochemical polymers to bio-based polymers has inflated into the branch of composite constituents. Throughout the last decade, it has become feasible to merge bio-based polymers with nature-based fillers to develop innovative substances referred to as bio-based composites. These materials have acquired global consciousness from both researchers and industries due to their diverse set of utilizations, including in aircraft components, vehicle parts, everyday products, athletic gear, and applications in the shipping and petroleum sectors (Rangappa et al., 2022). Composites that are bio-based are usually those types of substances that are combined with natural substances with non-biodegradable materials within a network. The most frequently utilized bio-based polymers include amylum, fiber, soy protein, lactic acid polymer, and polyhydroxyalkanoates (PHAs) (Lomeli-Ramirez et al., 2014). Sustainable filaments are mainly classified into two types: derived from flora and fauna. Flora-based fibers are typically cellulose-lignin composite, meaning they contain amorphous polysaccharides, diverse sugars, and pectin, while fauna containing fibers are made up of polypeptides, that includes silken fibers and fleece (Cheung et al., 2009). A considerable edge that bio-based composites acquired is that they need less energy and during their production, emission due to CO_2 is also very less if compared to conventional composites. They are also biodegradable, renewable, insubstantial, stronger and stiffer for their weight, extremely resilient towards electricity, cost-effective, and offer great heat and sound insulation (Jawaid, & Abdul Khalil, 2011). Bio-composites might suffer from some complications, such as being highly vulnerable to humidity, lacking durability and exhibiting fragile bonding between the base material and reinforcements (Thakur et al., 2014). Moreover, to address the current ecological troubles, there is a crucial need to create completely decomposable eco-friendly hybrids.

3.1.3. Environmental Friendliness

Green polymers are derived from sustainable materials (e.g., starch, cellulose, or PLA from corn) making them environment friendly. These biopolymers also assist to minimize the carbon footprint contrast to typical plastics. Few more properties of these biopolymers and their comparison with conventional polymers are also listed briefly in Table (1)

3.1.4. Thermal Properties

Biopolymers usually have lower heat resistance in contrast to artificial plastics because they emerge from earth's resources. Consequently, their usage is restricted in extreme heat intensive settings. Examining the thermal intensity of polymers helps us acknowledge their heat stability thoroughly (Singha, A.S., & Thakur, V.K. 2009; Yoichi, H., & Yutaka, T. 2010; Yoichi, H., & Yutaka, T. 2010; Singha, A.S., & Thakur, V.K. 2009; Nita, T., & Aurica, P.C. 2011). TGA involves the study of how materials break down and their effect on heat resistance under varied circumstances. It also helps to characterize the speed of physical and chemical changes happening within the sample material. Conversely, DTA entails measuring the specific heat deviation between a specimen and a constant reference material as both are subjected to heat.

3.1.5. Rheological Characteristics

Among grooming items, there are certain polymers like Xanthan gum and cellulose that help to stabilize and thicken the overall product or as drenching constituents, thereby intensifying the consistency of the product on the skin (like hyaluronic acid, guar derivatives) (Gruber, J.V. 1999). In self-care items, they are used to improve the product's capacity to hold moisture and deliver a smooth, uniform texture. Polysaccharide extracted from tamarind seeds are derived from the endosperm of its seed and *Tamarindus indica L.*, is a salient tree in the South Asian subcontinent for practical economic use (Rao et al., 1999). Polysaccharide sourced from tamarind seeds is a kind of xyloglucan (neutral and non-toxic cellulose) by incorporating β-(1,4)-D-glucan linkage as its main component. Few of the glucose units present in this chain are joined via α-D-xylopyranose at the O-6 position, and some units of this xylose are attached to β-D-galactose (Gidley et al., 1991). Water-based dispersions of xyloglucan are thick and allow an extensive range of pH levels (Nishinari et al., 2000). As a result, they are frequently used to enhance textural appearance and thicken the consistency of food, and pharmaceutical formulations (Sahoo et al., 2011).

Table 1. Shows properties of green polymers along with their brief description and their advantage over conventional polymers

Property	Description	Advantages Over Conventional Polymers
Ecofriendly	Emanate via sustainable materials, thereby reducing ecological impact of carbon assessment comparative to conventional plastics	Decreases dependence on non-renewable energy sources and lowers the release of greenhouse gases
Biodegradability	Breaks down naturally into water, CO_2, and biomass through microorganisms under suitable conditions	Eliminates long-term environmental pollution and microplastic accumulation
Mechanical Properties	Better strength and flexibility compared to traditional plastics, but depends on the type of polymer	Tunable properties while being environmentally safe at the end of the lifecycle
Thermal Properties	Lower thermal stability, limited applicability in high-temperature environments	Suitable for applications where lower thermal stability is not a concern
Water Sensitivity	Some are hydrophilic, affecting water resistance and durability	Decomposes faster in aquatic environments, aiding waste reduction
Composability	Meets stern degradation standards for industrial composting	Promotes sustainable waste management and nutrient recycling
Chemical Properties	Resistant to certain solvents and chemicals, but may be sensitive to acidic or basic environments	Safer disposal as they do not release hazardous toxins during degradation
Biocompatibility	Suitable for healthcare domains like administration of medicines	Non-toxic and ideal for use in biomedical and pharmaceutical applications
Durability	Shorter lifespan under environmental exposure compared to synthetic plastics	Avoids persistent waste problems caused by non-degradable synthetic materials
Cost-Effectiveness	Production costs are decreasing but still higher than conventional plastics	Long-term cost benefits due to reduced waste management expenses and eco-certifications

This table not only describes the properties of green polymers but also emphasizes their benefits, making them a more sustainable choice than conventional plastics.

4. SYNTHESIS AND METHOD OF PRODUCTION

4.1. Poly (Lactic Acid)

4.1.1. Synthesis and Method of Production of Poly (Lactic Acid)

Polylactic acid (PLA) is extensively used as a substitute for those plastics that are non-degradable. As polylactic acid is composed of imperishable, non-poisonous resources, PLA is considered as one of the crucial biopolymers (Singhvi et al., 2019). Before the early 2010s, PLA mostly operate in healthcare domain (Datta,

R., & Henry, M. 2006) because it possesses characteristics that allow it to be safely utilized within the human body. Later, in the late 2010s, polylactic acid became admired in the packaging industry as a sustainable material. Its popularity is due to key features such as the ability to break down naturally, offering durability, and being simple to manufacture (Godavitarne et al., 2017).

The making of lactic acid polymer starts with the breakdown of fermented carbohydrates obtained from renewable resources such as maize, cane sugar, and several edible ingredients, as shown in Figure 3. (Singhvi et al., 2019). The fundamental step of producing polylactic acid comprises of generating lactic acid (2-hydroxypropanoic acid), a universally familiarized hydroxycarboxylic acid initially discovered in 1970 by a Swedish scientist. The production of lactic acid can be either via anaerobic respiration or synthetic procedures. It is available organically in numerous edible products and is also generated through bacterial fermentation, examples of which include sour cabbage, curd, cultured buttermilk, leaven bread, and other inoculated foods. Lactic acid is one of the prime metabolic compounds that is present in several biological entities, from oxygen-free microorganisms to human beings (Datta, R., & Henry, M. 2006).

Figure 3. Polymerization process of the lactic acid synthesized through fermentation

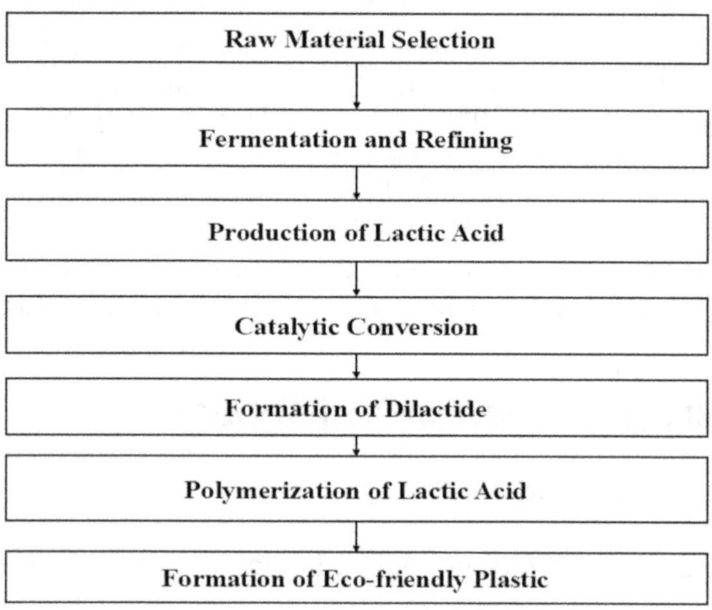

(Singhvi et al., 2019)

It has been conventionally utilized in the food sector for conservation, enhancing taste, regulating pH levels, or preventing the growth of bacteria in different types of canned foods (Schreurs, V.V, & Schaafsma, G. 2010). Industrial-grade lactic acid was additionally utilized for the manufacturing of leather. Lactic acid contains two different functional groups that can be used in multiple chemical transformations and such interactions are diverged into four varied categories i.e., reduction, condensation, esterification, and substitution of alcohol.

With the help of these chemical interactions and modifications, several products can be generated and used extensively in the industrial sector as well as commercial goods (Holten, C.H. 1971). When these chemical-based processes have been conducted there is the production of plastic-based fibers. Structural forms of lactic acid are categorized as L-lactic acid or D-lactic acid, illustrated in Figure 4. Within these different configurations, L-Lactic acid is preferred raw material for making lactic acid ester (Di-lactate) since it results in a better yield of lactate compound and a robust polymeric structure having higher crystallinity (Kister et al., 2000).

Figure 4. Structures of (A) L-lactic acid (B) D-lactic acid

The first stage of preparing polylactic acid is its polymerization (as in Figures 5–7) to make a bio-based polymer whose characteristics resemble conventional plastics. However, it is worth noting that, to synthesize polylactic acid, condensation method of polymerization proves to be cost-efficient but comes with certain drawbacks like difficulty in fabricating polymers that possess enhanced molecular mass and polymers that are free of solvents (Averous, L. 2008). Therefore, to fabricate high molecular weight polylactic acid (HM-PLA), there is a requirement of some cross-linking agents or esterification-aided chemicals. Minute quantities of plasticizers are also used as they interact with hydroxyl (-OH) and carboxyl (-COOH) end chain groups of polylactic acid.

Figure 5. Polymerization of lactic acid through direct condensation polymerization (A) low MW PLA (B) high MW PLA

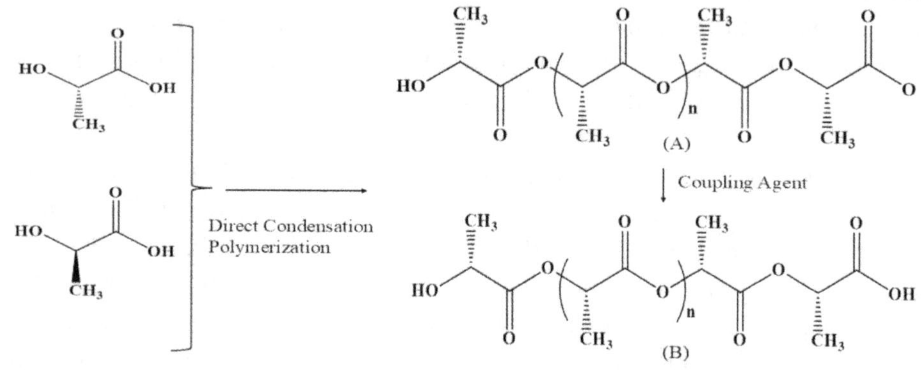

Figure 6. Polymerization of lactic acid through azeotropic dehydration condensation

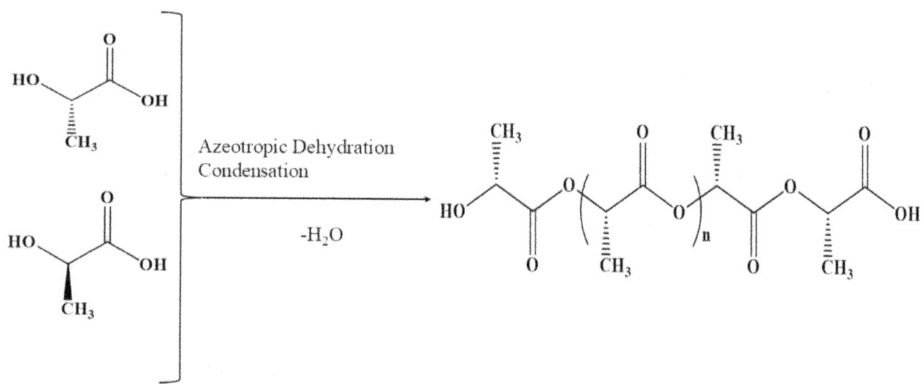

A drawback of chain lengthening agents is that after they are added, they disrupt the overall reaction by limiting its proceeding and some debris of these molecular linked agents remains within the PLA causing adulteration by introducing metal impurities. Additionally, a few of these chain lengthening agents drastically minimize the biodegradability of PLA. For instance, some of these extenders' agents are anhydrides, epoxides, and isocyanates. Moreover, introduction of isocyanates in PLA concluded to be dangerous, thereby limiting its consumption in food packaging industries.

One more vital method to adapt to making high molecular weight PLA is ring-opening polymerization. Moreover, the procedure is on the high-end side for industrial-level operations where producing scale is extremely high. The process involves breaking down of lactic acid (lactide) by cyclic dimerization method and then using a catalyst to speed up the process of polymerization. Condensation chain polymerization, polymer breakdown, and ring-opening reactions are well-known practices employed for polymer synthesis (Gupta et al., 2007). Metals such as Sn, Al, Pb, Zn, Bi, Fe, and Y are used to regulate the level of polymerization by adjusting factors like concentration, reaction duration, and heat (Singhhvi et al., 2019).

Figure 7. Polymerization of lactic acid through ring opening polymereization (A) PLA oligomer (B) lactide (C) high MW-PLA

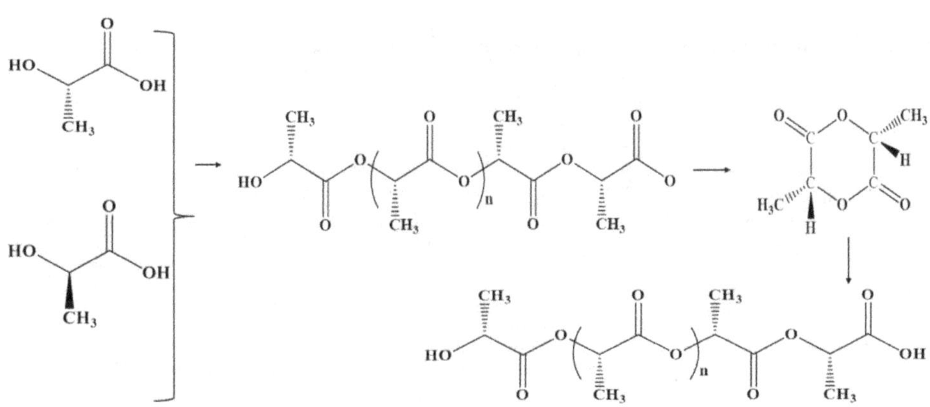

4.1.2. Synthesis and Method of Production of Polyhydroxyalkanoates (PHAs)

In recent years, polyhydroxyalkanoates (PHA) have gained a lot of recognition as being a bioplastic essentially obtained from microbes. The reason for their attentiveness is their unique physical and chemical characteristics and perspective to substitute non-biodegradable plastics. Studies on microbial reserve starts with the identification of fatty structures, which afterwards recognized as poly (3-hydroxybutyric acid). To begin with, PHA was originally thought to only contain 3-hydroxybutyrate (3HB). Further research revealed the presence of other hydroxy-alkanoates (HA) like 3-hydroxy-5-carbon carboxylate and 3-hydroxy-6-carbon carboxylate (3HHx), which inflate the possible uses of polyhydroxyalkanoates (Sudesh et al., 2000). A triad is seen in the evolution of polyhydroxyalkanoates. First, the discovery of intracellular granules of 3-hydroxybutyrate (3HB) in specific microorganisms initiated the emergence of polyhydroxyalkanoates experimentation. Then, identification of additional hydroxyalkanoate (HA) monomers expanded the potential uses of PHA (Dhania et al., 2022). The manufacturing of polyhydroxyalkanoates has been seen in several microorganisms, along with photoautotrophic species and a few archaebacterial strains (Muneer et al., 2022). This phase is concerned with identifying varied HA units and examining their biosynthesis processes. During the preparation of PHA, gene analysis and cloning reveals deeper insights into the molecular patterns. Recent studies try to explore three-dimensional and multi-subunit structures of PHA to know more about their catalytic activities and substrate preferences (Kaniuk, L., & Stachewicz, U. 2021; Stubbe, J., & Tian, J. 2003). The simplest synthesis process of PHA is shown in Figure 8.

Figure 8. Route for the biosynthesis of polyhydroxyalkanoates (PHAs)

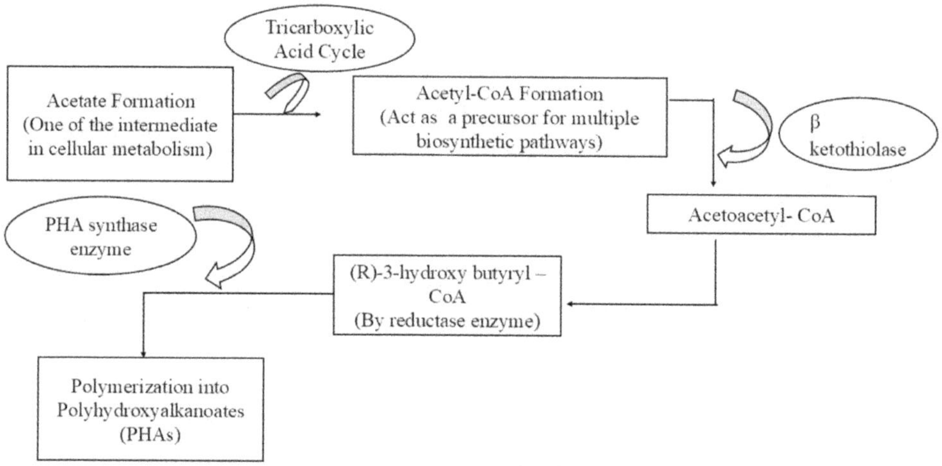

The extraction and processing of polyhydroxyalkanoates (PHA) emphasizes the requirement for alternative carbon resources to minimize reliance on food-based glucose and seed oils. Research is underway on synthesizing PHA within a biorefinery, primarily associated with bioethanol generation from biomass, lubricants, fats, cellulose, etc. (Winnacker, M. 2019). Advanced industrial biorefineries are anticipated to treat lignocellulosic biomass, that include farming and wood-derived byproducts, to generate sugars that can be transformed into PHAs. The challenges that come in recovering PHA are closely interrelated to microbial cell mass that requires few more steps for extraction and refinement. Although there are multiple approaches to recover PHA that includes techniques like solvent-based separation, enzyme-assisted breakdown, mechanical shearing, and supercritical fluid treatment (Madkour et al., 2013; Kannah et al., 2022; Kunasundari, B., & Sudesh, K. 2011). Focus has been shifted towards cost-effective and sustainable materials like unrefined oil, for creating biodegradable plastics. Chemical and catalytic techniques also attract attention to synthesize polyhydroxyalkanoates (Kumar et al., 2009; Jiang et al., 2016).

Another technique for producing PHA is by incorporating carbohydrates (Yalpani et al., 1991), such as simple sugars, short-chain sugars, and long-chain sugars. Complex sugars like cellulose, starch, and hemicellulose can be broken down into simple sugars or short chain sugars, which can undergo fermentation to generate PHAs. Bacteria like *Azotobacter vinelandii* and *Alcaligenes latus* can make PHAs from glucose and galactose found in sugar-rich sources or whey byproducts. Genetically modified *Escherichia coli* strains Have been employed for PHA synthesis from galactose.

4.1.3. Synthesize and Method of Production of Polycaprolactone (PCL)

Initially during the 1930s, a polymer named polycaprolactone (PCL) was made and commercialized thereafter due to its organically degradable nature (Finch, C. 1985). Several preparation techniques were used to synthesize PCL, one of them being ring cleavage polymerization or ROP. Polycaprolactone possesses characteristics like water-repelling, polymer with crystalline phase with effective dissolvability and modest liquefaction range. The innovation acquired interest in medical sector (Nair, L.S., & Laurencin, C.T. 2007). amid the growing use of bioresorbable polymers in the 1970s and 1980s. Its key benefits include adjustable breakdown rates, Superior mechanical features, and the ability to control drug release (Woodruff, M.A., & Hutmacher, D.W. 2010). In recent years, polycaprolactone (PCL) has acquired a resurgence in tissue regeneration technology elasticity and excellent flow which make it ideal for building scaffolds (Luciani et al., 2008). Moreover, cost-effectiveness along with regulatory clearance from the FDA and CE marking for certain pharmaceutical dispensing tools give it an edge in the market.

The production of polycaprolactone (PCL) begins with the chain formation of the precursor unit, as shown in Figure 9. During the process of transition, there is a reorganization of cyclohexyl alcohol into adipic acid (hexanedioic acid) via an oxidation procedure. The intermediate step in making of PCL involves 6-hydroxyhexanoic acid and ε-caprolactone (ε-CL). On an industrial scale, ε-CL is made by oxidizing cyclohexyl alcohol with peracetic acid (peroxyacetic acid) (Pyo et al., 2020; Labet & Thielemanas, 2009). However, there are two fundamental techniques to fabricate PCL. One involves polymerizing ε-CL through Ring Opening Polymerization (ROP) method while other used polycondensation of 6-hydroxycaproic acid (6-hydroxyhexanoic acid). There is another method where 6-hydroxyhexanoic acid was polymerized by creating a vacuum to make oligomers of PLC. Without any catalyst to speed up the process, reaction was concluded within 6 hours by gently elevating the temperature from 80 to 150 °C (Braud et al., 1998). Moreover, it is worth mentioning that polycondensation is often appraised than cyclopolymerization. The cyclopolymerization can be conducted using four vital catalysts that includes cationic, anionic, and monomer-initiated along with coordinated-coupled cyclic monomer polymerization. In anionic cyclopolymerization, there is establishment of negatively charged counterparts that interacts with the monomer's carbonyl carbon. This cleaves the ester linkage, triggering the emergence of an alcoholate (Stridsberg et al., 2002).

Figure 9. Synthesis pathway of Polycaprolactone (A) ε-caprolactone (B) polymerization technique (C) polycaprolactone

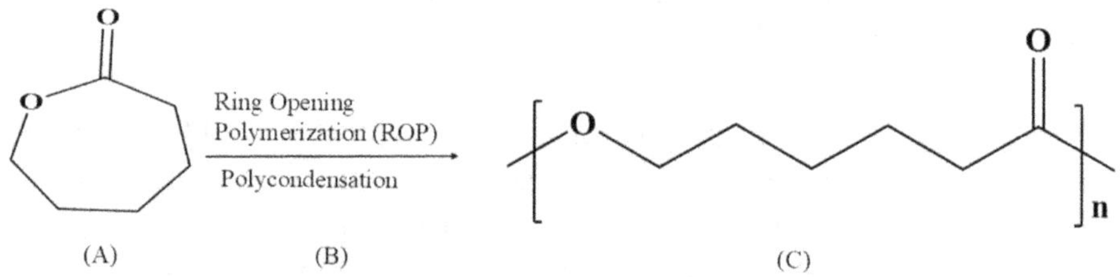

Moreover, one constraint of this process is the intramolecular reaction problem, where intra-molecular esterification process happens, leading to the development of petite-molecular polymers or cyclic compounds. Alternatively, cationic ROP depends on the formation of positive ions, which is subsequently targeted through oxygen functional group of the monomer's carbonyl group by a substation nucleophilic bimolecular reaction. The cyclopolymerization process is triggered by monomeric species, which are initially stimulated through an activator before interacting with the terminal extremity of the polymer strand. Coordination-based ring-opening polymerization, the most employed method, involves the monomer associating with the catalyst, followed by the inclusion of a single polymer in the metal oxide framework using an initiator.

In Ring-Opening Polymerization (ROP) of ε-caprolactone (ε-CL), when a starter and an activator have incorporated, there is an emergence of multiple undesirable byproducts (Riva et al., 2007; Detrembleur et al., 2000), namely cross-molecular alcoholysis depicted in Figure 10. An ester bond rearrangement within a single molecule, indicated (mechanism of ring polymerization) in Figure 11. These reactions typically materialize in the later stages of polymerization and under thermal settings. Therefore, the size of the polymers spread at a faster rate and disrupts the control of the polymerization mechanisms.

Figure 10. Possible ring opening of polymer caprolactam in presence of LDA

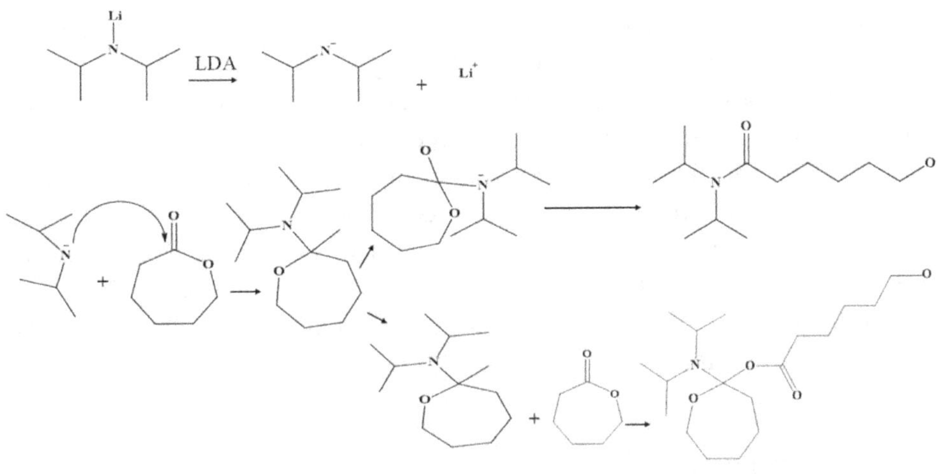

Figure 11. Mechanism of enzymatic ring opening polymerization of e-CL

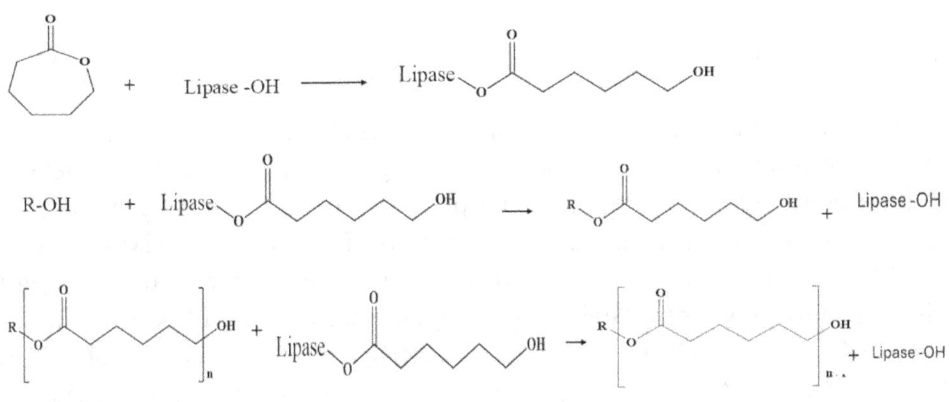

In ring-opening polymerization, catalysts that are alkali-metal derived shows certain activity. Moreover, because of some anionic interactions and ionic characteristics in ROP, adjusting the reaction temperature to control the rate of polymerization is quite challenging. Furthermore, alkali compounds often create clumps, which decrease their solubility (Platel et al., 2008). Rather than relying on metal-based catalysts, enzymes can act as a viable option. Researchers have described a technique involving enzyme-driven ring-cleavage polymerization of epsilon-caprolactone mediated by cholesteryl ester hydrolase (Nobes et al., 1996; MacDonald et al., 1995; Kobayashi, S. 1999; Kobayashi et al., 2001).

In this process, the lipase first reacts together with the single polymer to generate a complex with epsilon-caprolactone. Then, ethanol interacts with this compound to pursue the reaction process. Studies on the enzyme-driven ring-cleavage polymerization of epsilon-caprolactone have shown that polymerization, polymer disintegration, and enzyme deactivation happen at the same time. Polycaprolactone can be mixed with polymers like polylactic acid or cellulose or strengthened with fibers to improve its mechanical robustness. For instance, incorporating starch makes it stiffer by expanding its storage modulus and improves damping characteristics, which is suitable for healthcare applications (Asadpour et al., 2018; Wan et al., 2009). The ester bonds in PCL's structure, which can be broken down by hydrolysis, allow it to decompose over time.

5. APPLICATIONS

Green polymers and biodegradable plastics have found applications across diverse sectors, addressing the threats to nature raised by standard plastics while meeting the needs of modern industries. Below are key application areas

5.1. Food Packaging

Biobased plastics like polyhydroxyalkanoates (PHA) and polylactic acid (PLA) are extensively utilized for food packaging, packaging films, and single-use utensils, ensuring food protection while minimizing plastic pollution. Combining two polymer types can elevate the characteristics of the material. For instance, a mixture of PLA and TPS in a relevant proportion improves the attributes of the matter, making it suitable for producing compostable trays. Similar blends, such as PLA/PHB (Burgos et al., 2017), PLA/MCC/Ag (Fortunati et al., 2010), and PLA/PU (Ali et al., 2014), show significant potential in the packaging industry. Combinations of PHAs with dairy byproducts and gelatin (PHAs/CW/Gelatin) (Arrieta et al., 2017), and PHAs with keratin nanoparticles (Fabra et al., 2016), form thin films, making them a Viable substitute for packaging of processed snacks and supplementary plastic-enclosed food items. The combination of ZnO/PHB (Diez-Pascual & Diez-Vicente, 2014), PLA/PHB (Arrieta et al., 2014), PHBV/PLA (Jost & Kopitzky, 2015), and PBAT/PHB (Lin et al., 2018) enhanced the longevity of dehydrated items by improving the protective qualities of the mixtures. The PCL/LDH nano-based composite amalgamation demonstrated a significant improvement in the ability to withstand impact was enhanced, and packaging formulations were created with different heat, strength, and defense characteristics by modifying the LDH content within the matrix. Therefore, these substances are ideal for various ranges of packaging requirements. Combinations of PCL/chitosan (Joseph et al., 2011), cornstarch/chitosan (Silva et al., 2015), corn starch/citric acid/carboxymethylcellulose (C.S/CA/CMC) (Ghanbarzadeh et al., 2011), cassava starch/natural extracts (Medina et al., 2017), nanocrystalline cellulose/alginate (Huq et al., 2012), and PVA/CMC/CNC (El Achaby et al., 2017) were recognized as outstanding substances for producing films in the food packaging sector. For instance, Nature Works LLC produces Ingeo™ PLA, used in food containers and films, Tetra Pak incorporates bio-based polymers (sugarcane-derived LDPE) in its packaging, Danimer Scientific manufactures Nodax™ PHA, used in biodegradable straws and packaging.

5.1.1 Agriculture

Mulching with carbon-based or non-carbon-based substances aims to cover the soil, creating a barrier that minimizes water evaporation, controls weed, keeps the humus intact, and safeguards crops from contamination. Organic mulches assist in preserving soil organic content and structure (Tindall et al., 1991) while providing nourishment and protection to earthworms and other advantageous soil microorganisms (Doran, J.W. 1980). Compostable mulch films made from eco-friendly polymers enhance soil quality and minimize microplastic pollution by decomposing after use. For instance, Bio Bag® produces compostable mulch films, BASF's ecovio® is a biodegradable polymer used in agricultural films, Mater-Bi® by Novamont is a starch-based mulch film used globally.

5.1.2. Medicine and Healthcare

The reliance on plastic items in healthcare devices and materials creates complications for the environment, especially when it comes to garbage disposal, as many of these materials are single-use, non-recoverable, and non-degradable. Plastic waste from healthcare consists of products such as needles, hoses, intravenous bags, wrappers, and other items. Healthcare institutions give rise to a significant amount of plastic waste. Safely managing hazardous medical waste seems challenging. Numerous medical items are designed for one-time use or require thorough sterilization and complicating recycling efforts. Another concern is the release of chemicals, as certain plastics hold toxic substances that may seep into the surroundings or engage with the device's components, possibly leading to health hazards (Hahladakis et al, 2018). Creating eco-friendly substitutes for oil-derived plastics for healthcare devices equipment isn't an immediate priority, but it remains a significant effort. So far, research has focused on eco-friendly polymers (Ranakoti et al., 2022), composite materials (Thomas et al., 2023), surface modifications, microbial cellulose, 3D printing using eco-friendly materials (Hassan et al., 2024), clinical studies, collaborations with the industry, compliance challenges, and sustainable packaging solutions for healthcare products. Biodegradable options like starch-sourced acid, polylactic-co-glycolic acid (PLGA), and polycaprolactone (PCL), have shown promising results concerning biological acceptability, security, and potency. For instance, Evonik produces RESOMER® PLGA for bioresorbable implants, Corbion manufactures PURASORB® for sutures and drug delivery systems, Poly-Med Inc. develops bioresorbable medical textiles.

5.1.3. Environment

Sustainable polymers and degradable plastics are crucial for environmental conservation as they help to minimize plastic pollution and conserve resources. These substances decompose naturally into non-toxic components, reducing the buildup of waste in garbage sites and seas. Furthermore, eco-conscious polymers made from green resources, such as vegetable starch or agricultural residue, lessens the reliance on petroleum-based fuels and minimizes carbon emissions. By fostering a regenerative economy and reducing ecological harm, these substances offer an enduring solution to the worldwide plastic issue (Yin, G.Z., & Yang, X.M. 2020). For instance, TIPA® makes fully compostable flexible packaging, Avani Eco produces cassava-based biodegradable bags, New light Technologies creates Air Carbon®, a PHA-based alternative to conventional plastics.

5.1.4. Automotive

In the Automotive sector, extensively used polymers are polyamide (PA), polyurethane (PU), polypropylene (PP), and acrylonitrile-butadiene-styrene (ABS). The substitution of oil-derived plastics with bio-based substances in the automobile sector has been achieved, and intellectual property rights were secured, particularly involving plant-based fibers. Bio-based polymers are utilized in components such as dashboard panels, door trims, lighting fixtures, and structural grids within the vehicle production industry (Kelly, A. 2012). For instance, Ford uses soy-based polyurethane foam for seat cushions and headrests and Toyota incorporates bioplastics like bio-based PET in vehicle interiors.

5.1.5. Cosmetics and Personal Care

Eco-friendly polymers and compostable plastics are increasingly adopted in the beauty sector to promote sustainability while taking care of the product integrity. These materials act as environmentally conscious by providing options like sustainable containers, replacing conventional plastic bottles for lotions, shampoos, and cosmetics with biopolymers. Polymers rank as the second most prevalent component in beauty product formulations. Within the beauty sector, polymers serve as agents for stabilization, destabilization, and modification. They act as viscosity enhancers, emulsifying agents, and preservatives. In the cosmetic industry, polymers function as stabilizers, destabilizers, and modifiers (Agustiany et al., 2022). For instance, L'Oréal collaborates with Carbios to develop enzymatically recycled PET for packaging and Albéa Group offers eco-friendly paper-based and bio-resin tubes for skincare products.

5.1.6. Textile and Fashion

Bio-based polymers also benefit the apparel and textile sector by decreasing energy usage in unprocessed material, minimizing the cost and resources involved in sourcing, manufacturing, and disposal (Younes, B. 2016). In fact, the textile sector is among the most polluting industries globally, second only to petroleum; the most significant ecological harm arises from textile comes from fabrication, refinement, and staining (de Silva et al., 2021). As such, this industry needs green substitutes more urgently than others, with biopolymers offering a sustainable solution. Biodegradable fabrics, that must incorporate a minimum of 20% sustainable carbon, consist of organic materials and a mix of plant-based, synthetic, or reconstituted fibers. Naturally sourced biopolymers are derived from carbohydrates (such as fiber, phenolics, etc.), in addition to plant or animal polypeptides and fats (Patti, A., & Acierno, D. 2022). Indeed, organic fibers are also obtained from botanical origins such as wool, linen, cotton etc. Among them, cotton is extensively utilized in garment manufacturing because of its visual attractiveness and usability (Patti, A., & Acierno, D. 2022). In contemporary times, the push for eco-friendly manufacturing has redirected the textile sector's focus not only to materials like chemical-free cotton (cultivated without pest control chemicals and growth enhancers) and furthermore to the development of eco-friendly engineered fibers, referred to as "sustainable nonwoven fabrics." In this context, biopolymers are predominantly utilized in the production of fibers within the contemporary biodegradable fabric sector comprised of butyric acid (PHB), polycaprolactone (PCL), valeric acid (PHV), polylactic acid (PLA), and others (Fico et al., 2022). Some other applications are also discussed below in Figure 12. For instance, Adidas & Parley for the Oceans create sportswear using ocean plastic-based polyester.

Figure 12. Applications of green polymers and biodegradable plastics

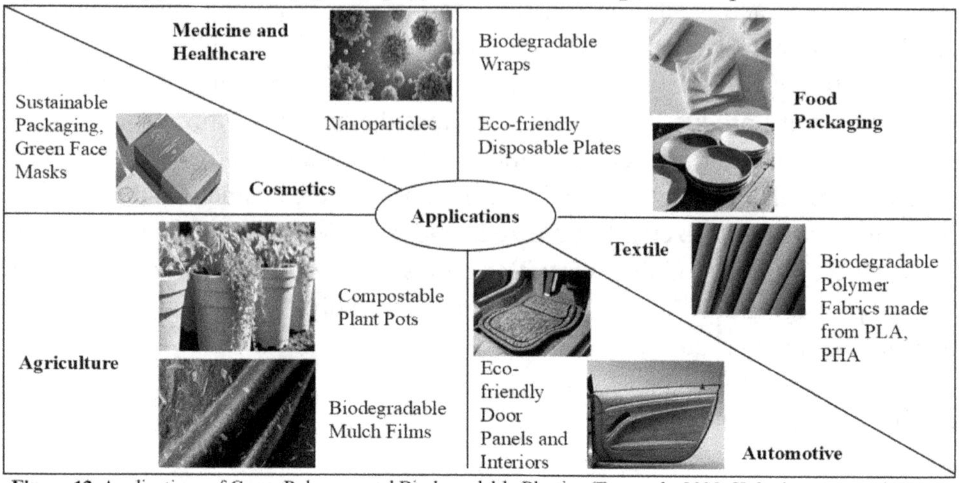

Figure 12. Applications of Green Polymers and Biodegradable Plastics (Fan et al., 2022; Kelaniyagama et al., 2024; Agustiany et al., 2022; Medina et al., 2017; Kelly, A. 2012).

(Fan et al., 2022; Kelaniyagama et al., 2024; Agustiany et al., 2022; Medina et al., 2017; Kelly, A. 2012)

6. CHALLENGES

Despite their potential to revolutionize material science and minimize environmental impact, green polymers and biodegradable plastics suffer numerous challenges that hinder their widespread adoption. These challenges arise from economic, technical, environmental, and societal factors, which must be addressed to unlock their full potential. The following table summarizes the key challenges associated with green polymers and biodegradable plastics, highlighting areas that require further innovation, investment, and awareness.

Table 2. Summarizes the key challenges associated with green polymers and biodegradable plastics.

Aspects	**Challenges**
High Production Cost	Manufacturing green polymers often requires innovative technologies and renewable feedstocks, making them more expensive compared to conventional plastics. For instance, PGA shows rapid degradation, brittleness, and expensive (Kim et al., 2005)
Limited Mechanical Properties	Green polymers may lack strength, durability, or flexibility, restricting their use in demanding applications like heavy-duty packaging or construction. For instance, Starch-based bioplastics may experience deteriorated mechanical properties and decreased heat resistance (Kim et al., 2005; Park, D.H., & Kim, B.S. 2011)
Dependence on Specific Conditions	Biodegradable plastics typically need commercial composting systems with controlled thermal settings, moisture, and microbial conditions for effective breakdown. For example, PBS shows good biodegradability but due to its low-mass compound and lower melting solidity significantly affects its functionality (Fan et al., 2022)
Recycling Challenges	Mixing biodegradable plastics with traditional plastics in recycling streams can contaminate the process, reducing the quality of recycled products
Lack of Infrastructure	Many regions lack the necessary composting or recycling facilities to process biodegradable plastics, limiting their environmental benefits
Consumer Awareness	Insufficient knowledge among consumers about the correct disposal methods for biodegradable plastics can result in improper handling and waste accumulation
Short Longevity	Compostable plastics may have a minuscule life span compared to regular plastics, limiting their use in long-term applications

7.1. Circular Economy

A circular economy is an approach of managing resources that aims to limit waste and use materials more efficiently. In contrast to the conventional linear model, which operates on an "extract, produce, and discard" system, a circular economy emphasizes environmental responsibility by ensuring products and materials remain in circulation for an extended period. It encourages designing products and packaging that can be reused, recycled, or composted at the end of their functional period, reducing their detrimental effects on the environment. Based on diverse investigations (UNEP, 2020; Abu et al., 2024; Sanchez et al., 2024), this change calls for redesigning products, choosing better materials, and improving waste handling methods (UNEP, 2020). There is an urgent need to set up frameworks like waste and pollution management that encourage suppliers to oversee their possessions throughout their operational period, covering activities like gathering, reprocessing, and getting rid of them. Circular economic efforts focus on using resources efficiently and cutting down on waste by using 3R strategy i.e., reducing, reusing, and recycling.

7.1.1. Reduction and Elimination of Single-Use Plastics

Cutting down plastic one-time is pivotal for bringing down plastic litter and pollution. This can be achieved by swapping items like single-use bags, plastic straws, and utensils with washable or eco-friendly choices. Encouraging individuals to change their behaviors, rigid enforcements, and supporting eco-friendly materials are successful methods to tackle the escalating crisis of disposable plastics (Jambeck et al., 2015). Companies and sectors are moving towards sustainable and rechargeable packaging options, cutting down on their dependence on one-time use plastics.

7.1.2. Collaboration and Partnerships

The studies examined highlighted that cooperation among administrations, businesses, non-governmental organizations (NGOs), and educational organizations fuels the progress in fostering broad transformations and supporting the progress of eco-friendly options (Rizos et al., 2017). Collaborations between the public and private sectors promote the transfer of expertise, pooling of assets, and unified efforts towards shared environmental objectives.

7.1.3. End-user Awareness and Involvement

Concerning the ecological impacts of petrochemical waste and the advantages of eco-friendly alternatives enables people to make well-informed decisions. Encouraging environmentally conscious living and adopting practices, like cutting down on waste, selecting green products, and endorsing sustainable and ethical brands, promotes widespread adoption of eco-friendly choices (Kumar et al., 2021).

8. MARKET AND DEMANDS

8.1. Current Scenario

Bio-based plastics captured 0.5% of the market in 2022 (Plastics Europe, 2023). Recently, polylactic acid (PLA) has become one of the top compostable plastics due to its robust physical characteristics, variety of raw material origins, and large-scale production capabilities (Li et al., 2024). It is the go-to material for 3D printing in consumer markets. PLA production grew from 200,000 tons in 2015 to 300,000 tons in 2019, making it the standard compostable plastic consumed in the market (Ghomi et al., 2021). Its manufacturing rate and market worth is expected to rise, potentially attaining a U.S. open market valuation of $3.2 billion by 2032, up from $1 billion in

2022 (Statista Research Department, 2024). As mentioned earlier, the high cost of large-scale PLA production is a significant hurdle to widespread global use. PLA also fails to stretch more than 10% before breaking, highlighting its brittleness and limited effectiveness in certain applications (Taib et al., 2022), particularly those requiring high stress resistance. An additional obstacle challenge in using PLA more extensively is its slow degradation, taking 3 to 5 years (Muthusamy, M.S., & Pramasivam, S., 2019). PLA is unsuitable for healthcare or medical applications because its slow breakdown within the body might result in negative repercussions (Mehmood et al., 2023). Another promising substitute for synthetic plastics, particularly in food packaging, is polyhydroxyalkanoate (PHA) (Vicente et al., 2023). PHA is a type of plastic made by bacteria and assembled from various derivatives such as sugar, botanical oils and corn. While PHA has lower heat processing potentiality than PLA, it meets the demand in biomedicine created by PLA's delayed breakdown speeds (Taib et al., 2022; Verma et al., 2024). Poly(3-hydroxybutyrate) stands out as the foremost form of PHA (Samir et al., 2022). It reveals analogous composition to poly (1-methylethylene), giving it similar physical characteristics such as resistance to water, durability, and melting temperature (Price et al., 2022).

Regardless of the surplus feedstock, wide-spread implementation is impeded due to elevated production expenses (Lepe et al., 2024). More than half of the total expenses associated with PHB manufacturing stem from heterotrophic fermentation employing an organic carbon substrate (Price et al., 2022). Utilizing climatic CO_2 for cyanobacterial PHB synthesis instead of the organic carbon source needed in heterotrophic fermentation may substantially lower PHB expenses during production (Lepe et al., 2024). Petroleum-based polymers like PP and HDPE can easily be replaced with bio-based poly (butylene succinate). Even though raw materials are plentiful, extensive use is limited due to the high production costs (Lepe et al., 2024). More than half of the manufacturing costs for PHB come from the use of carbon-based organic materials in heterotrophic microbial fermentation (Price et al., 2022). The production of cyanobacterial PHB with atmospheric carbon dioxide, in preference to the biological carbon required for non-photosynthetic fermentation, could drastically minimize the costs of making PHB. Bio-based poly (butylene succinate) has identified as a potential alternative to crude oil-based plastics like high-density polyethylene, and polypropylene (Lepe et al., 2024). It can be made from maize, beetroot, cane sugar, and tubers, all of them requires minimal amount of water and land (Platnieks et al., 2021). PBS being a polyester composed of dimethyl succinate, and few more building blocks (Platnieks et al., 2021; Rajgond et al., 2023). As the demand for biodegradable plastics like PBS has risen, export production of succinic acid has grown tremendously. The succinic acid economy is projected to hit $1.8 billion globally by 2025, growing at a yearly percentage of 27.4% (Nghiem et al., 2017).

8.1.1 Future Potential

Recently, in the latter half 2000s and beginning of 2010s, new biodegradable plastics have emerged in the market. Poly (ethylene 2,5-furandicarboxylate), made from 2,5-furandicarboxylic acid and ethylene glycol, is arousing curiosity due to its resemblance with PET (Loos et al., 2020). It not only shares similarities with PET, but it also has superior barrier qualities, including better resistance to gas and moisture resistance (Burgees et al., 2014; Burgess et al., 2014; Burgess et al., 2014; Sun et al., 2019). Initially patented by Celanese Corporation in 1946, PEF was least admired compared to effortlessly mass-produced bio-based plastics like PLA and PHA (Fei et al., 2022). Since 2009, a significant interest is sparked in producing PEF and attempts have been made to make it more affordable and scalable (Gandini et al., 2008; Gomes et al., 2011; Jiang et al., 2012). Recently, Avantium is working on expanding PEF's commercial use through their YXY Technology, aiming to replace PET-based products such as bottles, manufacturing fibers, and garment materials with PEF (Avantium, 2022; de Jong et al., 2012; de Jong et al., 2022). While this challenge is substantial, the beneficial properties of PEF could lead to its replacement of not only PET but also substances such as aluminum and glassware's across copious sectors, ranging from packaging to fabric production, could revolutionize these industries in the years to come.

9. RECOMMENDATIONS FOR FUTURE RESEARCH

- *Enhanced Mechanical Properties:* Focus on improving the robustness, pliability, and thermal stability of biodegradable plastics to expand their usability in intensive applications.
- *Energy-Efficient Processes:* Investigate low-energy manufacturing techniques to bring down carbon footprints.
- *Green Catalysis:* Develop non-toxic catalysts for the synthesis of biopolymers
- *Closed-Loop Systems:* Design systems to ensure bioplastics are recycled or composted effectively at the end of their lifecycle
- *Cost Reduction Strategies:* Develop economic models to minimize the production cost of green polymers, such as scaling up production or leveraging subsidies.
- *Policy Recommendations:* Study the impact of regulations and incentives on the adoption of biodegradable plastics in various industries.
- *Smart Bioplastics:* Investigate polymers with responsive properties (e.g., temperature, pH) for specialized applications.

- *Hybrid Materials:* Research composites of bioplastics with other sustainable materials for refined characteristics
- *Life Cycle Analysis (LCA):* Conduct LCAs to compare the overall environmental impact of green polymers versus traditional plastics.
- *Global Collaboration*: Promote interdisciplinary research and global partnerships to accelerate innovation in green polymers.

10. CONCLUSION

The transition to green polymers and biodegradable plastics represents a pivotal step toward achieving a sustainable future. These materials offer an eco-friendly substitute to conventional plastics by leveraging renewable resources, exhibiting enhanced biodegradability, and aligning with circular economy principles. The inspection of diverse feedstocks, innovative synthesis methods, and tailored material properties has propelled the development of green polymers into a viable solution for addressing global plastic pollution and environmental degradation. Despite their promise, challenges such as steep labor costs, narrow growth potential, and application-specific limitations persist. Addressing these issues requires a multidisciplinary approach encompassing material science, engineering, policymaking, and consumer engagement.

The unification of green polymers into multiple domains—including packaging, agriculture, medicine, cosmetics, and textiles—demonstrates their versatility and potential to transform industries. However, achieving long-term success will depend on fostering innovation, scaling up production, and ensuring compliance with environmental regulations. Public awareness, industry collaboration, and governmental incentives play vital roles in driving the adoption of biodegradable plastics and establishing a robust market for these sustainable replacements.

As global demand for sustainable materials is getting bigger, the future of green polymers lies in developing advanced, high-performance bioplastics that meet the stringent requirements of diverse applications. By prioritizing research, enhancing policy frameworks, and embracing sustainable practices, society can make significant strides toward curbing ecological impact and fostering a resilient, waste-reducing economy. Green polymers and biodegradable plastics are not just an alternative—they are a necessity for a greener, more sustainable world.

Funding: No external funding.

Conflicts of Interest: The authors declare no conflict of interest

REFERENCES

Agustiany, E. A., Ridho, M. R., Rahmi, M. D. N., Madyaratri, E. W., Falah, F., Lubis, M. A. R., & Solihat, N. N. (2022). Recent developments in lignin modification and its application in lignin-based green composites: A review. *Polymer Composites*, *43*(8), 4848–4865. DOI: 10.1002/pc.26824

Ali, F. B., Kang, D. J., Kim, M. P., Cho, C. H., & Kim, B. J. (2014). Synthesis of biodegradable and flexible polylactic acid-based thermoplastic polyurethane with high gas barrier properties. *Polymer International*, *63*(9), 1620–1626. DOI: 10.1002/pi.4662

Arrieta, M. P., López, M. D. M. C., Rayón, E., Barral-Losada, L. F., López-Vilariño, J. M., López, J., & González-Rodríguez, M. V. (2014). Plasticized poly (lactic acid)–poly(hydroxybutyrate) blends incorporated with catechin intended for active food-packaging applications. *Journal of Agricultural and Food Chemistry*, *62*(41), 10170–10180. DOI: 10.1021/jf5029812 PMID: 25255375

Arrieta, M. P., Peponi, L., López, D., López, J., & Kenny, J. M. (2017). Food packaging. In *Elsevier* (pp. 391–424). Academic Press.,

Asadpour, S., Yeganeh, H., Ai, J., Kargozar, S., Rashtbar, M., Seifalian, A., & Ghanbari, H. (2018). Polyurethane-polycaprolactone blend patches: Scaffold characterization and cardiomyoblast adhesion, proliferation, and function. *ACS Biomaterials Science & Engineering*, *4*(11), 4299–4310. DOI: 10.1021/acsbiomaterials.8b00848 PMID: 33418826

Avantium. (2022). FDCA flagship plant—YXY technology. Retrieved July 19, 2024, from https://avantium.com/technologies/yxy/

Bordes, P., Pollet, E., & Avérous, L. (2009). Nano-biocomposites: Biodegradable polyester/nanoclay systems.[[Doi:]. *Progress in Polymer Science*, *34*(2), 125–155. DOI: 10.1016/j.progpolymsci.2008.10.002

Braud, C., Devarieux, R., Atlan, A., Ducos, C., & Vert, M. (1998). Capillary zone electrophoresis in normal or reverse polarity separation modes for the analysis of hydroxy acid oligomers in neutral phosphate buffer. *Journal of Chromatography. B, Biomedical Sciences and Applications*, *706*(1), 73–82. DOI: 10.1016/S0378-4347(97)00468-4 PMID: 9544809

Brizga, J., Hubacek, K., & Feng, K. (2020). The Unintended Side Effects of Bioplastics: Carbon, Land, and Water Footprints. *One Earth*, *3*(1), 45–53. DOI: 10.1016/j.oneear.2020.06.016

Burgess, S. K., Karvan, O., Johnson, J. R., Kriegel, R. M., & Koros, W. J. (2014). Oxygen sorption and transport in amorphous poly (ethylene furanoate). *Polymer*, *55*(18), 4748–4756. DOI: 10.1016/j.polymer.2014.07.041

Burgess, S. K., Mikkilineni, D. S., Daniel, B. Y., Kim, D. J., Mubarak, C. R., Kriegel, R. M., & Koros, W. J. (2014). Water sorption in poly (ethylene furanoate) compared to poly (ethylene terephthalate). Part 2: Kinetic sorption. *Polymer*, *55*(26), 6870–6882. DOI: 10.1016/j.polymer.2014.10.065

Burgess, S. K., Mikkilineni, D. S., Daniel, B. Y., Kim, D. J., Mubarak, C. R., Kriegel, R. M., & Koros, W. J. (2014). Water sorption in poly (ethylene furanoate) compared to poly (ethylene terephthalate). Part 1: Equilibrium sorption. *Polymer*, *55*(26), 6861–6869. DOI: 10.1016/j.polymer.2014.10.047

Burgos, N., Armentano, I., Fortunati, E., Dominici, F., Luzi, F., Fiori, S., Cristofaro, F., Visai, L., Jiménez, A., & Kenny, J. M. (2017). Functional properties of plasticized bio-based poly (lactic acid)/poly(hydroxybutyrate) films for active food packaging. *Food and Bioprocess Technology*, *10*(4), 770–780. DOI: 10.1007/s11947-016-1846-3

Cheung, H. Y., Ho, M. P., Lau, K. T., Cardona, F., & Hui, D. (2009). Natural fibre-reinforced composites for bioengineering and environmental engineering applications. *Composites. Part B, Engineering*, *40*(7), 655–663. DOI: 10.1016/j.compositesb.2009.04.014

Datta, R., & Henry, M. (2006). Lactic Acid: Recent Advances in Products, Processes and Technologies—a Review. *Journal of Chemical Technology and Biotechnology*, *81*(8), 1119–1129. DOI: 10.1002/jctb.1486

de Jong, E., Dam, M. A., Sipos, L., & Gruter, G.-J. M. (2012). Furandicarboxylic acid (FDCA), A Versatile Building Block for a Very Interesting Class of Polyesters. In Smith, P. B., & Gross, R. A. (Eds.), *Biobased monomers, polymers, and materials* (Vol. 1105, pp. 1–13). American Chemical Society., DOI: 10.1021/bk-2012-1105.ch001

de Jong, E., Visser, H. R. A., Dias, A. S., Harvey, C., & Gruter, G.-J. M. (2022). The Road to Bring FDCA and PEF to the Market. *Polymers*, *14*(5), 943. DOI: 10.3390/polym14050943 PMID: 35267764

de Silva, C. J. G.Jr, de Medeiros, A. D. M., de Amorim, J. D. P., do Nascimento, H. A., Converti, A., Costa, A. F. S., & Sarubbo, L. A. (2021). Bacterial Cellulose Biotextiles for the Future of Sustainable Fashion: A Review. *Environmental Chemistry Letters*, *19*(4), 2967–2980. DOI: 10.1007/s10311-021-01214-x

Detrembleur, C., Mazza, M., Lou, X., Halleux, O., Lecomte, P., Mecerreyes, D., Hedrick, J. L., & Jérôme, R. (2000). New functional aliphatic polyesters by chemical modification of copolymers of ε-caprolactone with γ-(2-bromo-2-methylpropionate)-ε-caprolactone, γ-bromo-ε-caprolactone, and a mixture of β- and γ-ene-ε-caprolactone. *Macromolecules, 33*(21), 7751–7760. DOI: 10.1021/ma000488o

Dhania, S., Bernela, M., Rani, R., Parsad, M., Grewal, S., Kumari, S., & Thakur, R. (2022). Scaffolds the backbone of tissue engineering: Advancements in use of polyhydroxyalkanoates (PHA). *International Journal of Biological Macromolecules, 208*, 243–259. DOI: 10.1016/j.ijbiomac.2022.03.030 PMID: 35278518

Díez-Pascual, A., & Díez-Vicente, A. (2014). Poly(3-hydroxybutyrate)/ZnO bionanocomposites with improved mechanical, barrier, and antibacterial properties. *International Journal of Molecular Sciences, 15*(6), 10950–10973. DOI: 10.3390/ijms150610950 PMID: 24941255

Doran, J. W. (1980). Microbial changes associated with residue management with reduced tillage. *Soil Science Society of America Journal, 44*(3), 518–524. DOI: 10.2136/sssaj1980.03615995004400030016x

El Achaby, M., El Miri, N., Aboulkas, A., Zahouily, M., Bilal, E., Barakat, A., & Solhy, A. (2017). Processing and properties of eco-friendly bio-nanocomposite films filled with cellulose nanocrystals from sugarcane bagasse. *International Journal of Biological Macromolecules, 96*, 340–352. DOI: 10.1016/j.ijbiomac.2016.12.040 PMID: 27988293

Fabra, M. J., Pardo, P., Martínez-Sanz, M., Lopez-Rubio, A., & Lagarón, J. M. (2016). Combining polyhydroxyalkanoates with nanokeratin to develop novel biopackaging structures. *Journal of Applied Polymer Science, 133*(2), app.42695. Advance online publication. DOI: 10.1002/app.42695

Fan, P., Yu, H., Xi, B., & Tan, W. (2022). A review on the occurrence and influence of biodegradable microplastics in soil ecosystems: Are biodegradable plastics substitute or threat? *Environment International, 163*, 107244. DOI: 10.1016/j.envint.2022.107244 PMID: 35436719

Fan, P., Yu, H., Xi, B., & Tan, W. (2022). A review on the occurrence and influence of biodegradable microplastics in soil ecosystems: Are biodegradable plastics a substitute or threat? *Environment International, 163*, 107244. DOI: 10.1016/j.envint.2022.107244 PMID: 35436719

Fei, X., Wang, J., Zhang, X., Jia, Z., Jiang, Y., & Liu, X. (2022). Recent Progress on Bio-Based Polyesters Derived from 2,5-Furandicarbonxylic Acid (FDCA). *Polymers, 14*(3), 625. DOI: 10.3390/polym14030625 PMID: 35160613

Fico, D., Rizzo, D., Casciaro, R., & Esposito Corcione, C. (2022). A Review of Polymer-Based Materials for Fused Filament Fabrication (FFF): Focus on Sustainability and Recycled Materials. *Polymers, 14*(3), 465. DOI: 10.3390/polym14030465 PMID: 35160455

Fico, D., Rizzo, D., Casciaro, R., & Esposito Corcione, C. (2022). A Review of Polymer-Based Materials for Fused Filament Fabrication (FFF): Focus on Sustainability and Recycled Materials. *Polymers, 14*(3), 465. DOI: 10.3390/polym14030465 PMID: 35160455

Finch, C. (1985). Anionic polymerization to cationic polymerization. In Mark, H. F., Bikales, N. M., Overberger, C. G., & Menges, G. (Eds.), *Encyclopedia of Polymer Science and Engineering* (Vol. 2). Wiley-Interscience.,

Fortunati, E., Armentano, I., Iannoni, A., & Kenny, J. (2010). Development and thermal behaviour of ternary PLA matrix composites. *Polymer Degradation & Stability, 95*(11), 2200–2206. DOI: 10.1016/j.polymdegradstab.2010.02.034

Gandini, A., Silvestre, A. J. D., Neto, C. P., Sousa, A. F., & Gomes, M. (2008). The furan counterpart of poly (ethylene terephthalate): An alternative material based on renewable resources. *Journal of Polymer Science. Part A, Polymer Chemistry, 47*(1), 295–298. DOI: 10.1002/pola.23130

Ghanbarzadeh, B., Almasi, H., & Entezami, A. A. (2011). Improving the barrier and mechanical properties of corn starch-based edible films: Effect of citric acid and carboxymethyl cellulose. *Industrial Crops and Products, 33*(1), 229–235. DOI: 10.1016/j.indcrop.2010.10.016

Ghomi, E. R., Khosravi, F., Ardahaei, A. S., Dai, Y., Neisiany, R. E., Foroughi, F., Wu, M., Das, O., & Ramakrishna, S. (2021). The Life Cycle Assessment for Polylactic Acid (PLA) to Make It a Low-Carbon Material. *Polymers, 13*(11), 1854. DOI: 10.3390/polym13111854 PMID: 34199643

Gidley, M. J., Lillford, P. J., Rowlands, D. W., Lang, P., Dentini, M., Crescenzi, W., Edwards, M., Fanutti, C., & Reid, J. S. G. (1991). Structure and solution properties of tamarind seed polysaccharide. *Carbohydrate Research, 214*(2), 299–314. DOI: 10.1016/0008-6215(91)80037-N PMID: 1769022

Godavitarne, C., Robertson, A., Peters, J., & Rogers, B. (2017). Biodegradable Materials. *Orthopaedics and Trauma, 31*(6), 316–320. DOI: 10.1016/j.mporth.2017.07.011

Gomes, M., Gandini, A., Silvestre, A. J. D., & Reis, B. (2011). Synthesis and characterization of poly (2,5-furan dicarboxylate)s based on a variety of diols. *Journal of Polymer Science. Part A, Polymer Chemistry, 49*(16), 3759–3768. DOI: 10.1002/pola.24812

Gontard, N., Sonesson, U., Birkved, M., Majone, M., Bolzonella, D., Celli, A., Angellier-Coussy, H., Jang, G. W., Verniquet, A., Broeze, J., Schaer, B., Batista, A. P., & Sebok, A. (2018). A research challenge vision regarding management of agricultural waste in a circular bio-based economy. *Critical Reviews in Environmental Science and Technology, 48*(7–8), 614–654. DOI: 10.1080/10643389.2018.1471957

Green Dot Bioplastics. (2022). A Straightforward Explanation of Biodegradable vs. Compostable vs. Oxo-Degradable Plastics. Retrieved October 10, 2022, from https://www.greendotbioplastics.com/biodegradable-vs-compostable-vs-oxo-1479degradable-plastics-a-straightforward-explanation/

Gruber, J. V. (1999). Polysaccharide-Based Polymers in Cosmetics. In Goddard, E. D., & Gruber, J. V. (Eds.), *Principles of Polymer Science and Technology in Cosmetic and Personal Care* (Vol. 22, pp. 339–403). Marcel Dekker Inc.

Gupta, B., Revagade, N., & Hilborn, J. (2007). Poly (lactic acid) fiber: An overview. *Progress in Polymer Science, 32*(4), 455–482. DOI: 10.1016/j.progpolymsci.2007.01.005

Hahladakis, N. J., Costas, A. V., Weber, R., Iacovidou, E., & Purnell, P. (2018). An overview of chemical additives present in plastics: Migration, release, fate and environmental impact during their use, disposal and recycling. *Journal of Hazardous Materials, 344*, 179–199. DOI: 10.1016/j.jhazmat.2017.10.014 PMID: 29035713

Hassan, M., Mohanty, A. K., & Misra, M. (2024). 3D printing in upcycling plastic and biomass waste to sustainable polymer blends and composites: A review. *Materials & Design, 237*, 112558. DOI: 10.1016/j.matdes.2023.112558

Holten, C. H. (1971). *Lactic Acid: Properties and Chemistry of Lactic Acid and Derivatives*. Verlag Chemie. Retrieved from https://www.cabidigitallibrary.org/action/doSearch?do=Lactic+acid.+Properties+and+chemistry+of+lactic+acid+and+derivatives

Huq, T., Salmieri, S., Khan, A., Khan, R. A., Le Tien, C., Riedl, B., Fraschini, C., Bouchard, J., Uribe-Calderon, J., Kamal, M. R., & Lacroix, M. (2012). Nanocrystalline cellulose (NCC) reinforced alginate-based biodegradable nanocomposite films. *Carbohydrate Polymers, 90*(4), 1757–1763. DOI: 10.1016/j.carbpol.2012.07.065 PMID: 22944444

Ibrahim, N. I., Shahar, F. S., Sultan, M. T. H., Md Shan, A. U., Safri, S. N. A., & Yazik, M. H. M. (2021). Overview of Bioplastic Introduction and Its Applications in Product Packaging. *Coatings, 11*(11), 1423. DOI: 10.3390/coatings11111423

International Solid Waste Association (ISW). (2024). *Global Waste Management Outlook 2024: Beyond an Age of Waste, Turning Rubbish into a Resource*. Association ISW.

Jambeck, J. R., Geyer, R., Wilcox, C., Siegler, T. R., Perryman, M., Andrady, A., Narayan, R., & Lavender Law, K. (2015). Plastic waste inputs from land into the ocean. *Science, 347*(6223), 768–771. DOI: 10.1126/science.1260352 PMID: 25678662

Jawaid, M., & Abdul Khalil, H. P. S. (2011). Cellulosic/synthetic fibre reinforced polymer hybrid composites: A review. *Carbohydrate Polymers, 86*(1), 1–18. DOI: 10.1016/j.carbpol.2011.04.043

Jiang, G., Hill, D. J., Kowalczuk, M., Johnston, B., Adamus, G., Irorere, V., & Radecka, I. (2016). Carbon sources for polyhydroxyalkanoates and an integrated biorefinery. *International Journal of Molecular Sciences, 17*(7), 1157. DOI: 10.3390/ijms17071157 PMID: 27447619

Jiang, M., Liu, Q., Zhang, Q., Ye, C., & Zhou, G. (2012). A series of furan-aromatic polyesters synthesized via direct esterification method based on renewable resources. *Journal of Polymer Science. Part A, Polymer Chemistry, 50*(5), 1026–1036. DOI: 10.1002/pola.25859

Joseph, C. S., Prashanth, K. H., Rastogi, N., Indiramma, A., Reddy, S. Y., & Raghavarao, K. (2011). Optimum blend of chitosan and poly(ε-caprolactone) for fabrication of films for food packaging applications. *Food and Bioprocess Technology, 4*(7), 1179–1185. DOI: 10.1007/s11947-009-0203-1

Jost, V., & Kopitzky, R. (2015). Blending of polyhydroxybutyrate-co-valerate with polylactic acid for packaging applications: Reflections on miscibility and effects on the mechanical and barrier properties. *Chemical and Biochemical Engineering Quarterly, 29*(2), 221–246. DOI: 10.15255/CABEQ.2014.2257

Kaniuk, Ł., & Stachewicz, U. (2021). Development and Advantages of Biodegradable PHA Polymers Based on Electrospun PHBV Fibers for Tissue Engineering and Other Biomedical Applications. *ACS Biomaterials Science & Engineering, 7*(12), 5339–5362. DOI: 10.1021/acsbiomaterials.1c00757 PMID: 34649426

Kannah, R. Y., Kumar, M. D., Kavitha, S., Banu, J. R., Tyagi, V. K., & Rajaguru, P.. (2022). Production and recovery of polyhydroxyalkanoates (PHA) from waste streams: A review. *Bioresource Technology*, *366*, 128203. DOI: 10.1016/j.biortech.2022.128203 PMID: 36330969

Karande, R. D., Abitha, V. K., Rane, A. V., & Mishra, R. K. (2015). Preparation of polylactide from synthesized lactic acid and effect of reaction parameters on conversion. *Journal of Materials Science and Engineering: Advanced Technology*, *12*(1), 1–37.

Kelly, A. (Ed.). (2012). *Concise encyclopedia of composite materials*. Elsevier.

Kim, J. S., Lee, B. H., & Kim, B. S. (2005). Production of poly(3-hydroxybutyrate-co-4-hydroxybutyrate) by *Ralstonia eutropha*. *Biochemical Engineering Journal*, *23*(2), 169–174. DOI: 10.1016/j.bej.2005.01.016

Kister, G., Cassanas, G., Bergounhon, M., Hoarau, D., & Vert, M. (2000). Structural characterization and hydrolytic degradation of solid copolymers of D,L-lactide-co-ε-caprolactone by Raman Spectroscopy. *Polymer*, *41*(3), 925–932. DOI: 10.1016/S0032-3861(99)00223-2

Kobayashi, S. (1999). Enzymatic polymerization: A new method of polymer synthesis. *Journal of Polymer Science. Part A, Polymer Chemistry*, *37*(16), 3041–3056. DOI: 10.1002/(SICI)1099-0518(19990815)37:16<3041::AID-POLA1>3.0.CO;2-V

Kobayashi, S., Uyama, H., & Ohmae, M. (2001). Enzymatic polymerization for precision polymer synthesis. *Bulletin of the Chemical Society of Japan*, *74*(4), 613–635. DOI: 10.1246/bcsj.74.613

Kumar, P., Barrett, D. M., Delwiche, M. J., & Stroeve, P. (2009). Methods for pretreatment of lignocellulosic biomass for efficient hydrolysis and biofuel production. *Industrial & Engineering Chemistry Research*, *48*(8), 3713–3729. DOI: 10.1021/ie801542g

Kumar, R., Verma, A., Shome, A., Sinha, R., Sinha, S., Jha, P. K., Kumar, R., Kumar, P., Shubham, , Das, S., Sharma, P., & Vara Prasad, P. V. (2021). Impacts of plastic pollution on ecosystem services, sustainable development goals, and the need to focus on circular economy and policy interventions. *Sustainability (Basel)*, *13*(17), 9963. DOI: 10.3390/su13179963

Kunasundari, B., & Sudesh, K. (2011). Isolation and recovery of microbial polyhydroxyalkanoates. *Express Polymer Letters*, *5*(7), 620–634. DOI: 10.3144/expresspolymlett.2011.60

Labet, M., & Thielemans, W. (2009). Synthesis of polycaprolactone: A review. *Chemical Society Reviews*, *38*(12), 3484–3504. DOI: 10.1039/b820162p PMID: 20449064

Lepe, P. T., Heredia, K. V., Namur, E. C., Fabián, G. C. S., & García-Enríquez, S. (2024). Economics and commercialization of bioplastics. In *Bioplastics for Sustainability* (pp. 271–309). Elsevier., DOI: 10.1016/B978-0-323-95199-9.00018-4

Li, Y., Wang, S., Qian, S., Liu, Z., Weng, Y., & Zhang, Y. (2024). Depolymerization and Re/Upcycling of Biodegradable PLA Plastics. *ACS Omega*, *9*(12), 13509–13521. DOI: 10.1021/acsomega.3c08674 PMID: 38559974

Lin, X., Fan, X., Li, R., Li, Z., Ren, T., Ren, X., & Huang, T. S. (2018). Preparation and characterization of PHB/PBAT-based biodegradable antibacterial hydrophobic nanofibrous membranes. *Polymers for Advanced Technologies*, *29*(1), 481–489. DOI: 10.1002/pat.4137

Lomelí-Ramírez, M. G., Kestur, S. G., Manríquez-González, R., Iwakiri, S., De Muniz, G. B., & Flores-Sahagun, T. S. (2014). Bio-Composites of cassava starch-green coconut fiber: Part II—Structure and properties. *Carbohydrate Polymers*, *102*, 576–583. DOI: 10.1016/j.carbpol.2013.11.020 PMID: 24507321

Loos, K., Zhang, R., Pereira, I., Agostinho, B., Hu, H., Maniar, D., Sbirrazzuoli, N., Silvestre, A. J. D., Guigo, N., & Sousa, A. F. A. (2020). Perspective on PEF Synthesis, Properties, and End-Life. *Frontiers in Chemistry*, *8*, 585. DOI: 10.3389/fchem.2020.00585 PMID: 32850625

Luciani, A., Coccoli, V., Orsi, S., Ambrosio, L., & Netti, P. A. (2008). PCL microspheres-based functional scaffolds by bottom-up approach with predefined microstructural properties and release profiles. *Biomaterials*, *29*(35), 4800–4807. DOI: 10.1016/j.biomaterials.2008.09.007 PMID: 18834628

MacDonald, R. T., Pulapura, S. K., Svirkin, Y. Y., Gross, R. A., Kaplan, D. L., Akkara, J., Swift, G., & Wolk, S. (1995). Enzyme-catalyzed ε-caprolactone ring-opening polymerization. *Macromolecules*, *28*(1), 73–78. DOI: 10.1021/ma00105a008

Madkour, M. H., Heinrich, D., Alghamdi, M. A., Shabbaj, I. I., & Steinbüchel, A. (2013). PHA Recovery from Biomass. *Biomacromolecules*, *14*(10), 2963–2972. DOI: 10.1021/bm4010244 PMID: 23875914

Mehmood, A., Raina, N., Phakeenuya, V., Wonganu, B., & Cheenkachorn, K. (2023). The current status and market trend of polylactic acid as biopolymer: Awareness and needs for sustainable development. *Materials Today: Proceedings*, *72*, 3049–3055. DOI: 10.1016/j.matpr.2022.08.387

Muneer, F., Rasul, I., Qasim, M., Sajid, A., & Nadeem, H. (2022). Optimization, Production and Characterization of Polyhydroxyalkanoate (PHA) from Indigenously Isolated Novel Bacteria. *Journal of Polymers and the Environment, 30*(8), 3523–3533. DOI: 10.1007/s10924-022-02444-y

Muthusamy, M. S., & Pramasivam, S. (2019). Bioplastics—An Eco-Friendly Alternative to Petrochemical Plastics. *Current World Environment, 14*(1), 49–59. DOI: 10.12944/CWE.14.1.07

Nair, L. S., & Laurencin, C. T. (2007). Biodegradable polymers as biomaterials. *Progress in Polymer Science, 32*(8-9), 762–798. DOI: 10.1016/j.progpolymsci.2007.05.017

Nghiem, N. P., Kleff, S., & Schwegmann, S. (2017). Succinic Acid: Technology Development and Commercialization. *Fermentation (Basel, Switzerland), 3*(2), 26. DOI: 10.3390/fermentation3020026

Nishinari, K., Yamatoya, K., & Shirakawa, M. (2000). Xyloglucan. In Phillips, G. O., & Williams, P. A. (Eds.), *Handbook of hydrocolloids* (pp. 247–267). Woodhead Publishing.

Nita, T., & Aurica, P. C. (2011). Biopolymer-Based Green Composites: Mechanical, Thermal, and Physico-Chemical Characterization. *Journal of Polymers and the Environment, 19*(3), 546–556.

Nobes, G. A., Kazlauskas, R. J., & Marchessault, R. H. (1996). Lipase-catalyzed ring-opening polymerization of lactones: A novel route to poly(hydroxyalkanoates). *Macromolecules, 29*(15), 4829–4833. DOI: 10.1021/ma951774g

Park, D. H., & Kim, B. S. (2011). Production of poly(3-hydroxybutyrate) and poly(3-hydroxybutyrate-co-4-hydroxybutyrate) by *Ralstonia eutropha* from soybean oil. *New Biotechnology, 28*(6), 719–724. DOI: 10.1016/j.nbt.2011.01.007 PMID: 21333767

Patti, A., & Acierno, D. (2022). Towards the Sustainability of the Plastic Industry through Biopolymers: Properties and Potential Applications to the Textiles World. *Polymers, 14*(4), 692. DOI: 10.3390/polym14040692 PMID: 35215604

Plastics Europe. (2023). *Plastics—the fast facts 2023*. Retrieved May 29, 2024, from https://plasticseurope.org/knowledge-hub/plastics-the-fast-facts-2023/

Platel, R. H., Hodgson, L. M., & Williams, C. K. (2008). Biocompatible initiators for lactide polymerization. *Polymer Reviews (Philadelphia, Pa.), 48*(1), 11–63. DOI: 10.1080/15583720701834166

Platnieks, O., Gaidukovs, S., Thakur, V. K., Barkane, A., & Beluns, S. (2021). Bio-based poly (butylene succinate): Recent progress, challenges, and future opportunities. *European Polymer Journal*, *161*, 110855. DOI: 10.1016/j.eurpolymj.2021.110855

Price, S., Kuzhiumparambil, U., Pernice, M., & Ralph, P. (2022). Techno-economic analysis of cyanobacterial PHB bioplastic production. *Journal of Environmental Chemical Engineering*, *10*(3), 107502. DOI: 10.1016/j.jece.2022.107502

Pyo, S.-H., Park, J. H., Srebny, V., & Hatti-Kaul, R. (2020). A sustainable synthetic route for biobased 6-hydroxyhexanoic acid, adipic acid, and ε-caprolactone by integrating bio- and chemical catalysis. *Green Chemistry*, *22*(14), 4450–4455. DOI: 10.1039/D0GC01454K

Rajgond, V., Mohite, A., More, N., & More, A. (2023). Biodegradable polyester-polybutylene succinate (PBS): A review. *Polymer Bulletin*, *81*(7), 5703–5752. DOI: 10.1007/s00289-023-04998-w

Ranakoti, L., Gangil, B., Kumar Mishra, S., Singh, T., Sharma, S., Ilyas, R. A., & El-Khatib, S. (2022). Critical review on polylactic acid: Properties, structure, processing, biocomposites, and nanocomposites. *Materials (Basel)*, *15*(12), 4312. DOI: 10.3390/ma15124312 PMID: 35744371

Rangappa, S. M., Siengchin, S., Parameswaranpillai, J., Jawaid, M., & Ozbakkaloglu, T. (2022). Lignocellulosic fiber reinforced composites: Progress, performance, properties, applications, and future perspectives. *Polymer Composites*, *43*(2), 645–691. DOI: 10.1002/pc.26413

Rao, Y. S., Mary, K., Mathew, L., & Potty, S. N. (1999). Tamarind (Tamarindus indica L.) research—A Review. *Indian Journal of Arecanut, Spices and Medicinal Plants*, *1*(3), 127–145. https://www.cabidigitallibrary.org/action/doSearch?do=Indian+Journal+of+Arecanut%2C+Spices+and+Medicinal+Plants

Raza, S., Zhang, J., Ali, I., Li, X., & Liu, C. (2020). Recent trends in the development of biomass-based polymers from renewable resources and their environmental applications. *Journal of the Taiwan Institute of Chemical Engineers*, *115*, 293–303. DOI: 10.1016/j.jtice.2020.10.013

Riva, R., Schmeits, S., Jérôme, C., Jérôme, R., & Lecomte, P. (2007). Combination of ring-opening polymerization and "click chemistry": Toward functionalization and grafting of poly(ε-caprolactone). *Macromolecules*, *40*(3), 796–803. DOI: 10.1021/ma0624090

Rizos, V., Tuokko, K., & Behrens, A. (2017). The circular economy: A review of definitions, processes, and impacts. *CEPS Research Report, 2017/8*. Retrieved June 29, 2024, from http://aei.pitt.edu/id/eprint/85892

Sahoo, S., Sahoo, R., & Nayak, P. L. (2011). Tamarind Seed Polysaccharide: A Versatile Biopolymer for Mucoadhesive Applications. *Journal of Pharmaceutical and Biomedical Sciences*, *8*(1), 1–12.

Samir, A., Ashour, F. H., Hakim, A. A. A., & Bassyouni, M. (2022). Recent advances in biodegradable polymers for sustainable applications. *NPJ Materials Degradation*, *6*(1), 68. DOI: 10.1038/s41529-022-00277-7

Sarah, P., Julien, B. M. G., Melissa, I. G., Paolo, M., Adrienne, G., & Noémie, V. (2024). *Plastic Overshoot Day Report 2024*. EA-Earth Action.

Schreurs, V. V., & Schaafsma, G. (2010). Lactic acid and lactates: In Health and Wellness. *NUTRAfoods : International Journal of Science and Marketing for Nutraceutical Actives, Raw Materials, Finish Products*, *9*(1), 7–16. DOI: 10.1007/BF03223326

Shah, T. V., & Vasava, D. V. (2019). A glimpse of biodegradable polymers and their biomedical applications. *E-Polymers*, *19*(1), 385–410. DOI: 10.1515/epoly-2019-0041

Singha, A. S., & Thakur, V. K. (2009). Physical, Chemical, and Mechanical Properties of *Hibiscus sabdariffa* Fiber/Polymer Composite. *International Journal of Polymeric Materials*, *58*(4), 217–225. DOI: 10.1080/00914030802639999

Singha, A. S., & Thakur, V. K. (2009). Chemical Resistance, Mechanical, and Physical Properties of Biofibers-Based Polymer Composites. *Polymer-Plastics Technology and Engineering*, *48*(7), 736–742. DOI: 10.1080/03602550902824622

Singhvi, M., Zinjarde, S., & Gokhale, D. (2019). Polylactic acid: Synthesis and biomedical applications. *Journal of Applied Microbiology*, *127*(5), 1612–1626. DOI: 10.1111/jam.14290 PMID: 31021482

Statista Research Department. (2024). Distribution of Global Plastic Materials Production in 2022, by Region. Retrieved May 29, 2024, from https://www.statista.com/statistics/281126/global-plastics-production-share-of-various-countries-and-regions/

Stridsberg, K. M., Ryner, M., & Albertsson, A.-C. (2002). *Controlled ring-opening polymerization: Polymers with designed macromolecular architecture*. Springer.,

Stubbe, J., & Tian, J. (2003). Polyhydroxyalkanoate (PHA) homeostasis: The role of the PHA synthase. *Natural Product Reports*, *20*(5), 445–457. DOI: 10.1039/b209687k PMID: 14620841

Sudesh, K., Abe, H., & Doi, Y. (2000). Synthesis, structure and properties of polyhydroxyalkanoates: Biological polyesters. *Progress in Polymer Science*, *25*(10), 1503–1555. DOI: 10.1016/S0079-6700(00)00035-6

Sun, L., Wang, J., Mahmud, S., Jiang, Y., Zhu, J., & Liu, X. (2019). New insight into the mechanism for the excellent gas properties of poly (ethylene 2,5-furandicarboxylate) (PEF): Role of furan ring's polarity. *European Polymer Journal*, *118*, 642–650. DOI: 10.1016/j.eurpolymj.2019.06.033

Taib, N.-A. A. B., Rahman, R., Huda, D., Kuok, K. K., Hamdan, S., Bin Bakri, M. K., Bin Julaihi, M. R. M., & Khan, A. (2022). A review on poly lactic acid (PLA) as a biodegradable polymer. *Polymer Bulletin*, *80*(2), 1179–1213. DOI: 10.1007/s00289-022-04160-y

Thakur, V. K., Thakur, M. K., & Gupta, R. K. (2014). Review: Raw Natural Fiber–Based Polymer Composites. *IJPAC. International Journal of Polymer Analysis and Characterization*, *19*(3), 256–271. DOI: 10.1080/1023666X.2014.880016

Thomas, A. P., Kasa, V. P., Dubey, B. K., Sen, R., & Sarmah, A. K. (2023). Synthesis and commercialization of bioplastics: Organic waste as a sustainable feedstock. *The Science of the Total Environment*, *904*, 167243. DOI: 10.1016/j.scitotenv.2023.167243 PMID: 37741416

Tindall, J. A., Beverly, R. B., & Radcliffe, D. E. (1991). Mulch effect on soil properties and tomato growth using micro-irrigation. *Agronomy Journal*, *83*(6), 1028–1034. DOI: 10.2134/agronj1991.00021962008300060019x

United Nations Environment Program. (2020). *Single-use plastics: A roadmap for sustainability*. Retrieved June 29, 2024, from https://www.unep.org/resources/report/single-use-plastics-roadmap-sustainability

Verma, R., Vinoda, K., Papireddy, M., & Gowda, A. (2016). Toxic Pollutants from Plastic Waste—A Review. *Procedia Environmental Sciences*, *35*, 701–708. DOI: 10.1016/j.proenv.2016.07.069

Verma, S. K., Prasad, A., & Katiyar, V. (2024). State-of-the-art review on sustainable biodegradable polymers with a market overview for sustainability packaging. *Materials Today Sustainability*, *26*, 100776. DOI: 10.1016/j.mtsust.2024.100776

Vicente, D., Proença, D. N., & Morais, P. V. (2023). The Role of Bacterial Polyhydroalkanoate (PHA) in a Sustainable Future: A review on the biological diversity. *International Journal of Environmental Research and Public Health, 20*(4), 2959. DOI: 10.3390/ijerph20042959 PMID: 36833658

Wabnitz, C., & Nichols, W. J. (2010). Plastic Pollution: An Ocean Emergency. *Marine Turtle Newsletter, 129*, 1–4.

Wan, Y., Lu, X., Dalai, S., & Zhang, J. (2009). Thermophysical properties of polycaprolactone/chitosan blend membranes. *Thermochimica Acta, 487*(1-2), 33–38. DOI: 10.1016/j.tca.2009.01.007

Winnacker, M. (2019). Polyhydroxyalkanoates: Recent Advances in their Synthesis and Applications. *European Journal of Lipid Science and Technology, 121*(11), 1900101. DOI: 10.1002/ejlt.201900101

Woodruff, M. A., & Hutmacher, D. W. (2010). The return of a forgotten polymer—Polycaprolactone in the 21st century. *Progress in Polymer Science, 35*(10), 1217–1256. DOI: 10.1016/j.progpolymsci.2010.04.002

Yalpani, M., Marchessault, R. H., Morin, F. G., & Monasterios, C. J. (1991). Synthesis of poly(3-hydroxyalkanoate) (PHA) conjugates: PHA-carbohydrate and PHA-synthetic polymer conjugates. *Macromolecules, 24*(22), 6046–6049. DOI: 10.1021/ma00022a024

Yin, G. Z., & Yang, X. M. (2020). Biodegradable polymers: A cure for the planet, but a long way to go. *Journal of Polymer Research, 27*(2), 38. DOI: 10.1007/s10965-020-2004-1

Yoichi, H., & Yutaka, T. (2010). Polymer environment research. *Journal of Polymers and the Environment, 18*(1), 116.

Yoichi, H., & Yutaka, T. (2010). Mechanical, Thermal, and Physico-Chemical Characterization. *Journal of Polymers and the Environment, 18*(2), 326.

Younes, B. (2016). Classification, characterization, and the production processes of biopolymers used in the textiles industry. *Journal of the Textile Institute, 108*(5), 674–682. DOI: 10.1080/00405000.2016.1180731

Chapter 8
Hydrogen Storage Challenges and Solutions

Himadri Sekhar Das
https://orcid.org/0000-0002-3509-3388
Haldia Institute of Technology, India

Anjana Ghosh
Haldia Instiute of Technology, India

Shatarupa Bhowmick
Brainware University, Barasat, West Bengal, India

Sudipta Banerjee
https://orcid.org/0000-0003-0150-6794
Symbiosis Institute of Technology, Symbiosis International University, Pune, India

Heranmoy Maity
Ideal Institute of Engineering Kalyani, Nadia, India

Abhishek Bag
https://orcid.org/0009-0001-8390-9847
Haldia Instiute of Technology, India

Milan Hait
Dr. C.V. Raman University, Bilaspur, India

Mrinal Kanti Dash
Haldia Instiute of Technology, India

ABSTRACT

Abstract: Hydrogen is widely regarded as a promising clean energy carrier, yet its effective storage presents significant challenges that hinder widespread adoption. This chapter provides an in-depth exploration of various hydrogen storage technologies, examining their advantages, limitations, and potential solutions. We focus on physical, chemical, and material-based storage methods, evaluating key factors such as energy efficiency, safety, and economic feasibility. Additionally, we highlight emerging advancements in novel materials, hybrid storage systems, and policy frameworks that could play a pivotal role in addressing current storage challenges. The chapter concludes by discussing future directions and the need for continued research to overcome technical and economic barriers, ultimately enabling large-

DOI: 10.4018/979-8-3373-1409-9.ch008

scale hydrogen adoption as a cornerstone of a sustainable energy future.

INTRODUCTION:

Hydrogen is increasingly seen as a key component of future energy systems due to its high energy density and zero-emission potential. However, storing hydrogen efficiently and safely poses significant challenges. This chapter examines different hydrogen storage technologies, their limitations, and potential breakthroughs. Hydrogen has emerged as a pivotal element in the transition toward a sustainable energy future (A. Bag et al. (2023). As the world struggle with the dual challenges of reducing carbon emissions and meeting rising energy demands, hydrogen offers a compelling solution due to its high energy content and environmental benefits. Unlike conventional fossil fuels, hydrogen combustion produces only water as a by-product, making it an ideal candidate for clean energy applications. However, despite its promising attributes, hydrogen storage remains a significant technological and logistical challenge that must be addressed to enable its widespread adoption (G. Roymahapatra et al, (2022). One of the key advantages of hydrogen as an energy carrier is its superior gravimetric energy density-approximately three times that of gasoline. This characteristic makes it an attractive option for applications in transportation, power generation, and industrial processes. However, its low volumetric energy density poses a considerable obstacle to practical storage solutions. Hydrogen is the lightest element in the periodic table, and under standard conditions, it exists as a gas with extremely low density. As a result, storing hydrogen efficiently requires advanced methods that enhance its density while maintaining safety, economic feasibility, and ease of use. Hydrogen storage technologies can be broadly classified into three categories: physical storage, chemical storage, and materials-based storage (H. S. Das et al, (2025). Physical storage involves compressing hydrogen gas at high pressures or liquefying it at cryogenic temperatures to achieve higher densities. Compressed hydrogen storage is the most mature and widely used method, especially in fuel cell vehicles and industrial applications. Liquid hydrogen storage, on the other hand, offers greater energy density but requires extreme cooling (-253°C) and sophisticated insulation techniques to prevent boil-off losses. Chemical storage methods involve bonding hydrogen with other elements to form stable compounds, such as metal hydrides, ammonia, and liquid organic hydrogen carriers (LOHCs). These methods offer promising alternatives for safe and efficient hydrogen storage, as they allow for reversible hydrogen uptake and release under controlled conditions. However, challenges such as slow hydrogen release kinetics, high operational temperatures, and material degradation need to be overcome before widespread implementation. Materials-based storage focuses on novel approaches such as adsorption in porous materials like metal-organic frameworks (MOFs), carbon nanotubes, and zeolites. These materials have high surface areas and unique properties that enable efficient hydrogen capture and release at moderate temperatures and pressures. Research in this area is rapidly advancing, with the goal of developing lightweight, high-capacity storage systems that meet the stringent requirements of future hydrogen-powered applications. Despite significant progress, several challenges hinder the commercialization of hydrogen storage technologies. Safety concerns related to high-pressure storage, energy losses during compression and liquefaction, and the high cost of advanced materials remain critical barriers. Additionally, infrastructure development, including refuelling stations and distribution networks, is essential to facilitate large-scale hydrogen adoption. This chapter explores the current landscape of hydrogen storage technologies, analyzing their advantages, limitations, and recent breakthroughs (M. K. Dash et al, (2022). By examining emerging trends and innovative solutions, we aim to provide a comprehensive understanding of the future potential of hydrogen storage in enabling a clean and sustainable energy ecosystem. Fig.1 shows the different routes of hydrogen.

Figure 1. Different Routes of Hydrogen.

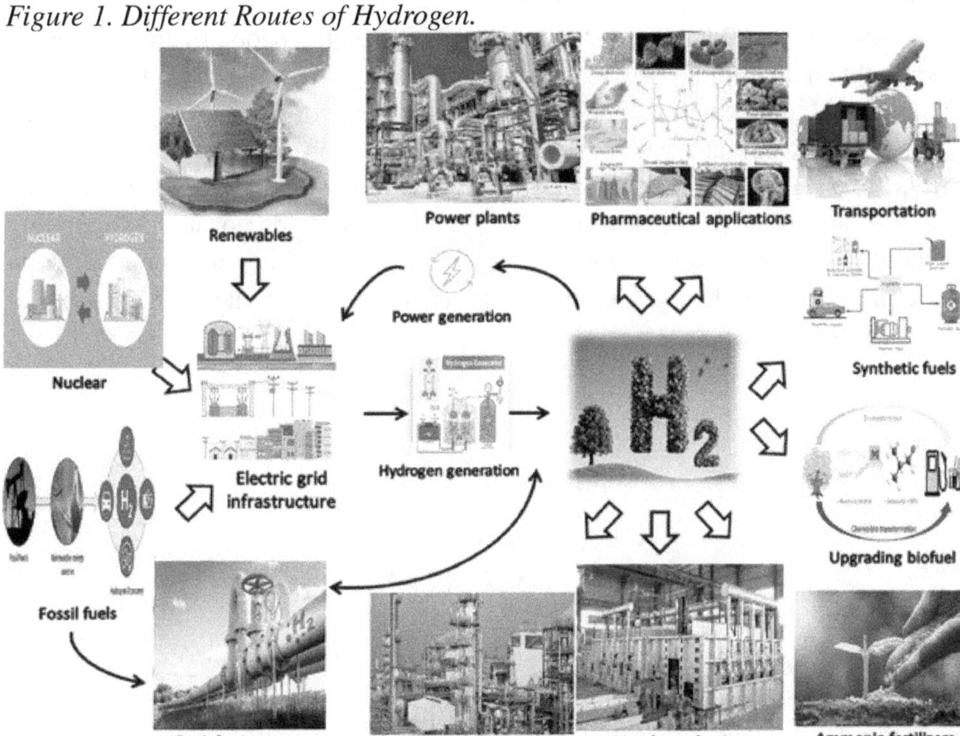

Hydrogen storage materials are essential for efficient energy applications, with different compounds offering varying storage capacities and operating conditions. Oxadiazole-x Li^+ complexes can store up to 19.93 wt% H_2 at or below 200 K, making them highly effective. Li-doped five-member aromatic heterocyclic complexes $(Ar\text{-}Li)^+$ can trap 9.8%-11.7% H_2 at 80 K, while lithiated triazine isomers store up to 17.39 wt% at 200 K. Alkali-metalated heterocyclic systems (M@C4H4N2) such as pyrazine and pyrimidine can accommodate 9 H_2 molecules with a 13% weight fraction. Li and Sc-doped C8N8 cages have a storage capacity of 8.32 wt%. Pyridine-lithium ion complexes store 8.50 wt% at 77 K, while lithium fluoride achieves a remarkable 43.48 wt% below 54 K. Lithium chloride and lithium bromide hold 32.00 wt% and 18.71 wt%, respectively, at temperatures below 51 K. Various lithium-doped boron hydrides can trap 11 H_2 molecules with a 17.7% weight fraction at temperatures below 100 K. Li12C60 is capable of storing 60 H_2 molecules at 9 wt%. Li-decorated metal-organic frameworks (MOFs) can store 18 H_2 molecules but only at 2.0 wt%. Li-functionalized [1,1]Para cyclophane achieves 13.42 wt% H_2 storage. Gold (I)-decorated isomeric triazine complexes store up to 4.07 wt% below 180 K, whereas Ag(I)/Au(I)-decorated five-member aromatic heterocyclic compounds store 3.66% to 5.45% below 180 K. Cu(I)-triazine complexes store up to 8.8 wt% below 150 K. An air-stable Cu(I) MOF exhibits good hydrogen storage under ambient conditions with a volumetric capacity of 10.4 g/L. Cu_2Bx (X=1-4) can hold 16 H_2 molecules at 13.80 wt% under ambient conditions. Cu(I)-decorated five-membered aromatic heterocyclic complexes exhibit storage between 6.38 to 7.15 wt% at 117 K. Ag(I)-decorated isomeric triazine complexes store between 5.08 wt% and 6.39 wt% below 160 K. These materials demonstrate significant potential in hydrogen storage applications, offering various options depending on temperature and weight capacity requirements. Table1 shows the different types of hydrogen storage materials.

Table 1. Different types hydrogen storage Materials

Sl. No.	Model Systems	Maximum Trapped H_2	Gravimetric Wt%	Temperature for spontaneity	Ref.
1.	[Oxadiazole-xLi+] (x = 1, 2)	$5H_2$ for mono $10H_2$ for di	11.49% (mono) and 19.93% (di)	at or below 200 K.	G. Roymahapatra et al, (2022)
2.	Li-doped five-member aromatic heterocyclic complexes (Ar-Li)+ [Ar = Imidazole, Pyrazole, Oxazole, Isoxazole, Thiazole, and Isothiazole].	$5H_2$ molecules	9.8%-11.7%	At 80 K temperature	M. K. Dash et al, (2022)
3.	lithiated Triazine isomers.	$5H_2$ for mono 10h2 for di	10.20.% (mono) and 17.39% (di)	200K temperature	M. K. Dash et al, (2021)
4.	The alkali (Li, Na, and K) metallated heterocyclic systems (M@ C4H4N2) Pyrazine, Pyrimidine, and Pyridazine.	$9H_2$ molecules	13%		R. Parida et al, (2021),
5.	Li and Sc doped novel C_8N_8 cage.		8.32 wt%		R. K. Sahoo et al, (2022)
6.	pyridine–lithium ion (1:1) complexes	$4H_2$ molecules	8.50 wt%	77K	S. Chattaraj et al, (2025)
7.	Lithium Fluoride.	$10H_2$ molecules	43.48 wt%	below 54 K.	M. K. Dash et al, (2022)
8.	Lithium Chloride/ Lithium Bromide.	$10H_2$ molecules	(32.00 wt%for LiCl and 18.71 wt%for LiBr),	below 51 K for LiCl and 49 K for LiBr	G. Roymahapatra et.al, (2022)
9.	Various Lithium— Doped Boron Hydrides	$11H_2$ molecules	17.7%	Below 100K	S Pan et.al,(2011),
10	Li12C60	$60H_2$ molecules	9 wt %		Q Sun et.al, (2006),
11	Li-decorated metal–organic framework 5	18 H_2 molecules	2.0 wt %		A. Blomqvist et. al, (2007)
12	Li functionalized [1,1]paracyclophane	8 H_2 molecules	13.42%		R Y Sathe et.al, (2020)
13	Au(I) Decorated Isomeric Triazine Complexes.	$5H_2$ for mono $10H_2$ for di	3.49 Wt% to 4.07 Wt%.	below 85 K for mono and below 180K for di	A. Bag et.al, (2024)

continued on following page

Table 1. Continued

Sl. No.	Model Systems	Maximum Trapped H_2	Gravimetric Wt%	Temperature for spontaneity	Ref.
14	Ag (I)/Au (I) Decorated Five-Member Aromatic Heterocyclic (AH) Compounds. [Ar = Imidazole, Pyrazole, Oxazole, Isoxazole, Thiazole, and Isothiazole].	$4H_2$ molecules	3.66% to 5.45%	Below 80K for Ag and below180K for Au	A. Bag et. al,(2024),
15	Isomeric Cu(I)-Triazine Complexes.	$5H_2$ for mono $10H_2$ for di	6.5 and 8.8 wt%	Below 150 K	A. Bag. et. al,(2023),
16	Air-Stable Cu(I) Metal–Organic Framework for Hydrogen Storage.	$6H_2$ Molecules	Good hydrogen storage capability under ambient conditions (Volumetric capacity 10.4 g/L, 233 K/100 bar to 296 K/5 bar)		D. Sengupta et. al, (2023),
17	Cu3Bx (X=1-4)	$16H_2$ Molecules	13.80 wt%	ambient conditions	Guo & Wang, (2020),
18	Cu (I)-Decorated Five-membered Aromatic Heterocyclic Complexes. [Ar = Imidazole, Pyrazole, Oxazole, Isoxazole, Thiazole, and Isothiazole].	$5H_2$ molecules	Between 6.38 to 7.15	117K	A. Bag et.al, (2025),
19	Ag(I) decorated isomeric triazine complexes.	$5H_2$ for mono $10H_2$ for di	between 5.08 wt% and 6.39 wt %.	Below 115K for mono and 160K for di	A. Bag et. al, (2025),

2. HYDROGEN STORAGE METHODS

2.1 Physical Storage

Physical storage methods involve compressing or liquefying hydrogen to increase its energy density (U. Eberle et al, (2009).

Compressed Gas Storage: Hydrogen is stored at high pressures (e.g., 350–700 bar). While mature, this method faces challenges related to energy consumption, material durability, and safety.

Liquefied Hydrogen Storage: Hydrogen is stored at cryogenic temperatures (~20 K). This method provides higher density but suffers from high energy losses due to liquefaction and boil-off losses.

Cryo-Compressed Storage: Combines advantages of compressed and liquid hydrogen storage, aiming to improve volumetric and gravimetric efficiency.

Material-Based Storage

Material-based hydrogen storage systems involve chemical interactions with various materials to store hydrogen in a compact and reversible form (Felderhoff, et al. (2007).

Metal Hydrides: Certain metals and alloys (e.g., MgH_2, TiFe) can reversibly absorb and release hydrogen. However, weight, absorption kinetics, and operating temperatures remain challenges.

Chemical Hydrides: Materials such as ammonia borane and sodium borohydride store hydrogen in chemical bonds but require efficient dehydrogenation methods.

Metal-Organic Frameworks (MOFs) and Porous Materials: High-surface-area materials that can adsorb hydrogen at relatively low pressures. Research is ongoing to improve storage capacity and reversibility.

Materials-Based Storage

Materials-based storage focuses on novel approaches such as adsorption in porous materials like metal-organic frameworks (MOFs), carbon nanotubes, and zeolites. These materials have high surface areas and unique properties that enable efficient hydrogen capture and release at moderate temperatures and pressures. Research in this area is rapidly advancing, with the goal of developing lightweight, high-capacity storage systems that meet the stringent requirements of future hydrogen-powered applications.

Despite significant progress, several challenges hinder the commercialization of hydrogen storage technologies. Safety concerns related to high-pressure storage, energy losses during compression and liquefaction, and the high cost of advanced materials remain critical barriers. Additionally, infrastructure development, including refueling stations and distribution networks, is essential to facilitate large-scale hydrogen adoption.

This chapter explores the current landscape of hydrogen storage technologies, analyzing their advantages, limitations, and recent breakthroughs. By examining emerging trends and innovative solutions, we aim to provide a comprehensive understanding of the future potential of hydrogen storage in enabling a clean and sustainable energy ecosystem.

2.2 Chemical Hydrogen Storage

Chemical storage involves binding hydrogen in molecules that release it through catalytic or thermal decomposition (Lavanya, M., et al. (2024).

Liquid Organic Hydrogen Carriers (LOHCs): Organic compounds such as toluene/methylcyclohexane store hydrogen reversibly but require high-energy input for release.

Ammonia as a Hydrogen Carrier: Ammonia has high hydrogen content and established transportation infrastructure, but its decomposition to release hydrogen requires efficient catalysts.

Chemical Storage

Chemical storage methods involve bonding hydrogen with other elements to form stable compounds, such as metal hydrides, ammonia, and liquid organic hydrogen carriers (LOHCs). These methods offer promising alternatives for safe and efficient hydrogen storage, as they allow for reversible hydrogen uptake and release under controlled conditions. However, challenges such as slow hydrogen release kinetics, high operational temperatures, and material degradation need to be overcome before widespread implementation (F. Schüth, (2009).Table.2 shows the comparison between different hydrogen storage methods. Fig. 2 shows the hydrogen Storage System.

Table 2. Comparative charts of different hydrogen storage methods:

Storage Method	Principle	Advantages	Challenges	Examples
Compressed Gas	Hydrogen is stored in high-pressure tanks (350–700 bar)	Mature technology, fast refueling, scalable	High energy loss in compression, requires strong tanks	Hydrogen fuel stations, fuel cell vehicles
Liquid Hydrogen	Hydrogen is cooled to -253°C and stored as a cryogenic liquid	High energy density, suitable for large-scale storage	High energy cost for liquefaction, boil-off losses	Space applications, heavy transport
Metal Hydrides	Hydrogen is absorbed and released by metal alloys	Safe, compact storage, moderate pressure	Heavy weight, slow absorption/desorption rates	Portable hydrogen storage, stationary applications
Chemical Hydrogen Storage	Hydrogen is stored in chemical compounds and released via reactions	High volumetric density, stable storage	Requires catalysts for release, regeneration can be energy-intensive	Ammonia, LOHCs (Liquid Organic Hydrogen Carriers)
Underground Storage	Hydrogen stored in geological formations like salt caverns	Large-scale storage, cost-effective	Limited to specific locations, risk of leakage	Grid energy storage, industrial hydrogen supply
Cryo-Compressed	Combination of liquid hydrogen and high pressure	Higher storage density than compressed or liquid H_2 alone	Complex thermal management, high cost	Advanced vehicle fuel storage, aerospace

Figure 2. Hydrogen Storage System.

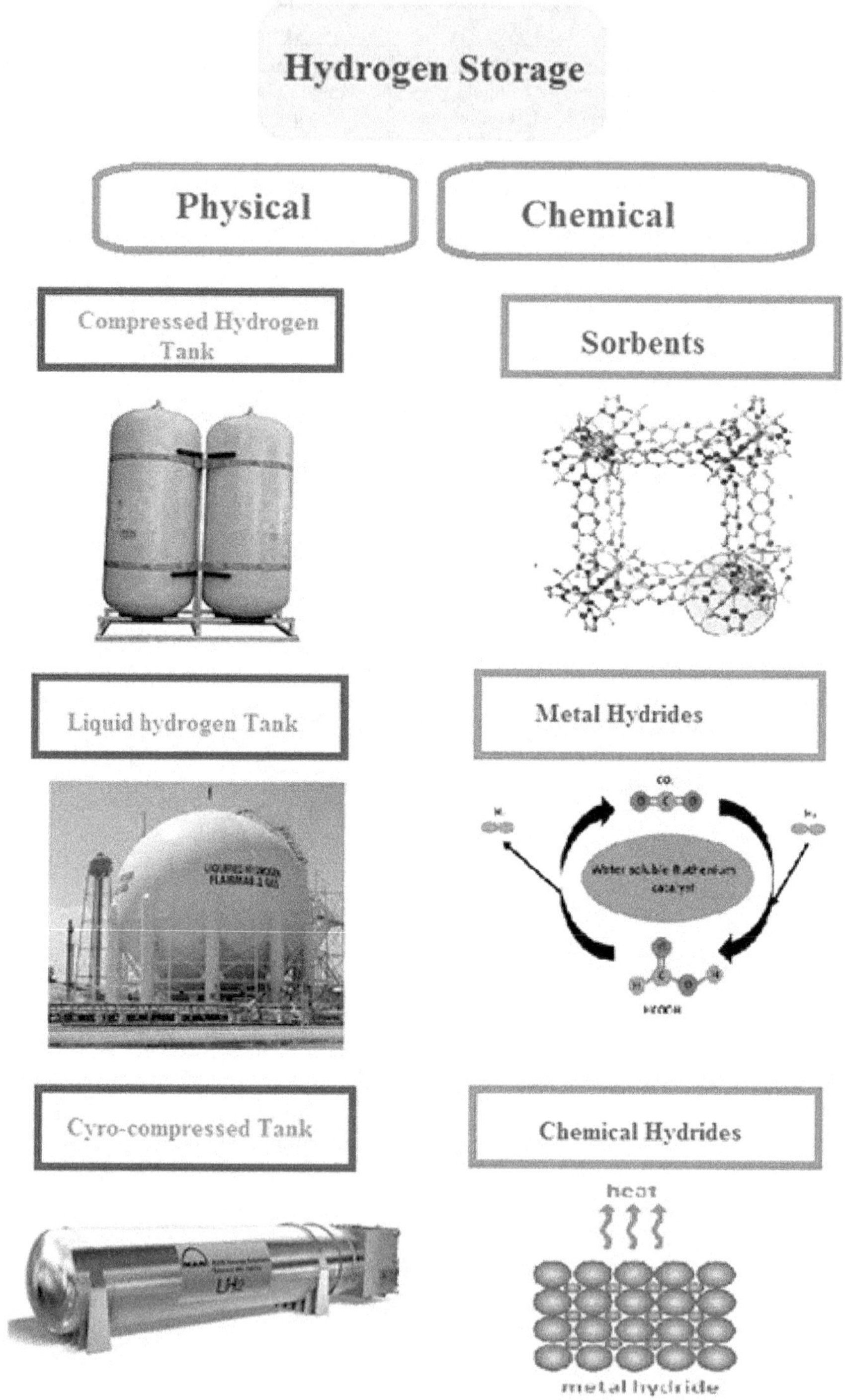

3.0 CHALLENGES IN HYDROGEN STORAGE

3.1 Energy Efficiency and Storage Density

Hydrogen has low volumetric energy density, making storage inefficient compared to conventional fuels.

Compression and liquefaction require significant energy input (~10–40% of stored energy).

Hydrogen's **low volumetric energy density** compared to conventional fuels poses a significant challenge for practical applications. Efficient storage solutions are essential to enhance its viability in transportation, energy storage, and industrial applications (Md Monjur Hossain Bhuiyan et al, (2025).

Compression and Liquefaction: Storing hydrogen as a compressed gas (350–700 bar) or liquid (-253°C) requires substantial energy, consuming approximately 10–40% of the stored energy in the process (Sadeq et al, 2024).

Metal Hydrides and Chemical Storage: While these methods improve volumetric density, they introduce weight and energy trade-offs due to the energy required for hydrogen absorption/desorption and system regeneration.

Material and Infrastructure Limitations: Advanced materials, such as lightweight composites for tanks or novel sorbents for solid-state storage, are crucial to balancing efficiency, safety, and cost.

3.2 Safety Concerns

The storage and handling of hydrogen involve several safety challenges due to its low molecular weight, high diffusivity, and flammability. Ensuring safe storage methods is critical for both industrial and commercial applications (M. Calabrese et al, (2024) and (Li et al, (2022).

High-Pressure Storage Risks: Storing hydrogen at 350–700 bar increases the risk of leaks, material fatigue, and potential explosions. Special composite tanks and leak detection systems are necessary to mitigate hazards (Abohamzeh et al, (2021).

Cryogenic Storage Challenges: Liquid hydrogen storage at -253°C suffers from boil-off losses, requiring advanced thermal insulation to minimize evaporation. Handling cryogenic hydrogen also demands specialized equipment to prevent frostbite hazards and pressure build-up.

Chemical Storage Considerations: Hydrogen carriers, such as metal hydrides, ammonia, and liquid organic hydrogen carriers (LOHCs), must address toxicity, reactivity, and controlled release to ensure safe operation. Proper containment and reaction control mechanisms are necessary to prevent unintended hydrogen release or hazardous by-products.

3.3 Cost and Infrastructure Limitations

The widespread adoption of hydrogen as an energy carrier is significantly hindered by cost-related challenges and insufficient infrastructure for storage and distribution.

High Capital and Operational Costs: The capital investment required for hydrogen storage systems-whether through compression, liquefaction, or chemical storage-remains high, with ongoing operational costs tied to energy consumption for compression or cooling. These high upfront costs make large-scale adoption difficult, particularly for industries requiring affordable solutions. For example, the installation of hydrogen refuelling stations and specialized storage tanks incurs substantial costs, limiting the ability to scale rapidly.

Lack of Infrastructure: The absence of a widespread hydrogen storage and distribution network further hinders market growth. Without an established infrastructure for transport, distribution, and refueling, the hydrogen economy faces delays in expanding into mainstream markets like transportation and energy storage.

Lack of integration with existing fuel infrastructure adds another barrier, as hydrogen needs specialized pipelines and storage facilities to be viable on a large scale.

4.0. EMERGING SOLUTIONS AND FUTURE PERSPECTIVES

4.1 Advanced Materials and Nanotechnology

The development of advanced materials and nanotechnology is the key for improving the efficiency and capacity of hydrogen storage systems, particularly in addressing the challenges of weight, density, and cost associated with traditional methods (Lavanya et al, (2024). **Lightweight, High-Capacity Materials:** The use of nano materials such as grapheme based sorbents and advanced MOFs shows significant promise for hydrogen storage. These materials can store hydrogen at higher volumetric and gravimetric densities compared to conventional methods.

Graphene-based Sorbents: Graphene, with its high surface area and tailored porosity, has been explored as a material for hydrogen adsorption at moderate pressures. The use of graphene oxide and graphene composites can enhance hydrogen storage capacity while keeping the system lightweight (Guo, (2015).

Metal-Organic Frameworks (MOFs): MOFs offer a highly porous structure and can be engineered to maximize hydrogen adsorption at both low and moderate pressures. Their flexibility and tunable properties make them ideal for improving storage densities and minimizing weight.

Nanotechnology for Enhanced Efficiency: By employing nano-structured materials and nano composites, researchers aim to enhance the kinetics of hydrogen absorption and desorption, improving the overall efficiency of storage systems. Nanotechnology also allows for improved thermal and mechanical stability, which can help overcome issues related to high pressure and cryogenic temperatures. These advancements in nanotechnology and material science offer the potential to dramatically enhance hydrogen storage, making it more viable for large-scale applications.

4.2 Hybrid Storage Approaches

Hybrid storage systems combine the strengths of multiple hydrogen storage technologies, aiming to optimize performance, capacity, efficiency, and safety. By integrating different methods, hybrid systems can address the limitations of individual approaches, offering enhanced overall hydrogen storage solutions for both stationary and transportation applications (Oladimeji Lawrence Oyewole et al, (2024).

Cryo-Compressed Hydrogen with MOFs: One promising hybrid approach combines cryogenic storage with compressed hydrogen at moderate pressures and metal organic frameworks (MOFs) for adsorption. The low temperature (cryogenic) helps to reduce the hydrogen volume, while the moderate pressure ensures that hydrogen can be stored more efficiently within the porous structure of MOFs. (Pamela Ramirez-Vidal et al, (2022)

This combination improves volumetric density compared to pure compression, and the MOFs enhance hydrogen retention even at higher temperatures. The result is an optimized hybrid system that balances energy efficiency and storage capacity.

Hybrid Metal Hydride Systems: Combining metal hydrides with compressed gas storage or cryogenic tanks can lead to hybrid systems that leverage the high volumetric density of metal hydrides for long-term storage while maintaining fast refuelling capabilities via compressed gas. This system could be particularly useful in applications like fuel cell vehicles where high efficiency and quick refuelling are required (M Davis Cortina et al, (2024).

Liquid Hydrogen and LOHCs (Liquid Organic Hydrogen Carriers): Another hybrid approach involves using liquid hydrogen in combination with liquid organic hydrogen carriers (LOHCs). LOHCs can absorb hydrogen at ambient conditions and release it through a chemical reaction when needed. By pairing LOHCs with liquid hydrogen storage, systems can optimize both volumetric density and safety, with LOHCs offering easier handling and lower risks of boil-off compared to cryogenic **storage** (Chu et al, (2023).

Cryo-Compression with Nanostructured Sorbents: Nanostructured materials, such as graphene and carbon nanotubes, are integrated into cryogenic storage systems to improve hydrogen adsorption capacity at low temperatures. This cryo-compressed hybrid system enhances the overall energy density while reducing boil-off losses. By combining various technologies, hybrid storage systems are poised to offer higher efficiency, better safety, and cost-effective solutions for large-scale and mobile hydrogen applications. Table. 3 shows the challenges and advantages of hydrogen storage technology:

Table 3. Challenges and advantages of hydrogen storage Technology:

Aspect	Challenges	Advantages
Storage Efficiency	Low volumetric energy density requiring large storage space.	High gravimetric energy density compared to fossil fuels.
Compression & Liquefaction	Requires high pressure (350–700 bar) or cryogenic temperatures (-253°C), increasing energy costs.	Enables long-term storage and transportation.
Material Selection	Limited availability of suitable materials for solid-state storage.	Metal hydrides and MOFs offer potential for safer storage.
Infrastructure	Limited refueling infrastructure and high initial investment costs.	Hydrogen can be stored in various forms (gas, liquid, solid).
Safety Concerns	Highly flammable and requires stringent safety measures.	Proper storage methods can mitigate risks effectively.
Durability & Lifetime	Storage tanks and materials degrade over time, requiring frequent maintenance.	Advanced materials research is improving longevity.
Energy Cost	Hydrogen storage involves energy-intensive processes, affecting overall efficiency.	Hydrogen can be produced from renewable sources for sustainability.
Scalability	Large-scale hydrogen storage is still in development stages.	Offers potential for large-scale energy storage solutions.
Mobility	High-pressure and cryogenic tanks add weight and complexity to vehicles.	Enables zero-emission fuel for transport applications.

4.3 Policy and Economic Considerations

Policy and Economic Considerations: The successful adoption and scaling of hydrogen storage technologies are closely tied to government policies and economic strategies that foster research, development, and integration into the broader energy system. Strategic policy frameworks, economic incentives, and the synergy with renewable energy sources are key to creating a sustainable hydrogen economy.

Government Incentives and Policies Supporting Hydrogen Storage R&D: Governments play a crucial role in stimulating innovation through funding programs, subsidies, and tax incentives aimed at advancing hydrogen storage technologies. Research and Development (R&D) support is vital to overcoming the technical and cost-related barriers to hydrogen storage. Policies that incentivize collaborative research, public-private partnerships, and demonstration projects can accelerate the commercial readiness of advanced storage solutions. Several countries have hydrogen roadmaps that allocate funds for hydrogen infrastructure development, including storage solutions, as part of their clean energy transition plans. For example, the EU Hydrogen Strategy and US Department of Energy's Hydrogen Program provide funding for projects aimed at improving energy efficiency, cost-effectiveness, and safety in hydrogen storage (Islam et al, (2024).

Integration with Renewable Energy Sources for a Sustainable Hydrogen Economy: One of the most promising pathways for hydrogen storage is its integration with renewable energy systems. Hydrogen can serve as an energy carrier that facilitates the storage of surplus renewable energy from solar, wind, or hydropower during periods of excess generation, storing it for later use when energy demand is high. This process is often referred to as Power-to-Gas (P2G). Hydrogen as a back-up storage medium for renewable can help address intermittency issues in energy generation. By coupling hydrogen storage with wind and solar plants, hydrogen can act as a flexible energy buffer, enhancing the reliability and stability of the overall energy grid. Policies promoting green hydrogen production (using renewable energy to produce hydrogen via electrolysis) can drive the market for both hydrogen storage and renewable energy integration, thus fostering a circular energy economy.

Economic Viability and Market Incentives: For hydrogen storage to become economically viable, the cost of production, storage, and distribution must be competitive with traditional energy storage systems. Economic incentives, such as carbon pricing or emissions trading systems, can help create a favourable economic environment for clean energy solutions. Governments may also support the development of hydrogen storage infrastructure by funding the construction of refueling stations and storage hubs, and promoting international hydrogen trade to reduce costs and improve access. Private investment in the sector will be critical for scaling up hydrogen storage technologies, as it will drive further innovation and cost reductions. Incentives that encourage investment in energy infrastructure, especially for hydrogen, can accelerate deployment at a large scale.

By aligning policy incentives with economic strategies, a hydrogen economy can emerge as a cornerstone of the global transition to clean, sustainable energy.

5. VARIOUS APPLICATION OF HYDROGEN

Fig.3 shows the different types of applications of hydrogen in different region like industry, transportation and energy generation.

Figure 3. Various Applications of Hydrogen.

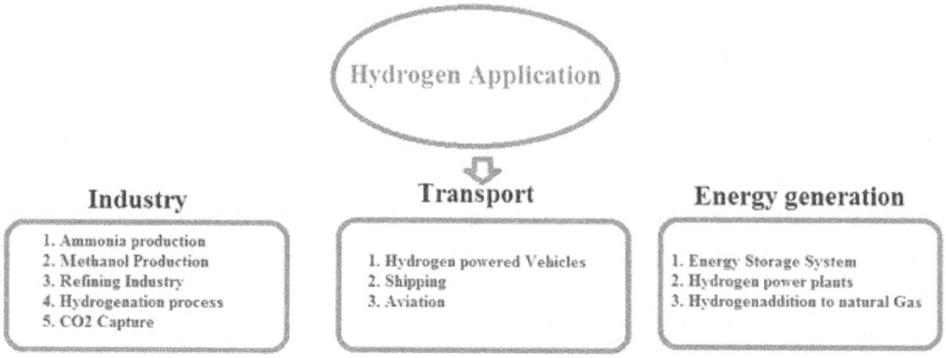

Hydrogen has a wide range of applications across various industries due to its high energy content, clean combustion properties, and versatility. Here's a categorized overview:

5.1. Energy & Power Generation:

Fuel Cells: Hydrogen is used in stationary and mobile applications for clean energy production.

Hydrogen Power Plants: It is used in gas turbines and combustion engines to generate electricity.

Energy Storage: Hydrogen acts as an energy carrier for storing excess renewable energy (Power-to-Gas systems).

5.2. Transportation:

Fuel Cell Vehicles (FCVs): It is used in hydrogen-powered cars, buses, trucks, trains, and even aircraft.

Maritime Applications: Hydrogen fuel cells are being explored for ships and submarines.

Aerospace & Space Exploration: Used in rocket propulsion and auxiliary power units.

5.3. Industrial Applications:

Ammonia & Fertilizer Production: Hydrogen is a key component in ammonia (NH_3) synthesis via the Haber-Bosch process.

Refinery & Petrochemical Industry: It is used in hydrocracking, desulfurization, and refining of crude oil.

Metal Production & Processing: It is used in reducing agents for metal refining and heat treatment processes.

5.4. Chemical Industry:

Methanol Production: Hydrogen is used as a feedstock for methanol synthesis.

Synthetic Fuels & Chemicals: Key component in producing e-fuels and synthetic hydrocarbons.

5.5. Domestic & Commercial Use:

Hydrogen Heating Systems: Being explored as a low-carbon alternative for residential and commercial heating.

Backup Power Supply: Fuel cells provide reliable power in off-grid and emergency situations.

5.6. Environmental Applications:

Carbon Capture & Utilization (CCU): Used in producing synthetic fuels while reducing CO_2 emissions.

Hydrogen for Grid Stability: Helps in balancing the electricity grid when using variable renewable energy sources.

5.7. Medical & Healthcare:

Hydrogen Therapy: Being studied for potential therapeutic applications in oxidative stress-related diseases.

Medical Gas Applications: It is used in specific controlled environments for research and healthcare.

6. CONCLUSION

Hydrogen storage continues to present a major challenge in the transition to a hydrogen-based economy. Despite the progress in developing physical, chemical, and material-based storage methods, each approach still faces distinct challenges related to energy efficiency, storage density, cost, and safety. However, ongoing research into novel materials, hybrid storage systems, and innovative policy frameworks offers promising solutions to these hurdles. Key advancements, such as the development of high-capacity materials like MOFs and graphene-based sorbents, and the integration of multiple storage technologies, have the potential to enhance hydrogen storage performance while reducing costs. Moreover, the combination of government incentives and public-private partnerships can significantly accelerate the deployment of advanced storage systems, making large-scale hydrogen storage more viable. In order to realize the full potential of hydrogen as a clean energy carrier, it is essential to continue addressing the technical and economic barriers that currently limit large-scale adoption. This includes further advancements in storage materials, improving system efficiency, and ensuring economic competitiveness with other energy storage solutions. Through continued innovation and collaboration, hydrogen storage can become a cornerstone in a sustainable and decarbonised energy future.

REFERENCES

Abohamzeh, E., Salehi, F., Sheikholeslami, M., Abbassi, R., & Khan, F. (2021). Review of hydrogen safety during storage, transmission, and applications processes. *Journal of Loss Prevention in the Process Industries*, *72*, 104569.

Bag, A., ... (2023) "Hydrogen Storage Efficiency of Isomeric Cu (I)-Triazine Complexes: In Quest of New Hydrogen Storage Material, Computational Studies: From Molecules to Materials.

Bag, A., ... (2023), CRC Press/Tylor & Francis Group, 2023. ISBN 9781032528540.

Bag, A., Chandra De, G., Bhattacharyya, S., Bepari, B., Das, H. S., Bandaru, S., & Roymahapatra, G. (2025). Ag(I) decorated isomeric triazine complexes as efficient hydrogen storage materials - A theoretical investigation. *Chemistry of Inorganic Materials.*, *2025*(5), 100093. DOI: 10.1016/j.cinorg.2025.100093

Bag, A., Dash, M. K., Tarif, E., Guo, J. Z., De, G. C., Maity, S., Ganguly, J., Bhattacharyya, S., & Roymahapatra, G. (2025). Cu(I)-Decorated Five-membered Aromatic Heterocyclic Complexes; A Potential H2 Storage System. *ES Chemistry and Sustainability*, *2025*(02), 1311. DOI: 10.30919/escs1311

Bag, A., Giri, S., Dhaiveegan, P., Jhalgam, R. T. T., De, G. C., Ganguly, J., & Roymahapatra, G. (2024). ES. *Energy & Environment*, *2024*(23), 1115. DOI: 10.30919/esee1115

Bag, A., Sinha, S., Das, H. S., Giri, S., De, G. C., Maity, S., Xu, B. B., Guo, Z., & Roymahapatra, G. (2024). Hydrogen Storage Efficiency of the Ag (I)/Au (I) Decorated Five-Member Aromatic Heterocyclic (AH) Compounds: A Theoretical Investigation. *Engineering and Science*, *2024*(28), 1062. DOI: 10.30919/es1062

Bhuiyan, M. M. H., & Siddique, Z. (2025). Hydrogen as an alternative fuel: A comprehensive review of challenges and opportunities in production, storage, and transportation. *International Journal of Hydrogen Energy*, *102*, 1026–1044.

Blomqvist, A., Araújo, C. M., Srepusharawoot, P., & Ahuja, R. (2007). Li-decorated metal–organic framework 5: A route to achieving a suitable hydrogen storage medium. *Proceedings of the National Academy of Sciences of the United States of America*, *104*(51), 20173–20176. www.pnas.org/cgi/doi/10.1073/pnas.0708603104. DOI: 10.1073/pnas.0708603104 PMID: 18093960

Calabrese, M., Portarapillo, M., Di Nardo, A., Venezia, V., Turco, M., Luciani, G., & Di Benedetto, A. (2024). Hydrogen Safety Challenges: A Comprehensive Review on Production, Storage, Transport, Utilization, and CFD-Based Consequence and Risk Assessment. *Energies, 2024*(17), 1350. DOI: 10.3390/en17061350

Chattaraj, S., Srinivasu, K., Mondal, S., & Ghosh, S. K. (2015). Hydrogen trapping ability of the pyridine–lithium+ (1: 1) complex. *The Journal of Physical Chemistry A, 119*(12), 3056–3063.

Chu, C., Wu, K., Luo, B., Cao, Q., & Zhang, H. (2023). Hydrogen storage by liquid organic hydrogen carriers: Catalyst, renewable carrier, and technology–a review. *Carbon Resources Conversion, 6*(4), 334–351.

Das, H. S.. (2025). *Advanced Nano-Structured Materials for Energy Storage Devices Title: Design.* Fabrication, and Significance of Advanced Nanostructured Materials., DOI: 10.4018/979-8-3693-5320-2.ch001

Dash, M. K.. (2022). S. Sinha, H. S. Das, G. C. De, S. Giri, G.Roymahapatra. *Sustainable Energy Technologies and Assessments, 2022*(52), 102235. DOI: 10.1016/j.seta.2022.102235

Dash, M. K., Chowdhury, S. D., Chatterjee, R., Maity, S., Roymahapatra, G., Huang, M., Bandyopadhyay, M. N., & Guo, Z. (2022). Can Lithium Fluoride (LiF) Be An Efficient Hydrogen Storage Material? An In-Silico Study. *Engineering and Science, 18.* Advance online publication. DOI: 10.30919/es8d658

Dash, M. K., Das, S., Giri, S., Chandra De, G., & Roymahapatra, G. (2021). Comprehensive in silico study on lithiated Triazine isomers and its H2 storage efficiency. *Journal of the Indian Chemical Society, 2021*(98), 100134. DOI: 10.1016/j.jics.2021.100134

Dash, M. K., Sinha, S., Das, H. S., De, G. C., Giri, S., & Roymahapatra, G. (2022). H2 storage capacity of Li-doped five member aromatic heterocyclic superalkali complexes; an in silico study. *Sustainable Energy Technologies and Assessments, 52*, 102235.

Davis Cortina, M., Romero de Terreros Aramburu, M., Neves, A. M., Hurtado, L., Jepsen, J., & Ulmer, U. (2024). The Integration of Thermal Energy Storage Within Metal Hydride Systems: A Comprehensive Review. *Inorganics, 2024*(12), 313. DOI: 10.3390/inorganics12120313

Eberle, U., Felderhoff, M., & Schüth, F. (2009). Chemical and physical solutions for hydrogen storage. *Angewandte Chemie International Edition, 48*(36), 6608–6630. DOI: 10.1002/anie.200806293 PMID: 19598190

Felderhoff, M., Weidenthaler, C., von Helmolt, R., & Eberle, U. (2007). Hydrogen storage: The remaining scientific and technological challenges. *Physical Chemistry Chemical Physics*, *9*(21), 2643–2653. DOI: 10.1039/b701563c PMID: 17627309

Gadipelli, S., & Guo, Z. X. (2015). Graphene-based materials: Synthesis and gas sorption, storage and separation. *Progress in Materials Science*, *69*, 1–60.

Guo, C., & Wang, C. (2020). The theoretical research of hydrogen storage capacities of Cu3Bx (X=1–4) compounds under ambient conditions. *International Journal of Hydrogen Energy*, *45*(46), 24947–24957. DOI: 10.1016/j.ijhydene.2020.06.089

Islam, A., Islam, T., Mahmud, H., Raihan, O., Islam, M. S., Marwani, H. M., & Awual, M. R. (2024). Accelerating the green hydrogen revolution: A comprehensive analysis of technological advancements and policy interventions. *International Journal of Hydrogen Energy*, *67*, 458–486.

Li, H., Cao, X., Liu, Y., Shao, Y., Nan, Z., Teng, L., & Bian, J. (2022). Safety of hydrogen storage and transportation: An overview on mechanisms, techniques, and challenges. *Energy Reports*, *8*, 6258–6269.

Lavanya, M.. (2024). An overview of hydrogen storage technologies–Key challenges and opportunities. *Materials Chemistry and Physics*, •••, 129710.

Lavanya, M., Shrivastava, S., Lakshmi, T., Sandadi, E. R., Gour, S., Thomas, N. A., & Sudhakar, K. (2024). An overview of hydrogen storage technologies–Key challenges and opportunities. *Materials Chemistry and Physics*, •••, 129710.

Oyewole, O. L., Nwulu, N. I., & Okampo, E. J. (2024). Optimal design of hydrogen-based storage with a hybrid renewable energy system considering economic and environmental uncertainties. *Energy Conversion and Management*, *300*, 117991.

Ramirez-Vidal, P., Sdanghi, G., Celzard, A., & Fierro, V. (2022). High hydrogen release by cryo-adsorption and compression on porous materials. *International Journal of Hydrogen Energy*, *47*(14), 8892–8915.

Pan, S., Giri, S., & Chattaraj, P. K. (2011). A computational study on the hydrogen adsorption capacity of various lithium—Doped boron hydrides. *Journal of Computational Chemistry*, *33*(4), 425–434. DOI: 10.1002/jcc.21985 PMID: 22121031

Parida, R., Dash, M. K., Giri, S., & Roymahapatra, G. (2021). Aromatic N-Heterocyclic superalkali M@C4H4N2 complexes (M=Li, Na, K); A promising potential hydrogen storage system. *Journal of the Indian Chemical Society*, *2021*(98), 100065. DOI: 10.1016/j.jics.2021.100065

Roymahapatra, G., Dash, M. K., Ghosh, A., Bag, A., Mishra, S., & Maity, S. (2022). Hydrogen Storage on Lithium Chloride/Lithium Bromide Surface at Cryogenic Temperature. *ES Energy & Environment*, *2022*(18), 90–100. DOI: 10.30919/esee8c756

Roymahapatra, G., Dash, M. K., Sinha, S., De, G. C., & Guo, Z. (2022). Theoretical Investigation of Hydrogen Adsorption Efficiency of [Oxadiazole- xLi+] Complexes (x = 1, 2): In Pursuit of Green Fuel Storage. *Engineering and Science*, *2022*(19), 114–124. DOI: 10.30919/es8d671

Sadeq, A. M., Homod, R. Z., Hussein, A. K., Togun, H., Mahmoodi, A., Isleem, H. F., & Moghaddam, A. H. (2024). Hydrogen energy systems: Technologies, trends, and future prospects. *The Science of the Total Environment*, *939*, 173622.

Sahoo, R. K., & Sahu, S. (2022, August 23). Reversible hydrogen storage capacity of Li and Sc doped novel C_8N_8 cage: Insights from density functional theory. *International Journal of Energy Research*, *46*(15), 22585–22600. DOI: 10.1002/er.8562

Sathe, R. Y., & Dhilip Kumar, T. J. (2020). Reversible hydrogen adsorption in Li functionalized [1,1]paracyclophane. *International Journal of Hydrogen Energy*, *2020*(45), 12940–12948. DOI: 10.1016/j.ijhydene.2020.03.009

Schüth, F. (2009). Challenges in hydrogen storage. *The European Physical Journal. Special Topics*, *176*(1), 155–166. DOI: 10.1140/epjst/e2009-01155-x

Sengupta, D., Melix, P., Bose, S., Duncan, J., Wang, X., Mian, M. R., Kirlikovali, K. O., Joodaki, F., Islamoglu, T., Yildirim, T., Snurr, R. Q., & Farha, O. K. (2023). Air-Stable Cu(I) Metal–Organic Framework for Hydrogen Storage. *Journal of the American Chemical Society*, *2023*(37), 20492–20502. Advance online publication. DOI: 10.1021/jacs.3c06393 PMID: 37672758

Sun, Q., Jena, P., Wang, Q., & Marquez, M. (2006). First-Principles Study of Hydrogen Storage on $Li_{12}C_{60}$. *Journal of the American Chemical Society*, *128*(30), 9741–9745. DOI: 10.1021/ja058330c PMID: 16866529

Chapter 9

Precision Agriculture Meets Sustainable Chemistry: Innovations for Eco-Friendly Farming

Azmirul Hoque
https://orcid.org/0009-0002-7220-9818
Assam University, Silchar, India

Mrutyunjay Padhiary
https://orcid.org/0000-0002-2236-568X
Assam University, Silchar, India

Suranjit Roy
https://orcid.org/0009-0004-1483-6447
Assam University, Silchar, India

ABSTRACT

Precision agriculture with the emergence of sustainable chemistry has modernized farming by enhancing efficiency, minimizing environmental impacts, and encouraging the conservation of resources. It makes use of more advanced techniques, such as drones, remote sensing, the Internet of things, and artificial intelligence. Sustainable chemistry focuses on biodegradable inputs, bio-based fertilizers, and globally inert insecticides. This chapter examines the entities between precision agriculture and supply chain, including AI-driven decision support systems, blockchain for traceability, and biopesticides for pest control. It also analyzes leading frameworks, financial incentives, and the development of digital networks. Future

DOI: 10.4018/979-8-3373-1409-9.ch009

research needs to concentrate on AI-driven precision agriculture, uses of nanotechnology, and climate-resilient agricultural approaches. Connecting precision agriculture with sustainable agriculture enhances the sustainability, productivity, and sustainability of farming systems, securing long-term food security and lowering agriculture's ecological footprint.

1. INTRODUCTION

1.1 Overview

Precision agriculture (PA) and sustainable chemistry are two fundamental concepts designed to enhance production and reduce environmental effects in agriculture. PA uses sophisticated technologies, such as artificial Intelligence (AI), internet of things (IoT), drones, and GIS, to meticulously monitor and manage agricultural processes, allowing farmers to enhance resource efficiency, increase crop yields, and minimize waste. The global PA market has shown substantial expansion, valued at \$9.3 billion in 2024 and anticipated to attain \$21.5 billion by 2033 (Bade and Tomomewo, 2024). IoT devices furnish real-time data regarding soil conditions, crop health, and meteorological patterns, allowing educated decision-making. In 2024, North America dominates PA adoption, capturing more than 51.2% of the market share (Foong et al., 2024). Sustainable chemistry (SC) in agriculture involves the development of products and processes that minimize the use and production of harmful substances, hence enhancing environmental health and resource efficiency. Green agrochemicals, including bio-based fertilizers and pesticides, aim to diminish the environmental impacts of agricultural practices (Dhuldhaj et al., 2023). Global governments are enacting laws to promote sustainable chemicals in agriculture, providing incentives for activities that match environmental conservation objectives.

1.2 Importance of Eco-Friendly Farming Practices

Ecologically sustainable farming practices are essential for sustaining food production, environmental preservation, and long-term agricultural sustainability (Padhiary & Kumar, 2024a). Traditional agricultural practices frequently result in soil erosion, low fertility, and desertification, which can be alleviated by sustainable techniques such as crop rotation, organic fertilizers, and conservation tillage (Diop et al., 2022). These techniques also reduce water contamination by promoting biofertilizers and integrated pest management (IPM). They also help reduce greenhouse gas (GHG) emissions, as agriculture accounts for 17–20% of global GHG emissions (Chataut et al., 2023). Eco-friendly agriculture increases food security and health by reducing toxic chemical residues, boosting nutritional quality, and nurturing biodiversity. It enhances farmers' profitability by eliminating reliance on costly chemical inputs, developing local and organic markets, and improving rural livelihoods.

Methods like cover cropping, decreased tillage, and composting enhance climate resilience by improving soil organic carbon and reducing reliance on chemical inputs. Furthermore, governments are imposing greater environmental restrictions, promoting low-impact agriculture via carbon credits and subsidies. Regulatory and consumer trends are evolving, with 65% of worldwide customers prepared to pay a premium for environmentally sustainable items (Velaoras et al., 2025). Developing sustainable agricultural techniques can secure a viable future for food production and the ecosystem.

1.3 Objectives

This chapter examines the convergence of PA and SC in environmentally sustainable farming practices. It outlines PA and its function in enhancing agricultural methodologies using AI, IoT, and remote sensing technologies. SC is used for environmentally friendly fertilizers, biopesticides, and soil management practices. PA lowers chemical overuse, soil degradation, and water contamination, whereas SC eliminates GHG emissions and safeguards biodiversity. AI and the IoT contribute to SA by minimizing resource waste and enabling real-time monitoring of soil health, crop development, and pest management. Case studies and practical applications and examine global policy frameworks and government programmes that promote sustainable agriculture. Challenges comprise elevated adoption expenses, data management problems, and regulatory limitations. Future innovations encompass nanotechnology, AI-based decision support systems, and blockchain for traceability.

2. FUNDAMENTALS OF PA

PA, also known as smart farming or digital farming, is a technology-based methodology aimed at enhancing farm management through the use of real-time data, automation, and artificial intelligence

2.1 Definition and Key Concepts of PA

PA is a technologically inclined approach to farming that uses real-time data, AI, IoT, and automation to improve agricultural efficiency, optimize resource utilization, and minimize the impact on the environment. It enables farmers to make data-driven decisions concerning crop management, irrigation, fertilization, insect control, and harvesting (Mishra and Mishra, 2024). Unlike conventional agriculture, which employs homogeneous inputs throughout a field, PA emphasizes site-specific management by examining soil, crop, and environmental conditions at a micro level, ensuring that resources such as fertilizers, water, and pesticides are used exclusively where and when necessary.

The fundamental principles of PA focus on efficiency, sustainability, and production. The basic foundation is variable-rate technology (VRT), which enables the accurate application of inputs according to soil and crop requirements, thereby minimizing waste and enhancing yield quality (Majumder et al., 2021). A fundamental concept is remote sensing, which uses satellites, drones, and multiple spectral bands to analyze crop health, identify nutritional deficiencies, and detect pest infestations prior to the onset of apparent symptoms. Besides, geographic information systems (GIS) and global positioning systems (GPS) are essential for mapping agricultural land and allowing farmers to assess field variability and formulate customized management strategies.

AI and machine learning (ML) improve PA by facilitating predictive analytics, wherein AI systems analyze extensive datasets, including sensors, climate models, and historical agricultural records to anticipate optimal planting periods, disease occurrences, and projected yield results (Manickam et al., 2022). The incorporation of IoT-enabled smart sensors in soils and crops facilitates real-time monitoring of moisture levels, temperature, and nutrient composition, enabling automated irrigation systems to offer accurate water volumes, minimize water waste, and enhance crop resilience to dry conditions. Automation and robots are shaping PA, with autonomous tractors, robotic weeders, and AI-driven harvesting devices increasing efficiency, decreasing labour expenses, and facilitating timely agricultural practices (Raza et al., 2024).

PA is a change from traditional farming methods to a more scientific, data-driven, and focused approach that promises higher yields, lower input costs, and fewer negative effects on the environment. Despite the difficulties of climate change, soil degradation, and water scarcity affecting global agriculture, the use of precision agricultural technologies is increasingly vital for maintaining sustainable and resilient food production systems.

2.2 Technologies in Precision Agriculture

PA utilizes a synthesis of sophisticated technology to improve farming efficiency, optimize resource utilization, and reduce environmental impact. Major technologies include GPS, IoT, AI, remote sensing, and unmanned planes (drones). These technologies collaborate to facilitate real-time data collection, analysis, and automation, enabling farmers to make informed choices and enhance productivity (Padhiary, Kumar, et al., 2024).

Global Positioning System (GPS): It is a fundamental technology in PA that facilitates accurate position monitoring of agricultural machinery, sensors, and equipment. It enables farmers to mark their fields, assess crop conditions, and apply inputs like water, fertilizer, and pesticides with precision. GPS-enabled autonomous tractors and sprayers ensure consistent resource application, reducing waste and enhancing efficiency (Padhiary, Tikute, et al., 2025). Likewise, GPS-guided VRT facilitates the precise application of inputs, enhancing plant development while reducing environmental effects.

Internet of Things (IoT) in smart farming: IoT in PA includes a network of linked sensors and devices that collect and relay real-time agricultural data. These sensors quantify soil moisture, temperature, humidity, nutrient levels, and crop vitality, assisting farmers in making data-informed decisions. IoT-enabled precision irrigation systems remotely modify water distribution according to soil moisture levels, decreasing water waste by as much as 50% (Ala'F et al., 2025). Additionally, IoT devices placed on livestock track their health, movements, and feeding behaviors, enhancing herd management and disease prevention. Cloud-based IoT technologies facilitate farmers' access to remote monitoring systems through Smartphone's or PCs, hence enhancing farm management.

Artificial Intelligence (AI) and Machine Learning: AI is essential for predictive analytics, automation, and decision-making in predictive analytics. AI-powered ML models evaluate extensive datasets from IoT sensors, satellite images, and historical meteorological records to predict ideal planting periods, and pest outbreaks, and yield results. AI-driven computer vision facilitates the automatic identification of plant diseases, weed infestations, and nutritional deficits in crops through drone or satellite footage (Aziz et al., 2025; Padhiary & Kumar, 2024b). The fusion of robotics with AI enhances PA by automating the weeding, harvesting, and sorting of products, hence reducing labour costs and increasing productivity. AI aids in the creation of digital twin models, wherein virtual simulations predict the impacts of various agricultural methods on crop development and soil vitality.

Remote Sensing and Satellite Imaging: Remote sensing technology uses satellites, drones, and multispectral sensors to obtain high-resolution photographs of agricultural land, offering critical insights about soil conditions, crop vitality, and environmental alterations. Remote-sensing methodologies, such as the normalized difference vegetation index (NDVI), assess plant health by estimating chlorophyll levels, thereby identifying initial indicators of illness, drought stress, or nutrient shortages (Kogan, 2022). Satellite-based weather forecasts assist farmers in organizing irrigation schedules, pest management, and harvesting to mitigate climate-related hazards. Moreover, remote sensing is essential for monitoring carbon footprints and certifying SA by evaluating land utilization and conservation initiatives.

Drones in Precision Agriculture: Drones are shifting agriculture through real-time aerial monitoring of agricultural fields. Fitted with high-resolution cameras and multispectral sensors, they collect precise field data, allowing farmers to identify crop stress, pest infestations, and nutrient deficits quickly (Laveglia et al., 2024). They can identify soil diversity for precise sowing and fertilization; automate pesticide application; and assess livestock health. The mixture of AI and drone technology improves real-time decision-making through the automation of image processing and predictive analytics.

2.3 Role of Big Data and AI in Farm Management

The synthesis of big data and AI in agriculture management has turned conventional farming methodologies through the facilitation of data-driven decision-making, predictive analytics, and automation. These technologies enhance crop yield, resource efficiency, pest management, and supply chain operations, resulting in improved efficacy, sustainability, and profitability. Big Data in agriculture covers the collection, processing, and analysis of extensive volumes of structured and unstructured data from diverse sources, including IoT sensors, satellite imaging, meteorological forecasts, machinery data, market trends, and genomic information (Wysel, 2023).

AI is essential for analyzing big data through the application of ML models, neural networks, and deep learning (DL) algorithms, delivering real-time insights and automated solutions for agricultural management. Potential uses of AI in agriculture comprise predictive analytics for crop planning and yield forecasting, AI-driven pest and disease identification, smart irrigation and water management, autonomous farming utilizing AI and robots, and AI in supply chain and market analysis.

Yet difficulties include high operational expenses, data privacy issues, and the necessity for technological proficiency. Progress in cloud computing, edge AI, and blockchain-based data security is addressing these concerns (Albshaier et al., 2025). AI-powered climate resilience models to deal with the effects of climate change, blockchain integration for open and safe data sharing, and AI-powered digital twins for simulating and making decisions about farming in the virtual world. **Figure 1** reveals the incorporation of big data and AI technologies within current farm management systems and expands agricultural practices via real-time decision support, predictive analytics, early disease detection, supply chain efficiency, and sustainable resource utilization, resulting in increased productivity and environmental sustainability.

Figure 1. Integration of big data and AI in farm management

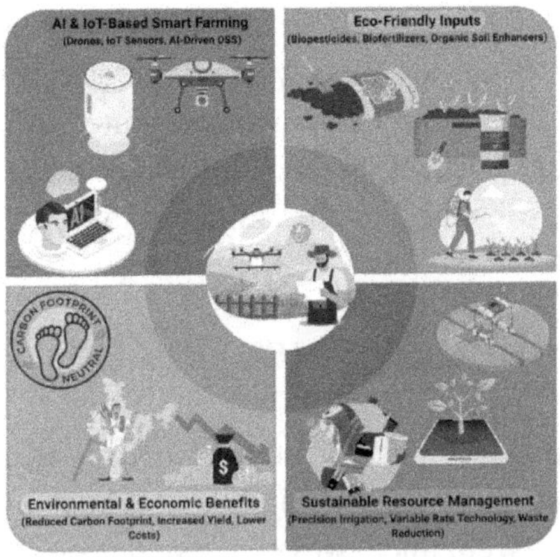

2.4 Challenges and Opportunities in PA Adoption

PA offers the ability to enhance agricultural efficiency, sustainability, and prosperity; though, it faces several hurdles, such as high initial expenditures, technical complexities, and data management complications. Problems include the major initial expenses associated with modern technology such as GPS-guided equipment, drones, IoT sensors, AI-driven software, and big data analytics platforms, which small and medium-sized farms frequently cannot afford.

Technical complexity and insufficient expertise are notable barriers to the adoption of PA, as some farmers lack the necessary education to proficiently run and maintain these systems (Pandeya et al., 2025). Data management and cybersecurity concerns present significant challenges, as PA produces huge amounts of data from sensors, satellites, and agricultural gear, necessitating effective storage, processing, and analysis. In rural agricultural areas, AI-driven decision support systems, IoT sensors, and remote monitoring technologies can't be used in real-time because of problems with connectivity and infrastructure (Redhu et al., 2022). Resistance to change and issues with adoption may also constitute an impediment. Prospects for PA adoption involve AI and ML for intelligent farming, the IoT and smart sensors for real-time monitoring, governmental support and incentive initiatives, blockchain for data security and transparency, and sustainable, climate-resilient agricultural practices.

AI-driven models can forecast yield, improve irrigation, automate pest identification, and enhance soil health management, facilitating data-informed decision-making and increasing production. IoT-based sensors provide real-time data on soil conditions, crop health, water use, and weather patterns, facilitating the adoption of precision farming practices and enhancing resource efficiency for small-scale farmers (Dhanaraju et al., 2022). Governments and international organizations offer financial incentives, subsidies, and training packages to facilitate PA adoption.

3. SUSTAINABLE CHEMISTRY (SC) IN AGRICULTURE

SC in agriculture promotes the application of environmentally friendly chemical procedures to enhance crop yield, soil vitality, and ecological preservation while reducing toxic compounds, chemical waste, and pollution (Wahab et al., 2024). The main ideas include making farming chemicals and methods that are better for the environment, supporting a circular economy, using fewer harmful agrochemicals, and making the soil healthier and more effective at using nutrients. Recently, there have been improvements in environmentally friendly chemistry. These include bio-based and nano-formulated agrochemicals, cold plasma technology for long-term control of diseases and pests, and enzyme-based solutions for keeping crops safe. Challenges include higher production expenses, insufficient farmer awareness, and rules and regulations. Prospective opportunities incorporate governmental legislation, advancements in AI and the IoT, and increased utilization of biodegradable and bio-based agrochemicals for enhanced environmental sustainability. **Table 1** compares traditional agricultural methods with SC-based strategies, recognizing advantages such as higher environmental safety, less dependence on chemicals, and increased resource efficiency. Sustainable practices rely on bio-based inputs, PA technologies, and circular economy principles to encourage environmentally friendly farming systems, hence improving resource efficiency.

Table 1. Comparison of conventional and sustainable chemistry approaches in agriculture

Aspect	Conventional Approach	Sustainable Chemistry Approach	Benefits	References
Fertilizers	Synthetic nitrogen-based	Biofertilizers, nano-fertilizers	Improves soil health, reduces runoff	(Arora et al., 2024)
Pesticides	Chemical-based (e.g., organophosphates)	Biopesticides (e.g., neem, microbial)	Less toxic, eco-friendly	(Archana et al., 2022)
Weed control	Herbicides (glyphosate)	Cover crops, biological control	Reduces chemical dependency	(Mennan et al., 2020)
Soil health management	Intensive tillage	Organic amendments, biochar	Enhances microbial activity	(Ahmad et al., 2022)
Waste management	Landfilling, incineration	Circular economy (composting, biogas)	Reduces pollution, improves sustainability	(Hidalgo et al., 2019)
Water management	Overuse of water, runoff	Smart irrigation, AI-based monitoring	Conserves water resources	(Krishnan et al., 2022)
Crop protection	Synthetic fungicides	Cold plasma, enzyme-based protection	Safe for human health and the environment	(Leskovac and Petrović, 2023)
Packaging	Plastic-based	Biodegradable, bio-based packaging	Reduces plastic waste	(Stark and Matuana, 2021)
Energy use	Fossil fuel-dependent	Renewable energy in farming	Lowers carbon footprint	(Hadj, 2021)
Technological integration	Traditional chemical application	AI-driven precision chemical use	Optimizes inputs, minimizes waste	(Sakkaravarthy et al., 2024)

3.1 Definitions and Principles of Sustainable Chemistry (SC)

SC, or green chemistry (GC), is a method that prioritizes the design of chemical products and processes to minimize environmental effects, decrease the use of hazardous compounds, and improve resource efficiency. SC in agriculture includes the use of nontoxic fertilizers, herbicides, and fertilizers that enhance crop yield while protecting soil ethics, water quality, and biodiversity (Rashmi et al., 2022). It supports biodegradable, non-toxic, and renewable substitutes for traditional agrochemicals, establishing a balance between agricultural productivity and environmental sustainability.

SC in agriculture reinforces waste and pollution reduction, optimized chemical reactions, and the adoption of safer chemicals and biodegradable inputs. It promotes the use of nano-formulated nutrients by substituting hazardous pesticides and synthetic fertilizers with biodegradable alternatives. Using waste-to-resource techniques, solar-powered AI-driven precision farming methods, and agrochemicals derived from plants and microbes all help promote renewable feedstock and a circular economy while also making energy use more efficient. The environmental persistence of chemicals can be reduced by the creation of agrochemicals that decompose spontaneously, leaving no hazardous residues. Enzyme-based treatments are employed for pest and disease management (Zhang et al., 2024). Advanced technologies such as AI, IoT, and remote sensing are employed to assess soil conditions and enhance chemical application efficiency.

3.2 Green Chemistry (GC) Approaches in Agriculture

GC in agriculture attempts to reduce environmental impact, improve resource efficiency, and encourage the use of safer, renewable inputs in place of conventional chemical-intensive practices. This methodology mixes biotechnology, nanotechnology, AI, and SC to enhance agricultural yield while preserving environmental equilibrium.

Bio-based fertilizers and nano-fertilizers have been promoted for sustainable nutrient provision by reducing soil deterioration, water contamination, and GHG emissions. Bio-based fertilizers contain microbial biofertilizers, compost, and organic amendments, as well as nano-fertilizers such as nano-urea, nano-phosphorus, and nano-zinc (Garg et al., 2023). These strategies reduce excessive fertilizer discharge, promote soil microbial health and fertility, and improve plant uptake efficiency. GC provides biopesticides and natural pest management by utilizing bacteria, fungi, or viruses to specifically target pests. Natural insect control involves neem oil, pyrethrin, rotenone, RNAi-based pest management, and predatory insects such as ladybirds, lacewings, and parasitic wasps (Nawaz and Ahmad, 2015). In conjunction with PA and AI-optimized input management, GC reduces chemical abuse through real-time monitoring and AI-driven decision-making. Essential technologies in PA involve IoT soil sensors, AI and big data analytics, drone-assisted spraying, and cold plasma technology for pathogen and enzyme management.

GC supports the development of bioherbicides like essential oils, allelopathic crops, and fungal bioherbicides that kill weeds effectively and sustainably. Cold plasma technology inactivates germs and spoilage enzymes in food and agricultural goods, effectively controlling E. coli, Salmonella, and Listeria. AI-driven smart irrigation strategies, including drip irrigation with AI monitoring and the application of hydrogen and biochar soil additives, encourage smart irrigation and water conservation. These methods conserve 30%–50% more water than flood irrigation, decrease fertilizer runoff in aquatic ecosystems, and promote climate resilience in agriculture (Sikka et al., 2022). Enzyme-based solutions for food preservation are aided by GC, which utilizes enzyme inhibitors to prevent enzymatic browning and spoilage; cold plasma-enhanced preservation to decrease microbial spoilage without chemical additives; and antimicrobial peptides to inhibit bacterial proliferation in fresh produce and dairy products. These methods diminish the need for artificial preservatives, naturally improve food quality and safety, and prolong shelf life without losing nutritional integrity.

3.3 Biodegradable Inputs, Bio-Based Fertilisers, and Eco-Friendly Pesticides

SA depends on biodegradable inputs, bio-based fertilizers, and environmentally harmless pesticides to reduce environmental pollution, improve soil fertility, and encourage safe food production. These options adhere to GC principles by reducing toxicity, enhancing biodegradability, and decreasing the carbon footprint of agricultural practices. Biodegradable inputs contain biodegradable mulch films, seed coatings, and plant-derived packaging materials that break down naturally into non-toxic substances (Acharjee et al., 2023). Also, they promote a circular bioeconomy through the use of plant-derived products.

Bio-based fertilizers originate from natural, renewable sources such as plant waste and animal excrement and are beneficial to microbes (Mandal et al., 2024). They facilitate nitrogen cycling, enhance soil organic matter, and reduce chemical runoff into aquatic systems. Categories of bio-based fertilizers encompass microbial biofertilizers, compost and vermicompost, seaweed and algae-derived fertilizers, and nano-biofertilizers. These products decrease reliance on chemical fertilizers, moderate soil deterioration, enhance microbiological health, improve nutrient availability, and also decrease eutrophication and water pollution.

Eco-friendly pesticides, or biopesticides, originate from natural components, including plants, microbes, and minerals (Vlaiculescu and Varrone, 2022). They quickly decompose, reduce toxicity hazards, and precisely target pests. Examples of environmentally sustainable pesticides involve microbial pesticides utilizing beneficial bacteria, fungi, or viruses for pest and plant disease management; botanical pesticides that interfere with insect metabolism; RNAi-based biopesticides employing RNA interference technology to inhibit crucial genes in target pests; and predatory insects, along with biological control agents that leverage beneficial insects to naturally decrease pest populations. **Figure 2** evaluates the application of biodegradable inputs, bio-based fertilisers, and environmentally friendly insecticides in SA to eliminate plastic waste, enhance soil health, and promote biodiversity.

Figure 2. Sustainable agricultural inputs: biodegradable inputs, bio-based fertilisers, and eco-friendly pesticides

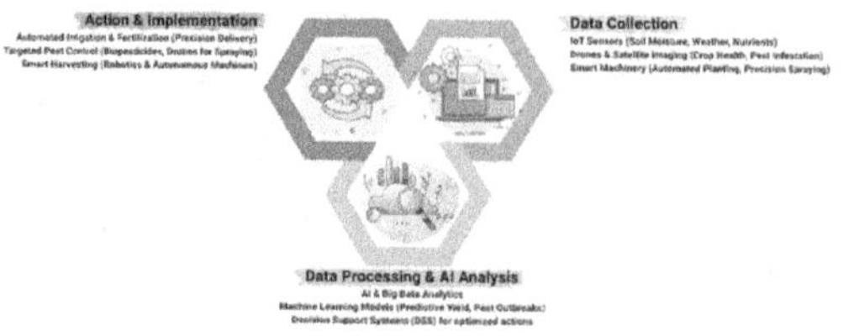

3.4 Comparative Analysis of Conventional vs. Sustainable Chemical Applications

The change from traditional chemical applications to SC in agriculture is driven by the necessity to reduce environmental effects, lower health hazards, and enhance long-term soil fertility. **Table 2** shows a comparison of traditional and sustainable chemical uses in agriculture. It shows important differences in where they come from, how they affect soil health and water pollution, how well they protect against pests, how safe food is, how much they cost, and how they affect the environment. It shows that sustainable chemical applications, such as biobased fertilizers, biopesticides, and biodegradable inputs, provide enduring advantages, including enhanced soil fertility, diminished chemical residues, and reduced environmental hazards compared to traditional synthetic chemicals. It clarifies the impact of precision application technologies and AI-driven optimizations in improving the efficiency of SA while adhering to legislation and consumer preferences.

Table 2. Comparison of conventional and sustainable chemical applications in agriculture

Parameter	Conventional Chemical Applications	Sustainable Chemical Applications	Impact on Agriculture & Environment	References
Source of chemicals	Synthetic fertilizers, pesticides, and herbicides	Bio-based fertilizers, biopesticides, and biodegradable inputs	Reduces reliance on fossil fuel-derived chemicals	(Yong et al., 2024)
Soil health impact	Depletes soil microbiome, leads to acidification and nutrient imbalance	Enhances soil biodiversity, improves nutrient retention	Promotes long-term soil fertility	(Tripathi et al., 2020)
Water contamination	Causes runoff, leading to eutrophication and groundwater pollution	Minimizes runoff with targeted application methods	Protects water bodies and ecosystems	(Haque, 2021)
Pest and disease resistance	Increases resistance due to overuse of synthetic pesticides	Uses natural predators, RNAi, and microbial pesticides	Reduces pest resistance buildup	(Suhag et al., 2021)
Residue in food products	High chemical residues, potential health risks	Low to no residue, safer for consumption	Improves food safety and consumer health	(Munir et al., 2024)
Application efficiency	Requires repeated applications, higher input costs	Uses AI, precision dosing, and slow-release formulations	Reduces input wastage and costs	(Kumar, 2025)
Environmental impact	High carbon footprint, biodiversity loss	Low emissions, eco-friendly formulations	Supports climate-friendly farming	(Chen et al., 2023)
Regulatory compliance	Stricter regulations on chemical residues and environmental effects	Encouraged under green and organic farming standards	Aligns with sustainability policies	(Nicastro et al., 2024)
Cost over time	Initial cost is lower, but long-term degradation leads to higher expenses	Higher initial cost, but long-term benefits improve ROI	Sustainable approach is more cost-effective in the long run	(Aghaei et al., 2022)
Consumer preference	Growing concerns over chemical residues and safety	Increasing demand for organic and eco-friendly products	Market shift towards sustainable agriculture	(Barwant et al., 2025)

4. PRECISION AGRICULTURE WITH SUSTAINABLE CHEMISTRY

PA and SC have modified agriculture by increasing efficiency, reducing environmental impact, and strengthening long-term sustainability. PA technology, including GPS, IoT sensors, AI, and remote sensing, facilitates the accurate application of fertilizers, pesticides, and water, thereby minimizing waste and environmental pollution (Getahun et al., 2024). SC substitutes synthetic agrochemicals with bio-based alternatives, ensuring safer and more efficient use of inputs. AI-driven decision-making for sustainable applications extensive agricultural data to propose optimal chemical application techniques, ensuring the precise quantity of bio-based fertilizers or environmentally friendly pesticides is administered at the appropriate time and location. Remote sensing for real-time monitoring facilitates the precise application of sustainable chemicals exclusively where necessary, thus reducing the overutilization of agricultural inputs, decreasing expenses, and improving the balance of nature.

PA and SC enable farmers to assess soil microbiota and modify chemical inputs to enhance beneficial microbes while circumventing harmful synthetic chemicals (Chen et al., 2023). Precision soil mapping facilitates techniques such as cover cropping, compost application, and biochar utilization, thereby enhancing soil structure and promoting long-term fertility. PA-guided smart irrigation, AI-optimized nutrient management, and sustainable inputs markedly diminish carbon emissions, chemical leaching, and non-biodegradable waste.

4.1 Role of AI and IoT in Optimizing Chemical Inputs

The pairing of AI and IoT in agriculture has markedly enhanced the efficacy and accuracy of chemical input applications, including fertilizers, insecticides, herbicides, and soil amendments. These technologies reduce waste, limit environmental effects, and improve crop productivity through real-time data collection, predictive analytics, and automated decision-making (Padhiary, Roy, et al., 2024). AI improves the accuracy and efficacy of chemical applications through the analysis of extensive datasets obtained from satellites, drones, soil sensors, and climate monitoring systems (Han et al., 2024). ML algorithms analyze this data to determine the precise quantity, variety, and timing of chemical applications required for optimal crop development. Predictive analytics models evaluate soil health, pest infestations, and nutrient deficits, assuring that fertilizers and pesticides are used exclusively when and where they are necessary.

IoT-enabled smart monitoring delivers real-time data on soil composition, moisture levels, temperatures, and insect activity, facilitating dynamic modifications in chemical input applications. Soil nutrient sensors quantify nitrogen, phosphorous, and potassium concentrations, directing farmers to apply fertilizer only in insufficient regions (Silva et al., 2024). Wireless sensor networks monitor environmental variables, facilitating precise pesticide administration. AI and IoT for Variable Rate Application (VRA) make it possible to accurately and location-specifically deploy chemical inputs based on changes in the field in real-time. Zone-based fertilization systems utilize AI to evaluate soil maps and administer nutrients exclusively in the required areas, thereby minimizing runoff and enhancing efficiency. Automated pest management systems engage pesticide sprayers only upon AI detection of pest presence via sensor data, hence reducing redundant chemical exposure. The melding of AI with IoT yields environmental and economic advantages, such as minimizing environmental degradation, decreasing GHG emissions, preserving biodiversity, and reducing production expenses.

4.2 Precision in the Application of Fertilizers and Pesticides

PA involves the meticulous application of fertilizers and pesticides, limiting their usage to specific locations, times, and required quantities. This methodology enhances agricultural productivity, lowers input expenses, and reduces ecological damage by curbing overutilization and runoff. Cutting-edge technologies, such as GPS, AI, IoT, remote sensing, and drone technology, have improved the utilization of these inputs in current agriculture (Nahiyoon et al., 2024). Precision fertilization maximizes the effectiveness of nitrogen, phosphorus, and potassium fertilizers, assuring that crops have the appropriate nutrients while minimizing surplus application. Technologies such as soil and nutrient sensors, variable rate technologies, smart irrigation systems, and AI-driven decision-making systems facilitate this.

Precision pesticide application ensures the effective utilization of herbicides, insecticides, and fungicides, minimizing chemical residues and protecting beneficial organisms. Essential technologies include drones and robotic sprayers, AI for pest and disease identification, IoT sensor networks, and electrostatic as well as controlled droplet application (CDA). Precision application offers advantages such as environmental conservation, financial efficiency, enhanced crop production and quality, and sustainability (Shah and Wu, 2019). It lowers excessive chemical application, averts soil degradation and water pollution, and encourages the judicious use of agrochemicals, thereby decreasing negative residues in food and the environment.

4.3 Sustainable Irrigation Management Using PA Tools

PA technologies serve in sustainable irrigation management to optimize water application, improve agricultural yield, and reduce environmental effects. PA-based systems provide accurate water distribution at the appropriate times and locations, thereby averting over-irrigation, soil deterioration, and nutrient leaching. Soil moisture sensors, remote sensing, satellite images, IoT-enabled smart irrigation systems, VRI technology, and AI-driven predictive irrigation models are some of the most important tools for precision agriculture (Nuwarapaksha et al., 2025). These technologies enhance irrigation efficiency by monitoring real-time soil moisture levels, evaluating plant water stress, and modifying irrigation schedules according to soil and meteorological circumstances. VRI employs GPS and AI algorithms to allocate water to various zones according to soil and crop needs, thereby minimizing water wastage and preventing over-irrigation in saturated regions. AI-driven predictive irrigation algorithms use past meteorological patterns, soil moisture metrics, and crop requirements to anticipate irrigation needs (Dhanke et al., 2024). Pennsylvania-based sustainable irrigation provides water conservation, energy efficiency, enhanced crop output, less soil degradation, and climate resilience. By conserving water use, optimizing pumping and watering schedules, and maintaining regular moisture levels, PA-based irrigation systems can mitigate water waste, degrade soil, and improve resilience to droughts and severe weather events. **Figure 3** depicts PA technologies used in sustainable irrigation management, encompassing soil moisture sensors, IoT-integrated smart irrigation systems, remote sensing, VRI, and AI-driven predictive models, facilitating efficient water utilization, minimizing water wastage, and enhancing crop productivity.

Figure 3. Precision agricultural tools for sustainable irrigation management

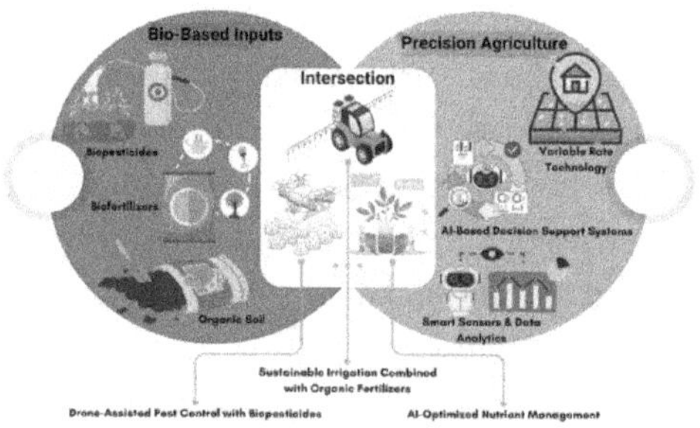

4.4 Case Studies of Integrated Approaches in Different Crop Systems

PA and SC have considerably enhanced resource efficiency, crop output, and environmental sustainability across diverse agricultural systems. Case studies demonstrate the successful deployment of AI-driven irrigation, bio-based fertilizer, drone-assisted pest management, and real-time soil monitoring in various agricultural environments. In India's rice cultivation, IoT-enabled soil moisture sensors and AI-driven predictive models were employed to enhance water efficiency, achieving a 30% decrease in water usage while sustaining high yields (Ashwini et al., 2025). In France, the implementation of AI-driven pest management in vineyards has resulted in a 40% reduction in chemical pesticide usage (Mwaka et al., 2024). In the United States, VRT was used in maize cultivation to regulate fertilizer delivery according to soil nutrient concentrations, thereby improving nutrient retention and increasing crop yield. Brazilian soybean fields implemented drone-assisted pesticide spraying alongside AI-driven pest monitoring, resulting in a 35% reduction in pesticide usage, less environmental contamination, and enhanced farm profitability (Upadhyay et al., 2025).

Automated greenhouse systems in tomato cultivation decreased water consumption by 40% and improved fruit quality by reducing chemical residues (Wu et al., 2023). Australian wheat fields employed remote sensing and AI-driven soil analysis to enhance organic fertilizer delivery, leading to a 20% rise in nitrogen usage efficiency (Padhiary, Saha, et al., 2024). AI-assisted irrigation in sustainable cotton agriculture resulted in a 25% reduction in water usage and improved fibre quality (Jiang et al., 2024). Apple cultivation utilized AI-driven disease prediction models and microbial-based bio fungicides to control fungal infections, resulting in a 50% reduction in fungicide usage (Shaikh et al., 2022). Chinese tea estates employed NDVI-based remote sensing and precise fertigation to improve tea leaf quality and decrease synthetic fertilizer applications by 30% (Ye et al., 2022). In Argentina, integrated crop-livestock systems used sensor-based grazing and organic manure recycling to improve soil fertility and decrease synthetic inputs, hence supporting long-term sustainability. **Table 3** provides an overview of real-world case studies showing the successful integration of PA technology with SC across several crop systems. It recognizes crop variety, working PA equipment, sustainable chemical methodologies, and significant results, illustrating how contemporary agricultural approaches improve efficiency, productivity, and environmental sustainability.

Table 3. Combination of PA and SC in different crop systems

Crop System	Implemented PA Tools	Sustainable Chemical Practices	Key Outcomes	References
Rice farming (India)	IoT-based soil moisture sensors, AI-driven irrigation scheduling	Bio-based fertilizers	30% reduction in water use, improved soil health	(Aghababaei et al., 2025)
Vineyards (France)	AI-based pest detection, drones, hyperspectral imaging	Eco-friendly biopesticides	40% reduction in pesticide use, preservation of beneficial insects	(Samal et al., 2024)
Corn fields (USA)	Variable Rate Technology (VRT), GPS mapping	Biochar-enriched organic fertilizers	15% increase in crop yield, reduced nitrogen runoff	(Saleem et al., 2023)

continued on following page

Table 3. Continued

Crop System	Implemented PA Tools	Sustainable Chemical Practices	Key Outcomes	References
Soybean cultivation (Brazil)	Drone-assisted pesticide application, AI-driven pest monitoring	Biopesticides	35% reduction in pesticide use, lower environmental contamination	(Istiak et al., 2023)
Tomato greenhouses (Netherlands)	Automated greenhouse climate control, AI-based irrigation	Organic nutrient dosing	40% reduction in water use, improved fruit quality	(Reza et al., 2025)
Wheat production (Australia)	Remote sensing, AI-powered soil nutrient analysis	Organic fertilizers	20% increase in nitrogen use efficiency, lower GHG emissions	(Padhiary et al., 2025)
Cotton farming (Egypt)	AI-based irrigation scheduling, IoT soil sensors	Biodegradable pesticides	25% water savings, improved fiber quality	(Maraveas, 2022)
Apple orchards (USA)	AI-powered disease prediction, microclimate sensors	Microbial-based biofungicides	50% reduction in fungicide use, cost savings	(Kaya, 2025)
Tea plantations (China)	NDVI-based remote sensing, precision fertigation	Reduced synthetic fertilizers	30% reduction in synthetic fertilizer use, enhanced leaf quality	(Al Aasmi et al., 2022)
Crop-livestock systems (Argentina)	Sensor-based grazing management, manure recycling	Organic soil enrichment	Improved soil fertility, reduced synthetic input dependency	(Bhattacharyya et al., 2020)

5. INNOVATIONS DRIVING ECO-FRIENDLY FARMING

Innovative technologies improve eco-friendly farming by enhancing sustainability, reducing environmental effects, and increasing productivity. These cover PA, AI, the IoT, robotics, and sustainable chemistry. AI-driven analytics and ML algorithms analyze extensive datasets from sensors, satellites, and drones to enhance irrigation, fertilization, and pest management techniques (Hoque & Padhiary, 2024). IoT-enabled smart farming uses IoT-based sensors to monitor soil moisture, temperature, nutrient levels, and weather conditions, facilitating automated precision irrigation and targeted nutrient application. Remote sensing and drone technologies utilize drones fitted with multispectral and thermal cameras to evaluate crop health, pest infestations, and soil conditions in real time, thereby minimizing chemical applications and enhancing efficiency. Innovations in biodegradable fertilizers, bio-based insecticides, and organic soil additives supplant synthetic chemicals, enhancing soil biodiversity and lowering pollution.

Sustainable irrigation management technologies, such as drip irrigation, AI-driven scheduling, and real-time soil moisture monitoring enhance water efficiency, minimize waste, and ensure adequate hydration for crops. Robotics and automation in SA assist in weeding, planting, and harvesting; hence, it decreases reliance on chemicals and labour expenses. The incorporation of renewable energy in agriculture, including solar-powered irrigation systems, wind energy, and biogas-based farming operations, reduces reliance on fossil fuels and lowers operational expenses. Vertical farming and controlled environmental agriculture (CEA) technologies permit continuous crop production throughout the year while minimizing water and land use, decreasing chemical pollution, and enhancing yield efficiency (Saha et al., 2025). Blockchain technology improves supply chain transparency and sustainability certification, enabling equitable trading practices.

5.1 Smart Sensors for Soil Health Monitoring

Digital sensors modify soil health assessment by delivering real-time, data-driven insights into soil conditions (Zeng et al., 2025). These sensors enhance water application, fertilizer management, and agricultural productivity, thereby promoting sustainable, environmentally friendly farming practices. A variety of smart sensors include moisture sensors, pH sensors, nutrition sensors, temperature sensors, and electrochemical sensors. The IoT and AI are crucial to soil health monitoring, as IoT-enabled sensors collect continuous soil data and relay it to cloud-based platforms for analysis. Analytics driven by AI can predict soil health trends, enhance irrigation schedules, and provide specific fertilization strategies. Wireless networks and cloud computing allow remote monitoring and data-informed decision-making for agricultural operators (Hoque et al., 2025). The benefits of smart soil sensors include precise irrigation, water saving, improved nutrient usage, higher soil sustainability, and enhanced yield and production. Automated irrigation systems eliminate water loss by as much as 50%, while evaluation of organic matter and microbial activity ensures sustained soil fertility. **Figure 4** describes the application of smart sensors for monitoring soil moisture, pH, temperature, and nutrient levels and employing IoT-enabled sensors for real-time data transfer and AI analytics for accurate suggestions on irrigation, fertilization, and soil management.

Figure 4. Smart soil sensors for real-time soil health monitoring

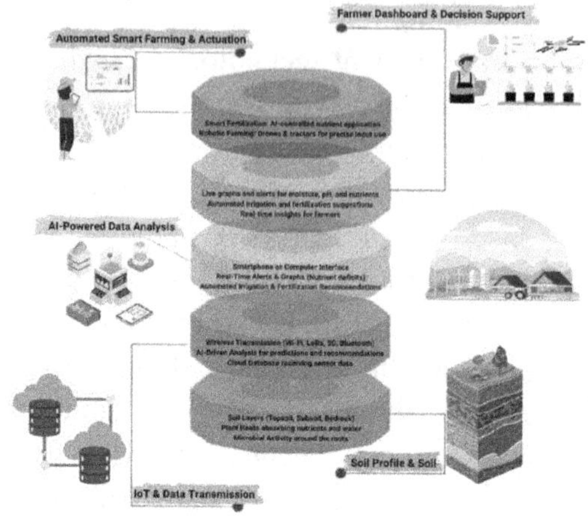

5.2 AI-Driven Decision Support Systems (DSS)

AI-enhanced Decision Support Systems (DSS) have also contributed to modern agriculture by combining big data analytics, ML, IoT, and remote sensing to enhance agricultural decision-making (Babbar et al., 2025). These technologies empower farmers to make data-informed, immediate decisions about crop management, irrigation, fertilization, pest control, and overall farm productivity, resulting in increased efficiency, cost savings, and environmental sustainability. Some of the most important features of AI-driven decision support systems are combining data from different sources, using predictive analytics and forecasting, making decisions automatically, monitoring from afar with real-time alerts, and being accessible from the cloud. Applications of AI-driven decision support systems in PA consist of precision irrigation management, fertilizer optimization, soil health management, pest and disease control, yield prediction, and crop planning, as well as climate change evolution (Roy et al., 2025). The use of AI in agricultural decision support systems has multiple benefits, such as improved decision-making, resource tuning, heightened farm productivity, less environmental impact, and financial savings for farmers. Limitations involve an extensive initial investment, complex data administration, farmer training and expertise, as well as scalability and adaptability issues.

Despite these obstacles, AI-driven decision support systems are anticipated to become increasingly accessible, economical, and efficient for farmers globally, owing to persistent advances in AI, IoT, and remote sensing technologies. The future of AI-driven agriculture lies in automated, data-informed decision-making systems that enhance production and promote environmental sustainability. AI-driven decision support systems are disrupting agriculture by delivering data-informed insights, enhancing resource efficiency, encouraging sustainable practices, and minimizing input expenses (Prasad et al., 2025). Challenges persist, such as significant initial investment, complex data management, and the need for farmer education and training. It anticipates AI-driven decision support systems to become increasingly accessible, economical, and efficient for farmers globally, despite these limitations.

5.3 Role of Biopesticides and biofertilizers

Biopesticides and biofertilizers are increasingly adopted in modern agriculture as an environmentally sustainable substitute for synthetic chemicals. Microbial solutions, covering bacteria, fungi, and viruses, are essential for boosting soil fertility and naturally managing pests. These methods enhance soil health, increase crop productivity, and reduce environmental harm, complying with SA principles and precision farming. Biopesticides are pest control products created from natural microorganisms, plant extracts, or helpful bacteria and are biologically based (Ghose et al., 2025). Target adaptation, biodegradability, and environmental friendliness characterize them. The benefits of microbial biopesticides include little impact on non-target organisms and biodiversity, lower chemical residues, and sustained efficacy. Biofertilizers are living microorganisms that promote plant growth by fixing nitrogen, solubilizing phosphorus, and aiding root development. These microbial treatments decrease our reliance on synthetic fertilizers and enhance soil fertility. Microbial biofertilizers include bacteria that fix nitrogen, like rhizobium, azotobacter, and azospirillum; microorganisms that dissolve phosphate and potassium; mycorrhizal fungi (AMF); and rhizobacteria that promote plant growth (PGPR) (Chakraborty and Akhtar, 2021; Nosheen et al., 2021). Benefits of microbial biofertilizers include enhanced soil fertility, elevated agricultural yields, and reduced environmental impact. Drawbacks incorporate storage and shelf life, effective field application techniques, regulatory permissions, and farmer wisdom. Biopesticides and biofertilizers will only work in the future if AI-driven PA technologies like smart sensors, drone-assisted spraying, and IoT-enabled soil monitoring are used (Hoque, 2024). These technologies can make farming more efficient and valuable, and they can also make sustainable farming more scalable.

5.4 Blockchain for Traceability in Sustainable Farming

Blockchain technology has reinforced SA by offering a safe, transparent, and irreversible system for monitoring food production, processing, and distribution. It enables farmers, suppliers, retailers, and consumers to obtain reliable information regarding agricultural products, thereby ensuring food safety, quality assurance, and equal trading practices. Essential features of the blockchain in SA include decentralization, immutability, transparency, smart contracts, and real-time monitoring. Blockchain technology is used extensively in numerous aspects of sustainable agriculture, improving transparency, accountability, and efficiency (Panwar et al., 2023). Farm-to-fork traceability represents a principal application of blockchain technology in agriculture, enabling consumers to verify the origin, quality, and sustainability of their food by scanning a QR code on the packaging. This feature enables retailers to confirm that organic bananas were cultivated without synthetic pesticides and transported under appropriate conditions. Blockchain was defined by supply chain accessibility and fair commerce (Kshetri, 2021). It removes superfluous intermediaries, facilitating direct interactions between producers and purchasers and thereby affecting exploitation and payment delays. Blockchain decreases food fraud by confirming and permanently storing certifications, ensuring that only verified organic and sustainably cultivated products are delivered to customers.

The inclusion of blockchain with IoT sensors and AI facilitates monitoring resource use, including water usage, chemical applications, and carbon emissions, thereby ensuring farmers adhere to environmental regulations while improving inputs to minimize waste. A vineyard adopting IoT-enabled irrigation devices may document real-time water use data on a blockchain ledger, thereby assuring sustainable water management (Adamashvili et al., 2024). Smart contracts for compliance and payments are autonomous agreements recorded on the blockchain, enabling automatic verification of compliance and immediate transactions. These contracts ensure farmers adhere to sustainability standards before receiving wages, thus decreasing the danger of fraud and delays in manual processing. However the adoption of blockchain in SA encounters constraints, including elevated initial expenses, technical intricacies, data privacy apprehensions, and scalability constraints. The future of blockchain in SA depends on its connection with AI, IoT, and ML to improve automation, real-time decision-making, and scalability (Khan et al., 2023). As an increasing number of governments and corporations implement blockchain-based traceability standards, it will emerge as a prevalent instrument for assuring food safety and sustainability adherence (Rajababu et al., 2025). **Figure 5** explains that blockchain technology promises transparency and traceability within the agricultural supply chain, beginning at the farm level with a ledger detailing growing techniques and sustainability certificates. It securely records transactions using QR codes and smart contracts and incorporates IoT sensors and artificial intelligence.

Figure 5. Blockchain-enabled traceability systems in sustainable farming

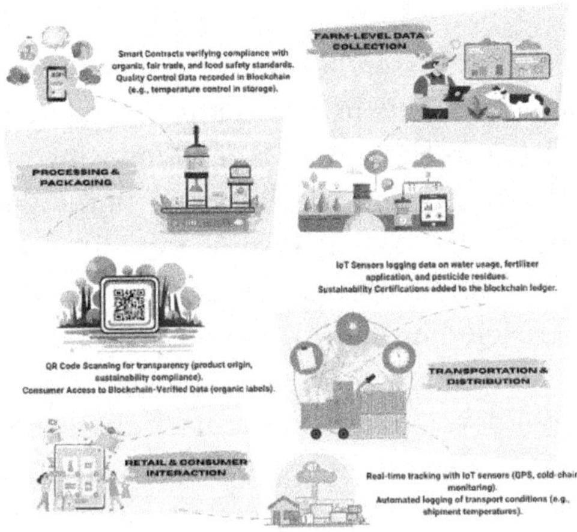

6. CHALLENGES AND FUTURE PERSPECTIVES

6.1 Technical, Economic, and Policy Challenges

The execution of PA and SC meets numerous obstacles. Technical challenges entail the large expense of sophisticated technology, like AI, IoT, and remote sensing, which restricts accessibility for small-scale farmers. The absence of regulated data collecting and interoperability among various platforms causes problems with implementation. Economic challenges include high initial investment expenses, continuous maintenance, and the requirement for proficient labour (Kim and Park, 2020). A multitude of farmers, especially in developing areas, face financial limitations and restricted access to subsidies or financing for the use of PA technologies. Policy problems include the lack of clearly articulated norms and incentives for sustainable agriculture. Governments frequently fail to establish comprehensive legislation to encourage consumption of environmentally SA inputs, restricting their broad adoption. Similarly, global inequalities in infrastructure and digital connectivity accentuate the disparity in technology adoption among areas.

6.2 Regulatory Frameworks for Sustainable Agricultural Chemistry

Regulatory frameworks are essential for enabling the safe and effective application of sustainable chemical inputs in agriculture. Governments and international organizations, like the FAO, WHO, and EPA, are imposing more stringent controls on chemical pesticides and fertilizers, promoting the shift towards biopesticides and biofertilizers (Fenibo et al., 2022). Certification initiatives such as the EU's Green Deal and the USDA Organic Standards advocate for the use of environmentally sustainable inputs while excluding detrimental elements. Even the absence of global standardization in regulatory rules results in variations in compliance across various locations. Also, numerous developing countries lack enforcement tools for verifying compliance with sustainable practices. Governments must implement incentives, like tax reductions, subsidies, and research grants, to promote the extensive adoption of GC in agriculture.

6.3 Future Trends in PA and GC

The future of PA and GC will be pushed by technical developments and sustainability-orientated technologies. AI and ML will continue to improve real-time decision-making in agricultural management by analyzing extensive datasets regarding soil health, meteorological conditions, and crop development (Delfani et al., 2024). Automation and robotics will further refine the application of fertilizers and pesticides, minimize chemical waste, and improve efficiency. In GC, biodegradable inputs, microbial-based fertilizers, and nanotechnology-enhanced biopesticides will replace conventional agrochemicals. The growth of carbon farming and regenerative agriculture will promote the use of sustainable methods. Governments and corporations are expected to invest in circular economy models that advocate waste-to-resource strategies. Moreover, the combination of blockchain and IoT will enhance food traceability and adherence to sustainability norms.

6.4 Potential of AI and Nanotechnology in Eco-Friendly Farming

The synthesis of AI and nanotechnology in agriculture offers the capacity to advance sustainable farming practices. AI-driven DSS can enhance irrigation, procedure, and pest management through the analysis of real-time data. ML algorithms can forecast crop illnesses and soil nutrient deficits, enabling farmers to adopt proactive remedies. Nanotechnology makes farming inputs like nanofertilizers and nanopesticides more effective by allowing for targeted delivery and controlled release while lowering the damage to the environment (Yadav et al., 2023). Nanosensors can identify soil moisture, nutrient concentrations, and pathogens, facilitating PA techniques. Furthermore, AI-powered autonomous drones and robotics will improve sustainable agricultural practices by minimizing chemical applications and labour expenses. These improvements will markedly enhance resource efficiency, environmental preservation, and food security in the following decades.

7. CONCLUSION

PA and SC are effective methodologies for enhancing agricultural output, conserving the environment, and optimizing resource usage. PA technologies, including AI, IoT, remote sensing, and drones, provide accurate monitoring and management of soil health, irrigation, and chemical inputs, thereby minimizing waste and enhancing output. SC promotes the use of biodegradable inputs, bio-based fertilizers, and sustainable pesticides, thereby reducing the adverse ecological effects of traditional agrochemicals. Even so, challenges such as elevated adoption costs, the absence of regulatory harmonization, restricted access to technology in rural regions, and resistance to change limit extensive implementation. For faster adoption of PA and SC, policy recommendations involve financial incentives and subsidies; standardized global policies for bio-based inputs, nanotechnology, and AI-driven agricultural management; capacity building and training; investment in digital infrastructure; collaboration with research institutions and agri-tech firms; as well as sustainability certification and market incentives. Future research should concentrate on AI and big data for PA, advancements in GC, the integration of AI and IoT for smart agriculture, blockchain for transparency in the agri-supply chain, and climate-resilient agricultural methodologies. Such approaches seek to raise agricultural productivity, prevent waste, and enhance resource efficiency while nurturing sustainable practices.

REFERENCES

Acharjee, S. A., Bharali, P., Gogoi, B., Sorhie, V., Walling, B., & Alemtoshi, . (2023). PHA-Based Bioplastic: A Potential Alternative to Address Microplastic Pollution. *Water, Air, and Soil Pollution*, *234*(1), 21. DOI: 10.1007/s11270-022-06029-2 PMID: 36593989

Adamashvili, N., Zhizhilashvili, N., & Tricase, C. (2024). The integration of the internet of things, artificial intelligence, and blockchain technology for advancing the wine supply chain. *Computers*, *13*(3), 72. DOI: 10.3390/computers13030072

Aghababaei, A., Aghababaei, F., Pignitter, M., & Hadidi, M. (2025). Artificial Intelligence in Agro-Food Systems: From Farm to Fork. *Foods*, *14*(3), 411. DOI: 10.3390/foods14030411 PMID: 39942003

Aghaei, M., Fairbrother, A., Gok, A., Ahmad, S., Kazim, S., Lobato, K., Oreski, G., Reinders, A., Schmitz, J., Theelen, M., Yilmaz, P., & Kettle, J. (2022). Review of degradation and failure phenomena in photovoltaic modules. *Renewable & Sustainable Energy Reviews*, *159*, 112160. DOI: 10.1016/j.rser.2022.112160

Ahmad, S., Hussain, I., Ghaffar, A., Rahman, M. H. U., Saleem, M. Z., Yonas, M. W., Hussnain, H., Ikram, R. M., & Arslan, M. (2022). Organic amendments and conservation tillage improve cotton productivity and soil health indices under arid climate. *Scientific Reports*, *12*(1), 14072. DOI: 10.1038/s41598-022-18157-0 PMID: 35982152

Al Aasmi, A., Alordzinu, K. E., Li, J., Lan, Y., Appiah, S. A., & Qiao, S. (2022). Rapid estimation of water stress in choy sum (Brassica chinensis var. Parachinensis) using integrative approach. *Sensors (Basel)*, *22*(5), 1695. DOI: 10.3390/s22051695 PMID: 35270842

Ala'F, K., Alqammaz, A., Khasawneh, A. M., Abualigah, L., Darabkh, K. A., & Zinonos, Z. (2025). An environmental remote sensing and prediction model for an IoT smart irrigation system based on an enhanced wind-driven optimization algorithm. *Computers & Electrical Engineering*, *122*, 109889. DOI: 10.1016/j.compeleceng.2024.109889

Albshaier, L., Almarri, S., & Albuali, A. (2025). Federated Learning for Cloud and Edge Security: A Systematic Review of Challenges and AI Opportunities. *Electronics (Basel)*, *14*(5), 1019. DOI: 10.3390/electronics14051019

Archana, H. R., Darshan, K., Lakshmi, M. A., Ghoshal, T., Bashayal, B. M., & Aggarwal, R. (2022). Biopesticides: A key player in agro-environmental sustainability. In *Trends of Applied Microbiology for Sustainable Economy* (pp. 613–653). Elsevier. https://www.sciencedirect.com/science/article/pii/B9780323915953000215

Arora, P. K., Tripathi, S., Omar, R. A., Chauhan, P., Sinhal, V. K., Singh, A., Srivastava, A., Garg, S. K., & Singh, V. P. (2024). Next-generation fertilizers: The impact of bionanofertilizers on sustainable agriculture. *Microbial Cell Factories*, *23*(1), 254. DOI: 10.1186/s12934-024-02528-5 PMID: 39304847

Ashwini, A., Sriram, S. R., Prabhakar, J. M., & Kadry, S. (2025). Farming 4.0: Cultivating the Future with Internet of Things Empowered on Smart Agriculture Solutions. In Dhanaraj, R. K., Sathyamoorthy, M., Balasubramaniam, S., & Kadry, S. (Eds.), *Networked Sensing Systems* (1st ed., pp. 247–271). Wiley., DOI: 10.1002/9781394310890.ch10

Aziz, D., Rafiq, S., Saini, P., Ahad, I., Gonal, B., Rehman, S. A., Rashid, S., Saini, P., Rohela, G. K., Aalum, K., Singh, G., Gnanesh, B. N., & Nabila Iliya, M. (2025). Remote sensing and artificial intelligence: Revolutionizing pest management in agriculture. *Frontiers in Sustainable Food Systems*, *9*, 1551460. DOI: 10.3389/fsufs.2025.1551460

Babbar, H., Rani, S., Soni, M., Keshta, I., Prasad, K. D. V., & Shabaz, M. (2025). Integrating remote sensing and geospatial AI-enhanced ISAC models for advanced localization and environmental monitoring. *Environmental Earth Sciences*, *84*(4), 1–12. DOI: 10.1007/s12665-025-12121-7

Bade, S. O., & Tomomewo, O. S. (2024). A review of governance strategies, policy measures, and regulatory framework for hydrogen energy in the United States. *International Journal of Hydrogen Energy*, *78*, 1363–1381. DOI: 10.1016/j.ijhydene.2024.06.338

Barwant, M. M., Ogidi, O. I., Yogita, C., & Munje, R. (2025). Study of Consumers Choices on Pesticides Use and Sustainability. In Babaniyi, B. R., & Babaniyi, E. E. (Eds.), *The Interplay of Pesticides and Climate Change* (pp. 469–516). Springer Nature Switzerland., DOI: 10.1007/978-3-031-81669-7_18

Bhattacharyya, P., Pathak, H., & Pal, S. (2020). *Climate smart agriculture: Concepts, challenges, and opportunities*.

Chakraborty, T., & Akhtar, N. (2021). Biofertilizers: Characteristic Features and Applications. In Inamuddin, M. I. Ahamed, R. Boddula, & M. Rezakazemi (Eds.), *Biofertilizers* (1st ed., pp. 429–489). Wiley. DOI: 10.1002/9781119724995.ch15

Chataut, G., Bhatta, B., Joshi, D., Subedi, K., & Kafle, K. (2023). Greenhouse gases emission from agricultural soil: A review. *Journal of Agriculture and Food Research, 11*, 100533. DOI: 10.1016/j.jafr.2023.100533

Chen, S., Wang, F., & Haroon, M. (2023). The impact of green economic recovery on economic growth and ecological footprint: A case study in developing countries of Asia. *Resources Policy, 85*, 103955. DOI: 10.1016/j.resourpol.2023.103955

Chen, W., Modi, D., & Picot, A. (2023). Soil and phytomicrobiome for plant disease suppression and management under climate change: A review. *Plants, 12*(14), 2736. DOI: 10.3390/plants12142736 PMID: 37514350

Delfani, P., Thuraga, V., Banerjee, B., & Chawade, A. (2024). Integrative approaches in modern agriculture: IoT, ML and AI for disease forecasting amidst climate change. *Precision Agriculture, 25*(5), 2589–2613. DOI: 10.1007/s11119-024-10164-7

Dhanaraju, M., Chenniappan, P., Ramalingam, K., Pazhanivelan, S., & Kaliaperumal, R. (2022). Smart farming: Internet of Things (IoT)-based sustainable agriculture. *Agriculture, 12*(10), 1745. DOI: 10.3390/agriculture12101745

Dhanke, J. A., Srivastava, D., Menaga, D., Raj, R., Kumar, K. V., Jangir, P., & Mani, P. (2024). Climate-Based AI-Powered Precision Irrigation: Sustainably Smart Agriculture Frameworks for Maximum Crop Yields. *Remote Sensing in Earth Systems Sciences*, •••, 1–12.

Dhuldhaj, U. P., Singh, R., & Singh, V. K. (2023). Pesticide contamination in agro-ecosystems: Toxicity, impacts, and bio-based management strategies. *Environmental Science and Pollution Research International, 30*(4), 9243–9270. DOI: 10.1007/s11356-022-24381-y PMID: 36456675

Diop, M., Chirinda, N., Beniaich, A., El Gharous, M., & El Mejahed, K. (2022). Soil and water conservation in Africa: State of play and potential role in tackling soil degradation and building soil health in agricultural lands. *Sustainability (Basel), 14*(20), 13425. DOI: 10.3390/su142013425

Fenibo, E. O., Ijoma, G. N., & Matambo, T. (2022). Biopesticides in Sustainable Agriculture: Current Status and Future Prospects. In Mandal, S. D., Ramkumar, G., Karthi, S., & Jin, F. (Eds.), *New and Future Development in Biopesticide Research: Biotechnological Exploration* (pp. 1–53). Springer Nature Singapore., DOI: 10.1007/978-981-16-3989-0_1

Foong, S. Y., Chan, Y. H., Yek, P. N. Y., Lock, S. S. M., Chin, B. L. F., Yiin, C. L., Lan, J. C.-W., & Lam, S. S. (2024). Microwave-assisted pyrolysis in biomass and waste valorisation: Insights into the life-cycle assessment (LCA) and techno-economic analysis (TEA). *Chemical Engineering Journal, 491*, 151942. DOI: 10.1016/j.cej.2024.151942

Garg, D., Sridhar, K., Stephen Inbaraj, B., Chawla, P., Tripathi, M., & Sharma, M. (2023). Nano-biofertilizer formulations for agriculture: A systematic review on recent advances and prospective applications. *Bioengineering (Basel, Switzerland), 10*(9), 1010. DOI: 10.3390/bioengineering10091010 PMID: 37760112

Getahun, S., Kefale, H., & Gelaye, Y. (2024). Application of Precision Agriculture Technologies for Sustainable Crop Production and Environmental Sustainability: A Systematic Review. *TheScientificWorldJournal, 2024*(1), 2126734. DOI: 10.1155/2024/2126734 PMID: 39421732

Ghose, S., Bhattacharjee, B., Rynjah, D., & Laloo, D. (2025). Pesticides and Allergens. In Odoh, U. E., Gurav, S. S., & Chukwuma, M. O. (Eds.), *Pharmacognosy and Phytochemistry* (1st ed., pp. 315–334). Wiley., DOI: 10.1002/9781394203680.ch16

Hadj, T. B. (2021). Nonlinear impact of biomass energy consumption on ecological footprint in a fossil fuel–dependent economy. *Environmental Science and Pollution Research International, 28*(48), 69329–69342. DOI: 10.1007/s11356-021-15521-x PMID: 34296408

Han, H., Liu, Z., Li, J., & Zeng, Z. (2024). Challenges in remote sensing based climate and crop monitoring: Navigating the complexities using AI. *Journal of Cloud Computing (Heidelberg, Germany), 13*(1), 34. DOI: 10.1186/s13677-023-00583-8

Haque, S. E. (2021). How effective are existing phosphorus management strategies in mitigating surface water quality problems in the US? *Sustainability (Basel), 13*(12), 6565. DOI: 10.3390/su13126565

Hidalgo, D., Martín-Marroquín, J. M., & Corona, F. (2019). A multi-waste management concept as a basis towards a circular economy model. *Renewable & Sustainable Energy Reviews, 111*, 481–489. DOI: 10.1016/j.rser.2019.05.048

Hoque, A. (2024). Artificial Intelligence in Post-Harvest Drying Technologies: A Comprehensive Review on Optimization, Quality Enhancement, and Energy Efficiency. *International Journal of Science and Research (Raipur, India), 13*(11), 493–502. DOI: 10.21275/SR241107163717

Hoque, A., & Padhiary, M. (2024). Automation and AI in Precision Agriculture: Innovations for Enhanced Crop Management and Sustainability. *Asian Journal of Research in Computer Science*, *17*(10), 95–109. DOI: 10.9734/ajrcos/2024/v17i10512

Hoque, A., Padhiary, M., Prasad, G., & Kumar, K. (2025). Real-Time Data Processing in Agricultural Robotics. In Bora, D. J., & Bania, R. K. (Eds.), *Computer Vision Techniques for Agricultural Advancements* (pp. 431–468). IGI Global., DOI: 10.4018/979-8-3693-8019-2.ch014

Istiak, M. A., Syeed, M. M., Hossain, M. S., Uddin, M. F., Hasan, M., Khan, R. H., & Azad, N. S. (2023). Adoption of Unmanned Aerial Vehicle (UAV) imagery in agricultural management: A systematic literature review. *Ecological Informatics*, *78*, 102305. DOI: 10.1016/j.ecoinf.2023.102305

Jiang, T., Wang, M., Zhang, W., Zhu, C., & Wang, F. (2024). A comprehensive analysis of agricultural non-point source pollution in China: Current status, risk assessment and management strategies. *Sustainability (Basel)*, *16*(6), 2515. DOI: 10.3390/su16062515

Kaya, C. (2025). Optimizing Crop Production With Plant Phenomics Through High-Throughput Phenotyping and AI in Controlled Environments. *Food and Energy Security*, *14*(1), e70050. DOI: 10.1002/fes3.70050

Khan, A. A., Laghari, A. A., Rashid, M., Li, H., Javed, A. R., & Gadekallu, T. R. (2023). Artificial intelligence and blockchain technology for secure smart grid and power distribution Automation: A State-of-the-Art Review. *Sustainable Energy Technologies and Assessments*, *57*, 103282. DOI: 10.1016/j.seta.2023.103282

Kim, J., & Park, C. (2020). Education, skill training, and lifelong learning in the era of technological revolution: A review. *Asian-Pacific Economic Literature*, *34*(2), 3–19. DOI: 10.1111/apel.12299

Kogan, F. (2022). New Remote Sensing Vegetation Health Technology. In F. Kogan, *Remote Sensing Land Surface Changes* (pp. 121–148). Springer International Publishing. DOI: 10.1007/978-3-030-96810-6_5

Krishnan, S. R., Nallakaruppan, M. K., Chengoden, R., Koppu, S., Iyapparaja, M., Sadhasivam, J., & Sethuraman, S. (2022). Smart water resource management using Artificial Intelligence—A review. *Sustainability (Basel)*, *14*(20), 13384. DOI: 10.3390/su142013384

Kshetri, N. (2021). Blockchain and sustainable supply chain management in developing countries. *International Journal of Information Management*, *60*, 102376. DOI: 10.1016/j.ijinfomgt.2021.102376

Kumar, P. (2025). Nanofertilizers, Nanopesticides for Individual's (Farmer's) Economy. In Kumar, P., & Dubey, R. C. (Eds.), *Nanofertilizers for Sustainable Agriculture* (pp. 383–406). Springer Nature Switzerland., DOI: 10.1007/978-3-031-78649-5_16

Laveglia, S., Altieri, G., Genovese, F., Matera, A., & Di Renzo, G. C. (2024). Advances in Sustainable Crop Management: Integrating Precision Agriculture and Proximal Sensing. *AgriEngineering*, *6*(3), 3084–3120. DOI: 10.3390/agriengineering6030177

Leskovac, A., & Petrović, S. (2023). Pesticide use and degradation strategies: Food safety, challenges and perspectives. *Foods*, *12*(14), 2709. DOI: 10.3390/foods12142709 PMID: 37509801

Majumder, D., Akhter, J., Mandal, A., Roy, R., Mondal, D., Bhatt, R., & Hossain, A. (2021). Precision Input Management for Minimizing and Recycling of Agricultural Waste. In Bhatt, R., Meena, R. S., & Hossain, A. (Eds.), *Input Use Efficiency for Food and Environmental Security* (pp. 567–603). Springer Nature Singapore., DOI: 10.1007/978-981-16-5199-1_19

Mandal, M., Roy, A., Das, S., Rakwal, R., Agrawal, G. K., Singh, P., Awasthi, A., & Sarkar, A. (2024). Food waste-based bio-fertilizers production by bio-based fermenters and their potential impact on the environment. *Chemosphere*, *353*, 141539. DOI: 10.1016/j.chemosphere.2024.141539 PMID: 38417498

Manickam, P., Mariappan, S. A., Murugesan, S. M., Hansda, S., Kaushik, A., Shinde, R., & Thipperudraswamy, S. P. (2022). Artificial intelligence (AI) and internet of medical things (IoMT) assisted biomedical systems for intelligent healthcare. *Biosensors (Basel)*, *12*(8), 562. DOI: 10.3390/bios12080562 PMID: 35892459

Maraveas, C. (2022). Incorporating artificial intelligence technology in smart greenhouses: Current State of the Art. *Applied Sciences (Basel, Switzerland)*, *13*(1), 14. DOI: 10.3390/app13010014

Mennan, H., Jabran, K., Zandstra, B. H., & Pala, F. (2020). Non-chemical weed management in vegetables by using cover crops: A review. *Agronomy (Basel)*, *10*(2), 257. DOI: 10.3390/agronomy10020257

Mishra, H., & Mishra, D. (2024). AI for Data-Driven Decision-Making in Smart Agriculture: From Field to Farm Management. In *Artificial Intelligence Techniques in Smart Agriculture* (pp. 173–193). Springer. https://link.springer.com/chapter/10.1007/978-981-97-5878-4_11

Munir, S., Azeem, A., Zaman, M. S., & Haq, M. Z. U. (2024). From field to table: Ensuring food safety by reducing pesticide residues in food. *The Science of the Total Environment*, *922*, 171382. DOI: 10.1016/j.scitotenv.2024.171382 PMID: 38432369

Mwaka, O., Mwamahonje, A., Nene, W., Rweyemamu, E., & Maseta, Z. (2024). Pesticides use and its effects on grape production: A review. *Sustainable Environment, 10*(1), 2366555. DOI: 10.1080/27658511.2024.2366555

Nahiyoon, S. A., Ren, Z., Wei, P., Li, X., Li, X., Xu, J., Yan, X., & Yuan, H. (2024). Recent development trends in plant protection UAVs: A journey from conventional practices to cutting-edge technologies—A comprehensive Review. *Drones (Basel), 8*(9), 457. DOI: 10.3390/drones8090457

Nawaz, A., & Ahmad, J. N. (2015). Insect Pest Management in Conservation Agriculture. In Farooq, M., & Siddique, K. H. M. (Eds.), *Conservation Agriculture* (pp. 133–155). Springer International Publishing., DOI: 10.1007/978-3-319-11620-4_6

Nicastro, R., Papale, M., Fusco, G. M., Capone, A., Morrone, B., & Carillo, P. (2024). Legal Barriers in Sustainable Agriculture: Valorization of Agri-Food Waste and Pesticide Use Reduction. *Sustainability (Basel), 16*(19), 8677. DOI: 10.3390/su16198677

Nosheen, S., Ajmal, I., & Song, Y. (2021). Microbes as biofertilizers, a potential approach for sustainable crop production. *Sustainability (Basel), 13*(4), 1868. DOI: 10.3390/su13041868

Padhiary, M., Hoque, A., Prasad, G., Kumar, K., & Sahu, B. (2025). Precision Agriculture and AI-Driven Resource Optimization for Sustainable Land and Resource Management. In Ruiz-Vanoye, J. A., & Díaz-Parra, O. (Eds.), *Smart Water Technology for Sustainable Management in Modern Cities* (pp. 197–232). IGI Global., DOI: 10.4018/979-8-3693-8074-1.ch009

Padhiary, M., Hoque, A., Prasad, G., Kumar, K., & Sahu, B. (2025). Precision Agriculture and AI-Driven Resource Optimization for Sustainable Land and Resource Management. In *Smart Water Technology for Sustainable Management in Modern Cities* (pp. 197–232). IGI Global Scientific Publishing., https://www.igi-global.com/chapter/precision-agriculture-and-ai-driven-resource-optimization-for-sustainable-land-and-resource-management/370442 DOI: 10.4018/979-8-3693-8074-1.ch009

Padhiary, M., & Kumar, R. (2024a). Assessing the Environmental Impacts of Agriculture, Industrial Operations, and Mining on Agro-Ecosystems. In Azrour, M., Mabrouki, J., Alabdulatif, A., Guezzaz, A., & Amounas, F. (Eds.), *Smart Internet of Things for Environment and Healthcare* (Vol. 1165, pp. 107–126). Springer Nature Switzerland., DOI: 10.1007/978-3-031-70102-3_8

Padhiary, M., & Kumar, R. (2024b). Enhancing Agriculture Through AI Vision and Machine Learning: The Evolution of Smart Farming. In Thangam, D. (Ed.), (pp. 295–324). Advances in Computational Intelligence and Robotics. IGI Global., DOI: 10.4018/979-8-3693-5380-6.ch012

Padhiary, M., Kumar, R., & Sethi, L. N. (2024). Navigating the Future of Agriculture: A Comprehensive Review of Automatic All-Terrain Vehicles in Precision Farming. *Journal of The Institution of Engineers (India): Series A, 105*(3), 767–782. DOI: 10.1007/s40030-024-00816-2

Padhiary, M., Roy, P., Dey, P., & Sahu, B. (2024). Harnessing AI for Automated Decision-Making in Farm Machinery and Operations: Optimizing Agriculture. In Hai-Jew, S. (Ed.), (pp. 249–282). Advances in Computational Intelligence and Robotics. IGI Global., DOI: 10.4018/979-8-3693-6230-3.ch008

Padhiary, M., Saha, D., Kumar, R., Sethi, L. N., & Kumar, A. (2024). Enhancing precision agriculture: A comprehensive review of machine learning and AI vision applications in all-terrain vehicle for farm automation. *Smart Agricultural Technology, 8*, 100483. DOI: 10.1016/j.atech.2024.100483

Padhiary, M., Tikute, S. V., Saha, D., Barbhuiya, J. A., & Sethi, L. N. (2025). Development of an IOT-Based Semi-Autonomous Vehicle Sprayer. *Agricultural Research, 14*(1), 229–239. DOI: 10.1007/s40003-024-00760-4

Pandeya, S., Gyawali, B. R., & Upadhaya, S. (2025). Factors influencing precision agriculture technology adoption among small-scale farmers in Kentucky and their implications for policy and practice. *Agriculture, 15*(2), 177. DOI: 10.3390/agriculture15020177

Panwar, S., Duggirala, K. S., Yadav, P., Debnath, N., Yadav, A. K., & Kumar, A. (2023). Advanced diagnostic methods for identification of bacterial foodborne pathogens: Contemporary and upcoming challenges. *Critical Reviews in Biotechnology, 43*(7), 982–1000. DOI: 10.1080/07388551.2022.2095253 PMID: 35994308

Prasad, G., Padhiary, M., Hoque, A., & Kumar, K. (2025). AI-Driven Personalized Nutrition Apps and Platforms for Enhanced Diet and Wellness. In Alhussaini Hamad, A. M., & Soni, R. (Eds.), *Food in the Metaverse and Web 3.0 Era* (pp. 125–158). IGI Global., DOI: 10.4018/979-8-3693-9025-2.ch006

Rajababu, D., Surya, S., Padhiary, M., & Modi, H. (2025). Blockchain for Cybersecurity_ Securing Data Transactions and Enhancing Privacy in Digital Systems. 2025 First International Conference on Advances in Computer Science, Electrical, Electronics, and Communication Technologies (CE2CT), 1426–1430. https://doi .org/DOI: 10.1109/CE2CT64011.2025.10939863

Rashmi, I., Karthika, K. S., Roy, T., Shinoji, K. C., Kumawat, A., Kala, S., & Pal, R. (2022). Soil Erosion and Sediments: A Source of Contamination and Impact on Agriculture Productivity. In Naeem, M., Bremont, J. F. J., Ansari, A. A., & Gill, S. S. (Eds.), *Agrochemicals in Soil and Environment* (pp. 313–345). Springer Nature Singapore., DOI: 10.1007/978-981-16-9310-6_14

Raza, A., Hu, Y., Acharki, S., Buttar, N. A., Zubair, M., Syed, N. R., Vishwakarma, D. K., & Elbeltagi, A. (2024). Role of Artificial Intelligence in Revolutionizing Agricultural Technology. In Pandey, K., Kushwaha, N. L., Pande, C. B., & Singh, K. G. (Eds.), *Artificial Intelligence and Smart Agriculture* (pp. 121–145). Springer Nature Singapore., DOI: 10.1007/978-981-97-0341-8_7

Redhu, N. S., Thakur, Z., Yashveer, S., & Mor, P. (2022). Artificial intelligence: A way forward for agricultural sciences. In *Bioinformatics in Agriculture* (pp. 641–668). Elsevier. DOI: 10.1016/B978-0-323-89778-5.00007-6

Reza, M. N., Lee, K.-H., Karim, M. R., Haque, M. A., Bicamumakuba, E., Dey, P. K., Jang, Y. Y., & Chung, S.-O. (2025). Trends of Soil and Solution Nutrient Sensing for Open Field and Hydroponic Cultivation in Facilitated Smart Agriculture. *Sensors (Basel), 25*(2), 453. DOI: 10.3390/s25020453 PMID: 39860823

Roy, P., Padhiary, M., Hoque, A., Sahu, B., Roy, D., & Kumar, K. (2025). Machine Learning for Precision Agriculture and Crop Yield Optimization: Techniques and Applications. In Emara, T., Hassan, E., Trinh, T., Li, G., & Saber, A. (Eds.), *The Role of Artificial Intelligence in Advancing Applied Life Sciences* (pp. 189–234). IGI Global., DOI: 10.4018/979-8-3693-9208-9.ch007

Saha, D., Padhiary, M., Hoque, A., & Prasad, G. (2025). 3D printing technology for valorization of food processing wastes and byproducts: A systematic review. *Waste Management Bulletin, 3*(4), 100192. DOI: 10.1016/j.wmb.2025.100192

Sakkaravarthy, S., Jano, N. A., & Vijayakumar, A. (2024). Overcoming Challenges in Traditional Waste Water Treatment Through AI-Driven Innovation. In Garg, M. C. (Ed.), *The AI Cleanse: Transforming Wastewater Treatment Through Artificial Intelligence* (pp. 53–81). Springer Nature Switzerland., DOI: 10.1007/978-3-031-67237-8_3

Saleem, S. R., Zaman, Q. U., Schumann, A. W., & Naqvi, S. M. Z. A. (2023). Variable rate technologies: Development, adaptation, and opportunities in agriculture. In *Precision Agriculture* (pp. 103–122). Elsevier., https://www.sciencedirect.com/science/article/pii/B9780443189531000106 DOI: 10.1016/B978-0-443-18953-1.00010-6

Samal, I., Bhoi, T. K., Mahanta, D. K., Komal, J., & Pradhan, A. K. (2024). Sustainable Plant Protection Measures in Regenerative Farming. In Kumar, S., Meena, R. S., Sheoran, P., & Jhariya, M. K. (Eds.), *Regenerative Agriculture for Sustainable Food Systems* (pp. 387–421). Springer Nature Singapore., DOI: 10.1007/978-981-97-6691-8_12

Shah, F., & Wu, W. (2019). Soil and crop management strategies to ensure higher crop productivity within sustainable environments. *Sustainability (Basel), 11*(5), 1485. DOI: 10.3390/su11051485

Shaikh, T. A., Rasool, T., & Lone, F. R. (2022). Towards leveraging the role of machine learning and artificial intelligence in precision agriculture and smart farming. *Computers and Electronics in Agriculture, 198*, 107119. DOI: 10.1016/j.compag.2022.107119

Sikka, A. K., Alam, M. F., & Mandave, V. (2022). Agricultural water management practices to improve the climate resilience of irrigated agriculture in India. *Irrigation and Drainage, 71*(S1), 7–26. DOI: 10.1002/ird.2696

Silva, F. M., Queirós, C., Pereira, M., Pinho, T., Barroso, T., Magalhães, S., Boaventura, J., Santos, F., Cunha, M., & Martins, R. C. (2024). Precision Fertilization: A critical review analysis on sensing technologies for nitrogen, phosphorous and potassium quantification. *Computers and Electronics in Agriculture, 224*, 109220. DOI: 10.1016/j.compag.2024.109220

Stark, N. M., & Matuana, L. M. (2021). Trends in sustainable biobased packaging materials: A mini review. *Materials Today Sustainability, 15*, 100084. DOI: 10.1016/j.mtsust.2021.100084

Suhag, A., Yadav, H., Chaudhary, D., Subramanian, S., Jaiwal, R., & Jaiwal, P. K. (2021). Biotechnological interventions for the sustainable management of a global pest, whitefly (*Bemisia tabaci*). *Insect Science, 28*(5), 1228–1252. DOI: 10.1111/1744-7917.12853 PMID: 32696581

Tripathi, S., Srivastava, P., Devi, R. S., & Bhadouria, R. (2020). Influence of synthetic fertilizers and pesticides on soil health and soil microbiology. In *Agrochemicals detection, treatment and remediation* (pp. 25–54). Elsevier., https://www.sciencedirect.com/science/article/pii/B9780081030172000027 DOI: 10.1016/B978-0-08-103017-2.00002-7

Upadhyay, A., Chandel, N. S., Singh, K. P., Chakraborty, S. K., Nandede, B. M., Kumar, M., Subeesh, A., Upendar, K., Salem, A., & Elbeltagi, A. (2025). Deep learning and computer vision in plant disease detection: A comprehensive review of techniques, models, and trends in precision agriculture. *Artificial Intelligence Review*, *58*(3), 1–64. DOI: 10.1007/s10462-024-11100-x

Velaoras, K., Menegaki, A. N., Polyzos, S., & Gotzamani, K. (2025). The role of environmental certification in the hospitality industry: Assessing sustainability, consumer preferences, and the economic impact. *Sustainability (Basel)*, *17*(2), 650. DOI: 10.3390/su17020650

Vlaiculescu, A., & Varrone, C. (2022). Sustainable and eco-friendly alternatives to reduce the use of pesticides. In *Pesticides in the natural environment* (pp. 329–364). Elsevier. DOI: 10.1016/B978-0-323-90489-6.00014-8

Wahab, A., Muhammad, M., Ullah, S., Abdi, G., Shah, G. M., Zaman, W., & Ayaz, A. (2024). Agriculture and environmental management through nanotechnology: Eco-friendly nanomaterial synthesis for soil-plant systems, food safety, and sustainability. *The Science of the Total Environment*, *926*, 171862. DOI: 10.1016/j.scitotenv.2024.171862 PMID: 38527538

Wu, Y., Si, W., Yan, S., Wu, L., Zhao, W., Zhang, J., Zhang, F., & Fan, J. (2023). Water consumption, soil nitrate-nitrogen residue and fruit yield of drip-irrigated greenhouse tomato under various irrigation levels and fertilization practices. *Agricultural Water Management*, *277*, 108092. DOI: 10.1016/j.agwat.2022.108092

Wysel, M. (2023). Data Sharing Platforms: How Value Is Created from Data Produced by Smart Agriculture. In Zhang, Q. (Ed.), *Encyclopedia of Digital Agricultural Technologies* (pp. 257–267). Springer International Publishing., DOI: 10.1007/978-3-031-24861-0_250

Yadav, A., Yadav, K., & Abd-Elsalam, K. A. (2023). Nanofertilizers: Types, delivery and advantages in agricultural sustainability. *Agrochemicals*, *2*(2), 296–336. DOI: 10.3390/agrochemicals2020019

Ye, J., Wang, Y., Kang, J., Chen, Y., Hong, L., Li, M., Jia, Y., Wang, Y., Jia, X., Wu, Z., & Wang, H. (2022). Effects of long-term use of organic fertilizer with different dosages on soil improvement, nitrogen transformation, tea yield and quality in acidified tea plantations. *Plants*, *12*(1), 122. DOI: 10.3390/plants12010122 PMID: 36616251

Yong, W. T. L., Thien, V. Y., Misson, M., Chin, G. J. W. L., Hussin, S. N. I. S., Chong, H. L. H., Yusof, N. A., Ma, N. L., & Rodrigues, K. F. (2024). Seaweed: A bioindustrial game-changer for the green revolution. *Biomass and Bioenergy*, *183*, 107122. DOI: 10.1016/j.biombioe.2024.107122

Zeng, Y., Verhoef, A., Vereecken, H., Ben-Dor, E., Veldkamp, T., Shaw, L., Van Der Ploeg, M., Wang, Y., & Su, Z. (2025). Monitoring and Modeling the Soil-Plant System Toward Understanding Soil Health. *Reviews of Geophysics, 63*(1), e2024RG000836. DOI: 10.1029/2024RG000836

Zhang, M., Ma, W., Tao, R., Fan, Q., Zhang, M., Qin, D., Cao, X., Li, J., Xiong, R., & Huang, C. (2024). Nanomaterials: Recent advances in plant disease diagnosis and treatment. *Nano Today*, *57*, 102326. DOI: 10.1016/j.nantod.2024.102326

Chapter 10
Sustainable Chemistry and Pioneering Green Engineering Solutions:
Green Chemical Processes for Industrial Application

Shrikant M. Ghodse
https://orcid.org/0000-0002-3082-7733
Principal K.M. Kundnani College of Pharmacy, Mumbai, India

Sakshi M. Gurav
Principal K.M. Kundnani College of Pharmacy, Mumbai, India

Gauraw R. Chopade
Principal K.M. Kundnani College of Pharmacy, Mumbai, India

Aamir M. Patel
Principal K.M. Kundnani College of Pharmacy, Mumbai, India

Kulbhushan A. Sasane
Independent Researcher, India

Hemant N. Akolkar
https://orcid.org/0000-0003-0882-1324
Abasaheb Marathe Arts and New Commerce, Science College, Ratnagiri, India

ABSTRACT

Green synthesis has become a reliable, sustainable, and ecological protocol for the synthesis of active pharmaceutical ingredients and their intermediates. Sustainable approaches include solvent free, catalysis, biocatalyst with start of art modified green methods explained. This chapter overview the various sustainable approaches for synthesis of various examples industrial important compounds. In addition to green approach case studies on pharmaceutical and agrochemical products and current updates were also explain.

DOI: 10.4018/979-8-3373-1409-9.ch010

1. INTRODUCTION TO GREEN SYNTHESIS

1.1 History of Green synthesis

A significant turning point in the global economic evolution was the expanding industrialization process. The environmental effects that the expansion of industrial activity could have on our planet were not taken into consideration by global government programs, despite their contribution to the improvement of quality of life. (Zygmunt *et al.*, 2009) (Marco *et al.*, 2019)

Due to excessive industrialization and increasing food production brought on by the population's fast growth, pollution, and resource depletion intensified. Natural resources were thus used as though there were no consequences for environmental issues.(Zygmunt *et al.*, 2009) (Marco *et al.*, 2019)

The novel "Silent Spring," which was published in the 1960s, sparked a modern environmental movement. The historical novel has generated significant government initiatives characterized by concern about the dangers of over-exploitation of natural resources and has increased understanding of ecological awareness. John Kenneth Galbraith ranked the book as one of the most significant works of Western literature, while Robert Downs described it as "the book that changed America."(Lutts, 2014) (Marco *et al.*, 2019)

Barry Trost, a chemistry professor at Stanford, was one of the pioneers of green chemistry. He initially proposed the idea of "atom economy" in 1973. Trost maintained that those in charge of creating chemicals should strive for elegant efficiency, which uses the maximum percentage of input atoms in the usable output, ideally leaving zero waste, rather than deeming a chemical process successful if it yields a usable product at a reasonable cost. When first offered, this idea looked idealistic, but more and more chemical processes, including biocatalysis, are currently being suggested that produce results that are somewhat like that objective. (Woodhouse, 2015) (Marco *et al.*, 2019)

The world first became aware of the potential harm that ecosystem degradation could do to humans in 1972 when the Stockholm Conference took place in Sweden. Several international conferences on the environment took place in the 1980s. To produce a report on global development and the environment, the UN established the World Commission on Environment and Development in 1983 following an assessment of the ten years of the planned activities at the Stockholm Conference. The world environment was under tremendous pressure at the time this commission was formed, and it was becoming increasingly clear that an extensive amount of development was unsustainable. (Marco *et al.*, 2019)

The 1987 publication of the "Brundtland Report," which introduced social and environmental concerns into harmony, established the idea of sustainable development for the first time without risking future generations. (Marco *et al.*, 2019)

The "Alternative Synthetic Routes for Pollution Prevention" program, started by the US Environmental Protection Agency (EPA) in 1991, reported a new philosophy and policy on minimizing the risks of toxic chemical products to prevent problems with such substances. It emphasized that the correct course of action would be to avoid producing these products in the first place. (Marco *et al.*, 2019)

The term "Green Chemistry" was likely first used in a 1990 paper title by Cathcart, who discussed the expansion of the Irish chemical sector. But it wasn't until 1996 that Anastas and Williamson's first publication addressed green chemistry with the current approach. (Marco *et al.*, 2019)

The Presidential Green Chemistry Challenge (PGCC) program was introduced by the US government in 1995. It considers the progress that has been made in the chemical industry to reduce waste generation across several production areas. Each year, five distinct categories are used to reward the works: Alternative Synthetic Routes, Small Business, Academics, Reactive Alternative Conditions, and Safer Chemical Designs. (Cann, 1999)(Marco *et al.*, 2019)

The Green Chemistry Institute (GCI) was established in 1997 as a non-profit organization to expand the use of green chemistry by using the expertise, experience, and ability of chemical companies to move towards sustainability. (Marco *et al.*, 2019)

Another significant breakthrough in green chemistry was the publication of the ground-breaking book Green Chemistry: Theory and Practice, which was co-authored by Paul Anastas and John C. Warner in 1998 and included 12 Green Chemistry principles. The 12 Principles of Green Chemistry are addressed in detail in the book, along with a philosophy that has always pushed businesses and university scientists to take ecologically responsible actions. (Marco *et al.*, 2019)

To facilitate and promote green chemistry and green engineering in the pharmaceutical industries, the American Chemical Society's Green Chemistry Institute (GCI) and multinational pharmaceutical companies organized a panel discussion in 2005. According to the panel discussion, "continuous processing" is essential to advancing "the green." Between 1997 and 2011, the International Union of Pure and Applied Chemistry (IUPAC) organized four conferences on green chemistry in collaboration with ACS and GCI. The conferences covered subjects such as environmentally friendly products and processes, energy production, renewable chemical waste sources, green policy adoption, and green chemistry education. (Marco *et al.*, 2019)

Even if sustainable practices have been incorporated into chemistry and ecological engineering research over time, it will be crucial to keep funding industrial methods and regulations to carry out environmental improvements. (Marco *et al.*, 2019)

The design and application of ecologically friendly chemical processes that seek to minimize the use of hazardous conditions, solvents, and toxic reagents is known as "green synthesis."(Anastas & Warner., 2000)

This strategy is based on the ideas of green chemistry, which emphasizes the creation of environmentally friendly, effective, and sustainable chemical processes. By using renewable resources, cutting waste, and increasing energy efficiency, green synthesis aims to reduce the environmental impact of producing chemicals in laboratories and industries.(A. Kumar, 2020)

Energy generation, materials research, nanotechnology, and medicines are just a few of the industries that use green synthesis techniques. In addition to addressing environmental issues, these techniques seek to increase the affordability and accessibility of chemical synthesis. The goal of green synthesis is to employ renewable energy sources for the power supply, minimize energy waste, and maximize consumption. (K. Betts, 2015)

Green chemistry focusses on synthesis processes following its 12 guiding principles, minimizing or eliminating adverse effects on the environment and human health. (Clarke *et al.*, 2018)

1.2 Twelve Principles of Green chemistry

1. *Prevention:* It is better to prevent waste than to treat or clean up waste after it has been created.
2. *Atom Economy:* Synthetic methods should be designed to maximize the incorporation of all materials used in the process into the final product.
3. *Less Hazardous Chemical Syntheses:* Wherever practicable, synthetic methods should be designed to use and generate substances that possess little or no toxicity to human health and the environment.
4. *Designing Safer Chemicals:* Chemical products should be designed to affect their desired function while minimizing their toxicity.
5. *Safer Solvents and Auxiliaries:* The use of auxiliary substances (e.g., solvents, separation agents, etc.) should be made unnecessary wherever possible and innocuous when used.
6. *Design for Energy Efficiency:* Energy requirements of chemical processes should be recognized for their environmental and economic impacts and should be minimized. If possible, synthetic methods should be conducted at ambient temperature and pressure.
7. *Use of Renewable Feedstocks:* A raw material or feedstock should be renewable rather than depleting whenever technically and economically practicable.

8. *Reduce Derivatives:* Unnecessary derivatization should be minimized or avoided, if possible because such steps require additional reagents and can generate waste.
9. *Catalysis:* Catalytic reagents are superior to stoichiometric reagents.
10. *Design for Degradation:* Chemical products should be designed so that at the end of their function they break down into innocuous degradation products and do not persist in the environment.
11. *Real-time Analysis for Pollution Prevention:* Analytical methodologies need to be further developed to allow for real-time, in-process monitoring and control before the formation of hazardous substances.
12. *Inherently Safer Chemistry for Accident Prevention:* Substances and the form of a substance used in a chemical process should be chosen to minimize the potential for chemical accidents, including releases, explosions, and fires.

Figure 1. Twelve Principles of green Chemistry

1.3 Types of Greener Synthesis Approach

Organic synthesis on a large scale involves the use of basic chemical ingredients from the petrochemical sector and catalysts; and after the end of the reaction, separation, purification, storage, packaging, distribution, etc. Conventional methods of organic synthesis usually need longer heating time, and tedious apparatus setup, which result in higher cost of process and the excessive use of solvents/reagents. During these processes, there are many problems of health and safety for workers in addition to the environmental problems caused by their use and disposition as waste. Green Chemistry would like to increase the efficiency of synthetic methods, use less toxic solvents, reduce the stages of the synthetic routes, and minimize waste as far as practically possible. (Clark & Macquarrie,. 2008)

There are some approaches listed below -

1.3.1 Physical Approach

Ball Mill
Microwave irradiation
Magnetic field-assisted synthesis
Ultrasound-assisted synthesis

1.3.2 Chemical Approach

Solvent free Synthesis
Greener solvent Synthesis
Catalysis

1.3.1 Physical Approach

Mechanosynthesis, microwave-assisted reactions (and their combination), ultrasound-assisted processes, UV-irradiation of the reaction system, and other physical techniques are currently being considered as well because they meet the requirements of green chemistry, do not require hazardous solvents, prevent pollution, and speed up synthesis processes. A few of these approaches are less widely used, such as magnetic field-assisted synthesis. (Kharissova *et al.*, 2019)

Ball milling, a type of mechanochemical synthesis that is utilized extensively in organic chemistry and has conventional application in inorganic solid-state processes, is an effective way to drive chemical reactions using mechanical energy. In addition to increasing energy efficiency, ball milling helps us stay away from hazardous solvents and chemicals.(Kharissova *et al.*, 2019)

These processes, or reactions, occur at room temperature and without the need for solvents. Although ball milling is regarded as a green tool for chemistry due to its simplicity and environmental friendliness, scientists do not frequently employ it, despite its enormous potential.(Do, 2017)

Examples -Conversion of benzyl alcohol to benzaldehyde, *Wang, Jing-Fei, et al* developed a facile, solvent free green protocol for oxidation of benzyl alcohol to benzaldehyde using $NaNO_3/P_2O_5$ under high-speed ball milling (Wang *et al.*, 2022) (Figure 2)

Figure 2. Conversion of benzyl alcohol to benzaldehyde

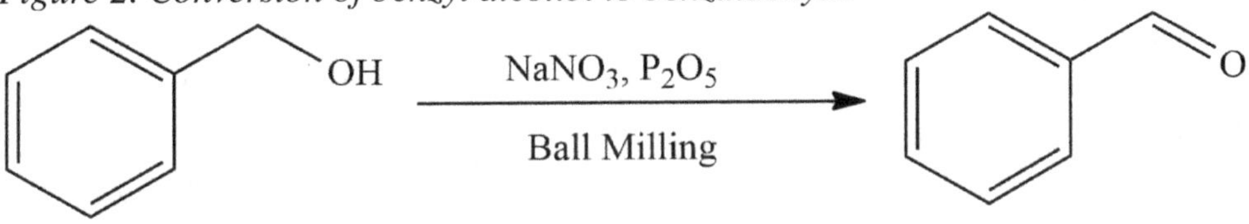

Microwave irradiation: Organic synthesis frequently uses microwave irradiation, a type of electromagnetic irradiation with mobile electric charges. Ionic conduction and dipolar polarisation are the two primary mechanisms that account for the heating mechanism of microwave radiation.(Kharissova *et al.*, 2019)

Conventionally and MW-heated simples have distinct temperature profiles. All of the samples experience uniform, volumetric heating; thermal conduction inside the sample reduces the energy needed for heat transmission. MW heating offers benefits like pure products, quick process speeds, reduced heat loss, high efficiency, waste reduction, low operating costs, and fewer potential side products. Ideal solvent-free conditions prevent explosive explosions. (Kitchen *et al.*, 2014)

Examples – Synthesis of *N*-alkyl phthalimides, *N*-Alkylations under microwave irradiation using phase transfer catalysts occupy a unique place in organic chemistry. The synthesis of *N*-alkyl phthalimides using phthalimide, alkyl halides, potassium carbonate, and TBAB; giving products in 45–98% yield. (Figure 3)(Bogda et al., 1996)

Figure 3. Synthesis of N-alkyl phthalimides

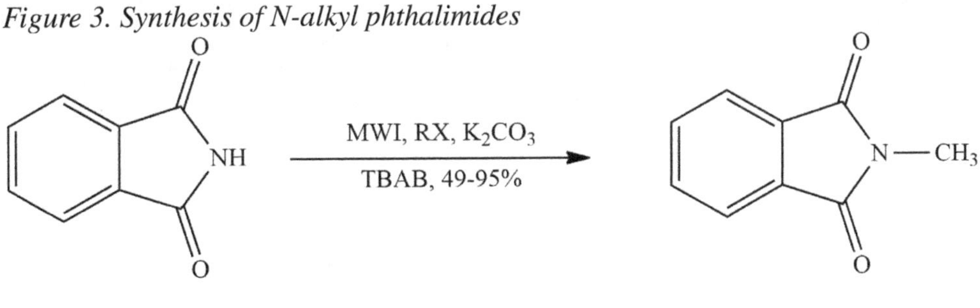

b) Synthesis of Pyrido[3,2,1-jk] carbazoles

Pyrido[3,2,1-jk] carbazoles were produced via the domino reactions of 2-(3-oxo-1,3-diarylpropyl)-1-cyclo-hexanones with phenyl hydrazine hydrochloride in water by Chitra, Selvam, et al A library of matching pyrido carbazoles can be prepared in aqueous medium using the cascade technique, which offers a very atom-efficient green approach with little purification required. (Figure 4) (Kitchen *et al.*, 2014) (Gawande *et al.*, 2014)

Figure 4. Synthesis of Pyrido[3,2,1-jk] carbazoles

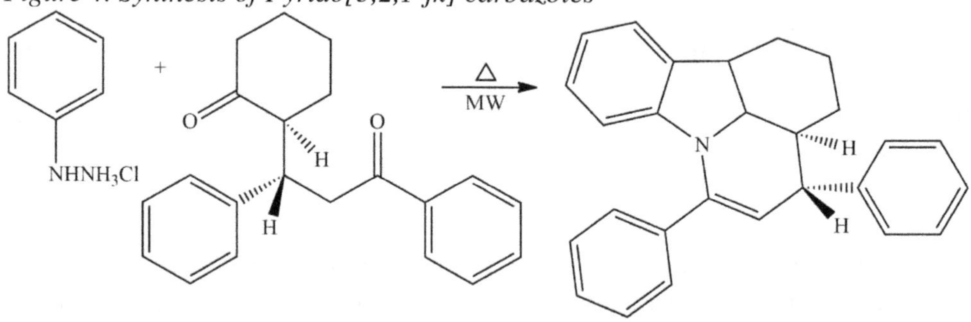

c) Synthesis of pyrazoles and diazepines

MW has been used to produce pyrazoles and diazepines without the use of solvents or catalysts; a full conversion takes place at 120 °C in 5–15 minutes. This method was developed by Vaddula, B. R., Varma, R. S., & Leazer, J. (Figure 4) (Figure 5)(Vaddula *et al.*, 2013)

Figure 5. Synthesis of pyrazoles

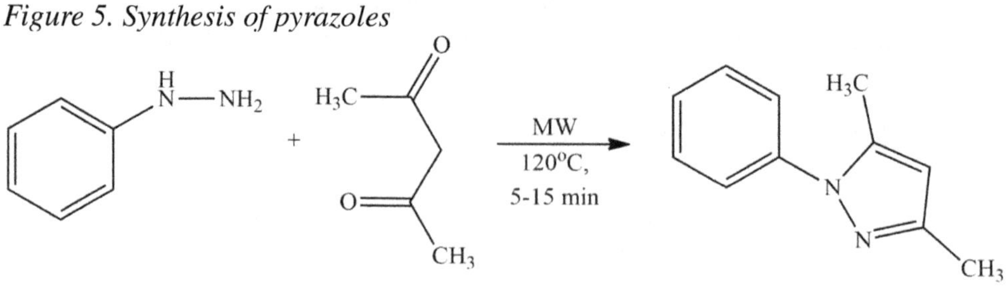

Figure 6. Synthesis of Diazepines

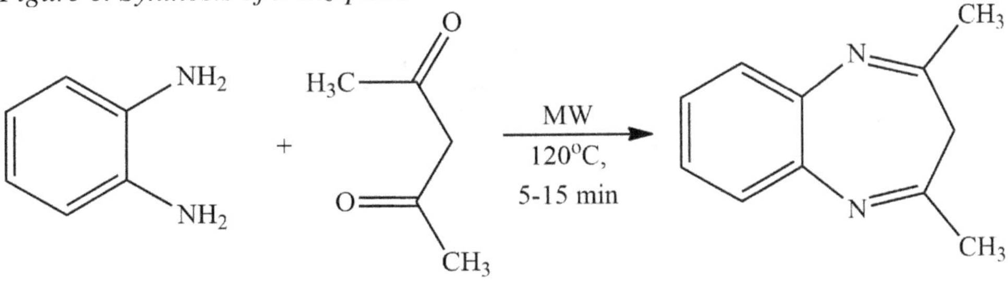

Magnetic-field-assisted synthesis: A chemical process known as magnetic-field-assisted (MFA) synthesis produces materials by applying an external magnetic field. Conventional techniques involve the use of hazardous solvents or extra energy-intensive procedures that may leave behind undesirable residues.(Kharissova *et al.*, 2019)

Magnetic fields can be used to skip these additional processes while synthesizing the required substance. An effective technique for enhancing electrocatalytic performance that is adaptable and compatible with current electrocatalytic devices is the use of an external magnetic field. Furthermore, magnetic fields can influence the catalytic reaction process and aid in catalyst production. (Krajewski, 2017)

Ultrasound-assisted synthesis: Sonochemistry is a potent and environmentally friendly method that uses ultrasonic technology to encourage chemical reactions and speed up the synthesis of organic molecules. (Mason, 1997)

The portion of the sonic spectrum that falls between roughly 20 KHz and 10 MHz is known as ultrasound. (Sci *et al.*, 2016)

Under ultrasonic irradiation, a wide range of chemical reactions may be conducted under moderate circumstances, with good yields and fast reaction times. Acoustic cavitation is the mechanism by which high-energy chemistry is produced by ultrasound. In both homogeneous and heterogeneous systems, ultrasound is a crucial source for the start or acceleration of catalytic processes. (Gharat & Rathod, 2020)

Examples –a) Synthesis of 2-imidazolines, *Sant Anna, Gabriela da S., et al.* synthesized a series of sixteen 2-substituted-2-imidazolines by the reaction of the substituted-aldehydes and ethylenediamine by ultrasound irradiation with NBS in water in high yields (80–99%). (Figure 7)(Gharat & Rathod, 2020)

Figure 7. Synthesis of 2-imidazolines

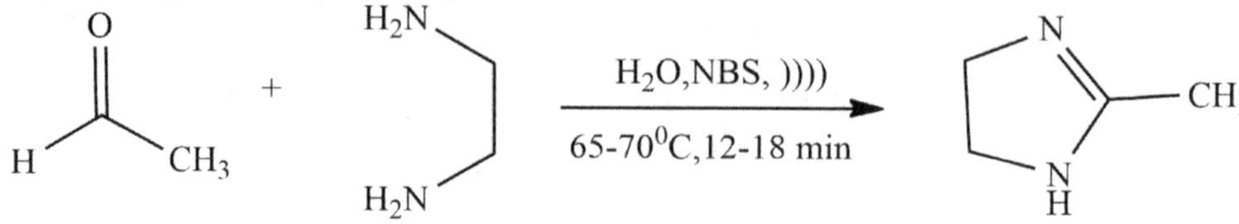

b) Formation of 4-methyloxazole-5-carbonitrile (Gharat & Rathod, 2020)

Figure 8. Formation of 4-methyloxazole-5-carbonitrile

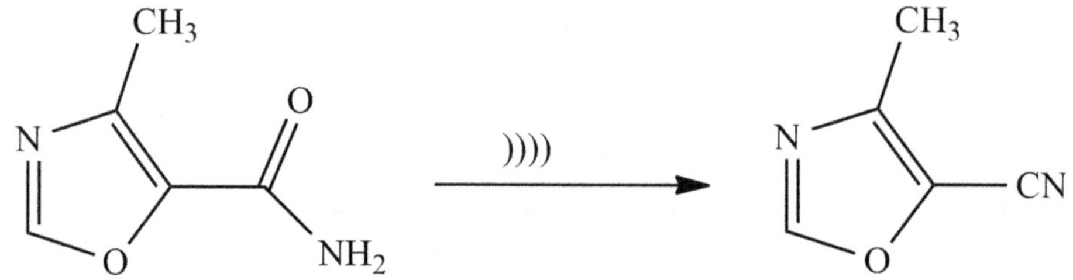

1.3.2 Chemical Approach

Solvent-free synthesis: Since solvents are frequently hazardous, costly, and difficult to use and remove, solvent-free methods are strategically advantageous. Solvent-free techniques are quite popular because they can modernize traditional processes and make them safer, easier, and cleaner. These methods can also make it possible to conduct studies without the use of powerful mineral acids, such as HCl or H_2SO_4, which can lead to issues with corrosion, safety, manipulation, and pollution as waste. Solid, recyclable acids like clays can effectively replace these acids. Solvent-free techniques are quite popular because they can modernize traditional processes and make them safer, easier, and cleaner. (Top *et al.*, 2016)

Examples –a) Synthesis of Chalcone(Marvaniya *et al.*, 2011)

Figure 9. Synthesis of Chalcone

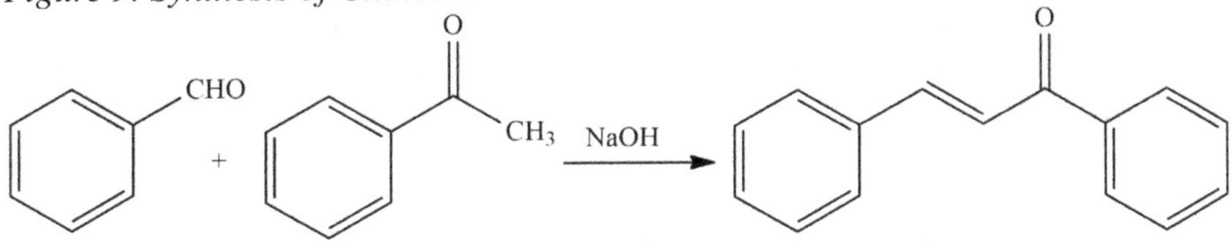

b) Synthesis of Dihydropyrimidinone(Marvaniya *et al.*, 2011)

Figure 10. Synthesis of Dihyropyrimidinones

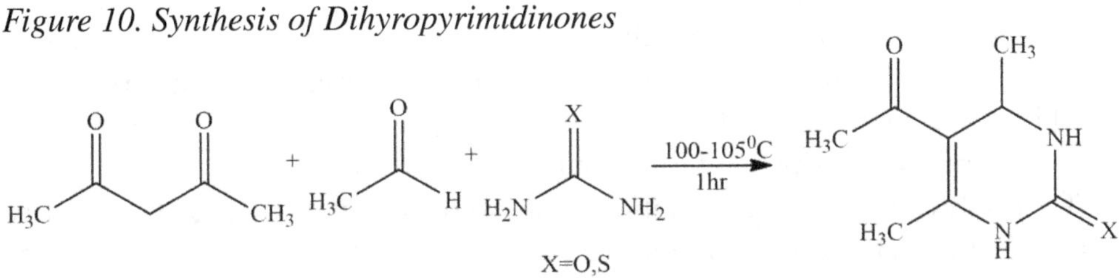

Greener Solvent synthesis: There are numerous toxic, hazardous, and environmentally damaging organic solvents. Their usage therefore puts the environment and human health in danger.(Kharissova *et al.*, 2019)

Volatile organic compounds are the most prevalent solvents among this waste and are consequently a significant environmental problem since they may produce smog and low-level ozone through free radical air oxidation processes. Many solvents have been graded according to their environmental, safety, and health (ESH) attributes as an understanding of solvent properties is essential to sustainable development. The solvent selection guide's fundamental idea is based on these rankings.(Kharissova *et al.*, 2019)

Water and other replacement inorganic solvents don't discharge into the atmosphere and damage the ozone layer, making them much less damaging to the environment. Some compounds cannot dissolve in water, thus supercritical fluid, a by-product of other operations, is used as an alternative to organic solvents. It is non-toxic and does not contribute to climate change. (Díaz-álvarez *et al.*, 2014)

Supercritical fluid technologies, such as supercritical water or CO_2, are advancing green nchemistry in nanomaterial synthesis and addressing environmental issues. (Wood *et al.*, 2004)

They are also used in extraction, cleaning, fractionation, drying, polymerization, hydrothermal reactions, plating, biomass conversion, and dyeing, offering solvent-free media, high yields, ease of use, and no wastewater or secondary pollution. (Kharissova *et al.*, 2019)

2. CATALYSIS FOR SUSTAINABLE MANUFACTURING

The Swedish chemist Jons Jacob Berzelius first proposed the idea of catalysis more than 180 years ago. According to Berzelius, a catalyst is any material that accelerates a chemical process without being consumed.(Ludwig *et al.*, 2017)

In the rate-determining phase, the catalyst affects the reaction rate, which is determined by how frequently the reactants come into touch with one another. In general, catalysts provide the reaction product with a different reaction channel, which speeds up the process and lowers the activation energy. (Madan, 2020)

This is accomplished by the catalysts reacting with one or more reactants to create intermediates, which in turn yield the end products.

The catalyst undergoes regeneration in the last stage. Catalysis is used in a variety of organic transformations to produce a large number of organic synthons. These days, organic chemical processes focus on improving selectivity and making better use of resources.

Catalysts are occasionally employed in chemical reactions to alter the target product's selectivity. (Madan, 2020)

In practice, many chemical processes do not continue without a catalyst since the observed increase in pace is frequently large enough.

It is hardly surprising that catalysis has become commonplace in our daily lives given its many benefits, including lower energy requirements, less of an impact on the environment, and total financial gains. Catalytic transformations are involved in almost 90% of modern chemical processes, which means that they account for the majority of the commodities generated. (Zhou, 2015) .

2.1 Classification of Catalysts

In the chemical and pharmaceutical sectors, as well as in refinement and environmental protection, catalysis is very significant and frequently used. A range of reactor types are used in catalytic processes, which include single-phase homogeneous liquid, two-phase heterogeneous liquid-solid, gas-solid (the most common), and three-phase heterogeneous liquid–solid–gas.(Hagemeyer and Clara, 2016)

The number of stages in the catalytic reaction that is conducted for a synthesis determines whether homogeneous or heterogeneous catalysis might be used. (Przegl and Pow, 2018)

Commercially, a wide variety of catalysts and catalyst types used are,

Homogeneous catalyst
Heterogeneous
Photocatalyst
Micellar catalyst
Biocatalyst

Homogeneous Catalyst: The molecules of a homogeneous catalyst are distributed in the same phase as the reactant. Several advantages of homogeneous catalysts include lowering reaction temperatures, conserving energy, and increasing selectivity by lowering the production of unwanted byproducts. Pharmaceutical synthesis makes considerable use of homogeneous catalysts due to their precision and efficacy. They are crucial to the production of pharmaceuticals because they encourage ecologically friendly and sustainable medication manufacturing methods. However, it has some limitations which include the need to separate the catalyst, the expense of recovering the catalyst, and the substantial release of heavier transition metals into the biosphere, all of which harm the environment. (Przegl & Pów, 2018)

Examples –a) Synthesis of Pyrano[4,3,-b]pyran, Khaligh *et.al* developed an environmentally friendly method of synthesizing Pyrano[4,3,-b]pyran derivatives by using Solar Energy under Solvent-Free Conditions employing Succinimide-N-sulfonic acid as an efficient Brönsted acid catalyst. (Figure 11) (Khaligh *et al.*, 2016)

Figure 11. Synthesis of Pyrano[4,3,-b]pyran

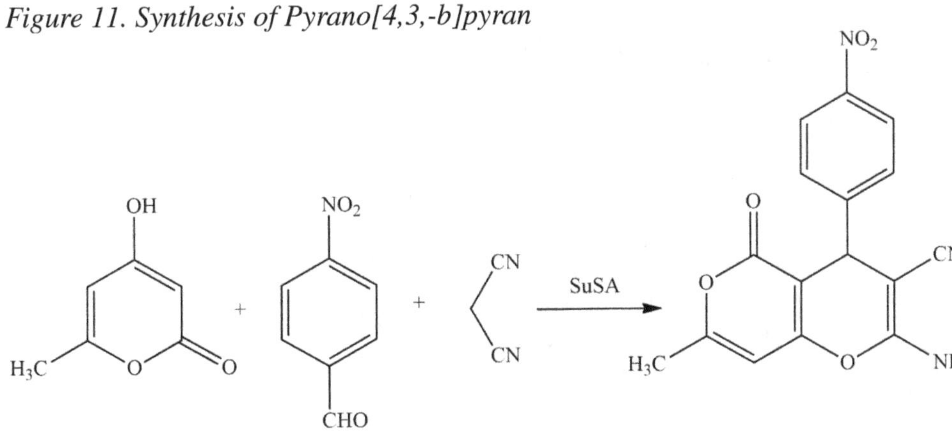

b) Synthesis of *L*-Dopa, Knowles at Monsanto invented the asymmetric hydrogenation of derivatives of cinnamic acid. *L*-dopa, a chiral amino acid used to treat Parkinson's disease, can be produced synthetically and used on a large scale. The nitrogen atom's acetyl protective group is removed to generate *L*-Dopa. A cationic rhodium compound with optically active phosphine ligands introduces the asymmetry. For the complexation of an olefin, asymmetric chelate ligands work especially well at creating an asymmetric coordination core. There are two diastereomeric variants of the resultant complex that vary in how the alkene is coordinated. Along with Noyori and Sharpless (for asymmetric epoxidation), Knowles shared the 2001 Nobel Prize in Chemistry. (Figure 12) (International, 2016)

Figure 12. Synthesis of L-Dopa

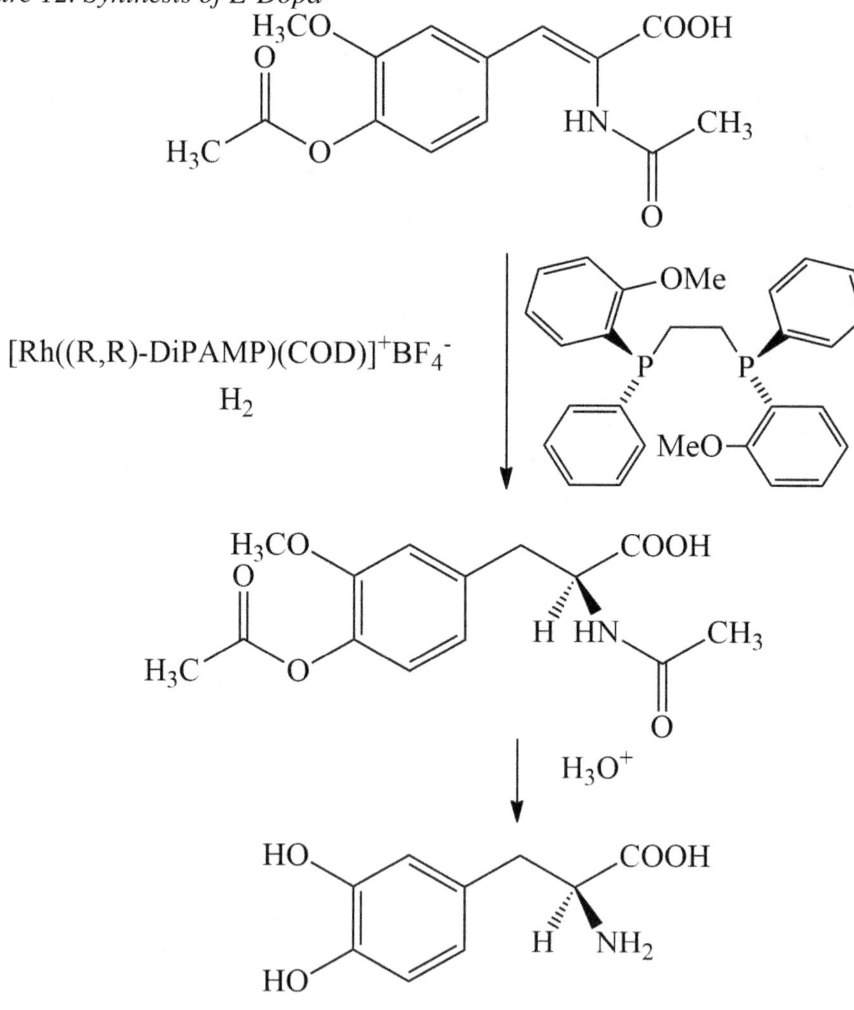

Heterogeneous catalysis plays a crucial role in the production of vital products in vital sectors such as agriculture, pharmaceuticals, and materials science, including the manufacture of polymers and various essential materials. Heterogeneous catalysis refers to systems where the reactions occur in distinct stages. Reactants are in the gaseous or liquid phase, whereas the catalyst is mostly in the solid phase. Heterogeneous catalysts offer unique advantages over reactants, such as simplicity in separation and the potential for catalytic recycling.(Przegl and Pow, 2018)

They operate in a distinct phase as well. Catalysts are often solid materials having active spots that facilitate the desired chemical processes. Despite its many advantages, heterogeneous catalysis can operate throughout a broad pH range, whereas homogenous catalysis is frequently pH-dependent, occasionally less effective, and less environmentally friendly. Furthermore, heterogeneous catalysis accounts for up to 80% of catalytic processes, making it a thriving area of chemistry. (Hutchings, 2009)(Lal & Parewa, 2021)

Examples –a) Synthesis of Niacinamide, Lonza AG, a Swiss-based company, is a leading provider of this product and has established two distinct production methods. The traditional approach is utilized at their Swiss facility, whereas a novel catalytic process is employed at their Guangzhou site in China, operated by Lonza Guangzhou. The classical route begins with acetaldehyde and ammonia, which are transformed into 5-ethyl-2-methylpyridine through a Tschitschibabin pyridine synthesis, and then subsequently oxidized with nitric acid. Beginning with 2-methylpentanediamine, the catalytic pathway cycles it over an acidic catalyst to 3-methylpiperidine. Over palladium, 3-methylpiperidine is dehydrogenated to 3-picoline. It uses a catalyst containing vanadium oxide to undergo an Ammon oxidation reaction to 3-cyanopyridine. Lastly, pharma-grade niacinamide is produced by selective enzymatic hydrolysis of nitrile. (Figure 13)(Heveling, 2012)

Figure 13. Synthesis of Niacinamide

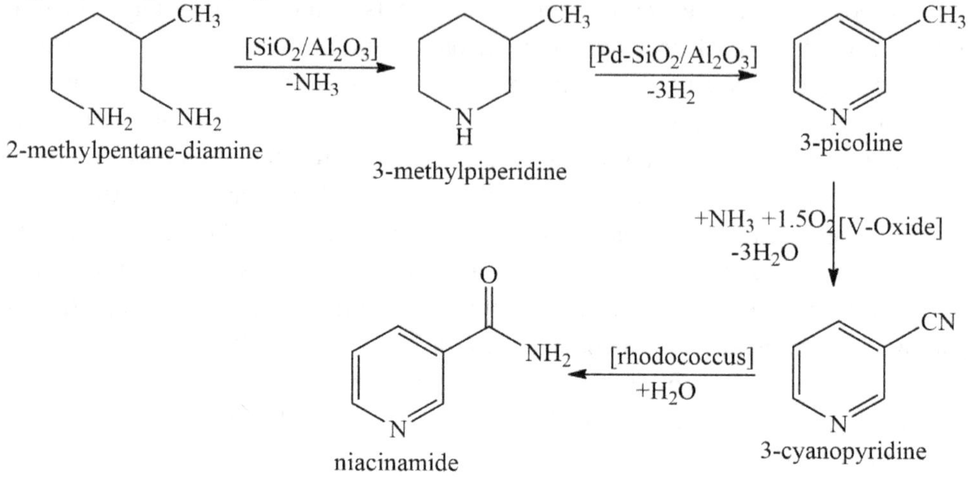

b) Synthesis of Biphenylacetic Acids

At Schering-Plough, *Gala et al.* revealed the use of Pd/C for the one-step synthesis of biphenyl acetic acids from aryl bromides and boric acids.

Although the intended transformation could be effectively carried out by techniques employing $Pd(OAc)_2$, development work required a less expensive, more process-friendly substitute. By diluting with an IPA/H_2O/NaOH combination and filtration to eliminate the Pd catalyst, the authors discovered that employing either dry or wet 5% Pd/C resulted in full conversion that could be separated. (Figure 14) (Suhelen *et al.*, 2021)

Figure 14. Synthesis of Biphenylacetic Acids

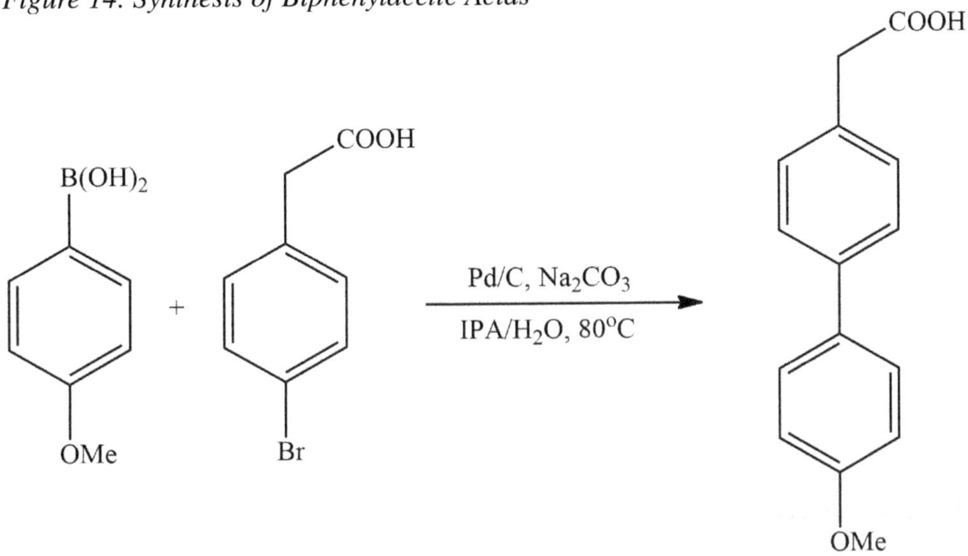

c) Synthesis of hexahydro-1,3,5-triazine derivatives

Dandia et.al used melamine to create graphitic carbon nitride (g-C_3N_4), which was then functionalized by treating it with H_2SO_4 with the aid of ultrasound. The author used a variety of analytical techniques to characterize the prepared functional carbon nitride (Sg–C_3N_4), which was then employed as a reusable catalyst to prepare 1,3,5-trisubstituted hexahydro-1,3,5-triazine derivatives using formaldehyde and aryl amines as a reactant in aqueous media under ultrasound irradiation. Several such tests demonstrate that the catalyst's SO_3H functionality plays a crucial role in this metal-free method. Under ultrasonic irradiation, Sg–C_3N_4 exhibits 21-fold greater catalytic activity compared to the traditional procedure. (Figure 15)(Lal and Parewa, 2021)

Figure 15. Synthesis of hexahydro-1,3,5-triazine derivatives

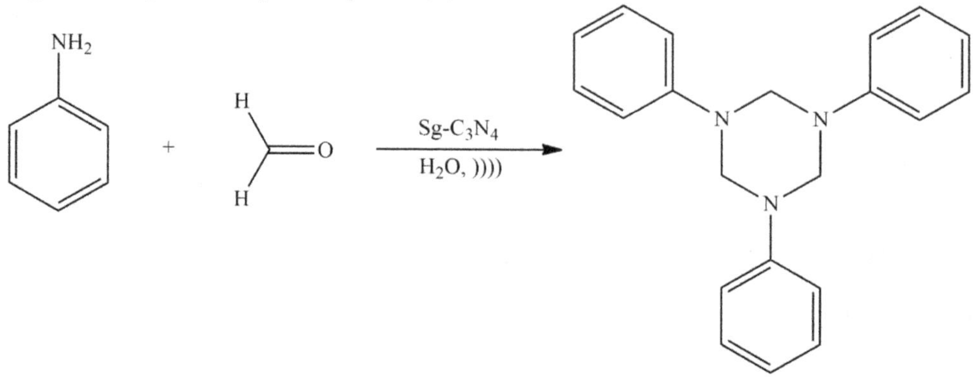

Photocatalyst

The phrase ***"photocatalyst"*** combines the terms "photo," which refers to a photon, and "catalyst," which refers to a material that, when present, changes the rate of a process. Consequently, substances known as photocatalysts alter a chemical reaction's rate when exposed to light. Under UV light, photochemical reactions are regarded as green chemistry interactions. They are based on electronic excitation, which affects the reagents' chemical reactivity during organic synthesis. (Zhu *et al.*, 2020)

Examples – a) Synthesis of substituted Furan (Crespi *et al.*, 2018)

Figure 16. Synthesis of substituted Furan

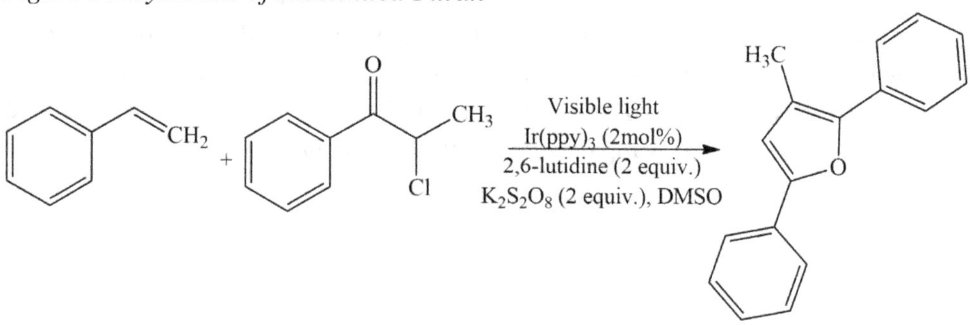

b) Synthesis of polysubstituted 1,4- dihydropyridines, *Zhu et.al* reported synthesis of 1,4-dihydropyridines using Ir(ppy)3 as a catalyst and dicumyl peroxide (DCP) as an oxidant, a visible light photocatalytic cascade cyclization process involving glycine derivatives and β-ketoesters At room temperature, a succession of N-aryl glycine esters easily progressed with the cyclization with β-ketoesters, yielding the required 1,4-dihydropyridines in a sufficient quantity. Figure 17 (Chen *et al.*, 2016)

Figure 17. Synthesis of polysubstituted 1,4- dihydropyridines

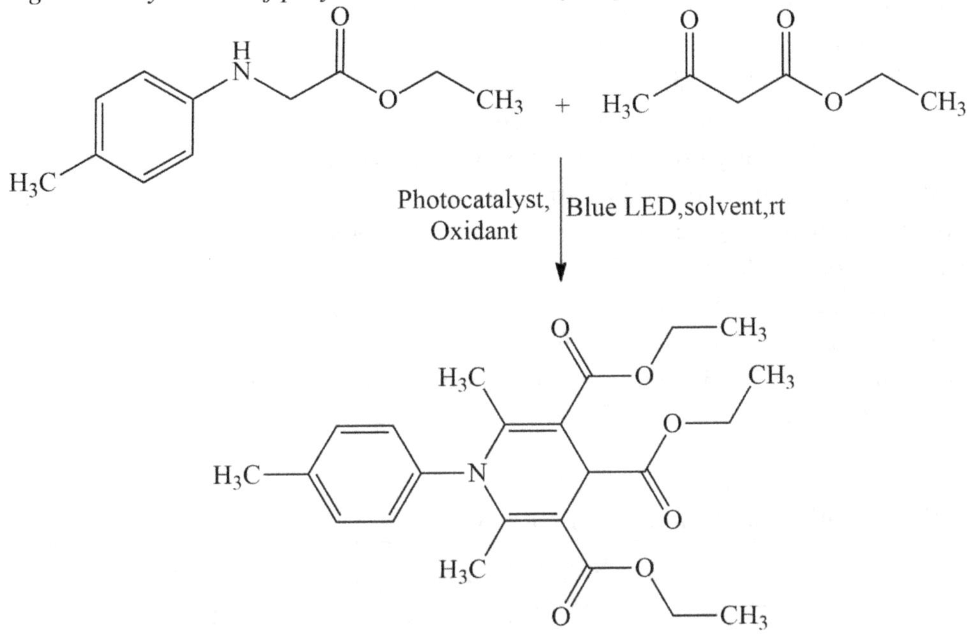

Biocatalysis For the production of active pharmaceutical ingredients (APIs), biocatalysis is continuously offering new possibilities. Aside from medication development and manufacturing, biocatalysis also contributes to drug discovery and can help with the early synthesis of several active ingredients to build up whole scaffolds in a focused and preparatory environment. Due to its many benefits, biocatalysis has become a vital technique for API synthesis.(Kharissova *et al.*, 2019)

As biocatalysts, optimized enzymes usually exhibit higher enantio-, stereo-, and regio-selectivity. Second, because biocatalysis permits reactions and reaction sequences that are not possible with traditional chemical methods, it can greatly reduce the length of multi-step synthesis pathways. Thirdly, biocatalysis is typically linked to benign reaction conditions, such as avoiding high pressure or temperature or the use of harmful chemicals. When these benefits are taken together, they often result in less problematic waste and fewer byproducts.(Kinner *et al.*, 2022)

Phyto nanotechnology is a medically useful and scalable method that uses microorganisms as natural nanofactories to convert metal ions into cero-valent nanoparticles. Biomass, bacteria, and mycosynthesis are used for simple, reliable, and potentially larger-scale production. Viruses have been used to create inorganic nanoparticles for semiconductor and quantum dot applications, including CdS, ZnS, Fe2O3, and SiO2. Yeasts generate metal nanoparticles, while algae are used for Au, Pd, and Pt nanoparticles. Proteins can act as a stabilizing agent, shape-control modifier, and reductant when used to create tetrahedral, decahedral, and icosahedral Au nanoparticles and bimetallic Ag/Ag nanoparticles. (Kharissova *et al.*, 2019)

The importance of enzymatic characteristics is the reason why many sectors, including chemical, food, textile, paper, biofuels, pharmaceuticals, agriculture, waste water treatment, healthcare, etc., are drawn to bio-catalysts (either enzymes or whole cells). With high selectivity, activity, and specificity, the enzymatic method produces more pure end-product streams while requiring moderate reaction conditions, minimal energy consumption, and fewer problems with isomerization and rearrangement.(Kate *et al.*, 2022)

Enzymatic bio-catalysts are needed for the development of sustainable processes, such as green catalysts, which enable reaction conduction even at lower pressures and temperatures, leading to productive, specific, and efficient process flow.(Federsel *et al.*, 2021)

Table 1. Application of enzymes in synthesis

Sr. No	Enzymes	Application	Reference
1	Alpha amylase	Biofuels	(Wiley, 2006)
2	Penicillin acylases	Synthesis of antibiotic drugs	(Fernandez-lafuente *et al.*, 1991)
3	Chitinase	Biopeptides	(Yasir *et al.*, 2021)
4	Lipase B	Antioxidant	(Chigorimbo-murefu *et al.*, 2009)
5	Lipase AK Amano	Synthesis of Chiral compounds	(Kasama *et al.*, 2020)
6	Resinase	Paper processing	(Singh *et al.*, 2016)
7	Protease	Esterification	(Riva *et al.*, 1988)

Additionally, the enzymes' immobilization allows for the use of small reaction capacities while reducing the chance of product contamination via bio-catalyst filtrates.

Numerous proposed works that illustrate immobilization techniques for biological, inert, and cross materials that result in a more active and stable bio-catalyst have been published in various scientific literatures. The examples and uses of the immobilized enzymes are shown in Table 2 above. (Luetz *et al.*, 2008)

Examples –a) Synthesis of Quinoxaline derivative Kamal and his team developed the reaction for the synthesis of quinoxaline derivatives by reacting1,2-diamine with a hydroxy ketone in the presence of *L*-proline as a catalyst and solvent like water (H2O) at 100 °Cover a duration of 12 to 14 hours with a consistently high yield. (Figure 18)(Karmakar and Mukhopadhyay, 2024)

Figure 18. Synthesis of Quinoxaline derivative

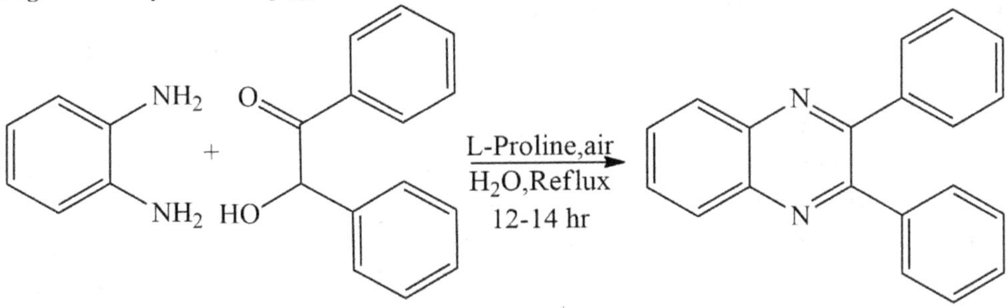

b) Synthesis of Nitriles: A classic example employing biocatalysis is the synthesis of nitriles. Nitriles are an important class of compounds that have a wide range of applications such as bulk and fine chemicals, pharmaceuticals, solvents, etc. There are various methods available for nitrile synthesis but they have some drawbacks such as the use of hazardous agents, employing harsh reaction conditions, long reaction times, etc. Therefore, Biocatalysis using the enzyme aldoxime dehydratase provides a green and cyanide-free synthesis of nitriles making it a versatile process. (Figure 19)(Hinzmann *et al.*, 2021)

Figure 19. Synthesis of Nitriles

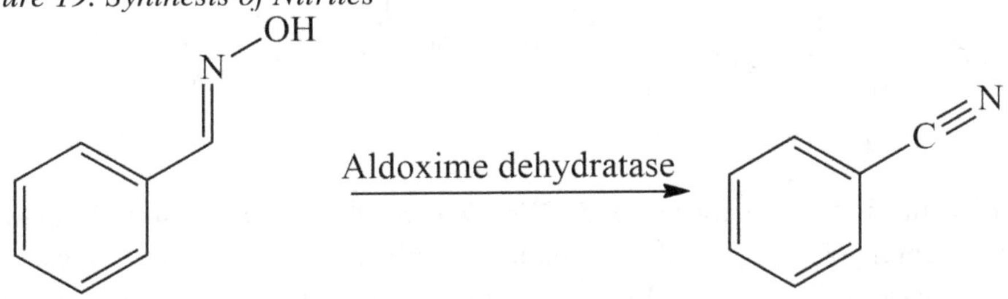

Micellar Catalysis: According to several sustainability matrices, including atom economy, process mass intensity, and E-factor, modern synthetic organic chemistry processes need to undergo a substantial shift from a sustainability perspective. Organic solvents have been identified as the primary responsible agent, and these elements have exposed the inherent inefficiencies inherent in standard techniques. As a result, using water as the main reaction medium for organic transformations has become a desirable substitute. When micellar catalysis emerged, this strategy took off. Amphiphilic compounds with both hydrophilic and hydrophobic domains are known as surfactants. (Ge *et al.*, 2021) The surfactant's distinct amphiphilic structure, which provides essentially mixed regions of different polarity, aids in the micelle-enabled organic processes. The micelle aqueous solution exhibits a very quick dynamic equilibrium between the surfactant monomers and the micelle aggregates, which promotes frequent organic compound and catalyst exchanges between water and micelles, hence speeding up the organic process. (Ge *et al.*, 2021)

The interfacial energy linked to the presence of hydrophobic derivatives in an aqueous environment can be significantly decreased by them. The association colloid, a self-assembled structure formed by amphiphilic molecules in water above a particular concentration known as the critical micellar concentration, is the fundamental idea behind micellar catalysis. Aggregates created in water by the hydrophobic effect function as nanoreactors. (Ge *et al.*, 2021) Higher local concentrations with rate enhancement and particular solvation effects result from this, improving the chemo-, regio-, and stereo-selectivity of the product.(S. Kumar *et al.*, 2020)

Examples–a) Synthesis of indolyl benzothiazole, *Kumar et.al* described eco-friendly production of indolyl benzothiazoles by using TPGS 750M facilitated nanomicellar catalysis in water. The reactions do not need extra catalysts and/or oxidants and occur at ambient temperature. (Figure 20)(S. Kumar *et al.*, 2020)

Figure 20. Synthesis of indolylbenzothiazole

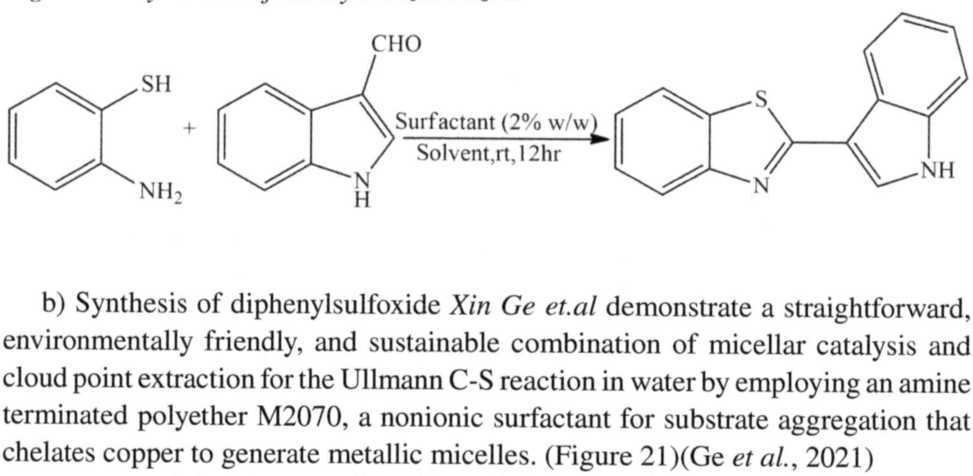

b) Synthesis of diphenylsulfoxide *Xin Ge et.al* demonstrate a straightforward, environmentally friendly, and sustainable combination of micellar catalysis and cloud point extraction for the Ullmann C-S reaction in water by employing an amine terminated polyether M2070, a nonionic surfactant for substrate aggregation that chelates copper to generate metallic micelles. (Figure 21)(Ge *et al.*, 2021)

Figure 21. Synthesis of diphenyl sulfoxide

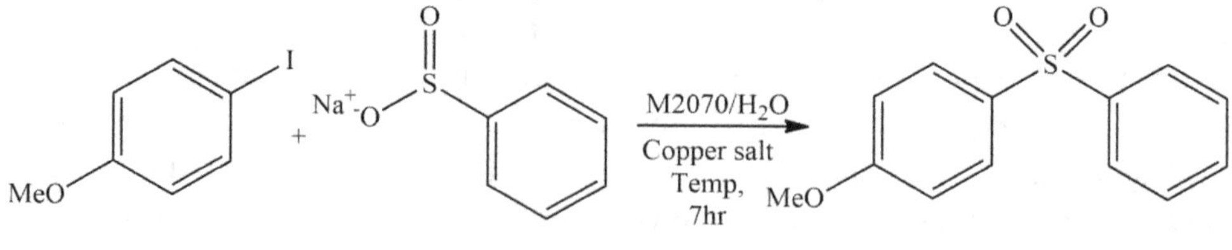

c) Synthesis of 2-Arylimidazo[1,2-a]pyridines

Bhutia.*et.al* in 2020, performed a reaction for the synthesis of 2-Arylimidazo[1,2-a]pyridines using sodium dodecyl sulfate as a micellar catalyst from acetophenone and 2 aminopyridyne. (Figure 22)(Bhutia *et al.*, 2020)

Figure 22. Synthesis of 2-Arylimidazo[1,2-a]pyridines

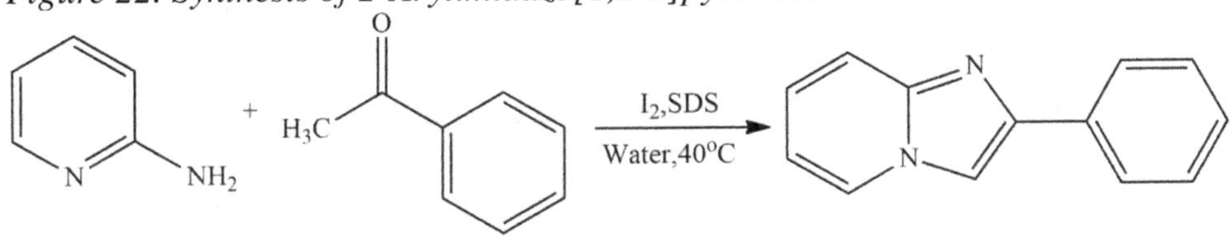

3. AQUEOUS-BASED PROCESSES

Aqueous-Based Processes: The fifth green chemistry principle—"safer solvents and auxiliaries"—"the use of the auxiliary substance"—was well described by Anastas and Warner in 2000 (C. Li & Zhou, 2019). For millions of years, water, the sole naturally occurring solvent on Earth, has served as a solvent for a variety of inorganic and biological interactions. In nature, both water-soluble and insoluble substrates can transform water; the latter often takes place at the active sites of enzymes, which are typically pockets or grooves.(Jun Li & Chan., 2007) Water has a high heat capacity and a big dielectric constant, among other special physical characteristics. Some other important properties of water include hydrogen-bonding effect and hydrophobic effect (the extrusion of nonpolar organics from the water phase). These physical and chemical properties bestow upon it some special advantages as a 'green solvent' for organic transformations (Manabe & Kobayashi, 2002) (C. Li *et al.*, 2006)

Chemistry research has been prompted to develop products and synthetic protocols from renewable resources due to the growing environmental concerns in society, which are primarily related to the increase in pollution, climate change, and resource depletion. These factors are affecting the health of people and animals. The chemical industry uses solvents extensively. The choice of solvent is crucial since it influences heat and mass transmission in addition to dissolution efficiency. Solvents can improve reagent solubility during catalytic conversions and assist product isolation by changing solvent conditions after the reaction. Examples that are currently available include CO_2-expanded liquids, supercritical fluids, fluorous biphasic systems, and partially miscible fluids.(C. Li & Zhou, 2019) Although they are rare, tuneable solvent systems, solvents whose solvation characteristics may be altered in situ by varying pressure and temperature are very desirable for chemical process design. Water is the most readily available and least expensive solvent to utilize in a process. It is nontoxic and nonflammable, and organic substrates can be separated using a straightforward phase separation method if a biphasic reaction system is employed.(Hailes, 2007)(Gru, 2020)

Advantages of employing water as a solvent (Jun Li & Chan., 2007)

- Gentler reaction circumstances: Reaction conditions can be made gentler by using water.
- Reduced residual concern: Using green solvents in the last stages of medication manufacturing reduces the likelihood of residuals in the API.
- Environmental friendliness: Water is a solvent that is good for the environment.
- Water can increase chemical selectivity, which lowers contaminants.
- Safe handling: It is okay to handle water.

Examples –a) Synthesis of 5,6,7-Trimethoxy-1-H-quinazoline-2,4-dione

Nitration of 2,3,4-trimethoxy-5-nitrobenzoic acid in presence of palladium on carbon as a catalyst and the water as solvent give product 5,6,7-Trimethoxy-1-H-quinazoline-2,4-dione. (Figure 23)(Gru, 2020)

Figure 23. Synthesis of 5,6,7-Trimethoxy-1-H-quinazoline-2,4-dio

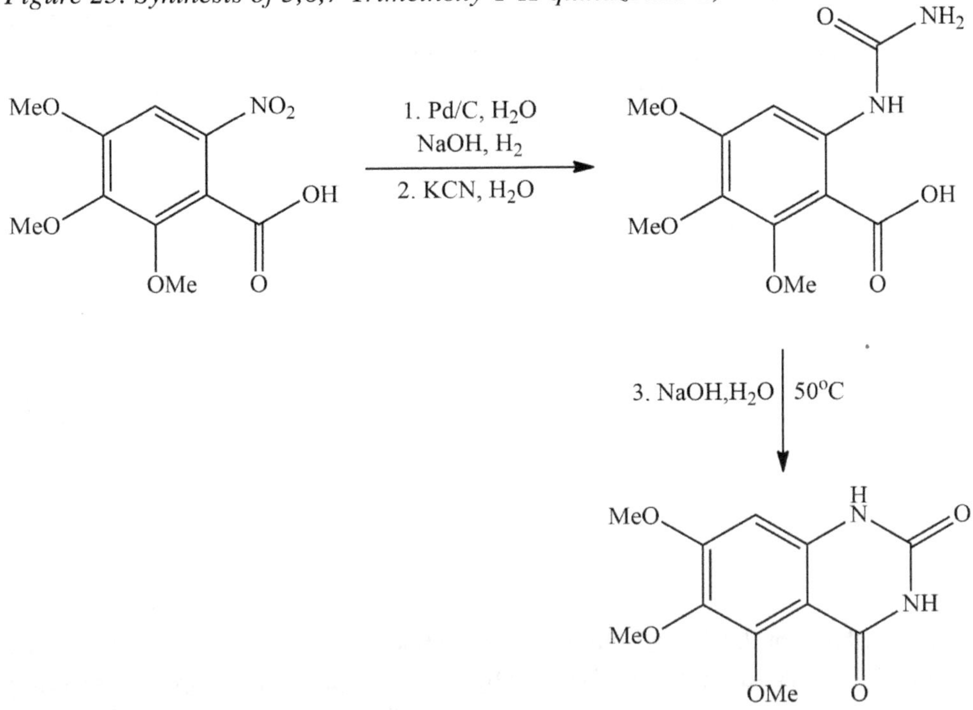

b) synthesis of pyranopyrazole derivatives: An effective and environmentally friendly multicomponent reaction system between an aldehyde, malononitrile, ethyl acetoacetate and hydrazine hydrate utilizing various base and water as solvent resulted in the formation of various pyranopyrazole derivatives.(Figure 24)(Vasuki & Kumaravel, 2008)

Figure 24. synthesis of pyranopyrazole derivatives

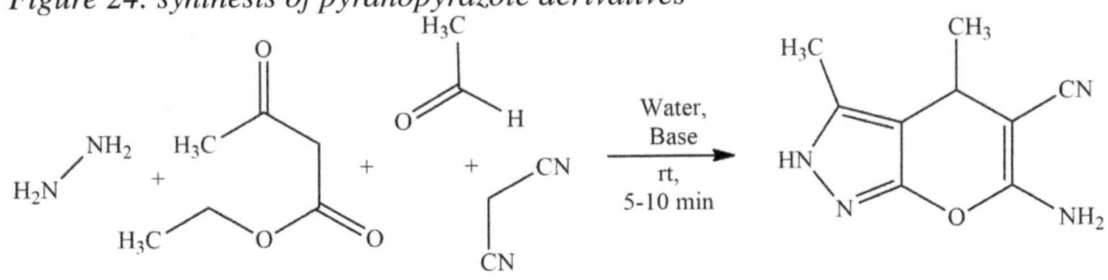

c) Synthesis of tetrahydrodipyrazolopyridine derivatives: An effective and environmentally friendly multicomponent reaction system between an aldehyde, ethyl acetoacetate, hydrazine hydrate, and ammonium acetate at room temperature in water as solvent and catalyst-free conditions resulted in formation of various tetrahydrodipyrazolopyridine derivatives. (Figure 25)(Keihanfar *et al.*, 2022)

Figure 25. synthesis of tetrahydrodipyrazolopyridine derivatives

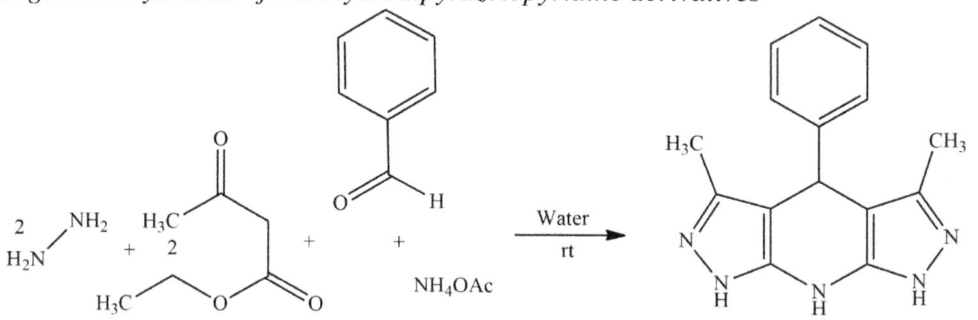

d) Synthesis of 2,4-Diaryl-quinoline derivative: Martins et al, 2021 synthesizes various 2,4-Diaryl-quinoline derivatives by using one pot multicomponent reaction between phenylacetylene, various aldehyde and aniline derivatives in water along with hydrochloric acid as the promoter in the reaction. (Figure 26)(Michelão *et al.*, 2021)

Figure 26. Synthesis of 2,4-Diaryl-quinoline derivative

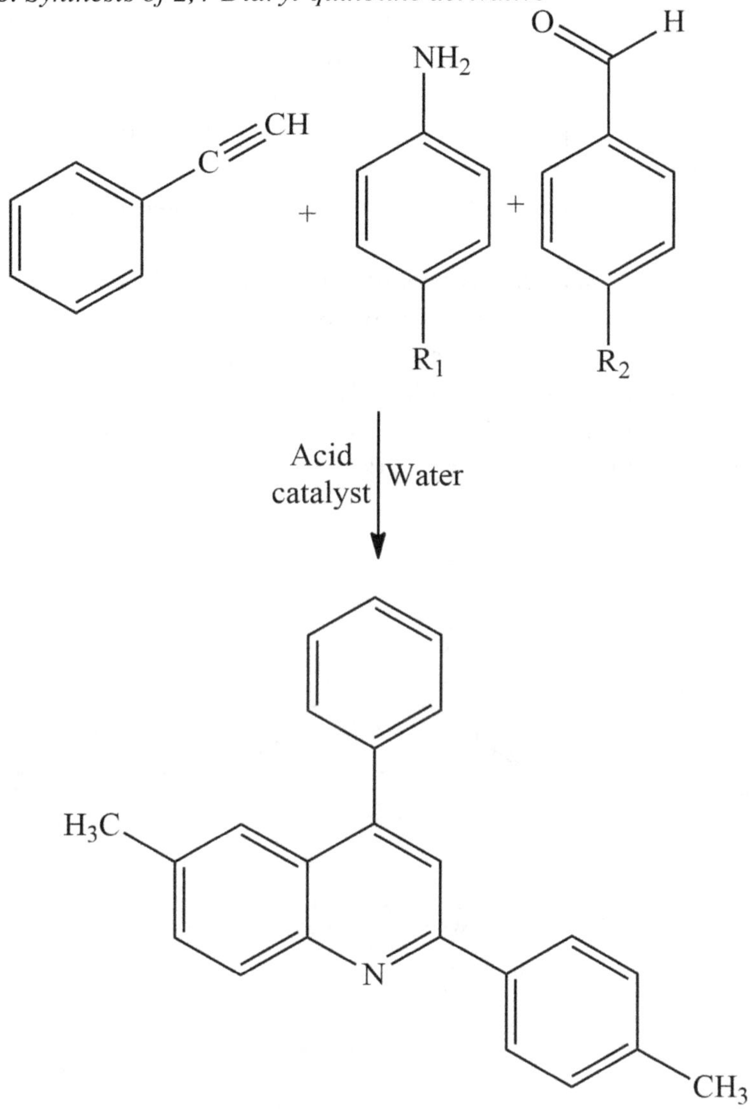

4. SOLVENT-FREE PROCESS

Crushed grapes produce wine through fermentation, while dried grapes do not lead to the creation of wine. While milk can sour and shake it can turn it into cheese, dried milk can be kept without alteration. Likewise, dried meat can be preserved for an extended period, whereas meat soup spoils quickly if left out. In the liquid state, but not in the solid state, it is evident that a material can change into another. In summarising these findings, Aristotle, one of the most well-known ancient Greek philosophers, concluded that "No Coopora nisi Fluida," or "No reaction occurs in the absence of solvent," applies. These ideas had a significant impact on the development of contemporary European sciences, which is one of the historical reasons why the majority of organic reactions have been investigated in solution.(Tanaka & Toda, 2000)

Yet, it is noteworthy that chemists continue to conduct their reactions in solution, even when there isn't a specific justification for using a solvent.

To produce a steady mixture of reactants and guarantee a smooth reaction, traditional organic chemical reactions primarily rely on organic solvents.

As a solvent, any liquid can be employed. There are very few solvents that are frequently used, though.(Nagendrappa, 2002)

Figure 27. classification of solvents

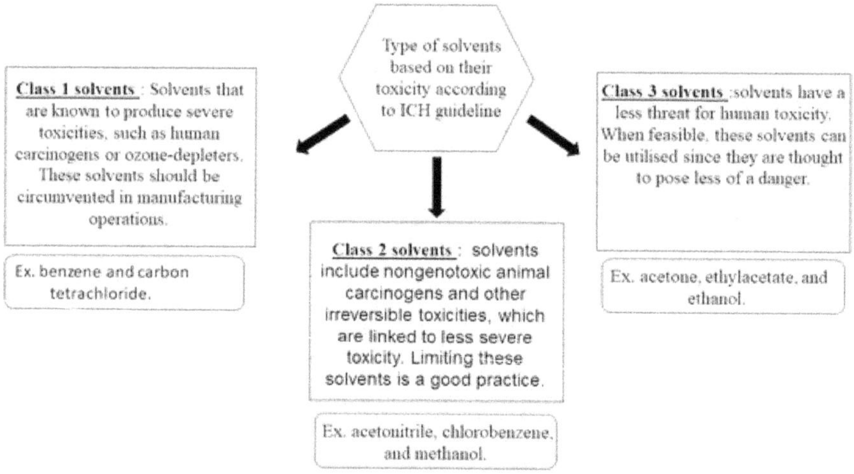

These solvents do, however, have several drawbacks, including flammability, high volatility, environmental contamination, toxicity, and explosion potential. Additionally, using them in chemical manufacturing necessitates extensive post-processing, which raises production costs.(M. Li *et al.*, 2024) Solvent-free reactions, a key component of green chemistry, have attracted considerable attention due to their environmentally friendly working methods, and these reactions have made significant advances over the past decades.(Zangade & Patil, 2020) Reactions without solvents are more efficient and allow for greater reaction concentrations than reactions with solvents. The famous chemist Sheldon famously said, “The best catalyst is no catalyst, and the best solvent is no solvent.” (M. Li *et al.*, 2024)

For the synthesis of physiologically active molecules, solid-state and solvent-free processes or methods are frequently employed. Reactions without volatile and poisonous organic solvents provide several benefits, including shorter reaction times, easier workup processes, higher product yields, and the elimination of the requirement for organic solvents. The development of environmentally friendly processes is further aided by the coupling of solvent-free reactions with different supported catalysts.(Zangade & Patil, 2019)

In certain instances, organic synthesis can be carried out without the use of a solvent throughout the reaction and product separation steps.(Tanaka & Toda, 2000) Using a chiral host molecule for the solid-state reaction of a prochiral guest allows for the execution of enantioselective thermal and photochemical reactions. The most promising enantioselective photosynthesis involves irradiating chiral crystals of achiral substances, where the molecules are organised chirally.(Nagendrappa, 2002)

However, there are currently very few examples of this phenomena.

Thermal organic synthesis without the use of solvents appears to be a very practical method, particularly.

Not every organic synthesis, nevertheless, can be completed without a solvent.

Solvent is helpful in these situations to moderate the pace of reaction.

Lastly, selecting the ideal conditions for organic synthesis is always crucial. Solid-state reactions would be a preferable option for reactions that proceed modestly in a water suspension or without a solvent.(Tanaka & Toda, 2000)

It would be preferable to conduct a solution reaction in a benign solvent for reactions that progress rapidly in the solid state.(Himaja *et al.*, 2011)

Examples –Synthesis of N-sulfonylformamidines, *Chandna et al.*by using direct condensation of N,N-dimethylformamide dimethyl acetal and sulfonamide derivatives at room temperature, created a solvent-free, environmentally friendly Method of Producing N-sulfonylformamidines. (Figure 28)(Chandna *et al.*, 2013)

Figure 28. Synthesis of N-sulfonyl formamidines

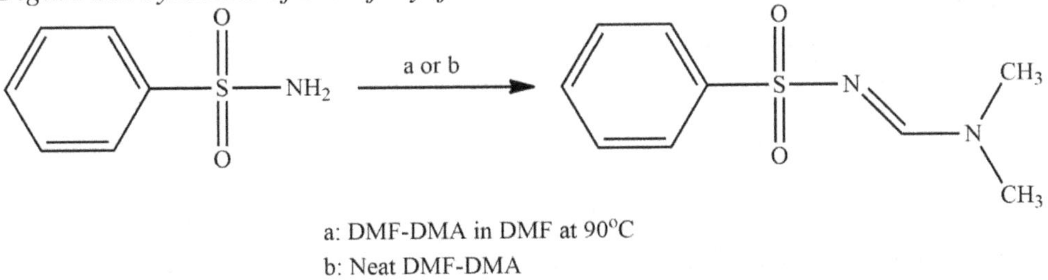

Metal catalysts and dangerous solvents, which are prohibited in the production of pharmaceuticals, are not used in the reactions or product isolation processes of the technique described.

Examples –a) synthesis of chalcone, *Saraci et.al* developed a combination of solvent-free synthesis and mechano-chemical techniques to gain traction to follow the 12 principles of green chemistry, continuously lowering the environmental impact in comparison to solvent-based procedures for the synthesis of chalcone. The method includes mechano-chemical processes mediated by $Mg(HSO_4)_2$. (Figure 29) (Argese & Zanoni, 2023)

Figure 29. synthesis of chalcone

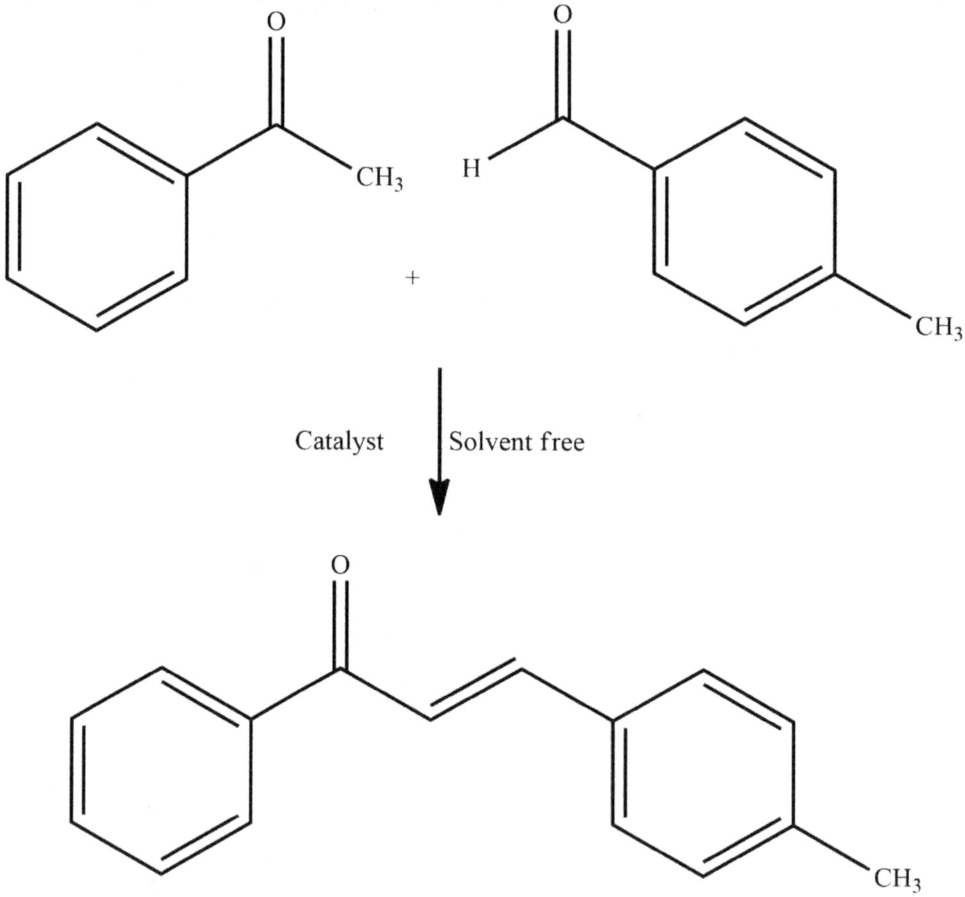

b) synthesis of quinazolin-4(3H)-ones: *Xue Ma et.al* by using 1-butyl-3-methylimida-zolium tetrafluoride ([Bmim]BF4) as an ionic liquid catalyst, condensation cyclization of 2-aminobenzamides and aldehydes developed a solvent-free and metal-free technique for the synthesis of quinazolin-4(3H)-ones. Without the need for further catalysts or solvents, [Bmim]BF4 functions as both a catalyst and a solvent in this process. This process produces several desirable products in moderate to excellent yields and shows good functional group tolerance in substrates. Furthermore, it is significant that even when [Bmim]BF4 is recycled at least four times, the reaction yield remains at 87%. (Figure 30)(Ma *et al.*, 2024)

Figure 30. synthesis of quinazolin-4(3H)-ones

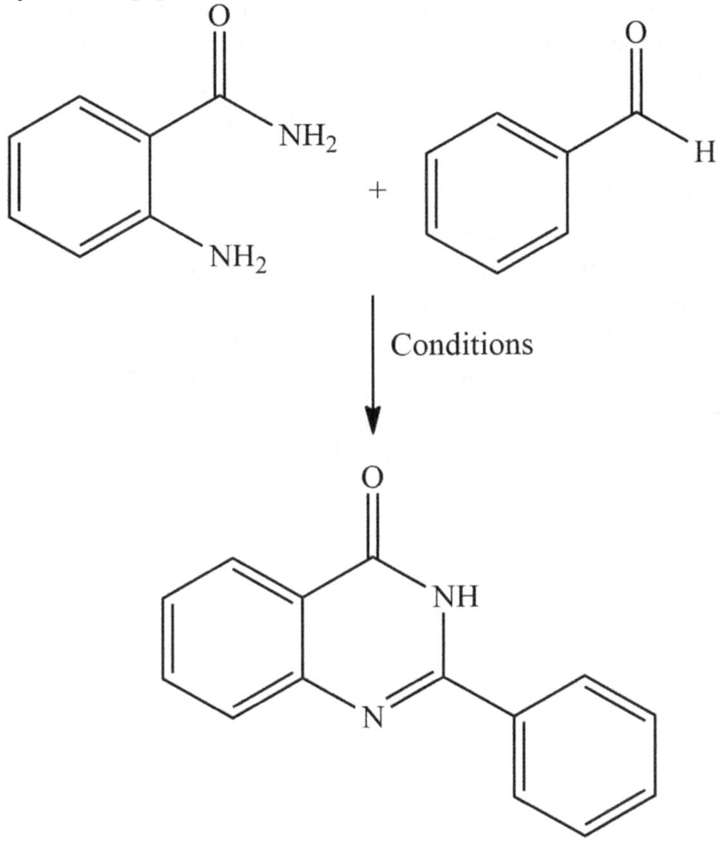

5. CASE STUDY

4.1 Pharmaceutical Applications

1. *Enzymatic preparation of Sitagliptin*

Previously with the aid of a chiral rhodium-based catalyst, an enamine undergoes asymmetric hydrogenation under high pressure to yield sitagliptin. This chemical process necessitates additional purification steps, which can lower the overall yield, to enhance both the enantiomeric excess and chemical purity due to inadequate stereoselectivity and the presence of rhodium impurities in the product stream.

A transaminase scaffold along with various protein engineering techniques were employed to establish an enzymatic process that greatly enhanced the production efficiency of sitagliptin. Substrate walking, modeling, and mutation techniques were utilized to create a transaminase with limited activity for synthesizing chiral amines, starting from an enzyme that initially exhibited no activity toward the prositagliptin ketone but possessed the necessary catalytic machinery to perform the desired reaction. Directed evolution was subsequently applied to further refine this selected enzyme variant for application in a manufacturing setting. The resulting biocatalysts demonstrated a broad potential for the synthesis of chiral amines, which were previously attainable only through the process of enzymatic resolution. (Figure 31)

Figure 31. Enzymatic preparation of Sitagliptin

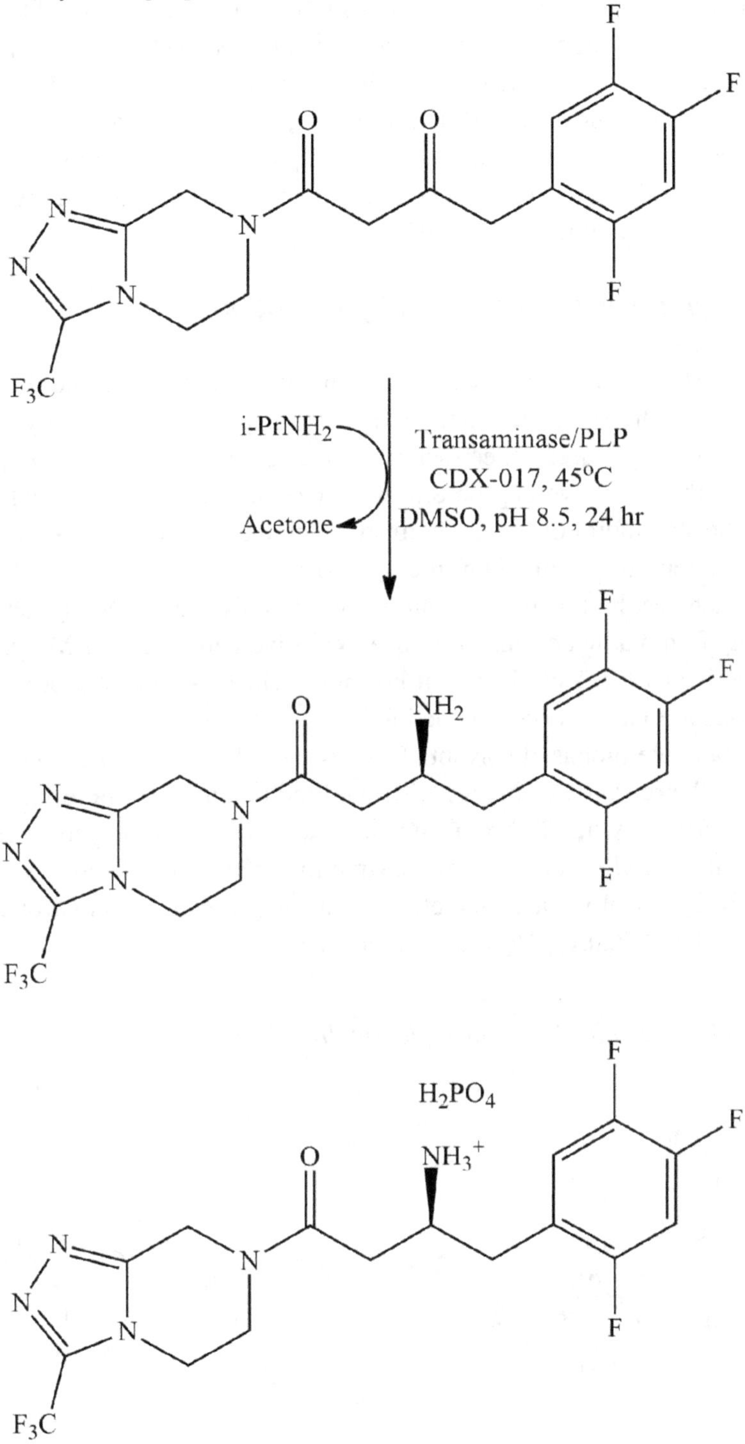

Under ideal circumstances, the best variation produced sitagliptin from 200 g/l prositagliptin ketone in >99.95%, or by using 6 g/l enzyme in 50% DMSO, with a 92% yield after the process. The biocatalytic process produced sitagliptin with a 10–13% increase in overall yield, a 53% increase in productivity, a 19% reduction in total waste, the removal of all heavy metals, and a lower total manufacturing cost when compared to the rhodium-catalyzed process. The enzymatic reaction is carried out in multipurpose vessels, negating the need for specialized high-pressure hydrogenation equipment. (Patel, 2018)

2. *Green synthesis of Bupropion hydrochloride*

Bupropion hydrochloride is made via brominating m-chloropropiophenone first, then amination with tert-butylamine, and finally precipitation to get the hydrochloride salt. This approach has already been acknowledged as a viable organic lab experiment.

However, there are several problems with the synthesis of bupropion hydrochloride from the standpoint of green chemistry. The REACH rule currently prohibits the use of the reprotoxic solvent N-methylpyrrolidinone (NMP). Dichloromethane (DCM), which has the potential to cause cancer, is also used. Diethyl ether, which is extremely flammable and may produce explosive peroxides, 12 M hydrochloric acid, which is caustic, and elemental bromine, which is deadly if consumed, are additional hazardous materials needed for this operation.

The sustainable biobased solvents Cyrene and ethyl acetate have been utilized to substitute the reprotoxic solvent NMP and the potentially carcinogenic solvent DCM, respectively, while N-bromosuccinimide has been employed as an alternative to bromine. Additionally, a novel extraction method has been developed that employs 1 M hydrochloric acid and ethyl acetate in place of 12 M hydrochloric acid and diethyl ether. (Figure 32)(Andrew *et al.*, 2022)

Figure 32. Green synthesis of Bupropion hydrochloride

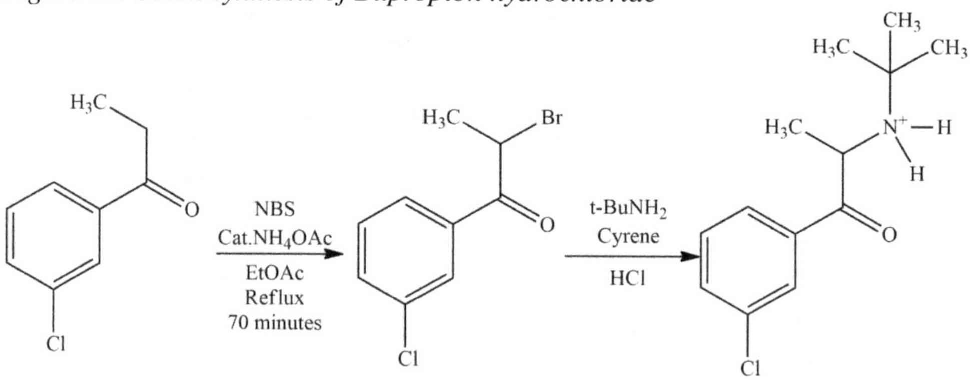

3. *Sustainable synthesis of Ibuprofen*

Due to the exhaustion of fossil fuels, the amount of energy required for chemical processes has become an environmental issue. A recent innovation has involved the use of satellite dishes as solar reflectors capable of producing heat through solar radiation. The potential of utilizing this solar reflector as the sole heat source for the synthesis of the widely used drug ibuprofen has been investigated. Furthermore, the current production of ibuprofen relies on chemicals that are not particularly environmentally sustainable (Figure 33)(Agee *et al.*, 2016)

Figure 33. Sustainable synthesis of Ibuprofen

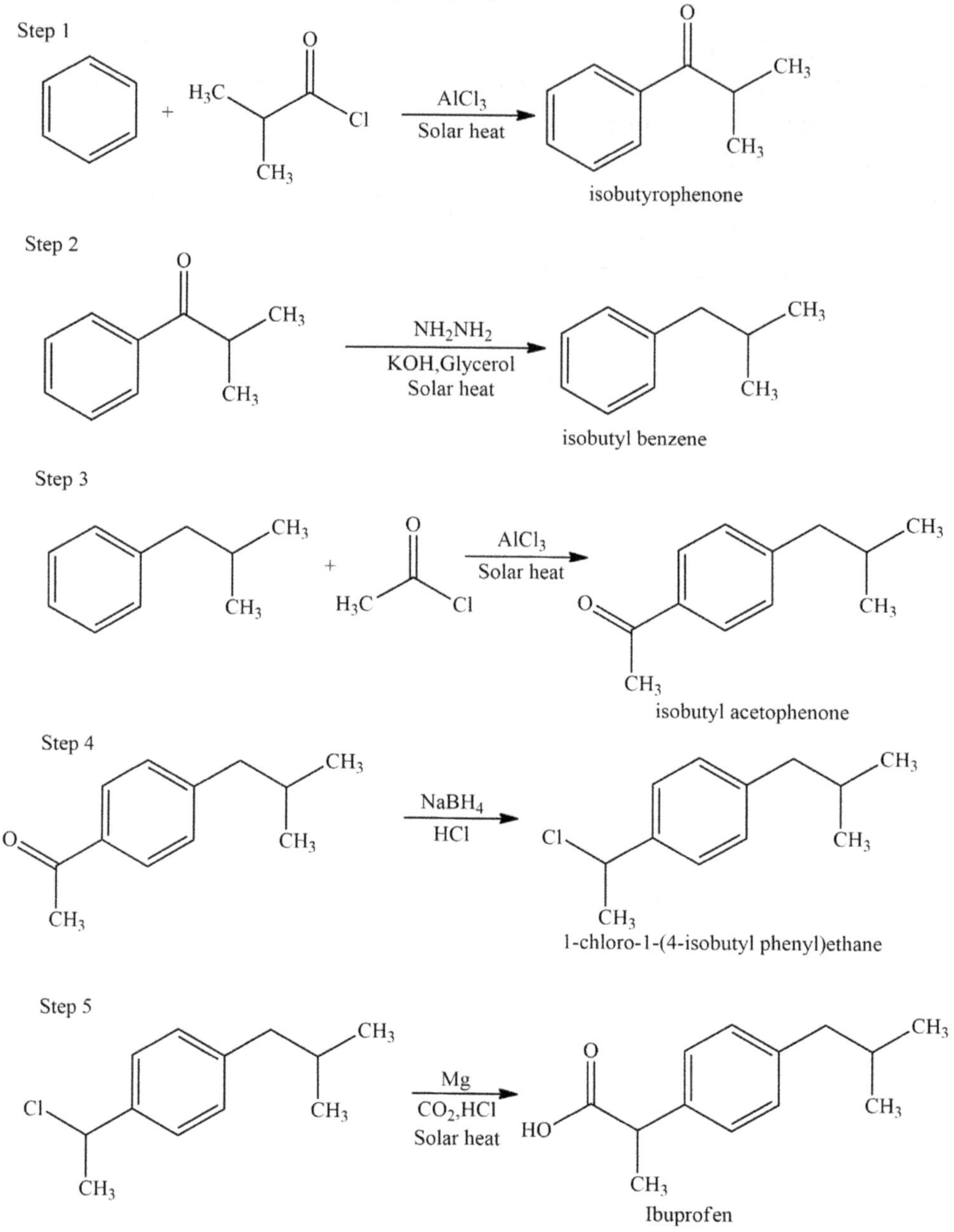

4. *Synthesis of Porphyrins*

A variety of metalloenzymes, such as heme, chlorophyll, bacteriochlorophyll, siroheme, and factor F430, contain the naturally occurring colored porphyrin. Synthetic porphyrinoids and their metal-containing versions have numerous applications, including acting as catalysts for the oxidation of organic molecules and water-splitting reactions, serving as potential elements in molecular electronics, functioning as dyes in dye-sensitized solar cells, contributing to supramolecular chemistry, and being utilized in photodynamic therapy (PDT).

A novel approach to the synthesis of porphyrins has been created. It's a straightforward two-step procedure. In the first stage, pyrrole and aldehyde are condensed using HCl in an H_2O–MeOH combination. The first step's precipitate was dissolved in reagent-grade dimethylformamide (DMF), refluxed for one and a half hours, and then stirred for the whole night at room temperature. Pure porphyrins were formed as a result of further purification using column chromatography or crystallisation. (Figure 34)

Figure 34. Synthesis of Porphyrins

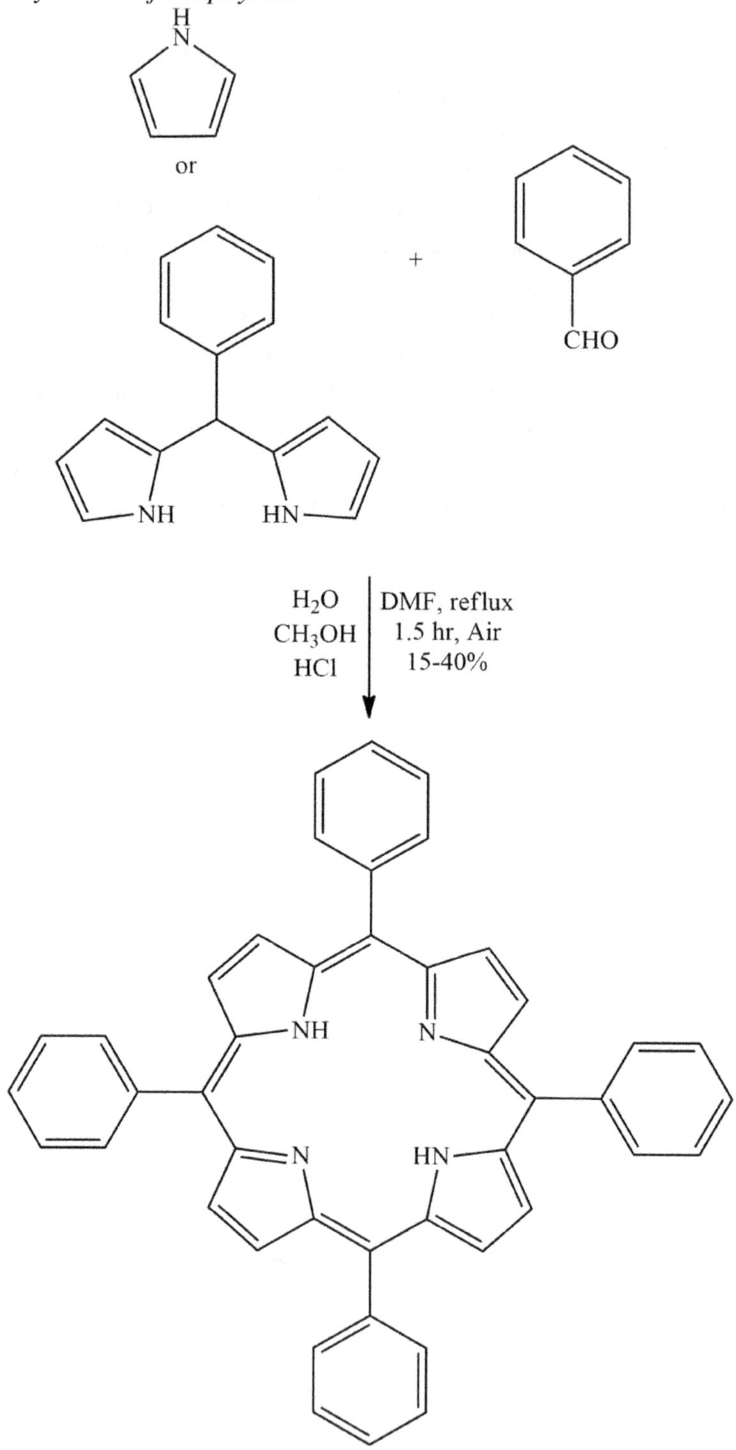

One benefit of this approach is that it eliminates the need for costly oxidising agents like chloranil and 2,3-dichloro-5,6-dicyanobenzoquinone (DDQ). Additionally, a significant amount of dry chlorinated solvents is not needed for this reaction. In contrast to previously published methods that are often ineffectual for producing porphyrins on a gram-scale, the current approach precisely. (Mondal *et al.*, 2021)

5. *Synthesis of Aspirin*

Aspirin is a commonly used medication that is traditionally made via an acid-catalyzed technique. Concentrated sulphuric and phosphoric acids are the most commonly used catalysts for this reaction. Because they are very corrosive, both acids need to be handled carefully. Recently, a different technique was described that allowed for the preparation of aspirin utilizing a low-temperature esterification process using acetyl chloride-pyridine as the acetylating agent. However, these chemicals are not appropriate for use in large-scale industrial processes due to their offensive and annoying smells. It has also been found that various derivatives including keggin-type heteropolyacids are effective catalysts for the esterification process.

Heravi *et al.* describe the use of a new and recyclable solid acid catalyst called the Preyssler catalyst, which has unique qualities that outperform those of Keggin hetero polyacids, for the quick and highly selective liquid-phase O-acetylation of salicylic acid to create aspirin at room temperature. (Bamoharram *et al.*, 2007)

6. *Synthesis of sildenafil citrate*

The sildenafil citrate process's development and scaling up were documented by Dunn *et al.* During the creation of sildenafil citrate, notable accomplishments were (i) identifying a convergent, effective synthetic pathway (ii) creating a seven-step process that eliminated extractive work-up at every stage (iii) putting effective solvent recovery into place early in the product's commercial life. With just 6 kg of waste per kilogramme of product, the work's E-factor is extremely low when compared to the industry norm of 25–100 kg. The sildenafil citrate method earned Pfizer the 2003 UK Award for Green Chemical Technology. (Figure 35) (Best method category). (Dunn *et al.*, 2004)

Figure 35. Synthesis of sildenafil citrate

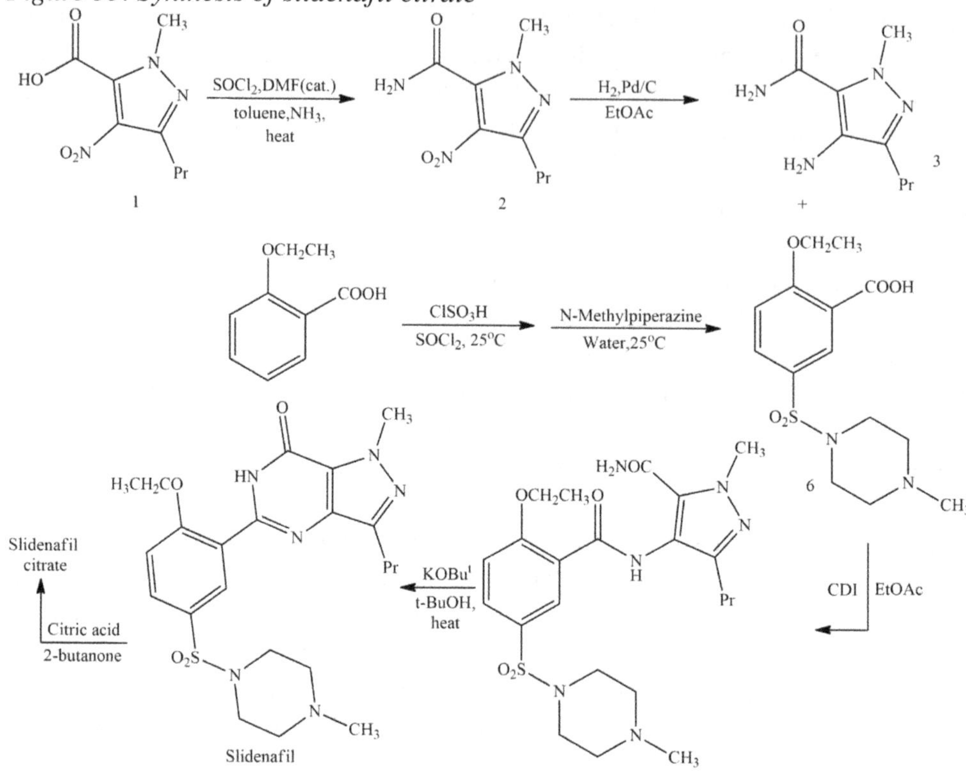

Case study agrochemical applications

1.*Bacillus thuringiensis*

The agricultural chemical industry generally views biopesticides as more recent. In contrast to many synthetic pesticides, Bacillus thuringiensis has a far longer history. One great example of a natural product that may be used as a green insecticide is Bacillus thuringiensis.

Since the US EPA registered Bacillus thuringiensis in 1961, it has been used for almost 40 years.Products based on Bt make up a significant portion of the worldwide biopesticide market.Because of its high prevalence of insect resistance and toxicity, its use has increased since synthetic products were deregulated. Bacillus thuringiensis was placed fourth among all pesticides in 1997 based on the number of fruit crops treated per acre; Bt has demonstrated durability in terms of performance and service behavior against recently produced synthetic products that exhibit resistance within five years (USEPA 2004, 2007, 2011). (Kulkarni, 2016)

2. *Synthesis of Inpyrfluxam*

To create inpyrfluxam, a modified Skraup cyclization of aniline and acetone is first catalyzed by lanthanides. The ring nitrogen of the resulting trimethylated 1,2-dihydroquinoline is then acetylated, which is then hydrogenated with hydrogen and palladium on charcoal to produce the tetrahydroquinoline derivative.12,13 This intermediate is then transformed via rearrangement in an acidic environment to 1,1,3-trimethylindan-4-amine.12,13 Following the chiral resolution of this racemic indane derivative into its (R)-enantiomer, this stereoisomer is then amidated with the acid chloride of the privileged SDHI acid 3-difluoromethyl-1-methyl-1H-pyrazole-4-carboxylic acid to produce inpyrfluxam. (Figure 36)(Jeschke, 2022)

Figure 36. Synthesis of Inpyrfluxam

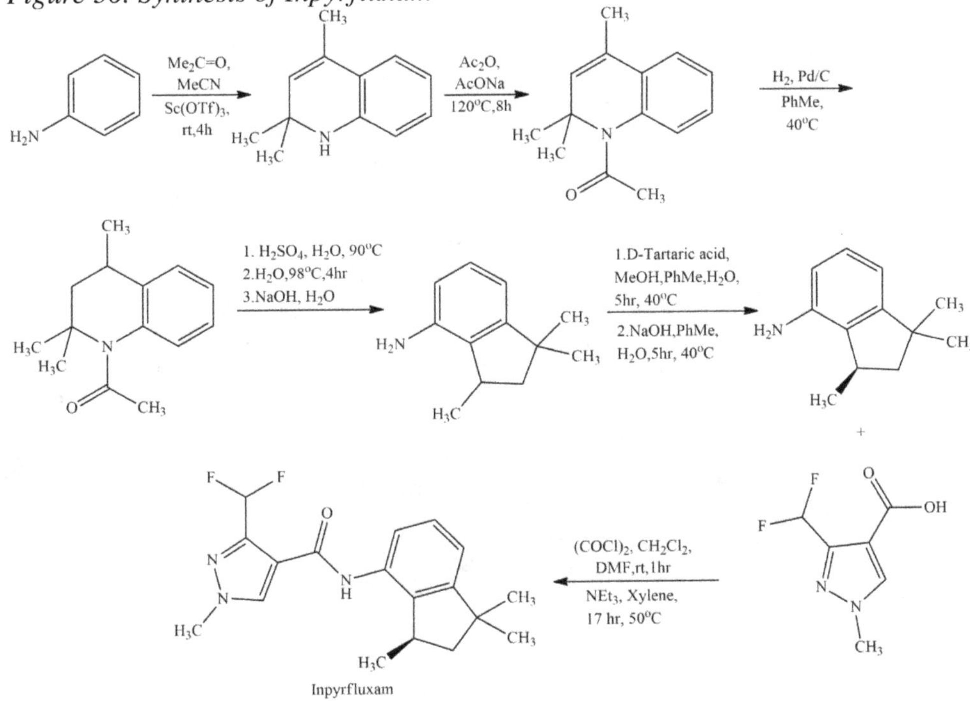

3. *Synthesis of fluoxapiprolin*

The trichlorinated and vinylated cyclohexene oxide, which provides the o,o'-disubstituted styrene through thermal aromatization, is the starting point for the synthesis of fluoxapiprolin. The α-chloroketo-substituted isoxazoline derivative is generated by converting the C–C double bond with chloroacetylhydroximinoyl chloride, which can be made by nitrosating 4-chloroacetoacetate, following sulfonylation to the equivalent mesylate. Two additional rings—a piperidine and a thiazole—are added during its condensation with piperidine- 4-carbothioamide, producing the piperidinium salt. This salt is then converted with chloroacetyl chloride to produce the chloroacetamide derivative. Fluoxapiprolin is obtained in the last stage by nucleophilically substituting 3,5-bis (difluoromethyl)pyrazole for this chlorine. (Figure 37) (Jeanmart *et al.*, 2021)

Figure 37. Synthesis of fluoxapiprolin

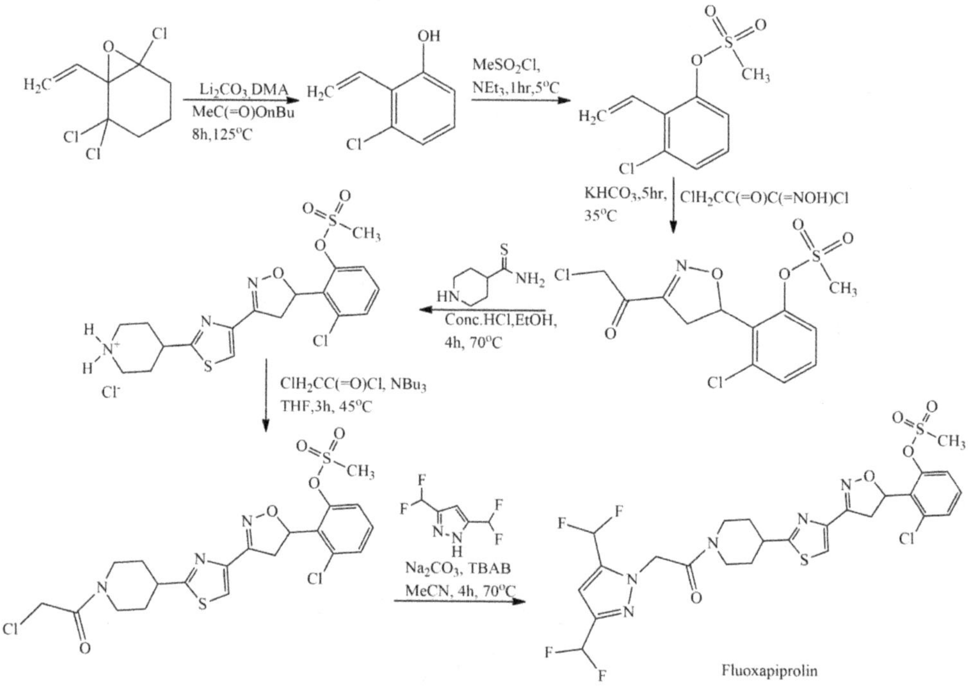

4. Synthesis of enantiomeric access of the trans-(1R,2S)indanylamine

In the presence of aluminum iso-propoxide, the trans-(1R,2S)indanylamine's enantiomeric access is prepared, coupled with N-cyano-guanidine, and cyclized with 2-fluoro-propionic acid methyl ester under basic conditions to form (1R,2S,6R)-indaziflam. The major (1R,2S,6R)-diastereoisomer and minor (1R,2S,6S)-diastereoisomer are present in the commercially available indaziflam in a 95:5 ratio. (Figure 38) (Jeanmart *et al.*, 2021)

Figure 38. Synthesis of enantiomeric access of the trans-(1R,2S)indanylamine

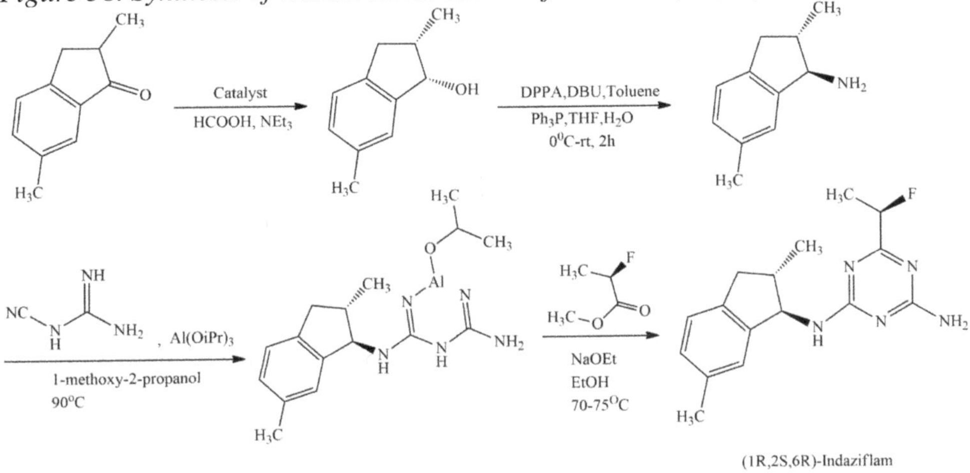

5. *Nano sensors*

Several insect pests, most notably stinkbugs, can affect soybean farms. In Brazil, phytophagous stink bugs, or Hemiptera, are the most important pests harming soybeans (Glycinemax (L.) Merrill).3 Up to 30% of plant development might be lost due to stink bug infestations, resulting in deformed and desiccated pods.4. Insects and plants mostly use chemical and physical interactions to communicate with their surroundings. They would not be able to exist without their extremely sensitive and specialized molecular recognition mechanisms. These substances that let creatures communicate with one another are called semiochemicals, and they are called pheromones when they interact with members of the same species.5. An alternative to the careless use of pesticides, which can have negative consequences on the environment and human health, is the direct use of semiochemicals for population monitoring or pest management. (Jeschke, 2022)

REFERENCES

Agee, B. M., Mullins, G., & Swartling, D. J. (2016). Progress towards a more sustainable synthetic pathway to ibuprofen through the use of solar heating. *Sustain Chem Process*, *4*(8), 1–9. DOI: 10.1186/s40508-016-0052-y

Anastas, P. T., & Warner, J. C. (2000). *Green chemistry: theory and practice*. Oxford university press., DOI: 10.1093/oso/9780198506980.001.0001

Andrew, O. B., Sherwood, J., & Hurst, G. A. (2022). A Greener Synthesis of the Antidepressant Bupropion Hydrochloride. *Journal of Chemical Education*, *99*(9), 3277–3282. DOI: 10.1021/acs.jchemed.2c00581

Argese, M., & Zanoni, G. (2023). Solvent-free synthesis of chalcones using $Mg(HSO_4)_2$. *RSC Sustainability*, *1*(3), 504–510. DOI: 10.1039/D3SU00003F

Bamoharram, F. F., Heravi, M. M., Roshani, M., & Jahangir, M. (2007). Catalytic Method for Synthesis of Aspirin by a Green, Efficient and Recyclable Solid Acid Catalyst (Preyssler ' s Anion) at Room Temperature. *Journal of the Chinese Chemical Society (Taipei)*, *54*(4), 1017–1020. DOI: 10.1002/jccs.200700146

Betts, K. (2015). How industrial applications in green chemistry are changing our world. *Journal of the American Chemical Society*, **•••**, 1–28.

Bhutia, Z. T., Panjikar, P. C., Iyer, S., Chatterjee, A., & Banerjee, M. (2020). Iodine Promoted E ffi cient Synthesis of 2 - Arylimidazo[1,2 - a]pyridines in Aqueous Media: A Comparative Study between Micellar Catalysis and an " On-Water " Platform. *ACS Omega*, *5*(22), 13333–13343. DOI: 10.1021/acsomega.0c01478 PMID: 32548520

Bogda, D., Pielichowski, J., & Boro, A. (1996). Remarkable fast microwave-assisted N-alkylation of phthalimide in dry media. *Synlett*, *1996*(09), 873–874. DOI: 10.1055/s-1996-5587

Breynaert, E., Houlleberghs, M., Radhakrishnan, S., Grübel, G., Taulelle, F., & Martens, J. A. (2020). Water as a tuneable solvent: A perspective. *Chemical Society Reviews*, *49*(9), 2557–2569. DOI: 10.1039/C9CS00545E PMID: 32238999

Cann, M. C. (1999). Bringing State-of-the-Art, Applied, Novel, Green Chemistry to the Classroom by Employing the Presidential Green Chemistry Challenge Awards. *Journal of Chemical Education*, *76*(12), 1639–1641. DOI: 10.1021/ed076p1639

Chandna, N., Chandak, N., Kumar, P., Kapoor, J. K., & Sharma, P. K. (2013). Metal- and solvent-free synthesis of N-sulfonylformamidines. *Green Chemistry, 15*(8), 2294–2301. DOI: 10.1039/c3gc40797g

Chen, J., Hu, X., Lu, L., & Xiao, W. (2016). Exploration of Visible-Light Photocatalysis in Heterocycle Synthesis and Functionalization : Reaction Design and Beyond. *Accounts of Chemical Research, 49*(9), 1911–1923. DOI: 10.1021/acs.accounts.6b00254 PMID: 27551740

Chigorimbo-murefu, N. T. L., Riva, S., & Burton, S. G. (2009). *Journal of Molecular Catalysis B :* Enzymatic Lipase-catalysed synthesis of esters of ferulic acid with natural compounds and evaluation of their antioxidant properties. *56*, 4, 277–282. https://doi.org/DOI: 10.1016/j.molcatb.2008.05.017

Clarke, C. J., Tu, W., Levers, O., Bro, A., & Hallett, J. P. (2018). Green and Sustainable Solvents in Chemical Processes. *Chemical Reviews, 118*(2), 747–800. DOI: 10.1021/acs.chemrev.7b00571 PMID: 29300087

Crespi, S., Fagnoni, M., & Ring, F. (2018). *Photocatalyzed Formation of Heterocycles.Topics in Heterocyclic Chemistry, 54*. Springer., DOI: 10.1007/7081_2017_13

Diaz-alvarez, A. E., Francos, J., Crochet, P., Cadierno, V. Recent advances in the use of glycerol as green solvent for synthetic organic chemistry. *Current Green Chem.*, 1, 51-65DOI: 10.2174/1570178620666230418093820

Do, J. (2017). Mechanochemistry: A Force of Synthesis. *ACS Cent. Sci.* 3, 1, 13–19https://doi.org/DOI: 10.1021/acscentsci.6b00277

Dunn, P. J., Hettenbach, K., Ct, U. K., Pharmaceuticals, I., Plant, R. A. P. I., & Box, P. O. (2004). The development of an environmentally benign synthesis of sildenafil citrate (Viagra ™) and its assessment by Green Chemistry metrics. *Green Chemistry, 6*(1), 43–48. DOI: 10.1039/B312329D

Federsel, H. J., Moody, T. S., & Taylor, S. J. (2021). Recent trends in enzyme immobilization—Concepts for expanding the biocatalysis toolbox. *Molecules (Basel, Switzerland), 26*(9), 2822. DOI: 10.3390/molecules26092822 PMID: 34068706

Fernandez-Lafuente, R., Rosell, C. M., & Guisán, J. (1991). Enzyme reaction engineering: synthesis of antibiotics catalysed by stabilized penicillin G acylase in the presence of organic cosolvents. *Enzyme and microbial technology, 13*(11), 898-905. https://doi.org/ rights and contentDOI: 10.1016/0141-0229(91)90106-KGet

Gawande, M. B., Shelke, S. N., Zboril, R., & Varma, R. S. (2014). Microwave-assisted chemistry: Synthetic applications for rapid assembly of nanomaterials and organics. *Accounts of Chemical Research, 47*(4), 1338–1348. DOI: 10.1021/ar400309b PMID: 24666323

Ge, X., Song, W., Chen, X., Qian, C., Zhou, S., & Liu, X. (2021). Tandem micellar catalysis and cloud point extraction process for CS coupling reaction in water. *Colloids and Surfaces. A, Physicochemical and Engineering Aspects, 616*, 126263. DOI: 10.1016/j.colsurfa.2021.126263

Gharat, N. N., & Rathod, V. K. (2020). Ultrasound-assisted organic synthesis. In *Green sustainable process for chemical and environmental engineering and science* (pp. 1–41). Elsevier., DOI: 10.1016/B978-0-12-819540-6.00001-2

Hagemeyer, A., & Clara, S. (2016). *Materials : Catalysts $. March 2015*, 1–8. https://doi.org/DOI: 10.1016/B978-0-12-803581-8.01121-8

Himaja, M., Poppy, D., & Asif, K. (2011). Green technique-solvent free synthesis and its advantages. *International Journal of Research in Ayurveda and Pharmacy, 2*(4).

Kasama, K., Aoyama, H., & Akai, S. (2020). Enantiodivergent Synthesis of Axially Chiral Biphenyls from σ-Symmetric 1, 1′-Biphenyl-2, 6-diol Derivatives by Single Lipase-Catalyzed Acylative and Hydrolytic Desymmetrization. *European Journal of Organic Chemistry, 2020*(6), 654–661. DOI: 10.1002/ejoc.201901583

Krajewski, M. (2017). Magnetic-field-induced synthesis of magnetic wire-like micro-and nanostructures. *Nanoscale, 9*(43), 16511–16545. DOI: 10.1039/C7NR05823C PMID: 29067381

Kulkarni, S. (2016). Sustainable agrochemicals for conservation of agriculture and climate change. *Conservation Agriculture:* An Approach to Combat Climate Change in Indian Himalaya, 135-157. https://doi.org/DOI: 10.1007/978-981-10-2558-7

Li, C. J., & Chan, T. H. (2007). *Comprehensive organic reactions in aqueous media*. John Wiley & Sons. DOI: 10.1002/9780470131442

Ma, X., Li, P., Chen, X., Li, S., Sun, W., & Zhong, Q. (2024). Recyclable ionic liquid-mediated synthesis of quinazolin-4 (3H)-ones under metal-free and solvent-free conditions. *Giant*, 18, 100255.Elsevier.https://doi.org/DOI: 10.1016/j.giant.2024.100255

Madan, Y. (2020). A mini review study on "the generation of bioactive heterocyclic motifs by using sustainable metal nanocatalysed organic synthetic routes". *Materials Today: Proceedings, 30*, 197–202. DOI: 10.1016/j.matpr.2020.05.742

Manabe, K., & Kobayashi, S. (2002). Catalytic asymmetric carbon–carbon bond-forming reactions in aqueous media. *Chemistry (Weinheim an der Bergstrasse, Germany)*, *8*(18), 4094–4101. DOI: 10.1002/1521-3765(20020916)8:18<4094::AID-CHEM4094>3.0.CO;2-G PMID: 12297999

Martins, L. M., Moreno, V. F., Rosario, I. S., de Oliveira Graeff, C. F., & Silva Filho, L. C. (2021). Bronsted Acid Mediated Facile Greener Multicomponent Synthesis of 2, 4-Diaryl-quinoline Derivatives in Water. *Orbital: Electron. J. Chem.*, *14*(1), 9. DOI: 10.17807/orbital.v14i1.1689

Marvaniya, H. M., Modi, K. N., & Sen, D. J. (2011). Greener reactions under solvent free conditions. *Int. J. Drug Dev. Res*, *3*(2), 42–51.

Mason, T. J. (1997). Ultrasound in synthetic organic chemistry. *Chemical Society Reviews*, *26*(6), 443–451. DOI: 10.1039/cs9972600443

Mondal, S., Pain, T., Sahu, K., & Kar, S. (2021). Large-scale Green synthesis of porphyrins. *ACS Omega*, *6*(35), 22922–22936. DOI: 10.1021/acsomega.1c03534 PMID: 34514263

Nagendrappa, G. (2002). Organic synthesis under solvent-free condition: An environmentally benign procedure—I. *Reson*, *7*(10), 59–68. DOI: 10.1007/BF02835544

Patel, R. N. (2018). Biocatalysis for synthesis of pharmaceuticals. *Bioorganic & Medicinal Chemistry*, *26*(7), 1252–1274. DOI: 10.1016/j.bmc.2017.05.023 PMID: 28648492

Riva, S., Chopineau, J., Kieboom, A. P. G., & Klibanov, A. M. (1988). Protease-catalyzed regioselective esterification of sugars and related compounds in anhydrous dimethylformamide. *Journal of the American Chemical Society*, *110*(2), 584–589. DOI: 10.1021/ja00210a045

Singh, G., Capalash, N., Kaur, K., Puri, S., & Sharma, P. (2016). Enzymes: applications in pulp and paper industry. In Agro-industrial wastes as feedstock for enzyme production (pp. 157-172). *Academic Press*. https://doi.org/DOI: 10.1016/B978-0-12-802392-1.00007-1

Tanaka, K., & Toda, F. J. C. R. (2000). Solvent-free organic synthesis. *Chemical Reviews*, *100*(3), 1025–1074. DOI: 10.1021/cr940089p PMID: 11749257

Vaddula, B. R., Varma, R. S., & Leazer, J. (2013). Mixing with microwaves: Solvent-free and catalyst-free synthesis of pyrazoles and diazepines. *Tetrahedron Letters*, *54*(12), 1538–1541. DOI: 10.1016/j.tetlet.2013.01.029

Vasuki, G., & Kumaravel, K. (2008). Rapid four-component reactions in water: Synthesis of pyranopyrazoles. *Tetrahedron Letters*, *49*(39), 5636–5638. DOI: 10.1016/j.tetlet.2008.07.055

Wang, J. F., Lu, L. Y., Zhang, P., Qin, Y. J., & Guo, Z. X. (2022). Selective oxidation of primary benzylic alcohols to aldehydes using NaNO3 under ball milling. *Journal of Chemical Research*, *46*(4), 17475198221101983. DOI: 10.1177/17475198221101983

Wood, C. D., Cooper, A. I., & DeSimone, J. M. (2004). Green synthesis of polymers using supercritical carbon dioxide. *Current Opinion in Solid State and Materials Science*, *8*(5), 325–331. DOI: 10.1016/j.cossms.2005.02.001

Yasir, M., Zeshan, B., Daud, N. H. A., Shahid, I., & Khalid, H. (2021). Characterization of bacteriocin and chitinase producing bacterial isolates with broad-spectrum antimicrobial activities. *SN Applied Sciences*, *3*(8), 1–12. DOI: 10.1007/s42452-021-04740-z

Zangade, S., & Patil, P. (2019). A review on solvent-free methods in organic synthesis. *Current Organic Chemistry*, *23*(21), 2295–2318. DOI: 10.2174/13852728 23666191016165532

Zhou, Q. L. (2015). Transition-metal catalysis and organocatalysis: Where can progress be expected? *Angewandte Chemie*, *55*(18), 5352–5353. DOI: 10.1002/anie.201509164 PMID: 26662619

Zhu, X., Zhu, Z. Q., Guo, D., Liu, S., Ji, J. J., Tang, J., Yuan, E., Xie, Z.-B., & Le, Z. G. (2020). Cascade cyclization of glycine derivatives with β-ketoesters for polysubstituted 1, 4-dihydropyridines by visible light photoredox catalysis. *Tetrahedron*, *76*(33), 131353. DOI: 10.1016/j.tet.2020.131353

Chapter 11 Waste Management and Circular Economy

Ahass Bin Sreenath
Amity Institute of Geoinformatics and Remote Sensing, Amity University, Noida, India

K. Dhanaraj
Amity Institute of Geoinformatics and Remote Sensing, Amity University, Noida, India

Prangya Rath
https://orcid.org/0000-0002-6764-1625
Amity Institute of Environmental Sciences, Amity University, Noida, India

Laxmi Kant Bhardwaj
https://orcid.org/0000-0001-7518-4199
Amity Institute of Environmental Toxicology, Safety, and Management, Amity University, Noida, India

Rakesh Choudhary
Department of Civil Engineering, National Institute of Technology Delhi, New Delhi, India

ABSTRACT

The prompt urbanization and expansion in Kozhikode Corporation, Kerala, have led to a surge in solid waste creation, posing environmental and public health challenges. This chapter aims to recognize an environmentally and developmentally appropriate spot for solid waste dumping using GIS and remote sensing technologies. By employing supervised classification, reclassification, and Euclidean distance analysis, key parameters such as land use, elevation, geology, water resources, and

DOI: 10.4018/979-8-3373-1409-9.ch011

proximity to infrastructure were analyzed. Weighted overlay analysis was applied to classify the suitability of sites into five categories: very high, high, moderate, low, and very low. Results indicate that the eastern wards, including Kovoor, Mayanad, and Chelavoor, are most suitable for waste disposal due to their sparse buildup and availability of barren lands, with approximately 2.52 hectares identified as very highly suitable. This research underscores the importance of sustainable waste management practices and provides actionable insights for reducing environmental and health risks in urban areas.

INTRODUCTION:

Waste is a material that has effects on the world and is generated from manufacturing processes, industries, and other human activities (Kan 2009). The waste creation amount depends on elements such as population mass, financial status, level of commercial action, culture, and region. Due to rapid development and excessive economic growth, the urban inhabitants are growing, producing additional garbage (Rajput et al. 2009; Öncel et al. 2017). The existence of waste is a warning of overconsumption, ineffective utilization of items, and inadequate waste-discarding processes.

Waste is created by several manufacturing, commercial, farming, and even local supplies. The risks of wastes can vary conditional on their categories and environmental settings and numerous short-term and long-term impacts ranging from acute to chronic are assumed. Earth waste creation is estimated to be ~ 27 billion tonnes annually by 2050, one-third of which will derived from Asia, with foremost supplies from China and India (Modak et al. 2010). The waste formed in urban zones of India is ~ 170,000 tonnes per day, equal to about 62 million tonnes annually, and this is predicted to upsurge by 5% annually owing to rises in residents and altering routines (Planning Commission, Government of India, 2014).

1.1. Types of Wastes: Municipal solid waste (MSW), agriculture waste, industrial waste, e-waste, biomedical waste, construction, and radioactive waste are the different types of waste.

- **Municipal Solid Waste (MSW):** MSW is described as a waste of robust things, nondurable things, pots and packaging, food scrapings, yard trimmings, and various inorganic garbage from residential, commercial, and industrial resources (Gupta et al. 2024). It may also include dangerous trash such as paints, pesticides, batteries, and depleted medicine (Ishchenko et al. 2018; Elbeshbishy and Okoye 2019). Compostable organics involve vegetables, food waste, and fruits. The amount of MSW created is influenced by living values, the level and kind of commercial action, intake practices, and period (Kaushal et al. 2012). The World Bank identifies MSW as "non-dangerous leftover produced in homes, industrial and business corporations, institutes, and non-dangerous industrial practice trashes, farming garbage, and sewage mud". Additional trash per capita is created in elevated-income nations than in low-income nations as illustrated in table 1 (World Bank report 2012).

As per the data of the Census of India (2011), the populace of India was 1252 million in 2013. Populace expansion is a chief reason for creating elevated quantities of MSW in India. As per the study of CPCB (2000), India creates ~ 133,760 tonnes of MSW per day, of which 91,152 tonnes are gathered and 25, 884 tonnes are cured. MSW creation per capita in India varies from ~ 0.17 kg per individual per day in small cities to ~ 0.62 kg per individual per day in cities (Kumar et al. 2009). The average constituents of MSW generated by Indian towns are ~ 41% organic, 40% inert, and 19% recyclable items (Sharholy et al. 2008). Largely organic surplus is created from homes, and inert surplus is created from structure, destruction, and road clearing.

Currently, the method of gathering, transport, and dumping of MSW in India can be measured to be disorderly, with unscientific approaches for dumping. In India, substantial municipalities accumulate around 70 to 90% of MSW created, in comparison to minor municipalities that collect < 50% of waste created. The worrying waste management procedure is that upwards of 91% of MSW accumulated is discarded and empty in illegal landfills. Table 2 shows the MSW created in main cities in India (CPCB 2011). Kumar et al. (2017) deliberate the MSW creation in diverse states and informed the extreme waste creation in Maharashtra, followed by Uttar Pradesh, Tamil Nadu, West Bengal, Andhra Pradesh, Kerala and Madhya Pradesh, Rajasthan, Gujarat, Karnataka and Mizoram. They also informed lower garbage creation in Jammu & Kashmir, Bihar, Jharkhand, Chhattisgarh, Orissa, Goa, Assam, Arunachal Pradesh, Meghalaya, Tripura, Nagaland and Manipur.

Table 1. Per Capita Urban Municipal Solid Waste (MSW) Generation as Per Income of Countries

S. No.	Country	Per Capita Urban Municipal Solid Waste (MSW) Generation (kg/day)	
		1999	2025
1	Low-income countries	0.45-0.90	0.60-1.00
2	Middle-income countries	0.52-1.10	0.80-1.50
3	High-income countries	1.10-5.07	1.10-4.54

Table 2. The Generation of Municipal Solid Waste (MSW) in the Major Cities in India

S. No.	Name of the City	Municipal Solid Waste (MSW) (Tonnes per day)
1	Ahmedabad	2,300
2	Bangalore	3,700
3	Chennai	4,500
4	Delhi	6,800
5	Hyderabad	4,200
6	Kanpur	1,600
7	Kolkata	3,670
8	Lucknow	1,200
9	Mumbai	6,500
10	Pune	1300

- **Industrial Waste:** It is created by manufacturing units, mills, workshops, and mines (Krishna et al. 2020). It may be dangerous or non-dangerous items (Bulucea et al. 2010). The hazardous trash that is produced from manufacturing units contains harmful compounds and heavy metals (Alam et al. 2023; Bhardwaj et al. 2023), while non-hazardous trash includes paper, wood, and glass objects. This type of waste is formed by modern civilizations as an obvious by-product of mining, industrial actions, and the requirements of today's purchasers (Artiola 2019). During the 19th century, a large amount of industrial waste was produced because this era is called the industrial developing period.
- **Agriculture Waste:** It contains consumed grains, grass, bagasse, molasses, husk (rice, maize, and wheat), skin (banana and avocado), shell (coconut, walnut, and groundnut), and harvest stalks (cotton) (Dai et al. 2018; Sharma

et al. 2020). This waste contains more nutrients (Ravindran et al. 2018). Food and Agriculture Organization (FAO) in 2013 stated that worldwide 250 million tons of agro-industrial trash annually is created from non-eatable plants during the management of separate produce (Heredia-Guerrero et al. 2017). India creates > 350 million tons of agro-industrial trash yearly (Madurwar et al. 2013). Usually, this unused is either burned in the grounds thus foremost to the creation of fume, poisonous gases (such as SO_2, CH_4, N_2O), carcinogens (for eg: polycyclic aromatic hydrocarbons (PAHs), furans, and dioxins) and greenhouse gases (GHGs) displaying harmful properties on human health and environment (Sharma et al. 2020).

- **E-waste:** Electronic utensils are being manufactured in massive quantities nowadays (Ilyas et al. 2007). E-waste is the quickest-rising waste type in the globe that holds abandoned electronic machines such as mobile, telephones, televisions, computers, etc (Bhardwaj et al. 2024; Bhardwaj et al. 2025). The international creation of e-waste is projected to be 20 to 25 million tonnes annually, with maximum e-waste being generated in the United States (US), Europe and Australia. This waste holds hazardous pollutants like mercury (Hg), lead (Pb), antimony (Sb), nickel (Ni), cadmium (Cd), polychlorinated biphenyls (PCBs), and polybrominated diphenyl ethers (PBDEs). After burning up, e-unused may create furans, dioxins, PAHs, and polyhalogenated aromatic hydrocarbons (PHAHs). E-waste is different from other forms of leftover; it includes both significant and dangerous constituents that need different supervision and recovering approaches to avoid environmental pollution and damaging properties on human health.
- **Biomedical Waste:** This type of waste can exist in gas, solid, and liquid forms and includes throwaway syringes, sanitary supplies, body solutions, bandages, and blood-holding cloths (Mathur 2012; Bhalla et al. 2013). It is believed dangerous because it may harbor pathogens and other kinds of bacteria. Inappropriate dumping of this trash in open zones or landfills can trigger the distribution of infections and water pollution. Radiographic image is a polyester layer and is used in therapeutic and dental examinations for some inquiries. ~ 2 billion radiographs are consumed around the globe annually, including chest X-rays, CT scans, etc. Conventionally, 94 to 98% of X-ray pictures are utilized in curative facilities (Khunprasert et al. 2008). Burning is the greatest technique to throw away these wastes.
- **Construction and Radioactive Waste:** Construction waste may be further described as either non-organic material or organic material that has a long-time cycle (Taurino et al. 2023). It includes construction debris such as bricks, rubble, glass, plastic containers, and concrete. Radioactive trash is generated by medical treatments, nuclear power plants, certain industrial

processes, and scientific research (Zhang et al. 2019; Deng et al. 2020). It is very dangerous and can cause skin cancer and mutation. So, it requires careful and specialized handling. It is the by-product of nuclear resource creation and other uses of nuclear fission (Rao et al. 2022). It encloses radionuclides which are unbalanced atoms of an element that release emission at the time of deterioration. It creates a major hazard to human health; thus, it must be controlled appropriately to ensure that radioactive revealing does not surpass values and principles (Poškas et al. 2019).

Kozhikode city is situated in Kerala, India. Initially, it was recognized as Calicut, and it is one of the first and most valuable cultural spots of India's old cultures. It has been divided into 75 wards. It comprises Kannur district in the northern region, Malappuram district in the southern region, Wayanad district in the eastern region, and the Arabian Sea in the western region. The geocoordinates of the city are 11° 8' 00" to 11° 50' 00" north latitude and 75° 30' 00" to 76° 8' 00" East longitude. According to 2011 census statistics, the total residents of Kozhikode City is roughly 5500440. Kozhikode Municipal Corporation was developed in 1962, and it is the second ancient municipal corporation in Kerala. The Corporation has four electorates namely Kozhikode North, Kozhikode South, Baypur, and Elathur. Kozhikode is the leading municipal corporation in the development sector of Kerala. Therefore, development and urbanization lead to the accumulation of enormous amounts of waste in this area. Mainly food packets plastic aluminum or thermocol, bottled water, cold drinks cane, chip's pouches, and polybags have caused a huge flow of garbage in this corporation.

The aim of this examine is to find an environmentally appropriate optimal location for solid waste dumping in Kozhikode Corporation. GIS and remote sensing are used to collect and analyze data with the help of various quantitative techniques. Here we can find a suitable location by providing a variety of parameters. This includes the assessment of many factors, including environmental footprint, distance to residential areas, walkability, lack of land, and potential contamination of water supplies. Based upon these parameters, the objective is to guarantee the choice of suitable locations that will reduce the possible effect on the environment and public health and to take into account the practical requirements for the waste management system. Evaluation is used to identify the right places where solid waste needs to be disposed of, leading to sustainability and suitability in waste management.

2. REVIEW OF LITERATURE:

The aim of the study "*The Municipal Solid Waste Disposal Site Suitability Analysis Using Multi-Criteria Evaluation in Assosa*" was to identify suitable landfill sites in Assosa municipality by commissioning multi-criteria decision analysis and GIS. Parameters containing the slope and land usage of the vicinity, as well as the proximity to sensitive sites, roads, rivers, airports, and boreholes, were applied. The GIS interface had created and standardized map layers for each requirement. It was categorized as highly suited, somewhat suitable, less suitable, and inappropriate (Saketa et al. 2023).

GIS, the Analytical Hierarchy Process (AHP), and remote sensing procedures are utilized in the study "*Solid Waste Disposal Site Selection with GIS and AHP Methodology: A Case Study in Senirkent – Uluborlu (Isparta) Basin, Turkey*" to pick a landfill location. There are ten distinct criteria: aspect, elevation, slope, lithology, aquifer, surface water, land application, groundwater depth, and lineaments. The categorizations were extremely suited, slightly suitable, and unsuitable (Sener et al. 2011).

The study "*Site Suitability for Urban Solid Waste Disposal Using Geoinformatics: A Case Study of Pune Municipal Corporation, Maharashtra, India*" uses a geoinformatics methodology to recognize suitable solid unused dumping spots. To do this, a number of themed coats were generated applying medium-resolution satellite information and topographical charts. Nine factors are taken into account when analyzing a site's potential for disposing of urban solid waste, including the road system, rivers, geology, lakes, population density, canals, slope, land use, and airport. It was characterized as extremely suitable, slightly appropriate, unsuitable, and less suitable (Jaybhaye et al. 2014).

The study "*Site Suitability Analysis for Municipal Solid Waste Disposal in Birnin Kebbi, Nigeria*" aimed to observe the appropriate site for the discarding of MSW in Birnin Kebbi by administration trash at its absolute removal surroundings. Tools such as remote sensing and GIS were applied in this study to efficiently analyze the examined area's location appropriateness. For this study, Landsat 8 OLI was used to create the essential data (Kaoje et al. 2016).

The study "*Remote Sensing and GIS-based Site Suitability Analysis for Solid Waste Disposal a Multi-Criteria Analysis of the Khulna City Area*" was designed to find a feasible site for trash dumping in Khulna City, Bangladesh. GIS and remote sensing technology were used in this study. The methodology utilized in this study consists of gathering information, arranging the data, and using Multi-Criteria Analysis (MCA) to the supplied data in order to verify the effective spot for Khulna's solid waste removal. An arithmetical model with a three-dimensional determination of

30 m, a Landsat picture with a resolution of 30 m, and GPS-composed land monitor spots were applied in this study (Khan et al. 2015).

In the study "*Landfill Site Suitability Analysis Using AHP for Solid Waste Management in the Guwahati Metropolitan Area, India*" AHP and GIS-based multi-criteria verdict investigation procedures were used to observe possible locations in the Guwahati, Assam, India. Seven factors were included the type of land used, the height, the slope, and the distance from wetlands, rivers, roads, and airports (Hazarika and Saikia 2020).

In the study "*Site Selection in Municipal Solid Waste Management with an Extended VIKOR Method Under a Fuzzy Environment*" the weights and ratings for the assortment criteria were evaluated using linguistic variables, which can be represented as trapezoidal fuzzy statistics. The choice maker's attitude was taken into consideration while converting the fuzzy decision matrix into crisp values using the ordered weighted averaging operator (Liu et al. 2014).

In the study "*The Land Use and Land Cover Change Database and Its Relative Studies in China*" remotely sensed Landsat Thematic Mapper pictures were orthorectified and georeferenced, and they were translated into land-use/land-cover (LULC) classes at a 1:100,000 scale using an entire digital software environment alteration. The LUCC database was successfully updated by contrasting the mid-1990s TM data with fresh data sources obtained from 1999 to 2000 and 1989 to 1990. The national LUCC research program in China has benefited tremendously from the LUCC database, and several related investigations are only partially discussed in this study (Jiyuan et al. 2002).

The study "*The Land Use/Land Cover Change and Statistical Modelling of Cultivated Land Change Drivers in Nigeria*" focused on the evolution to refined land use, analyzed the spatiotemporal strength of land usage, and explained changes in LULC that have occurred in Nigeria throughout a ten-year study period (2000 to 2010). It also identified the variables that drove these changes (Arowolo et al. 2018).

The study "*Spatio-Temporal Analysis of Land Use/Land Cover Pattern Changes in Arasbaran Biosphere Reserve: Iran*" focused on assessing quantitative and qualitative changes in this biosphere reserve's LULC throughout time. The LULC spatiotemporal alterations in the Arasbaran biosphere reserve were categorized as forest, agricultural, and barren areas and contrasted with coming spatial patterns. This investigation covered the entire region as well as every zone inside the Arasbaran biosphere reserve (Amini Parsa et al. 2016).

In order to properly manage solid waste, the study "*GIS-Based Sanitary Landfill Suitability Analysis for Sustainable Solid Waste Disposal*" introduced the incorporation of GIS with AHP for the purpose of choosing the best and most sustainable locations for sanitary landfills. For this study, a number of input criteria were employed, including soil, geology, urban areas, unoccupied and agricultural lands, rail lines, water bodies, roads, slopes, and elevations. The study's findings were divided into three categories: most, suitable, and inappropriate landfill locations (Mohammed et al. 2019).

The study "*GIS-Based Approach for Optimized Siting of Municipal Solid Waste Landfills*" used a multi-criteria decision investigation and overlay evaluation with GIS to address the site of a new landfill. Geology, water stream resources, land utilization, sensitive locations, air quality, and groundwater value were some of the parameters (Natesan and Sarkar 2008).

The study "*Site Suitability Analysis for a Decentralized Sustainable Solid Waste Management System for Coimbatore City Using GIS*" used weighted overlay analysis and GIS to choose decentralized processing unit sites zone-by-zone. In order to determine the greatest place for decentralized solid waste management (SWM), a weighted overlay analysis is performed using thematic layers such as LULC, soil, geology, geomorphology, lineaments, groundwater depth, remoteness from railway tracks, remoteness from water bodies, road density, drainage density, and remoteness from roads. The classification was very appropriate, high to moderately appropriate, moderately appropriate, poorly appropriate, and unsuitable (Poornima and Nithya 2022).

This study "*Land Suitability Analysis for Solid Waste Landfills in Kano Metropolis Using Multi-Criteria Evaluation Techniques*" focused on the goal of finding a good location and geographical dispersion for landfills. Advanced Spaceborne Thermal Emission and Reflection Radiometer (ASTER)- Global Digital Elevation Model (GDEM) from the United States Geological Survey (USGS) was utilized to generate slope and elevation, and road networks were extracted from satellite pictures obtained with Google Earth Pro. The USGS Landsat-7 ETM+ picture is used to classify LULC. The Ministry of Environment Kano State provides a list of dumpsite addresses along with their coordinates derived from fieldwork (Mohammed and Zubair 2022).

3. METHODOLOGY

3.1. Study Area: Kozhikode Municipal Corporation was the study area (figure 1). The geographical coordinates were 11° 9' 30" to11° 22' 00" North Latitude and 75° 44' 00" to 75° 52' 30" East Longitude.

Figure 1. Diagrammatical Representation of the Study Area (Kozhikode Corporation)

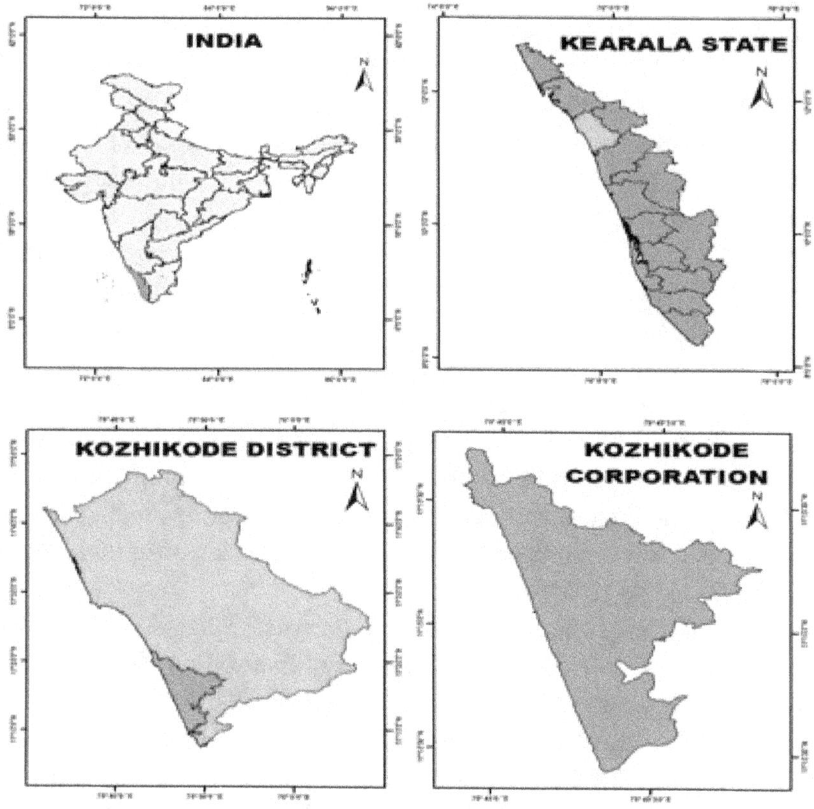

3.2. Geomorphology: Geomorphology is the science of landform and sediment deposition processes. It exposes the dynamic interaction of natural processes in the study area. The most obvious geomorphic features are pediment and pediplain formations, which are very extensive in the northeastern area. These characteristics are usually developed over very long timeframes of erosion and weathering. They are relevant to the structural and geological importance of the region. Coastal flats and floodplains are also prominent, highlighting the role played by marine and fluvial processes as part of landscape evolution. These wetland areas, located mostly in the rest of the locale, are of major value in maintaining biodiversity and natural flood control through bio-swamps. Learning these geomorphic features is critical to sustainable land-use planning and environmental protection in this geologically eclectic region. Figure 2 represents the geomorphology and transportation networks of Kozhikode Corporation in the study area.

Figure 2. (A) Geomorphology and (B) Transportation Network of the Study Area

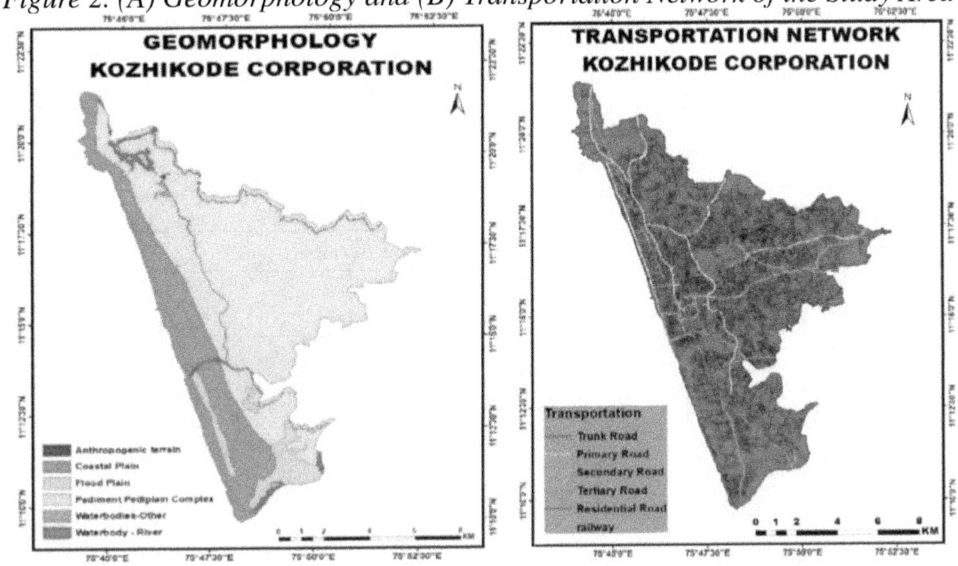

A B

3.3. Data and Methods: Different data, software, and procedures are used in this study. Most of the analysis was completed on a Geoinformatics software basis. Figure 3 depicts the flowchart showing the methodological overview. Here eight parameters are used to perform the weighted over analysis which is the main part (table 3). The main data used are Landsat Data and Sentinal Data, SRTM DEM.

Figure 3. The Flowchart of the Methodology Used in the Study

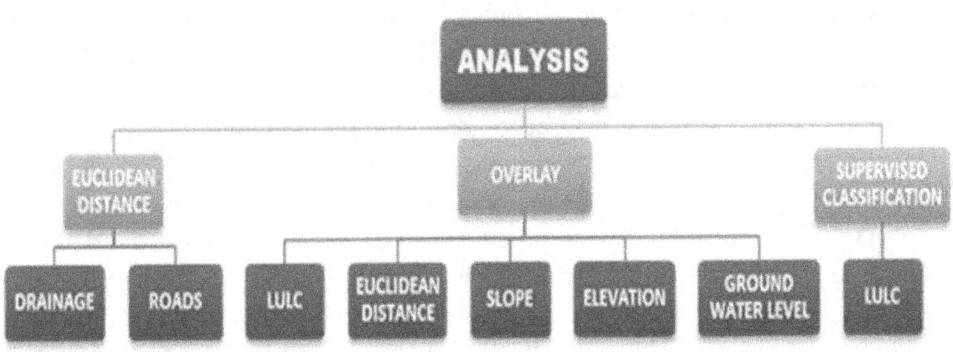

Table 3. Different Parameters, Satellite/Sensor, Source, and Date/Period

Parameters	Satellite/ Sensor	Source	Date/ Period
Geology		**GSI Bhukosh**	
Elevation	SRTM	USGS	2014
Slope	SRTM	USGS	2014
Land use land cover	Sentinal	Copernicus	2023
Euclidean distance (road)		OpenStreetMap data	
Euclidean distance (drainage)	Sentinal	Copernicus	
Groundwater level		India Water Resources Information System	
Euclidean distance (Builtup)	Sentinal	Copernicus	

4. METHODS

4.1. **Supervised Classification:** The technique of giving a set of grid compartments in a raster GIS record a class value based on the standards of comparable grid compartments that have already been identified as demonstrative of a class or group is known as supervised classification. The ERDAS IMAGINE software is used for supervised classification this has been analyzed into five main classifications. As part of supervised classification, the classification is classified as built-up, vegetation barrier land, water body, and agriculture about 100 training samples were taken for each class. Sentinal 2 data used for supervised classification. The recode method used in this analysis to compensate for the error caused by pixels.

4.2. **Reclassify:** Reclassification is applied to make easier the understanding of raster information by altering a single value to an innovative value range of values into particular values. Reclassification is the method of reselecting one or more values in a raster dataset to additional output values. The Reclassify instrument is accessible in the Spatial Analyst expansion in ArcMap. Geology and geomorphological features are reclassified here. Its data is mainly collected from the Geological Survey of India website. Here the geology is divided into ten classes grey fine sand, sand, clay, clayey sand, pebble bed, terri sand, sandstone, laterite, charnockite, and banded iron formation. Based on classification, charnockite rock is mostly found here. Similarly gray fine sand and clay laterite are very less. In reclassify, the highest weightage is given to charnockite. The same weightage is given to other rocks.

4.3. **Euclidean Distance:** Euclidean distance indicates the separation between the nearest source and each raster cell. Raster or feature class data can be used as the input source data. All of the source raster's cells with valid values make up the set of source cells when the input source data is a raster. The source set excludes any cells with no data values. A valid source is one with a value of 0. The extraction tools can be used to build a source raster. Before conducting the analysis, the source locations are internally transformed into a raster when the input source data is a feature class. Roads, water bodies, and built-up land elements are chosen for Euclidean distance analysis (table 4). Finding appropriate dumpsites adjacent to road water bodies built-up land characteristics might be aided by distance analysis. Areas away from all three of these features can be detected by it. Here, the higher value is assigned the highest weightage in each of the three attributes, and the lowest value is assigned the lowest weightage as well.

Table 4. Different Parameters with Classes and Weightage

Parameter	Classes	Weightage
Road	Below 0 - 343 343.1 - 685 685.1 - 1027 1027.1 - 1370 1370.1 - 1713 & above	1 2 3 4 5
Water bodies	Below 0 - 200 200 - 600 600 - 1200 1200 - 2000 2000 - 2491 & above	1 2 3 4 5
Built up land	Below 0 - 72.5 72.5 - 145 145.1 - 217.5 217.5 - 290 290.1 - 362.5 & above	1 2 3 4 5

4.4. **Analytic Hierarchy Process (AHP):** An overall system of measurement, the AHP is used to originate ratio scales from continuous and discrete paired assessments, which can be derived from real extents or from a fundamental scale reflecting the relative asset of feelings and preferences. The AHP is particularly concerned with deviations from uniformity in its measurement and on dependencies within and between the collections of essentials of its construction, its most common uses are in multicriteria decision-making, planning and resource distribution, and conflict steadfastness.

AHP is a still prevalent decision-making methodology that offers a systemized approach for comparing and ranking several aspects regarding a set of criteria. It is a process of disassembling complex decision problems into a hierarchy that is, the goal, criteria, sub-criteria, and alternatives. AHP enables decision-makers to make value judgments by comparing the elements and then transforming the comparisons into numerical values. By comparing ratio scales, AHP allows a more intrinsic insight into the relative importance of various factors (tables 5-8)

An important asset that AHP possesses is the ability to handle noise in the sequence of decisions and reveal relationships among different factors in a problem to be solved. Concentrating on inconsistencies AHP allows the decision-maker to assess and correct his/her preferences in order to maintain the integrity of the decision-making process. It has been widely used when several criteria have to be taken into account, e.g., resource allocation, planning, negotiation/conflict resolution. By their systematic nature, AHP enables clear justifiable, and efficient decision-making, therefore AHP is a valuable tool for making decisions at an individual level as well as at a group level.

Table 5. Comparison Matrix of Different Layers Used for Analytic Hierarchy Process (AHP) Analysis

	GEOLOGY	ELEVATION	SLOPE	LAND USE LAND COVER	EUCLIDEAN DISATANCE (ROAD)	EUCLIDEAN DISTANCE (DRAINGAE)	GROUND WATER LEVEL	EUCLIDEAN DISTANCE (BUILTUP)
GEOLOGY	1	2	2	2	2	3	2	2
ELEVATION	0.5	1	1	3	3	3	2	2
SLOPE	0.5	1	1	3	3	2	2	2
LAND USE LAND COVER	0.5	0.333333333	0.333333333	1	5	2	2	2
EUCLIDEAN DISATANCE (ROAD)	0.5	0.333333333	0.333333333	0.2	1	1	2	1
EUCLIDEAN DISTANCE (DRAINGAE)	0.333333333	0.333333333	0.5	0.5	1	1	2	2
GROUND WATER LEVEL	0.5	0.5	0.5	0.5	0.5	0.5	1	2
EUCLIDEAN DISTANCE (BUILTUP)	0.5	0.5	0.5	0.5	1	0.5	0.5	1
TOTAL	4.333333333	6	6.166666667	10.7	16.5	13	13.5	14

Table 6. Consistency Ratio Calculated for Analytic Hierarchy Process (AHP) Analysis

	D	CP	D*	E=D/CP	l max	CI	RI	CR=CI/RI
GEOLOGY	1.92935726	0.217512813	1.86814991	8.58868903	8.686314304	0.098044901	1.14	0.086004299
ELEVATION	1.646452553	0.185618565	1.591953744	8.576479098	8.700260771	0.100037253	1.14	0.087751976
SLOPE	1.56508458	0.176445263	1.506291021	8.536874234	8.72089105	0.102984436	1.14	0.090337224
LAND USE LAND COVER	1.104964781	0.124572054	1.155875055	9.278766891	8.757694414	0.108242059	1.14	0.094949175
EUCLIDEAN DISTANCE (ROAD)	0.621367359	0.070052014	0.623311927	8.89784447	8.627426294	0.089632328	1.14	0.078624849
EUCLIDEAN DISTANCE (DRAINGAE)	0.759835686	0.085662723	0.720873656	8.41525493	8.537286902	0.076755272	1.14	0.067329186
GROUND WATER LEVEL	0.648419777	0.073101863	0.637102987	8.715277047	8.598302889	0.085471841	1.14	0.074975299
EUCLIDEAN DISTANCE (BUILTUP)	0.594603558	0.067034704	0.568543359	8.48132873	8.48132873	0.068761247	1.14	0.060316884
TOTAL	8.870085554							

Consistency index $CI = \frac{(\lambda max - n)}{(n-1)}$

Consistency Ratio $CR = \frac{CI}{RI}$

where n and λmax represent the dimension of the Eigenvalue and matrix. The judgment evenness is checked by CR which is taken by dividing CI with Random Index (RI) from table 7 (Saaty 1987).

Table 7. Weightage and Rank for Different Parameters

	GEOLOGY	ELEVATION	SLOPE	LAND USE LAND COVER	EUCLIDEAN DISTANCE (ROAD)	EUCLIDEAN DISTANCE (DRAINGAE)	GROUND WATER LEVEL	EUCLIDEAN DISTANCE (BUILTUP)	Weightage (W)	% W	Rank (0-5)
GEOLOGY	0.230769	0.333333	0.324324	0.186916	0.12121	0.230769	0.1481	0.1429	1.718329	21.47912	5
ELEVATION	0.115385	0.166667	0.162162	0.280374	0.18182	0.230769	0.1481	0.1429	1.42818	17.85225	4
SLOPE	0.115385	0.166667	0.162162	0.280374	0.18182	0.153846	0.1481	0.1429	1.351257	16.89071	4
LAND USE LAND COVER	0.115385	0.055556	0.054054	0.093458	0.30303	0.153846	0.1481	0.1429	1.066334	13.32917	3
EUCLIDEAN DISTANCE (ROAD)	0.115385	0.055556	0.054054	0.018692	0.06061	0.076923	0.1481	0.0714	0.600792	7.509896	2
EUCLIDEAN DISTANCE (DRAINGAE)	0.076923	0.055556	0.081081	0.046729	0.06061	0.076923	0.1481	0.1429	0.688823	8.610289	2
GROUND WATER LEVEL	0.115385	0.083333	0.081081	0.046729	0.0303	0.038462	0.0741	0.1429	0.612224	7.652797	2
EUCLIDEAN DISTANCE (BUILTUP)	0.115385	0.083333	0.081081	0.046729	0.06061	0.038462	0.037	0.0714	0.534061	6.675765	1
TOTAL									8	100	

4.5. Weighted Overlay Analysis Ranking

Table 8. Ranking of Different Parameters Used in the Study

Parameters	Class /Value	Rank
Geology	Laterite	1
	Charnockite	5
	Grey fine sand	1
	Pebble bed	1
	Banded iron formation	1
	Clayey sand	1
	Sand	1
	Clay	1
	Terri sand	1
	Sandstone	1
Built up land	Below 0 - 72.5	1
	72.5 - 145	2
	145.1 - 217.5	3
	217.5 - 290	4
	290.1 - 362.5 & above	5
Water bodies	Below 0 - 200	1
	200 - 600	2
	600 - 1200	3
	1200 - 2000	4
	2000 - 2491 & above	5
Road	Below 0 - 343	1
	343.1 - 685	2
	685.1 - 1027	3
	1027.1 - 1370	4
	1370.1 - 1713 & above	5
Land use land cover (LULC)	Built up land	1
	Agriculture	1
	Barren land	5
	Vegetation	1
	Waterbodies	1

Parameters	Class /Value	Rank
Groundwater level	Below 3 - 5	5
	5 - 10	4
	10 - 15	3
	15 - 20	2
	20 - 31 & above	1
Slope	Below 0 – 2	1
	2.01- 4	3
	4.01 - 6	5
	6.01 – 10	4
	10.01 – 23 & above	2
Elevation	Below -9 - 0	1
	0 - 23	2
	23.01 - 35	5
	35.01 - 51	4
	51.01 – 95 & above	3

5. RESULT AND DISCUSSION

5.1. Land Use Land Cover (LULC): A LULC map as depicted in figure 4, is used to recognize the physical land type of the area. LULC has mainly classified the area into five classes as shown in table 9 (Barren land, Agricultural land, Built-up land, Vegetation land, and Waterbodies). Built-up land is seen more in the western fraction of the study region. Likewise, Vegetation and Agriculture features are present in the eastern part. Waterbodies are mostly found in the North and Southern parts. Being an urban area, there is very less concentration of Barren land.

The LULC classification can be useful for understanding the spatial scattering of various land cover types in the area of study which summarises the interaction between natural features and urban expansion. The dominance over built-up land in the western area highlights the dominance of urban structures and residential areas, which is suggestive of greater human activities and developmental pressures in the western area. By contrast, the east, due to the dominant coverage of vegetation and agricultural land, is a key area for ecological equilibrium and a promising site for green infrastructure development. There is a scattered distribution of water bodies in the north and south, which enhances the hydrological stability of the region and provides potential for sustainable water resource management. The relatively scarce amount of undeveloped land in this urbanized landscape shows the high degree of land use intensification, and planning for its development and environmental protection needs to be taken seriously.

Accuracy Assessment

Table 9. Accuracy Assessment of Land Use and Land Cover (LULC) Classification Images

Year	Accuracy	CLASSES					Overall Accuracy	Overall Kappa
		Builtup	Barren land	Agricultural	Vegetation	Water bodies		
2024	Reference Total	103	21	80	40	10	81.25	0.75
	Classified	70	60	80	37	9		
	Correctly classified	67	20	77	35	9		
	User's Accuracy (%)	95.71	33.33	96.25	94.59	100		
	Producers Accuracy (%)	65.05	95.24	96.25	87.5	90		
	Kappa	0.93	0.27	0.95	0.94	1		

Figure 4. Land Use and Land Cover (LULC) Classification Image of the Study Area

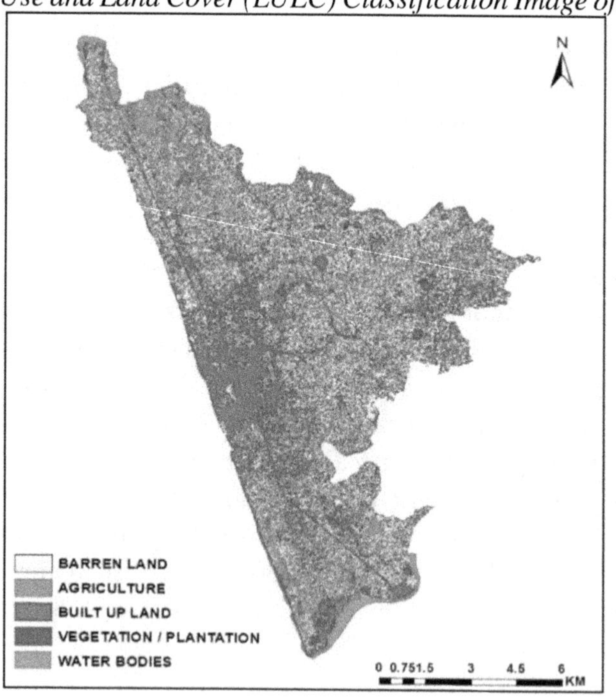

5.2. Geology: It is an area of unique importance in terms of geologic features. There are about ten different geological deposits here. Charnockite is the predominant geological formation in this region. Similarly, Sand and Terri Sand are also concentrated here. Charnockite is more abundant in the Northeastern part. Gray Fine Sand is found near the coastal area. The study area is geologically diverse and thus of exceptional interest from a point of geoscience view. Within the 10 types of geological structures, as depicted in figure 5, charnockite is the most represented structure, primarily in the northeastern part. Very hard, metamorphic rock is important for determining the structural stability of that region, in addition to its role in predicting how that region can be suitable for many developmental uses. Besides charnockite the occurrence of Sand and Terri Sand further showcases the geological potentiality of the region where these deposits could be useful for construction and land development purposes. Coastal deposits, in the form of Gray Fine Sand, represent the effects of marine and fluvial processes and highlight the dynamic interaction of geology and hydrology in landscape evolution. These characteristics emphasize the need to realize the value of geological context in urban design and sustainable practices.

Figure 5. Stratigraphy Map of the Study Area

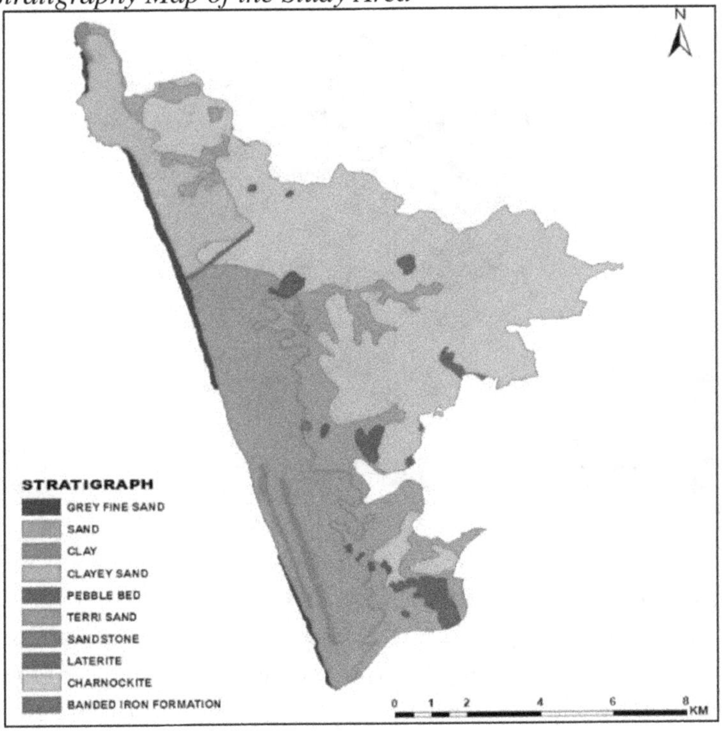

5.3. Elevation: Elevation refers to areas above sea level. The elevation range of this area is 9 m to 95 m. The northeastern part is higher being a coastal area. The elevation is very low in the western part. The elevation is classified into five categories. Land altitude is an important factor in determining land suitability for different purposes such as SWM, urbanization, and ecological preservation. Elevation in the study zone diverges from 9 m to 95 m above sea level and thus represents an area of considerable topographic diversity. Figure 6 depicts the elevation of Kozhikode derived from DEM. The northeastern region owing to its higher altitude, provides improved drainage and lower flood risks, thus it is more suitable for infrastructure development and waste management units. Alternatively, the western coastal lands are highly low and thus very susceptible to inundation and waterlogging, especially with major rainfall or storm surges. The elevation has been categorized into five different classes, and this gives a granular insight into the variation of topography which is necessary for planned and sustainable use of the land resources of the area.

Figure 6. Elevation of the Study Area Derived from DEM

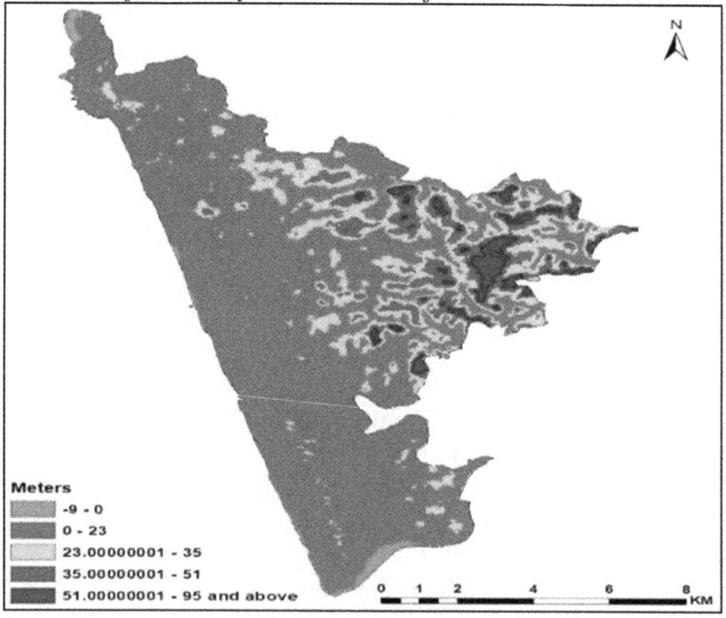

5.4. Slope: The slope range in this area is below 2° to 23° above. As observed from figure 7, most of the slopes are on the northeastern side. The slope decreases on the western side. Slope is generally low in this area. Slope features of the study area are of great importance in the assessment of land use suitability and environmental stability. The slope ranges from less than 2° to more than 23°, with steeper slopes mainly representing the Northeastern region. This steeper gradient in the northeast could present construction and waste management activities with some challenges but could at the same time be beneficial to natural drainage and erosion control. On the contrary, the slope is reduced drastically towards the western side, generating gentler areas favorable to urban expansion and road infrastructure improvement. Overall, the generally low slope throughout the area ensures accessibility and ease of land use while highlighting the need for targeted slope management strategies in steeper regions to prevent soil erosion and maintain stability.

Figure 7. Slope of the Study Area in Degrees

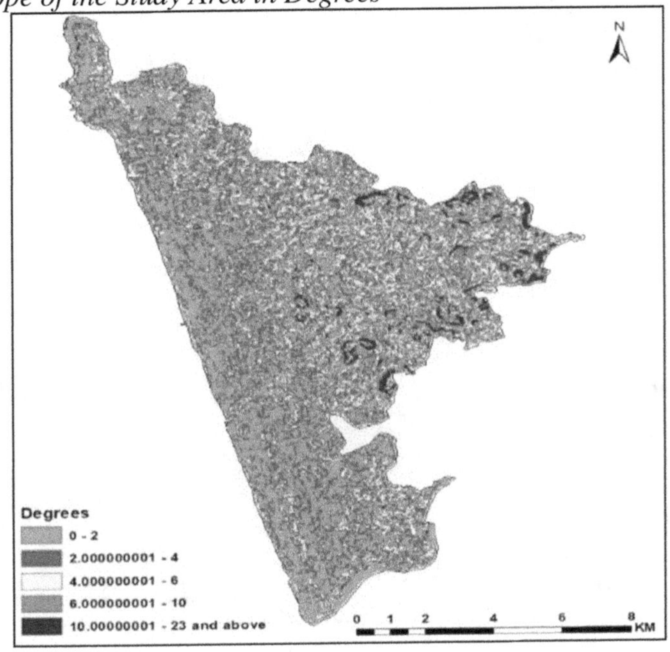

5.5. Ground Water: The depth of groundwater is about 3 m to 31 m and above. Figures 8-11 depicts the euclidean data of various parameters of the study area. The area near the coastal region has less depth. From the coastal side to the eastern part amount of depth increases. Data from 10 groundwater stations are represented here. The depth of groundwater in the region of study is highly heterogeneous and extends from 3 m down to more than 31 m, as a consequence of the wide variety of hydrogeologic properties of the region. Coastal zones have shallower groundwater table depths in relation to the sea and occur with higher water table exposures, increasing their vulnerability to contamination and/or saltwater intrusion. Conversely, as moving eastward the groundwater depth follows, i.e., being gradually reduced, i.e., the water table gradually diminishing. This trend plays an important role in groundwater extraction/recharge activity planning. Data from 10 groundwater stations give an overall picture of these changes, important for sustainable water resource management, safe drinking water, and prevention of overexploitation in the area.

Figure 8. Euclidean Distance from the Ground Water Locations

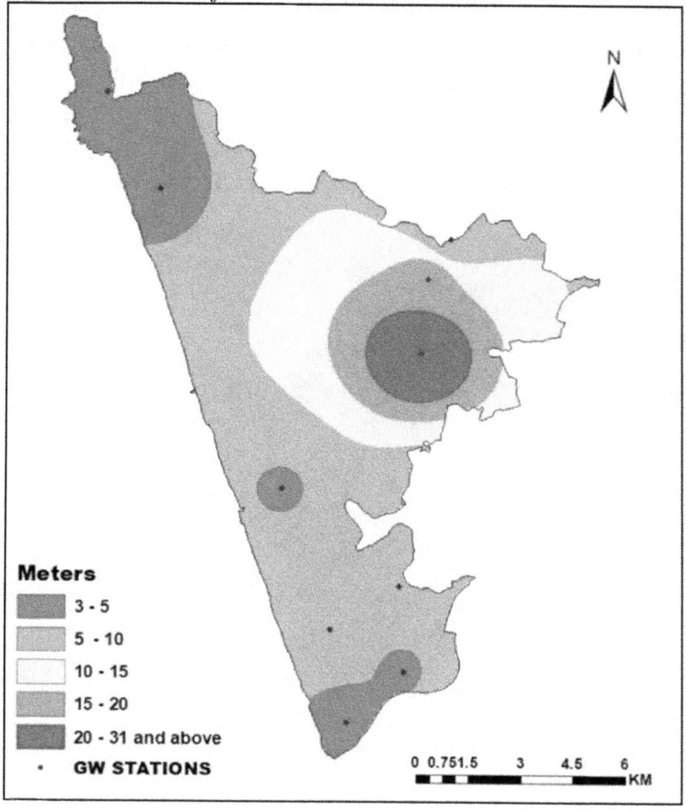

5.6. Euclidean Distance of Water Bodies

Figure 9. Euclidean Distance from the Locations of the Surface Water

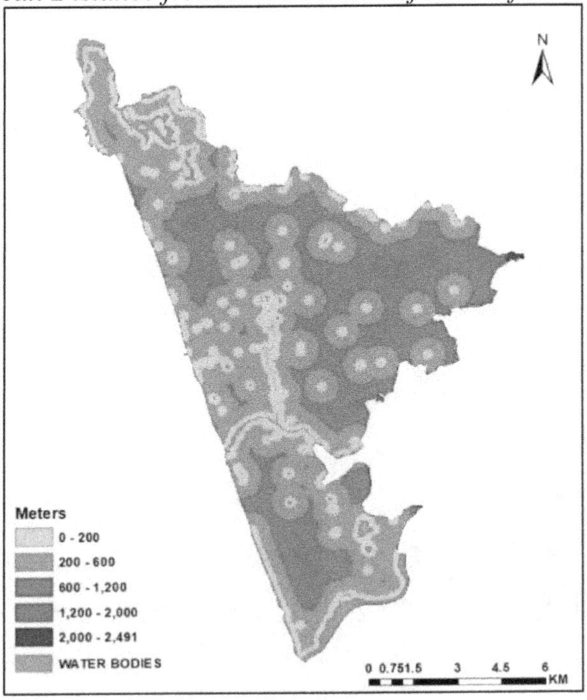

5.7. Euclidean Distance of Built-Up Land

Figure 10. Euclidean Distance from the Locations of Built-Up Patches

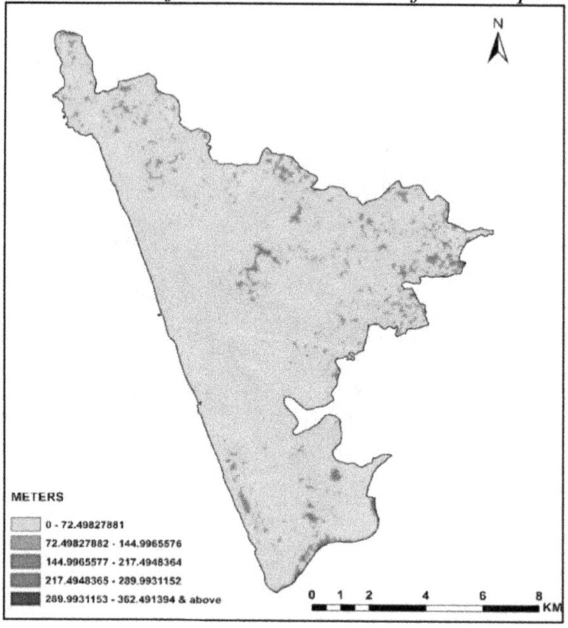

5.8. Euclidean Distance of Road

Figure 11. Euclidean Distance from the Locations of Roads

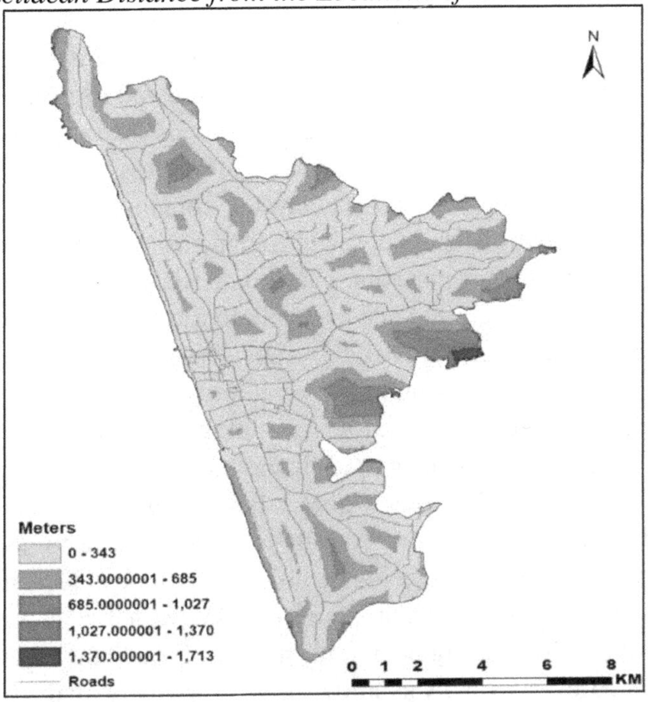

5.9. Weighted Overlay Analysis: Weighted Overlay analysis is used to find suitable solid waste dumpsites. Dumpsite is classified into five main classes: high, very high, low, moderate, and very low. The parameters used for weighted overlay analysis are LULC, Elevation, Slope, Groundwater level, Geology, Euclidean distance - road, Euclidean distance – drainage, and Euclidean distance - built-up land. Most of the solid waste deposit spots are located in the eastern wards of Kozhikode Corporation. Figure 12 represents the suitable sites for Solid Waste Disposal at Kozhikode Corporation. Comparatively, these wards are slightly behind in terms of buildup. Compared to other spots in the corporation, there is a different concentration in these wards. These are the parts that are not adjacent to the coastal area. About seven wards cover areas suitable mainly for dumpsites. Table 10 depicts area of solid waste disposal suitability in the study area. Kovoor is the most suitable ward for dumpsites as the concentration of dumpsites in this ward is higher than in other wards. A very high suitable site area will be around 2.52119 ha. Similarly, the highly suitable area is about 1778.932651 ha. Kovoor, Mayanad, Chelavoor, Moozhikkal,

Vellimadukunnu, Kommeri, and Kuttiyil Thazha have very high suitability for solid waste dumpsites and these wards have barren lands also.

The spatial ecology of suitable solid waste dump sites emphasizes the necessity for place of waste management strategies to coordinate with their natural and urban environment. The eastern wards of Kozhikode Corporation, which can be defined as the most appropriate areas, exhibit distinct geographic and infrastructural features that render them suitable to function as dumping grounds. The use of parameters like minimized groundwater depletion, optimum elevation level, and minimized slope gradient reduces the leaching risk of groundwater contamination and soil erosion. In addition, the Euclidean distances of roads, drainage, and built-up areas guarantee the accessibility of waste transportation but diminish the environmental impact on residential and commercial areas. Such a quantitative analysis guarantees that waste management infrastructure not only stands, runs, or works, but also functions well and that it has benign environmental and social impacts.

The availability of adequate dumpsites within the eastern wards constitutes an opportunity for linking waste management infrastructure with community development activities. Wards like Kovoor and associated regions, due to a high proportion of barren lands, can be planned strategically to provide space for advanced waste processing facilities including recycling facilities and composting plants. This would not only be a way of mitigating the problems of solid waste but also create employment opportunities and a strong economic structural foundation in such areas. In addition, weighted overlay analysis allows a data-driven, explainable process for site selection that can be extrapolated to find more locations as the city's waste management grows. For effective urban development and environmental sustainability, forward-looking schedule planning is important.

Table 10. Area of Solid Waste Disposal Suitability in the Study Area

Classes	Area (ha)
Very high	2.52119
High	1778.932651
Moderate	3739.484325
Low	5465.098486
Very low	699.116517

5.10. Suitable Solid Waste Dumpsite

Figure 12. Suitable sites for Solid Waste Disposal at Kozhikode Corporation

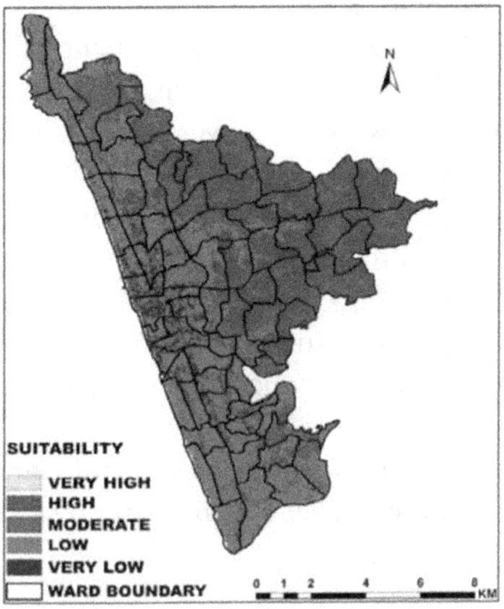

A B

6. Waste Management: In the 21st century, main emphasis is being levied upon environmental protection and worry regarding human health. In this application, SWM needs foremost consideration. Understanding in civilization is profusely necessary for the minimization of solid waste creation (Chatterjee et al. 2016). SWM is defined as the method of collection, separation, transportation, as well as disposal of solid wastes in a manner by which they become less toxic for humans, flora, fauna, and the environment. It simultaneously considers maintaining proper public health as well as aesthetic quality (Olanibi and Emmanuel 2022). Environmental issues have been reported to have a significant role in decision-making as to where to dump these solid wastes as it might lead to damaging the biophysical environment as well as the ecosystem of the region. The optimal sites for solid waste dump locations are spots with mild rolling and flat hills that are not prone to floods (Sener et al. 2011).

One of the most critical issues regarding the environment is waste management (Demirbas 2011). It consists of the activities dealing with the waste, including the collection, transportation, management, recovering, and dumping of waste objects. These can be in the form of solids, liquids, gases, and even radioactive materials, with each requiring its specific techniques and area of specialization. Different categories of waste can be stored separately. Waste management is also involved to retrieve sources from it (Yedla 2005). Liquid waste management involves wastewater handling, sewage treatment, and chemical and biochemical treating. Waste management methods are distinct for established and developing countries, urban and rural regions, as well as domestic and industrial manufacturers. The current practices in waste management in developing nations are not up to par with the needs of handling the enormous daily piles of waste. The goal of waste management is the minimization of the refusal effects of waste on human health, the environment, and society. A waste management idea involves the following objectives (Kan 2009):

(i) Decrease of the total volume of waste by decreasing and recovering refuse.
(ii) Reutilizing and re-introduction of appropriate groups of materials into manufacture cycles as secondary raw constituents.
(iii) Re-introduction of organic garbage into the natural rotation.
(iv) Finest-possible decrease of remaining waste masses, which are to be disposed of in "appropriate" landfills.
(v) Flexible theory regarding variations in waste capacities and the constitution of household waste. New progresses in the field of waste management must be add in in the organization.

The unofficial sector has a key responsibility in separating value from waste, with ~ 90% of remaining waste presently being abandoned rather than appropriately landfilled (Narayana 2009). There is a critical requirement to shift to a more balanced SWM, and this wants new management practices and waste management capabilities. Currently, SWM structures are not very efficient. This has been informed to initiate harmful effects on the physical condition of everyone, nature, as well as the economy (Biswas et al. 2010). The Waste Management and Handling Rules have been executed by the Ministry of Environment and Forests (MoEF) in India (MoEF 2015). The MoEF announced Management and Handling of Municipal Solid Waste Rules 2000 to check that appropriate waste management takes place in India. Recently, it has recently published new rules in its updated draft (MoEF 2015). Municipal authorities have been given the duty of implementing strict rules and simultaneously developing infrastructure that is responsible for collecting, storing, segregating, transporting, processing, as well as disposing of MSW. Rana et al. (2015) stated that the SWM rule was implemented first time in Chandigarh, India.

In India, the percentage of urban people was seen to be 37.7% in the year 2015 while it was 17.29% in the year 1951. This has been associated with rapid urbanization. It has been reported that it is important for developing and propagating an effective system for waste management by the government (Census of India 2011). In the majority municipalities, municipal litterbins have been fitted for collecting garbage; however, they do not get cleaned up repeatedly by the authorities of municipalities, thereby resulting in the dumping of household wastes by people. This, earlier, used to happen along roadsides, path corners, as well as empty lands thereby creating regional illegal and unclean landfills (Gidde et al. 2008).

The status of SWM in India is currently considered poor. It is because the usage of best and most appropriate methods, i.e. from waste collection to disposal is not being conducted. There is insufficiency of SWM training protocols and the availability of qualified professionals for wastes management is less. Current SWM systems lack accountability in India (Khajuria et al. 2010). Municipal authorities have been made responsible for MSW management in country. However, these do not have sufficient budgets that could be used to cover the costs related to developing appropriate wastage collection, storage, treatment, as well as disposal. Unavailability of strategic MSW plans, wastage collection/segregation, as well as government finance regulatory frameworks are considered as some of the major barriers that are responsible for hindering effective SWM in country. Less awareness associated to environmental problems that are combined with lesser motivation has been responsible for inhibiting innovations and simultaneously adopting new technologies which could transform waste management in country. Public attitude towards wastes is also one of the major barriers towards improving SWM in country.

6.1. Challenges in Waste Management: Infrastructure related to waste management has played a key responsibility in delivering sustainable growth. Waste management has been one of the major challenges that are being faced by many developing countries. It is difficult in a city with a greater population density to manage effectively (Kumar et al. 2017). The waste management infrastructure adds substantially to sustainable development. In developing nations, the issue of municipal solid waste management (MSWM) is especially critical because of the tendencies of urbanization, and increased population growth. There are some real barriers to managing the MSW in India that are actually complex in nature (Kumar et al. 2017). These include poor collection and transportation mechanisms, no source segregation, insufficient capacities for processing, and dependence on dirty dumps for disposal. In response, the Indian government has developed different measures to enhance source segregation and promote methods of recycling and composting (Mani and Singh 2016).

However, the problem of how to apply these measures on a commercial scale is still one of the most significant barriers. If these challenges are addressed, it will be possible for India to minimize the effects of environmental diseases as well as environmental degradation due to improper disposal of wastes. Although there has been gradual progress in recent years, further systems solutions are required in the Indian waste recycling market in terms of waste sorting, improvement of waste treatment capacities, and regularization of the irregular sector. The people of India and the government must practice a proper and effective waste management system. Sharing knowledge, experience, and skills can work together to create and implement new thinking to address these issues and improve waste management systems in the world thereby encouraging sustainability. It involves the exchange of information on relevant studies and technologies, joint implementation of the activities on waste minimization and recycling, as well as compliance with intergovernmental conventions and standards on the management and disposal of hazardous waste and its effects on the biosphere.

6.2. Effects of Waste on Humans and Environment: Waste dumps have been reported to generate adverse impacts on the health of the public as well as the environment (Rathi 2006). Open dumps have been reported to release methane gas (CH_4) that is generated from biodegradable waste decomposition that occurs in anaerobic conditions. CH_4 has been seen to cause fires as well as explosions. They are one of the major contributors to global warming (Sridevi et al. 2012). They cause problems like bad odor and leachate migration into water bodies (Srivastava et al. 2014). Smell is considered a major issue, especially throughout summer seasons which the average temperatures of India exceed 45°C (Dasgupta et al. 2013). Tires that are discarded at dumps have been observed to collect water, thereby helping mosquitoes to easily breed. This increases the risk of mosquito-related diseases like malaria, dengue, as well as West Nile fever. Burning of wastes in an uncontrollable manner at dump sites has been reported to release fine particles that lead to major respiratory diseases and smog (Sridevi et al. 2012). A study conducted in Mumbai has reported that open on fire of MSW and tires produces nearly 22,000 tonnes of different contaminants into the surroundings each year (Annepu 2012). The negative effects of inadequate waste management on the physical condition of the public have been well documented. They lead to increased incidences of infections in the throat and nose, difficulties in respiring, asthma, inflammation, microbial contaminations, anemia, decreased immunity, allergies, as well as other infections (CPCB 2000).

6.3. Waste Management and Circular Economy: The circular economy is considered a sustainable financial paradigm that seeks to reduce trash and pollution degrees, keep yields as well as materials in use, as well as rebuild natural systems. Unlike the traditional linear economy that circulates around "take, make, dispose of", the circular economy revolves around the concept of product lifecycles by means of its repair, reuse, as well as recycling. Incorporating principles of circular economy within environmentally friendly management is considered a transition towards an economy that emphasizes minimal carbon and less pollution. While the idea of circular economy is comparatively new, the concept of circular economy is closely linked with several other financial sustainability methods such as industrial ecosystem and industrial relationship, essentially targeting the circularization of linear value chains (Morseletto 2020). Successful implementations of circular economy can lead to superior sustainable development than many existent results.

6.3.1. Recycling and Material Recovery: Recycling operates as a main recycling technique in the circular economy. Waste management systems sort waste into raw materials which can then be reintegrated into the production cycle. Examples here are metals, plastics, and paper.

6.3.2. Composting and Biological Cycles: After composting of organic garbage like food or agricultural remains, the nutrients are restored to the soil and thus confirm recreating agriculture.

6.3.3. Extended Producer Responsibility (EPR): The Extended Producer Responsibility (EPR) strategies are intended to make producers responsible for the whole process of life-cycle administration, provoking designers to introduce revisions that would support recycling and reuse.

6.3.4. Waste-to-Energy: The process of combustion (incineration) or anaerobic digestion can convert non-recyclable waste to electricity, thereby reducing our dependence on landfills.

6.3.5. Reduction Strategies: Effective waste management includes reducing the generation of waste through education, policies, and innovation.

7. Significance of the Study: In this era where solid waste is a major issue, its proper disposal and collection are important. Kozhikode Corporation is one of the leading urban corporations in developing sectors of Kerala and hence does not lag behind in terms of population. There is excessive waste in the area as population and development go simultaneously. Various schemes are being implemented by the corporation to manage this issue in a proper manner. In this case, it is imperative to find a suitable dumpsite for solid waste disposal. Therefore, this study may help to find suitable dumpsites through geo-informatics based on satellite images and other digital data. This study is conducted using various important parameters so that the suitability of the solid waste dumping spot can be understood.

Because of the prompt growth in the populace of Kozhikode city and the urban infrastructure of Kozhikode, solid waste is generated in a great proportion, putting a challenge in the effective management and disposal of waste material. As urbanization increases, pressure on current waste management infrastructure has been elevated, calling for new solutions. Geo-informatics, which uses satellite imagery and GIS technologies, constitutes a useful method to locate and assess potential dumpsite areas of interest according to the risk factors in environmental, social, and economic aspects. This technology allows accurate assessment of land suitability suitable for decision-making on where to best locate waste disposal facilities by authorities.

Besides environmental aspects, this work also addresses aspects regarding the reach, legal restrictions, and acceptance in society of potential waste disposal facilities. It is considered regarding the effects of waste disposal on neighboring communities and the environment, and site selection is employed so that the chosen locations are sufficiently distanced from residential areas in order to minimize health and environmental adverse effects. Using the data of GIS and remote sensing, the study tries to establish a systematic framework to assist the Kozhikode Corporation in formulating an effective and sustainable SWM strategy for the city development which contributes to both the overall development of the city and public health and environmental problems.

8. RECOMMENDATIONS

The local waste management systems should be installed to guarantee local waste is reused within the region to decrease the measure of waste and create advantages for the residents.

Wet waste can be reused for biogas manufacture and the sludge treated by vermiculture plants can produce local organic manure.

E-waste can require monetary worth to residents upon deposit.

Authority and nationals should jointly work together to significantly implement and apply the 3R (Reduce, Reuse, and Recycle) philosophy.

The people should be educated about simple techniques to transform waste into beneficial services.

9. Future Prospects: The understanding achieved by literature reviews on SWM has helped to streamline some of the important aspects that would be beneficial for drafting future pathways toward managing different types of solid waste and simultaneously achieving sustainable development goals (SDGs). Some of these are:

- **Incorporating Recyclable Materials into Daily Life**

One method for addressing the solid waste problem is by utilizing recycled materials in our daily routines. Encouraging the usage of materials that are reusable, and recyclable would help in reducing waste. Patronizing recycling behavior not only saves natural resources but also reduces the reliance on garbage landfills. Incorporation of sustainable practice into routine activity will help to create a responsible environmental society.

- **Proper Administration by Municipal Authorities**

Efficient SWM needs strong and active administration by municipal authorities. There shall be strict regulation on waste segregation, collection, and processing, and such compliance must be enforced at all levels. Implementation of a clear surveillance system and sufficient resources are crucial to keeping the waste management process running smoothly. Municipal corporations are also required to work in partnership with private and community stakeholders in order to improve efficiency and develop waste management infrastructure. Raising awareness of solid waste issues has great significance in the treatment of solid waste problems. Knowing about the negative consequences of the wrong waste management and understanding the need for green practices can give the citizen the power to decide. Focusing on the economic and environmental advantages of waste management can generate a shared sense of ownership, thereby promoting an engaged approach to sustainable practices.

- **Incorporating Dumpsites into Town Planning**

Integration of dedicated landfills in urban design needs to be taken seriously in order to resolve the challenge of spatial constraints in waste management. Each ward in a municipal corporation must have a properly designed dumpsite for providing waste disposal facilities in a regulated manner and without environmental pollution risks or negative impact on the health of the public. Integration of waste management infrastructure in the town planning process can ensure that logistical efficiency is maximized, illegal dumping avoided, and waste collection processes optimized. Such strategic thinking promotes environmentally sustainable urban development with a low environmental footprint.

- **Regular Waste Collection and Processing**

The reliability of waste collection and processing is very important for hygiene and waste clogging. Municipal corporations are obliged to collect and process waste on a regular basis to prevent the outbreak of unsanitary conditions that are susceptible to infestation and spill-over infections. It can be effectively reduced through scheduling efficient collection intervals and capitalizing on innovative processing capabilities. In addition, routine waste management activities are also a beneficial factor to improve the overall efficacy of solid waste systems and guarantee a cleaner, healthier urban space.

- **Installation of Local Waste Management Systems**

Implementing point-of-waste-management-based systems plays a key role in improving the performance as well as the sustainability of waste recycling activities. This is achieved by ensuring that generated waste gets processed and then recycled within the desired territory. As a result, the requirement for transportation of waste gets minimized, and associated costs as well as carbon emissions are reduced. These systems can also make jobs and provide economic and social benefits to the local community.

- **Recycling Wet Waste for Biogas and Organic Fertilizer**

Wet waste is a great resource for sustainable recycling, especially for biogas production and organic fertilizer production. Biogas plants can be used to transform organic waste into a renewable energy resource, offering a cleaner energy solution for small consumption. Besides, the by-product sludge can be treated to perform vermiculture-based high-quality organic fertilizer production. These fertilizers have the potential to mediate local agricultural life, decreasing outside use of chemical fertilizers, and fostering environmentally benign agriculture. Through wet waste recycling, communities are well-equipped to complete the waste-energy-nutrient loop effectively.

- **Leveraging Dry and E-Waste for Monetary Benefits**

Dry and e-waste have the potential to be valuable materials when properly managed. With the introduction of systems providing monetary rewards for dry and e-waste deposits, people are motivated to become involved in recycling programmes in an active way. This method not only reduces pollution of the environment but also causes a return to the citizens, increasing the motivation to have a habit of sustainable disposal of waste.

- **Cooperative Efforts for Adopting the 3R Philosophy**

Efficient implementation of the 3R (i.e. Reduce, Reuse, Recycle) concept has been observed to depend on the joint collaborations and efforts of the government as well as citizens. Policymakers are required to implement effective legislation and/or financial incentives that would help encourage the reduction of waste and economic conservation. Community-led activities, including volunteer-led waste exchange programmes, repair and reuse workshops, and local material recovery centers, can facilitate the mainstreaming of the 3R concepts in everyday life.

- **Educating Citizens on Waste Conversion Techniques**

Workshops and training courses can be used to instruct citizens on how to make compost from kitchen scraps, build biogas systems, or recycle waste items into useful objects. These educational programs not only help to decrease the amount of waste going ultimately into the landfills but also allow people to use their waste in an everyday and economical way. Raising awareness promotes an innovative and autonomous culture in waste management.

- **Using scientific software tools**

By locating the dump site using accurate parameters, the unsanitary and other problems faced by local residents and public spaces can be solved. Given the parameters, the dump site will be located in a manner that is environmentally friendly. Spatial data can be combined and analyzed using GIS software. Thus, remote sensing can bring positive changes in waste management. Applying this analysis results in a well-defined rise for a developing region. Finding a suitable dump site may help to improve surface water contamination issues, groundwater contamination, atmosphere, soil, and human health in this region.

10. Conclusions: India encounters major conservational challenges linked with waste creation and inadequate waste gathering, transportation, management, and removal. A particular problem that significantly impacts environmental sustainability is solid waste. Thus, improper processing of the material could have an impact on future generations. Because solid waste is largely created in the area of urban development, a suitable disposal facility is crucial in an area that has the size of a municipal corporation. In Kozhikode Corporation's case, there is extremely little space for disposing of solid trash. It receives a suitable space of only seven wards in the corporation's eastern region. If solid wastes do not get systematically deposited, a number of hazards have been observed to occur.

This study highlights that addressing challenges associated with SWM in Kozhikode Corporation needs multi-pronged approaches that combine novel technologies, participation of communities as well as strong policy interventions. Awareness campaigns, training sessions, or workshops as well as community outreach could complement the experience, and knowledge gaps, as well as inspire communities to take action.

Sustainable waste disposal technologies will ideally focus on maximizing landfill use and on supporting the development of high-tech waste processing, including composting, recycling, and waste treatment to energy. Perhaps the most effective way to decrease the load on disposal facilities is to engage local communities through information campaigns and encourage waste sorting at the source. In addition, switching to a decentralized waste management architecture, where processing is decentralized (i.e., smaller) and more local (i.e., localized), will help overcome the geographical limitations in populous regions. Combining all these parameters, Kozhikode Corporation can take the first step towards an effective, eco-friendly waste management system, which will ensure environmental sustainability and the health of people for generations ahead.

11. ACKNOWLEDGMENT

The authors are deeply grateful to Amity University for providing us the platform to do this study.

12. REFERENCES

Alam, S., Bhardwaj, L. K., Mallick, R., & Rai, S. (2023). Estimation of heavy metals and fluoride ion in vegetables grown nearby the stretch of River Yamuna, Delhi (NCR), India. *Indian Journal of Environmental Protection*, *43*(1), 64–73.

Amini Parsa, V., Yavari, A., & Nejadi, A. (2016). Spatio-temporal analysis of land use/land cover pattern changes in Arasbaran Biosphere Reserve: Iran. *Modeling Earth Systems and Environment*, *2*(4), 1–13. DOI: 10.1007/s40808-016-0227-2

Annepu, R. K. (2012). Sustainable solid waste management in India. *Columbia University, New York, 2*(01), 1-89.

Arowolo, A. O., Deng, X., Olatunji, O. A., & Obayelu, A. E. (2018). Assessing changes in the value of ecosystem services in response to land-use/land-cover dynamics in Nigeria. *The Science of the Total Environment*, *636*, 597–609. DOI: 10.1016/j.scitotenv.2018.04.277 PMID: 29723833

Artiola, J. F. (2019). Industrial waste and municipal solid waste treatment and disposal. In *Environmental and pollution science* (pp. 377–391). Academic Press. DOI: 10.1016/B978-0-12-814719-1.00021-5

Bhalla, B., Saini, M. S., & Jha, M. K. (2013). Effect of age and seasonal variations on leachate characteristics of municipal solid waste landfill. *International Journal of Research in Engineering and Technology*, *2*(8), 223–232. DOI: 10.15623/ijret.2013.0208037

Bhardwaj, L. K., Rath, P., Jain, H., & Choudhury, M. (2025). Exploring the effects of e-waste on soil, water quality and human health. *Discover Civil Engineering*, *2*(1), 12. Advance online publication. DOI: 10.1007/s44290-025-00167-2

Bhardwaj, L. K., Rath, P., & Tokas, R. (2024). E-Waste Management in Developing Countries: Current Practices, Challenges, Disposal, and Impact on Human Health and Environment. *Sustainable Solutions for E-Waste and Development*, 90-102.

Bhardwaj, L. K., Sharma, S., & Jindal, T. (2023). Estimation of physico-chemical and heavy metals in the lakes of Grovnes & Broknes Peninsula, Larsemann Hill, East Antarctica. *Chemistry Africa*, *6*(5), 2677–2694. DOI: 10.1007/s42250-023-00668-6

Biswas, A. K., Kumar, S., Babu, S. S., Bhattacharyya, J. K., & Chakrabarti, T. (2010). Studies on environmental quality in and around municipal solid waste dumpsite. *Resources, Conservation and Recycling*, *55*(2), 129–134. DOI: 10.1016/j.resconrec.2010.08.003

Bulucea, C. A., Mastorakis, N. E., Bulucea, C. A., Boteanu, N. I. C. U. L. A. E., & Stinga, A. U. R. E. L. (2010). Systemic approach of hazardous and non-hazardous waste management. In *4th WSEAS International Conference on Energy Planning, Energy Saving, Environmental Education, EPESE* (Vol. 10, pp. 181-189).

Census of India. (2011). *Rural and Urban Distribution of population*. Government of India.

Chatterjee, D., Kuotsu, R., James Kikon, Z., Sarkar, D., Ao, M., Ray, S. K., Bera, T., & Deka, B. C. (2016). Characterization of vermicomposts prepared from agricultural solid wastes in north eastern hill region of Nagaland, India. *Proceedings of the National Academy of Sciences. India. Section B, Biological Sciences*, *86*(4), 823–833. DOI: 10.1007/s40011-015-0538-5

CPCB (Central pollution Control Board). 2000 Management of municipal solid waste Delhi. See http://www.cpcb.nic.in/divisionsofheadoffice/pcp/ MSW_Report.pdf (accessed 27 June 2015).

CPCB (Central pollution Control Board). (2011). *Status report on Municipal Solid waste management*. Central Pollution Control Board.

Dai, Y., Sun, Q., Wang, W., Lu, L., Liu, M., Li, J., Yang, S., Sun, Y., Zhang, K., Xu, J., Zheng, W., Hu, Z., Yang, Y., Gao, Y., Chen, Y., Zhang, X., Gao, F., & Zhang, Y. (2018). Utilizations of agricultural waste as adsorbent for the removal of contaminants: A review. *Chemosphere*, *211*, 235–253. DOI: 10.1016/j.chemosphere.2018.06.179 PMID: 30077103

Dasgupta, B., Yadav, V. L., & Mondal, M. K. (2013). Seasonal characterization and present status of municipal solid waste (MSW) management in Varanasi, India. *Advances in Environmental Research*, *2*(1), 51–60. DOI: 10.12989/aer.2013.2.1.051

Demirbas, A. (2011). Waste management, waste resource facilities and waste conversion processes. *Energy Conversion and Management*, *52*(2), 1280–1287. DOI: 10.1016/j.enconman.2010.09.025

Deng, D., Zhang, L., Dong, M., Samuel, R. E., Ofori-Boadu, A., & Lamssali, M. (2020). Radioactive waste: A review. *Water Environment Research*, *92*(10), 1818–1825. DOI: 10.1002/wer.1442 PMID: 32860717

Elbeshbishy, E., & Okoye, F. (2019). Improper disposal of household hazardous waste: Landfill/municipal wastewater treatment plant. *Municipal Solid Waste Management, 183*.

Gidde, M. R., Todkar, V. V., & Kokate, K. K. (2008, March). Municipal solid waste management in emerging mega cities: a case study of Pune city. In *Proceedings of Indo Italian Conference on Green and Clean Environment,* Pune, India *(*March 20-21*)*.

Gupta, P., Sharma, A., & Bhardwaj, L. K. (2024). Solid waste management (SWM) and its effect on environment & human health. Futuristic Trends in Agriculture Engineering & Food Sciences Volume 3; 91-101. IIP Proceedings. https://www.doi.org/10.58532/V3BCAG4P1CH7

Hazarika, R., & Saikia, A. (2020). Landfill site suitability analysis using AHP for solid waste management in the Guwahati Metropolitan Area, India. *Arabian Journal of Geosciences*, *13*(21), 1–14. DOI: 10.1007/s12517-020-06156-2

Heredia-Guerrero, J. A., Heredia, A., Domínguez, E., Cingolani, R., Bayer, I. S., Athanassiou, A., & Benítez, J. J. (2017). Cutin from agro-waste as a raw material for the production of bioplastics. *Journal of Experimental Botany*, *68*(19), 5401–5410. DOI: 10.1093/jxb/erx272 PMID: 28992177

Ishchenko, V., Pohrebennyk, V., Borowik, B., Falat, P., & Shaikhanova, A. (2018). Toxic substances in hazardous household waste.

Jaybhaye, R., Mundhe, N., & Dorik, B. (2014). Site suitability for urban solid waste disposal using geoinformatics: A case study of Pune Municipal Corporation, Maharashtra, India. *Int J Adv Remote Sens GIS*, *3*(1), 769–783.

Jiyuan, L., Mingliang, L., Xiangzheng, D., Dafang, Z., Zengxiang, Z., & Di, L. (2002). The land use and land cover change database and its relative studies in China. *Journal of Geographical Sciences*, *12*(3), 275–282. DOI: 10.1007/BF02837545

Kan, A. (2009). General characteristics of waste management: A review. *Energy Education Science and Technology, Part A. Energy Science and Research*, •••, 23.

Kaoje, I. U., Dankani, I. M., & Ishiaku, I. (2016). Site suitability analysis for municipal solid waste disposal in Birnin Kebbi, Nigeria. *IOSR Journal of Humanities and Social Science (IOSR-JHSS), 21*(7), 01-10.

Kaushal, R. K., Varghese, G. K., & Chabukdhara, M. (2012). Municipal solid waste management in India-current state and future challenges: A review. *International Journal of Engineering Science and Technology*, *4*(4), 1473–1489.

Khajuria, A., Matsui, T., Machimura, T., & Morioka, T. (2010). Assessment of the challenge of sustainable recycling of municipal solid waste management in India. *International Journal of Environmental Technology and Management, 13*(2), 171–187. DOI: 10.1504/IJETM.2010.034300

Khan, M. S., Hossain, S., & Jahan, I. (2015). *RS and GIS based site suitability analysis for solid waste disposal: A multi criteria analysis of Khulna City area. Proc. of the WasteSafe*. KUET.

Khunprasert, P., Grisdanurak, N., Thaveesri, J., Danutra, V., & Puttitavorn, W. (2008). Radiographic film waste management in Thailand and cleaner technology for silver leaching. *Journal of Cleaner Production, 16*(1), 28–36. DOI: 10.1016/j.jclepro.2006.06.010

Krishna, R. S., Mishra, J., Meher, S., Das, S. K., Mustakim, S. M., & Singh, S. K. (2020). Industrial solid waste management through sustainable green technology: Case study insights from steel and mining industry in Keonjhar, India. *Materials Today: Proceedings, 33*, 5243–5249. DOI: 10.1016/j.matpr.2020.02.949

Kumar, S., Bhattacharyya, J. K., Vaidya, A. N., Chakrabarti, T., Devotta, S., & Akolkar, A. B. (2009). Assessment of the status of municipal solid waste management in metro cities, state capitals, class I cities, and class II towns in India: An insight. *Waste Management (New York, N.Y.), 29*(2), 883–895. DOI: 10.1016/j.wasman.2008.04.011 PMID: 18595684

Kumar, S., Smith, S. R., Fowler, G., Velis, C., Kumar, S. J., Arya, S., Rena, , Kumar, R., & Cheeseman, C. (2017). Challenges and opportunities associated with waste management in India. *Royal Society Open Science, 4*(3), 160764. DOI: 10.1098/rsos.160764 PMID: 28405362

Liu, H. C., You, J. X., Chen, Y. Z., & Fan, X. J. (2014). Site selection in municipal solid waste management with extended VIKOR method under fuzzy environment. *Environmental Earth Sciences, 72*(10), 4179–4189. DOI: 10.1007/s12665-014-3314-6

Madurwar, M. V., Ralegaonkar, R. V., & Mandavgane, S. A. (2013). Application of agro-waste for sustainable construction materials: A review. *Construction & Building Materials, 38*, 872–878. DOI: 10.1016/j.conbuildmat.2012.09.011

Mani, S., & Singh, S. (2016). Sustainable municipal solid waste management in India: A policy agenda. *Procedia Environmental Sciences, 35*, 150–157. DOI: 10.1016/j.proenv.2016.07.064

Mathur, V. (2012). SCOPE OF RECYCLING MUNICIPAL SOLID WASTE IN DELHI AND NATIONAL CAPITAL REGION (NCR). *Integral Review. Journal of Management, 5*(2).

Ministry of Environment and Forests (MoEF). 2015 The Gazette of India. Municipal solid waste (Management and Handling) rules, New Delhi, India.

Modak, P., Jiemian, Y., Hongyuan, Y., & Mohanty, C. R. (2010). Municipal solid waste management: turning waste into resources. *Shanghai manual: a guide for sustainable urban development in the 21st century*, 1-36.

Mohammed, A., & Zubair, Z. L. (2022). Land Suitability Analysis for Solid Waste Landfills in Kano Metropolis Using Multi-Criteria Evaluation Techniques. *AFRIGIST JOURNAL OF LAND ADMINISTRATION AND ENVIRONMENTAL MANAGEMENT, 2*(1), 52–68. DOI: 10.54222/afrigist/ajlaem/v2i1.5

Mohammed, H. I., Majid, Z., & Yamusa, Y. B. (2019, February). GIS based sanitary landfill suitability analysis for sustainable solid waste disposal. [). IOP Publishing.]. *IOP Conference Series. Earth and Environmental Science, 220*, 012056. DOI: 10.1088/1755-1315/220/1/012056

Morseletto, P. (2020). Targets for a circular economy. *Resources, Conservation and Recycling, 153*, 104553. DOI: 10.1016/j.resconrec.2019.104553

Narayana, T. (2009). Municipal solid waste management in India: From waste disposal to recovery of resources? *Waste Management (New York, N.Y.), 29*(3), 1163–1166. DOI: 10.1016/j.wasman.2008.06.038 PMID: 18829290

Natesan, U., & Sarkar, C. (2008). GIS-based approach for optimized siting of municipal solid waste landfill. *Waste Management, 28*(11).

Olanibi, O. A., & Emmanuel, A. A. (2022). Site Suitability Analysis of Solid Waste Disposal in Ilesa, Nigeria. *European Journal of Development Studies, 2*(2), 49–59. DOI: 10.24018/ejdevelop.2022.2.2.60

Öncel, M. S., Bektaş, N., Bayar, S., Engin, G., Çalışkan, Y., Salar, L., & Yetiş, Ü. (2017). Hazardous wastes and waste generation factors for plastic products manufacturing industries in Turkey. *Sustainable Environment Research, 27*(4), 188–194. DOI: 10.1016/j.serj.2017.03.006

Planning Commission, Government of India. 2014 Report of the task Force on waste to energy (Volume I) in the context of integrated municipal solid waste management. See http:// planningcommission.nic.in/reports/genrep/rep_ wte1205.pdf (accessed 1 July 2015).

Poornima, P., & Nithya, R. (2022). Site suitability analysis for decentralized sustainable solid waste management system for Coimbatore city using GIS. *Sustainability, Agri, Food and Environmental Research, 10*.

Poškas, P., Kilda, R., Šimonis, A., Jouhara, H., & Poškas, R. (2019). Disposal of very low-level radioactive waste: Lithuanian case on the approach and long-term safety aspects. *The Science of the Total Environment*, *667*, 464–474. DOI: 10.1016/j.scitotenv.2019.02.373 PMID: 30833245

Rajput, R., Prasad, G. A. K. C., & Chopra, A. K. (2009). Scenario of solid waste management in present Indian context.

Rana, R., Ganguly, R., & Gupta, A. K. (2015). An assessment of solid waste management system in Chandigarh City, India. *The Electronic Journal of Geotechnical Engineering*, *20*(6), 1547–1572.

Rao, T. S., Panigrahi, S., & Velraj, P. (2022). Transport and disposal of radioactive wastes in nuclear industry. In *Microbial biodegradation and bioremediation* (pp. 419–440). Elsevier.

Rathi, S. (2006). Alternative approaches for better municipal solid waste management in Mumbai, India. *Waste Management (New York, N.Y.)*, *26*(10), 1192–1200. DOI: 10.1016/j.wasman.2005.09.006 PMID: 16288861

Ravindran, R., Hassan, S. S., Williams, G. A., & Jaiswal, A. K. (2018). A review on bioconversion of agro-industrial wastes to industrially important enzymes. *Bioengineering (Basel, Switzerland)*, *5*(4), 93. DOI: 10.3390/bioengineering5040093 PMID: 30373279

Saaty, R. W. (1987). The analytic hierarchy process—What it is and how it is used. *Mathematical Modelling*, *9*(3-5), 161–176. DOI: 10.1016/0270-0255(87)90473-8

Saketa, Y., Tamene, N., & Melknew, M. (2023). Municipal solid waste disposal site suitability analysis using multi-criteria evaluation in Assosa, Ethiopia. *International Journal of Environmental Science and Technology*, *20*(4), 3815–3830. DOI: 10.1007/s13762-022-04221-9

Şener, Ş., Sener, E., & Karagüzel, R. (2011). Solid waste disposal site selection with GIS and AHP methodology: A case study in Senirkent–Uluborlu (Isparta) Basin, Turkey. *Environmental Monitoring and Assessment*, *173*(1-4), 533–554. DOI: 10.1007/s10661-010-1403-x PMID: 20213053

Sharholy, M., Ahmad, K., Mahmood, G., & Trivedi, R. C. (2008). Municipal solid waste management in Indian cities–A review. *Waste Management (New York, N.Y.)*, *28*(2), 459–467. DOI: 10.1016/j.wasman.2007.02.008 PMID: 17433664

Sharma, P., Gaur, V. K., Kim, S. H., & Pandey, A. (2020). Microbial strategies for bio-transforming food waste into resources. *Bioresource Technology*, *299*, 122580. DOI: 10.1016/j.biortech.2019.122580 PMID: 31877479

Sridevi, V., Modi, M., Lakshmi, M. V. V. C., & Kesavarao, L. (2012). A review on integrated solid waste management. *Int. J. Eng. Sci. Adv. Technol*, *2*, 1491–1499.

Srivastava, R., Krishna, V., & Sonkar, I. (2014). Characterization and management of municipal solid waste: A case study of Varanasi city, India. *International Journal of Current Research and Academic Review*, *2*(8), 10–16.

Taurino, R., Bondioli, F., & Messori, M. (2023). Use of different kinds of waste in the construction of new polymer composites. *Materials Today Sustainability*, *21*, 100298. DOI: 10.1016/j.mtsust.2022.100298

World Bank Group (Ed.). World Development Indicators 2012. World Bank Publications; 2012 Apr 18.

Yedla, S. (2005). Modified landfill design for sustainable waste management. *International Journal of Global Energy Issues*, *23*(1), 93–105. DOI: 10.1504/IJGEI.2005.006412

Zhang, X., Gu, P., & Liu, Y. (2019). Decontamination of radioactive wastewater: State of the art and challenges forward. *Chemosphere*, *215*, 543–553. DOI: 10.1016/j.chemosphere.2018.10.029 PMID: 30342399

Chapter 12
Water Treatment and Sustainable Chemistry: A Review

Dattatraya N. Pansare
https://orcid.org/0000-0002-0419-3538
Deogiri College, India

Rohini Shelke
Radhabai Kale Mahila Mahavidyalaya, India

Devidas Bhagat
Government Institute of Forensic Science, Chhatrapti Sambhajinagar, India

Pravin Chavan
Doshi Vakil Arts College, India & GCUB Science and Commerce College, India

Dhanraj Kamble
S.B.E.S. College of Science, India

ABSTRACT

Water treatment is a cornerstone of public health and environmental protection, yet the challenges of increasing global water demand, pollution, and climate change necessitate innovative and sustainable solutions. This review explores advancements in sustainable chemistry for water treatment, emphasizing green methodologies and resource-efficient technologies. Key approaches include the development of biodegradable adsorbents, nanomaterials, and photocatalysts for contaminant removal. These innovations reduce reliance on harmful chemicals and improve the efficiency of processes such as adsorption, membrane filtration, and advanced oxidation. Emerging strategies like bio-inspired and waste-to-resource technologies highlight

DOI: 10.4018/979-8-3373-1409-9.ch012

the potential for circular economy integration in water treatment.

1. INTRODUCTION

Importance of water treatment in global sustainability

Water treatment plays a crucial role in advancing global sustainability by ensuring access to clean and safe water, promoting public health, supporting ecosystems, and enabling economic development. As the global population grows and environmental challenges intensify, effective water treatment systems become essential to achieving the United Nations Sustainable Development Goals (SDGs), particularly SDG 6, which focuses on clean water and sanitation for all. Access to clean water is fundamental to human health and well-being. Contaminated water is a leading cause of waterborne diseases, such as cholera and dysentery, which disproportionately affect vulnerable populations, especially in developing countries. According to the World Health Organization (WHO), over 2 billion people globally lack access to safe drinking water, contributing to significant health and economic burdens. Advanced water treatment technologies, such as filtration, disinfection, and desalination, remove harmful pathogens, pollutants, and contaminants, reducing the prevalence of disease and improving quality of life. By prioritizing water treatment, communities can safeguard public health and enhance resilience to health crises (United Nations, n.d.). Water treatment is also vital for preserving ecosystems and maintaining biodiversity. Pollutants from industrial discharges, agricultural runoff, and untreated sewage can degrade aquatic habitats, harm wildlife, and disrupt ecological balance. Effective treatment processes, including biological and chemical treatments, help mitigate these impacts by removing pollutants before they reach natural water bodies. Additionally, the reuse of treated wastewater can alleviate pressure on freshwater resources, promoting water conservation and supporting ecosystems during periods of scarcity (World Health Organization (WHO, 2019). Economic development is closely tied to water quality and availability. Industries such as agriculture, energy, and manufacturing rely heavily on water as a critical resource. Contaminated or inadequate water supplies can hinder industrial operations, reduce agricultural productivity, and limit energy generation. Water treatment enables the sustainable use of water resources by ensuring that water meets quality standards required for various applications. For example, in agriculture, treated wastewater can be used for irrigation, reducing the reliance on freshwater and enhancing food security. In urban settings, efficient water treatment systems support infrastructure development and attract investments, driving economic growth. Moreover, water treatment contributes to climate change mitigation and adaptation. As climate change exacerbates water scarcity and pollution, innovative treatment technologies can enhance water resource management. For instance, energy-efficient desalination methods and membrane filtration systems offer sustainable solutions to address water shortages. Additionally, treating and reusing water reduces the need for extracting and transporting freshwater, lowering greenhouse gas emissions associated with these activities (UNESCO, 2020). Despite its critical importance, achieving universal access to effective water treatment remains a challenge. Financial constraints, technological gaps, and inadequate governance hinder progress, particularly in low-income regions. To overcome these barriers, investments in infrastructure, research, and capacity building are essential. Collaborative efforts involving governments, private sector stakeholders, and international organizations can accelerate the deployment of advanced water treatment technologies and ensure equitable access to clean water (Gleick, 2014).

Overview of sustainable chemistry principles.

Sustainable chemistry, also known as green chemistry, seeks to design and implement chemical processes and products that reduce or eliminate environmental harm while enhancing efficiency. Rooted in the 12 principles developed by Paul Anastas and John Warner in 1998, this field promotes innovations that minimize waste, conserve resources, and prioritize safety for humans and ecosystems. The principles encourage the use of renewable feedstocks, the design of biodegradable products, and energy-efficient synthesis methods. For example, atom economy focuses on maximizing the incorporation of raw materials into the final product, reducing byproducts. Similarly, the use of safer solvents and reaction conditions decreases the reliance on hazardous chemicals. Catalysis is another key aspect, as it enhances reaction efficiency and minimizes energy consumption. Advancements in sustainable chemistry have enabled the production of bio-based plastics, green solvents like supercritical carbon dioxide, and renewable energy storage materials. Despite progress, challenges remain in balancing economic feasibility with sustainability goals. By fostering interdisciplinary collaboration, sustainable chemistry is integral to achieving the United Nations' Sustainable Development Goals (SDGs), particularly those addressing clean water, affordable energy, and climate action. As industries and academia align with these principles, they pave the way for a more sustainable future (Anastas & Warner, 1998).

Objectives of the review.

Assessing Current Water Treatment Techniques
Exploring Sustainable Chemistry Principles
Integrating Interdisciplinary Approaches

Promoting Circular Economy Models
Evaluating Environmental and Socioeconomic Impacts
Identifying Knowledge Gaps and Future Directions

2. ADVANCEMENTS IN WATER TREATMENT TECHNOLOGIES

Water treatment technologies have advanced significantly in recent years, addressing growing concerns over water scarcity and pollution. One notable innovation is **membrane filtration** systems, such as reverse osmosis (RO) and nanofiltration (NF), which offer high efficiency in removing contaminants, including heavy metals, pathogens, and dissolved solids. These technologies are increasingly used in desalination plants, providing potable water from seawater or brackish sources (Elimelech & Phillip, 2011). Another breakthrough is the adoption of **advanced oxidation processes (AOPs)**. By using reactive oxygen species, AOPs effectively degrade organic pollutants and pharmaceutical residues, making them particularly useful in treating industrial wastewater (Saharan et al., 2014). Biotechnology has also made strides with **biofilm reactors** and microbial fuel cells (MFCs). These systems harness microbial activity to degrade organic matter while generating bioenergy as a byproduct (Logan et al., 2006). Additionally, **smart sensors** integrated with artificial intelligence (AI) are revolutionizing real-time water quality monitoring, enabling quicker responses to contamination events (Krishnan et al., 2022). These advancements not only enhance efficiency and sustainability but also ensure the availability of clean water globally. Continued research and investment are crucial to scale these innovations and meet the growing demand for safe water.

3. SUSTAINABLE CHEMISTRY IN WATER TREATMENT

- Green chemistry principles applied to water treatment.

Green chemistry principles aim to minimize environmental impact and promote sustainability by designing processes and products that reduce the use of hazardous substances, energy, and waste. In water treatment, these principles guide the development of innovative methods to purify and manage water while minimizing environmental and health risks. Here's how green chemistry principles are applied to water treatment (Anastas & Warner, 1998; Shannon et al., 2008; Savage & Diallo, 2005).

Prevention of Waste

Traditional water treatment often generates sludge and secondary pollutants that require disposal. Green chemistry emphasizes processes that minimize waste. For instance, advanced oxidation processes (AOPs) like photocatalysis degrade pollutants into benign substances (e.g., CO_2 and H_2O) without generating secondary waste (Mauritz & Moore, 2004; Andrady, 2011).

Example: Using zero-valent iron nanoparticles to decompose organic pollutants directly reduces sludge formation.

Designing Safer Chemicals

Green chemistry promotes the use of safer chemicals in treatment processes. For example, the use of non-toxic coagulants like biopolymers (e.g., chitosan) instead of traditional alum-based coagulants reduces risks to aquatic ecosystems and human health.

Example: Chitosan is biodegradable and effectively removes suspended solids and heavy metals from water.

Energy Efficiency

Water treatment can be energy-intensive, particularly in processes like desalination. Green chemistry encourages energy-efficient alternatives, such as using solar-driven desalination or low-energy reverse osmosis membranes.

Example: Photocatalytic water treatment uses sunlight and photocatalysts like titanium dioxide (TiO_2) to degrade pollutants, significantly reducing energy demands.

Use of Renewable Feedstocks

Renewable materials are increasingly used in water treatment. For example, biochar, derived from agricultural waste, is utilized as an adsorbent for heavy metals and organic pollutants.

Example: Agricultural by-products like coconut shells are converted into activated carbon, offering a sustainable alternative for adsorption-based treatments.

Catalysis

Catalysts enhance the efficiency of chemical reactions without being consumed, aligning with the principle of reducing excess reagent use. In water treatment, catalysts like TiO_2 in photocatalytic processes are highly effective for degrading persistent organic pollutants (POPs).

Example: Nano-catalysts in advanced oxidation processes (e.g., Fenton-like reactions) efficiently break down contaminants under mild conditions.

Design for Degradation

Treatment systems are designed to break down pollutants into non-toxic byproducts. For instance, AOPs convert organic pollutants into CO_2 and H_2O, ensuring that no harmful residues remain in treated water.

Example: Ozonation and hydrogen peroxide treatments degrade pharmaceuticals and pesticides into non-toxic components.

Reducing Derivatives

Avoiding unnecessary derivatization steps reduces the number of chemicals used. Processes like direct electrochemical oxidation eliminate the need for added reagents, simplifying treatment systems.

Example: Electrochemical methods directly treat water without the need for intermediate compounds or reagents.

Safer Solvents and Auxiliaries

Solvent use is minimized or replaced with environmentally benign options like water or supercritical CO_2 in water treatment processes.

Example: Supercritical CO_2 is used for extracting oil and grease contaminants from wastewater, replacing organic solvents.

Real-Time Pollution Monitoring

Incorporating real-time monitoring systems prevents overuse of chemicals and optimizes treatment efficiency.

Example: Sensors for monitoring pH, turbidity, and organic contaminants allow dynamic adjustment of treatment protocols.

Holistic System Design

Integrating green chemistry with circular economy principles ensures that water treatment processes recycle and reuse resources. For instance, treated wastewater can be repurposed for agriculture or industrial processes.

Example: Membrane bioreactors combine biological treatment with filtration, producing high-quality reusable water.

- Development of eco-friendly materials and processes.

The increasing awareness of environmental degradation, resource depletion, and climate change has driven substantial research and innovation in the development of eco-friendly materials and processes. These advancements aim to minimize ecological impact, conserve resources, and promote sustainable development across various industries.

Eco-Friendly Materials

Eco-friendly materials are designed to reduce environmental harm throughout their life cycles, from raw material extraction to disposal. Key categories of such materials include (Bajpai et al., 2014; Anastas & Warner, 1998; Ashby, 2015):

Biodegradable Polymers: Derived from natural resources like starch, cellulose, or polylactic acid (PLA), biodegradable polymers decompose under environmental conditions, reducing plastic pollution. Applications range from packaging to medical implants.

Recycled Materials: Incorporating post-consumer and industrial waste into new products helps reduce resource consumption and landfill use. Examples include recycled concrete in construction and recycled plastics in manufacturing.

Bio-based Composites: Combining natural fibers (e.g., hemp, flax, or jute) with biodegradable resins creates sustainable alternatives to synthetic composites, widely used in automotive and construction industries.

Green Concrete: This innovative material uses industrial by-products such as fly ash and blast furnace slag to replace traditional cement, significantly lowering greenhouse gas emissions associated with concrete production.

Eco-Friendly Processes

In addition to materials, sustainable production processes are crucial in minimizing industrial environmental footprints. Key advancements include:

Green Chemistry: This approach emphasizes designing chemical processes that eliminate or reduce hazardous substances. For instance, solvent-free reactions and the use of catalysts derived from renewable sources are notable developments.

Energy-Efficient Manufacturing: Processes such as additive manufacturing (3D printing) reduce material waste and energy usage compared to traditional subtractive manufacturing techniques.

Waste Valorization: Transforming industrial waste into valuable by-products exemplifies circular economy principles. For example, food waste can be processed into bioenergy or bioplastics, reducing landfill dependence.

Water-Efficient Technologies: Processes that recycle and treat water used in industries, such as zero-liquid discharge systems, reduce water consumption and minimize pollution.

Challenges and Opportunities

Despite significant progress, challenges remain in scaling eco-friendly innovations due to cost, technical limitations, and market acceptance. However, government policies, corporate social responsibility, and consumer demand for sustainable products are fostering a favorable environment for their adoption.

Emerging technologies like nanotechnology, artificial intelligence, and biotechnology hold immense potential for further enhancing the sustainability of materials and processes. For example, AI can optimize supply chains to reduce waste, while biotechnological advances can engineer microbes to produce bio-based chemicals more efficiently.

- Case studies of sustainable practices.

4. INTEGRATION OF RENEWABLE ENERGY WITH WATER TREATMENT

The integration of renewable energy into water treatment processes offers a sustainable solution to address water scarcity, energy demands, and environmental concerns. By combining renewable energy technologies with water treatment systems, the industry can reduce its carbon footprint, improve energy efficiency, and enhance access to clean water in remote and underserved regions.

Importance of Renewable Energy in Water Treatment

Water treatment processes are often energy-intensive, requiring significant power for pumping, filtration, disinfection, and desalination. Conventional energy sources, such as fossil fuels, contribute to greenhouse gas emissions and are prone to supply fluctuations. Renewable energy sources, including solar, wind, and hydroelectric power, provide a cleaner, more reliable alternative that reduces operational costs and environmental impacts.

Key Renewable Energy Technologies in Water Treatment

1. **Solar Energy** Solar power is widely used to support decentralized water treatment systems, particularly in regions with abundant sunlight. Photovoltaic (PV) panels provide a stable energy source for powering pumps, UV disinfection systems, and membrane filtration technologies. Solar thermal systems can also support thermal desalination processes, such as multi-effect distillation (MED) and multi-stage flash distillation (MSF).
2. **Wind Energy** Wind turbines can generate electricity to power water treatment plants in coastal and open-area regions. Wind energy is effective for large-scale applications where sustained wind speeds can provide consistent power generation. Hybrid systems combining wind and solar technologies further enhance reliability.
3. **Hydropower** Small-scale hydropower systems can generate energy from rivers, canals, or water pipelines, providing localized power solutions for water treatment facilities. Hydropower is particularly effective in regions with reliable water flow and elevation differences.
4. **Biogas and Biomass Energy** Biogas generated from organic waste treatment can be harnessed to supply energy for water treatment facilities. Biomass energy systems can convert agricultural and forestry residues into heat or electricity, supporting various treatment processes.

Applications of Renewable Energy in Water Treatment

1. **Desalination** Renewable energy technologies are increasingly being used to power desalination plants, particularly in arid regions. Solar-powered reverse osmosis (RO) systems have demonstrated effective water purification while reducing energy costs.
2. **Wastewater Treatment** Renewable energy sources can support aerobic and anaerobic digestion systems in wastewater treatment plants. Solar panels and biogas recovery units can provide sustainable energy to drive pumps, aerators, and monitoring systems.
3. **Rural and Off-Grid Water Supply** Remote communities lacking grid infrastructure benefit significantly from renewable energy-driven water treatment systems. Standalone solar-powered filtration units and wind-powered pumping stations offer practical solutions for ensuring safe drinking water access.

Benefits of Integration

- **Energy Efficiency:** Renewable energy systems reduce dependency on fossil fuels, lowering energy costs and improving overall efficiency.
- **Environmental Impact:** By minimizing greenhouse gas emissions, renewable-powered water treatment contributes to climate change mitigation.
- **Economic Viability:** Although initial investments may be higher, long-term savings from reduced energy costs make renewable energy integration economically advantageous.
- **Resilience and Reliability:** Decentralized renewable energy systems improve energy security and water availability during power outages or emergencies.

Challenges and Future Outlook

Despite its advantages, integrating renewable energy into water treatment faces challenges such as fluctuating energy supply, high installation costs, and the need for storage solutions. Advances in energy storage systems, smart grid integration, and improved efficiency of renewable technologies are essential for expanding their role in water treatment.

In the future, combining renewable energy with innovative treatment technologies, such as advanced membrane systems and artificial intelligence-driven process optimization, will enhance water security and sustainability worldwide.

5. CIRCULAR ECONOMY IN WATER TREATMENT

The circular economy in water treatment focuses on minimizing waste, maximizing resource recovery, and promoting sustainable water management. Unlike the traditional linear model of "take, use, and dispose," the circular approach emphasizes reuse, recycling, and resource efficiency.

In water treatment, this concept involves reclaiming wastewater for non-potable applications such as irrigation, industrial processes, or groundwater recharge. Nutrients like phosphorus and nitrogen can be extracted from wastewater to produce fertilizers, while biosolids can be repurposed for soil conditioning or biogas production. Energy recovery is another key aspect, where anaerobic digestion processes convert organic waste into biogas, reducing energy demands.

By integrating circular economy principles, water treatment plants can reduce environmental impact, conserve resources, and lower operational costs. Technologies such as membrane filtration, advanced oxidation, and decentralized water systems play a vital role in enabling these sustainable practices. Additionally, collaboration between industries, municipalities, and environmental organizations is crucial to developing closed-loop water systems.

Overall, adopting a circular economy in water treatment enhances water security, mitigates pollution, and supports the transition to a more sustainable and resilient water infrastructure.

6. GLOBAL POLICY AND REGULATION

Sustainable water treatment is a critical aspect of environmental management that ensures water resources are conserved, protected, and safely reused. Global policies and regulations play a vital role in guiding sustainable practices. Key frameworks such as the **United Nations Sustainable Development Goal 6 (SDG 6)** promote universal access to clean water and sanitation, emphasizing resource efficiency and pollution control. International agreements like the **Water Convention (UNECE)** and the **Ramsar Convention** support cross-border cooperation for water resource management.

Countries are increasingly adopting stricter regulations to promote sustainable practices. For instance, the **European Union Water Framework Directive** enforces integrated water management plans, while the **Clean Water Act (USA)** regulates pollutant discharge to maintain water quality. Additionally, policies encourage the use of advanced treatment technologies such as membrane filtration, bioremediation, and energy-efficient desalination.

To achieve sustainable outcomes, policymakers promote water reuse, recycling, and circular economy principles. They also incentivize industries to adopt low-impact treatment methods that minimize chemical use and energy consumption. Effective global cooperation, combined with localized regulations, is essential for achieving long-term water sustainability while safeguarding ecosystems and human health.

7. CONCLUSION

In conclusion, water treatment is indispensable for global sustainability, addressing interlinked challenges in health, environment, economy, and climate. By prioritizing and investing in effective water treatment systems, societies can advance toward a more sustainable and equitable future. Ensuring access to clean water for all is not only a moral imperative but also a prerequisite for achieving long-term global development goals. Water treatment and sustainable chemistry are critical to addressing global challenges such as water scarcity, pollution, and climate change. This study highlights the significant advancements in these fields, focusing on innovative materials, energy-efficient processes, and integrated systems that ensure water security and environmental sustainability. Key achievements include the development of advanced filtration technologies, catalytic systems for pollutant degradation, and circular approaches that minimize waste generation while recovering valuable resources. By integrating interdisciplinary knowledge and leveraging technological innovations, we can build a more resilient and sustainable water infrastructure. The use of nanomaterials, bio-based polymers, and hybrid composites has enhanced the efficiency and specificity of water treatment systems. These materials exhibit superior adsorption, filtration, and catalytic properties, making them pivotal in addressing complex pollutants. Further research should focus on the development of scalable, cost-effective, and environmentally friendly materials that can address emerging contaminants and adapt to diverse water quality challenges.Continued exploration of hybrid processes that combine physical, chemical, and biological treatment methods is essential to improve overall efficiency and flexibility.

8. REFERENCES

Anastas, P. T., & Warner, J. C. (1998). *Green Chemistry: Theory and Practice*. Oxford University Press.

Andrady, A. L. (2011). Microplastics in the marine environment. *Marine Pollution Bulletin*, *62*(8), 1596–1605. DOI: 10.1016/j.marpolbul.2011.05.030 PMID: 21742351

Ashby, M. F. (2015). *Materials and Sustainable Development*. Butterworth-Heinemann.

Bajpai, P. K., Singh, J., & Madaan, A. (2014). Development and applications of bio-composites. *Journal of Reinforced Plastics and Composites*, *33*(20), 1854–1866.

Elimelech, M., & Phillip, W. A. (2011). The future of seawater desalination: Energy, technology, and the environment. *Science*, *333*(6043), 712–717. DOI: 10.1126/science.1200488 PMID: 21817042

Gleick, P. H. (2014). Water, Drought, Climate Change, and Conflict in Syria. *Weather, Climate, and Society*, *6*(3), 331–340. DOI: 10.1175/WCAS-D-13-00059.1

Logan, B. E., Hamelers, B., Rozendal, R., Uwe, S., & Jürg, K. (2006). Stefano, Freguia, Peter A., Willy, V., Korneel, R. Microbial fuel cells: Methodology and technology. *Environmental Science & Technology*, *40*(17), 5181–5192. DOI: 10.1021/es0605016 PMID: 16999087

Mauritz, K. H., & Moore, R. B. (2004). State of Understanding of Nafion. *Chemical Reviews*, *104*(10), 4535–4546. DOI: 10.1021/cr0207123 PMID: 15669162

Savage, N., & Diallo, M. S. (2005). Nanomaterials and Water Purification: Opportunities and Challenges. *Journal of Nanoparticle Research*, *7*(4-5), 331–342. DOI: 10.1007/s11051-005-7523-5

Shannon, M. A., Bohn, P. W., Elimelech, M., Georgiadis, J. G., Marĩas, B. J., & Mayes, A. M. (2008). Science and technology for water purification in the coming decades. *Nature*, *452*(7185), 301–310. DOI: 10.1038/nature06599 PMID: 18354474

Siva Rama Krishnan, M. K. (2022). Smart Water Resource Management Using Artificial Intelligence: A Review. *Sustainability*, *14*(20), 13384. DOI: 10.3390/su142013384

UNESCO. (2020). *The United Nations World Water Development Report 2020: Water and Climate Change*. UNESCO Publishing.

United Nations. Sustainable Development Goal 6: Ensure availability and sustainable management of water and sanitation for all. https://sdgs.un.org/goals/goal6

Virendra Kumar Saharan. (2014). *Dipak Pinjari, Parag R Gogate, Aniruddha Bhalchandra Pandit. Advanced Oxidation Technologies for Wastewater Treatment.* Industrial Wastewater Treatment, Recycling and Reuse., DOI: 10.1016/B978-0-08-099968-5.00003-9

World Health Organization (WHO). (2019). Drinking-water. Retrieved from https://www.who.int/news-room/fact-sheets/detail/drinking-water

Compilation of References

Abeng, F. E., & Anadebe, V. C. (2023). Combined electrochemical, DFT/MD-simulation and hybrid machine learning based on ANN-ANFIS models for prediction of doxorubicin drug as corrosion inhibitor for mild steel in 0.5 M H2SO4 solution. *Computational & Theoretical Chemistry*, *1229*, 114334. DOI: 10.1016/j.comptc.2023.114334

Abohamzeh, E., Salehi, F., Sheikholeslami, M., Abbassi, R., & Khan, F. (2021). Review of hydrogen safety during storage, transmission, and applications processes. *Journal of Loss Prevention in the Process Industries*, *72*, 104569.

Abo-Neima, S. E., Motaweh, H. A., & Elsehly, E. M. (2020). Antimicrobial activity of functionalised carbon nanotubes against pathogenic microorganisms. *IET Nanobiotechnology / IET*, *14*(6), 457–464. DOI: 10.1049/iet-nbt.2019.0342 PMID: 32755954

Acharjee, S. A., Bharali, P., Gogoi, B., Sorhie, V., Walling, B., & Alemtoshi, . (2023). PHA-Based Bioplastic: A Potential Alternative to Address Microplastic Pollution. *Water, Air, and Soil Pollution*, *234*(1), 21. DOI: 10.1007/s11270-022-06029-2 PMID: 36593989

Adamashvili, N., Zhizhilashvili, N., & Tricase, C. (2024). The integration of the internet of things, artificial intelligence, and blockchain technology for advancing the wine supply chain. *Computers*, *13*(3), 72. DOI: 10.3390/computers13030072

Agajie, T. F., Ali, A., Fopah-Lele, A., Amoussou, I., Khan, B., Velasco, C. L. R., & Tanyi, E. (2023). A Comprehensive Review on Techno-Economic Analysis and Optimal Sizing of Hybrid Renewable Energy Sources with Energy Storage Systems. *Energies*, *16*(2), 642. DOI: 10.3390/en16020642

Agee, B. M., Mullins, G., & Swartling, D. J. (2016). Progress towards a more sustainable synthetic pathway to ibuprofen through the use of solar heating. *Sustain Chem Process*, *4*(8), 1–9. DOI: 10.1186/s40508-016-0052-y

Aghababaei, A., Aghababaei, F., Pignitter, M., & Hadidi, M. (2025). Artificial Intelligence in Agro-Food Systems: From Farm to Fork. *Foods*, *14*(3), 411. DOI: 10.3390/foods14030411 PMID: 39942003

Aghaei, M., Fairbrother, A., Gok, A., Ahmad, S., Kazim, S., Lobato, K., Oreski, G., Reinders, A., Schmitz, J., Theelen, M., Yilmaz, P., & Kettle, J. (2022). Review of degradation and failure phenomena in photovoltaic modules. *Renewable & Sustainable Energy Reviews*, *159*, 112160. DOI: 10.1016/j.rser.2022.112160

Agustiany, E. A., Ridho, M. R., Rahmi, M. D. N., Madyaratri, E. W., Falah, F., Lubis, M. A. R., & Solihat, N. N. (2022). Recent developments in lignin modification and its application in lignin-based green composites: A review. *Polymer Composites*, *43*(8), 4848–4865. DOI: 10.1002/pc.26824

Ahmad, S., Hussain, I., Ghaffar, A., Rahman, M. H. U., Saleem, M. Z., Yonas, M. W., Hussnain, H., Ikram, R. M., & Arslan, M. (2022). Organic amendments and conservation tillage improve cotton productivity and soil health indices under arid climate. *Scientific Reports*, *12*(1), 14072. DOI: 10.1038/s41598-022-18157-0 PMID: 35982152

Ahmad, V., & Ansari, M. O. (2022). Antimicrobial Activity of Graphene-Based Nanocomposites: Synthesis, Characterization, and Their Applications for Human Welfare. *Nanomaterials (Basel, Switzerland)*, *12*(22), 4002. DOI: 10.3390/nano12224002 PMID: 36432288

Al Aasmi, A., Alordzinu, K. E., Li, J., Lan, Y., Appiah, S. A., & Qiao, S. (2022). Rapid estimation of water stress in choy sum (Brassica chinensis var. Parachinensis) using integrative approach. *Sensors (Basel)*, *22*(5), 1695. DOI: 10.3390/s22051695 PMID: 35270842

Ala'F, K., Alqammaz, A., Khasawneh, A. M., Abualigah, L., Darabkh, K. A., & Zinonos, Z. (2025). An environmental remote sensing and prediction model for an IoT smart irrigation system based on an enhanced wind-driven optimization algorithm. *Computers & Electrical Engineering*, *122*, 109889. DOI: 10.1016/j.compeleceng.2024.109889

Alam, S., Bhardwaj, L. K., Mallick, R., & Rai, S. (2023). Estimation of heavy metals and fluoride ion in vegetables grown nearby the stretch of River Yamuna, Delhi (NCR), India. *Indian Journal of Environmental Protection*, *43*(1), 64–73.

Alam, S., Khan, M. I., Fiaz, F., Iqbal, M. Z., Alam, F., Ahmad, Z., & Hegazy, H. H. (2023). Advancements in asymmetric supercapacitors: Material selection, mechanisms, and breakthroughs with metallic oxides, sulfides, and phosphates. *Journal of Energy Storage*, *72*, 108208. DOI: 10.1016/j.est.2023.108208

Albshaier, L., Almarri, S., & Albuali, A. (2025). Federated Learning for Cloud and Edge Security: A Systematic Review of Challenges and AI Opportunities. *Electronics (Basel)*, *14*(5), 1019. DOI: 10.3390/electronics14051019

Ali, F. B., Kang, D. J., Kim, M. P., Cho, C. H., & Kim, B. J. (2014). Synthesis of biodegradable and flexible polylactic acid-based thermoplastic polyurethane with high gas barrier properties. *Polymer International*, *63*(9), 1620–1626. DOI: 10.1002/pi.4662

Ali, M. Y. S., Anuradha, V., Abishek, R., Yogananth, N., & Sheeba, H. (2017). In Vitro Anticancer Activity of Green Synthesis Ruthenium Nanoparticle from Dictyota Dichotoma Marine Algae. *NanoWorld Journal*, *03*(4), 1. DOI: 10.17756/nwj.2017-049

Al-thani, H., Koç, M., & Isaifan, R. J. (2018). A review on the direct effect of particulate atmospheric pollution on materials and its mitigation for sustainable cities and societies. *Environmental Science and Pollution Research International*, *25*(28), 27839–27857. DOI: 10.1007/s11356-018-2952-8 PMID: 30128969

Amini Parsa, V., Yavari, A., & Nejadi, A. (2016). Spatio-temporal analysis of land use/land cover pattern changes in Arasbaran Biosphere Reserve: Iran. *Modeling Earth Systems and Environment*, *2*(4), 1–13. DOI: 10.1007/s40808-016-0227-2

Amir, M., Deshmukh, R. G., Khalid, H. M., Said, Z., Raza, A., Muyeen, S. M., Nizami, A.-S., Elavarasan, R. M., Saidur, R., & Sopian, K. (2023). Energy storage technologies: An integrated survey of developments, global economical/environmental effects, optimal scheduling model, and sustainable adaption policies. *Journal of Energy Storage*, *72*, 108694. DOI: 10.1016/j.est.2023.108694

Amooaghaie, R., Saeri, M. R., & Azizi, M. (2015). Synthesis, Characterization and Biocompatibility of Silver Nanoparticles Synthesized from Nigella Sativa Leaf Extract in Comparison with Chemical Silver Nanoparticles. *Ecotoxicology and Environmental Safety*, *120*, 400–408. DOI: 10.1016/j.ecoenv.2015.06.025 PMID: 26122733

Anastas, P. T., & Warner, J. C. (1998). *Green Chemistry: Theory and Practice*. Oxford University Press.

Anastas, P. T., & Warner, J. C. (2000). *Green chemistry: theory and practice*. Oxford university press., DOI: 10.1093/oso/9780198506980.001.0001

Anderson, J. O., Thundiyil, J. G., & Stolbach, A. (2012). Clearing the air: A review of the effects of particulate matter air pollution on human health. *Journal of Medical Toxicology; Official Journal of the American College of Medical Toxicology*, *8*(2), 166–175. DOI: 10.1007/s13181-011-0203-1 PMID: 22194192

Andrady, A. L. (2011). Microplastics in the marine environment. *Marine Pollution Bulletin*, *62*(8), 1596–1605. DOI: 10.1016/j.marpolbul.2011.05.030 PMID: 21742351

Andrew, O. B., Sherwood, J., & Hurst, G. A. (2022). A Greener Synthesis of the Antidepressant Bupropion Hydrochloride. *Journal of Chemical Education*, *99*(9), 3277–3282. DOI: 10.1021/acs.jchemed.2c00581

Annepu, R. K. (2012). Sustainable solid waste management in India. *Columbia University, New York, 2*(01), 1-89.

Archana, H. R., Darshan, K., Lakshmi, M. A., Ghoshal, T., Bashayal, B. M., & Aggarwal, R. (2022). Biopesticides: A key player in agro-environmental sustainability. In *Trends of Applied Microbiology for Sustainable Economy* (pp. 613–653). Elsevier. https://www.sciencedirect.com/science/article/pii/B9780323915953000215

Argese, M., & Zanoni, G. (2023). Solvent-free synthesis of chalcones using $Mg(HSO_4)_2$. *RSC Sustainability*, *1*(3), 504–510. DOI: 10.1039/D3SU00003F

Arora, P. K., Tripathi, S., Omar, R. A., Chauhan, P., Sinhal, V. K., Singh, A., Srivastava, A., Garg, S. K., & Singh, V. P. (2024). Next-generation fertilizers: The impact of bionanofertilizers on sustainable agriculture. *Microbial Cell Factories*, *23*(1), 254. DOI: 10.1186/s12934-024-02528-5 PMID: 39304847

Arowolo, A. O., Deng, X., Olatunji, O. A., & Obayelu, A. E. (2018). Assessing changes in the value of ecosystem services in response to land-use/land-cover dynamics in Nigeria. *The Science of the Total Environment*, *636*, 597–609. DOI: 10.1016/j.scitotenv.2018.04.277 PMID: 29723833

Arrieta, M. P., López, M. D. M. C., Rayón, E., Barral-Losada, L. F., López-Vilariño, J. M., López, J., & González-Rodríguez, M. V. (2014). Plasticized poly (lactic acid)–poly(hydroxybutyrate) blends incorporated with catechin intended for active food-packaging applications. *Journal of Agricultural and Food Chemistry*, *62*(41), 10170–10180. DOI: 10.1021/jf5029812 PMID: 25255375

Arrieta, M. P., Peponi, L., López, D., López, J., & Kenny, J. M. (2017). Food packaging. In *Elsevier* (pp. 391–424). Academic Press.,

Artiola, J. F. (2019). Industrial waste and municipal solid waste treatment and disposal. In *Environmental and pollution science* (pp. 377–391). Academic Press. DOI: 10.1016/B978-0-12-814719-1.00021-5

Arunachalam, S., Brantley, H., Barzyk, T. M., Hagler, G., Isakov, V., Kimbrough, E., Naess, B., Rice, N., Snyder, M. G., Talgo, K., & Venkatram, A. (2015). Assessment of port-related air quality impacts: Geographic analysis of population. *International Journal of Environment and Pollution*, *58*(4), 231–250. DOI: 10.1504/IJEP.2015.077455

Asadpour, S., Yeganeh, H., Ai, J., Kargozar, S., Rashtbar, M., Seifalian, A., & Ghanbari, H. (2018). Polyurethane-polycaprolactone blend patches: Scaffold characterization and cardiomyoblast adhesion, proliferation, and function. *ACS Biomaterials Science & Engineering*, *4*(11), 4299–4310. DOI: 10.1021/acsbiomaterials.8b00848 PMID: 33418826

Ashby, M. F. (2015). *Materials and Sustainable Development*. Butterworth-Heinemann.

Ashwini, A., Sriram, S. R., Prabhakar, J. M., & Kadry, S. (2025). Farming 4.0: Cultivating the Future with Internet of Things Empowered on Smart Agriculture Solutions. In Dhanaraj, R. K., Sathyamoorthy, M., Balasubramaniam, S., & Kadry, S. (Eds.), *Networked Sensing Systems* (1st ed., pp. 247–271). Wiley., DOI: 10.1002/9781394310890.ch10

Avantium. (2022). FDCA flagship plant—YXY technology. Retrieved July 19, 2024, from https://avantium.com/technologies/yxy/

Aziz, D., Rafiq, S., Saini, P., Ahad, I., Gonal, B., Rehman, S. A., Rashid, S., Saini, P., Rohela, G. K., Aalum, K., Singh, G., Gnanesh, B. N., & Nabila Iliya, M. (2025). Remote sensing and artificial intelligence: Revolutionizing pest management in agriculture. *Frontiers in Sustainable Food Systems*, *9*, 1551460. DOI: 10.3389/fsufs.2025.1551460

Azizi-Lalabadi, M., Hashemi, H., Feng, J., & Jafari, S. M. (2020). Carbon nanomaterials against pathogens; the antimicrobial activity of carbon nanotubes, graphene/graphene oxide, fullerenes, and their nanocomposites. *Advances in Colloid and Interface Science*, *284*, 102250–102267. DOI: 10.1016/j.cis.2020.102250 PMID: 32966964

Babar, Z. B., & Shareefdeen, Z. (2014). Management and control of air emissions from electronic industries. *Clean Technologies and Environmental Policy*, *16*(1), 69–77. DOI: 10.1007/s10098-013-0594-6

Babbar, H., Rani, S., Soni, M., Keshta, I., Prasad, K. D. V., & Shabaz, M. (2025). Integrating remote sensing and geospatial AI-enhanced ISAC models for advanced localization and environmental monitoring. *Environmental Earth Sciences*, *84*(4), 1–12. DOI: 10.1007/s12665-025-12121-7

Bade, S. O., & Tomomewo, O. S. (2024). A review of governance strategies, policy measures, and regulatory framework for hydrogen energy in the United States. *International Journal of Hydrogen Energy*, *78*, 1363–1381. DOI: 10.1016/j.ijhydene.2024.06.338

Bag, A., ... (2023) "Hydrogen Storage Efficiency of Isomeric Cu (I)-Triazine Complexes: In Quest of New Hydrogen Storage Material, Computational Studies: From Molecules to Materials.

Bag, A., ... (2023), CRC Press/Tylor & Francis Group, 2023. ISBN 9781032528540.

Bag, A., Chandra De, G., Bhattacharyya, S., Bepari, B., Das, H. S., Bandaru, S., & Roymahapatra, G. (2025). Ag(I) decorated isomeric triazine complexes as efficient hydrogen storage materials - A theoretical investigation. *Chemistry of Inorganic Materials.*, *2025*(5), 100093. DOI: 10.1016/j.cinorg.2025.100093

Bag, A., Dash, M. K., Tarif, E., Guo, J. Z., De, G. C., Maity, S., Ganguly, J., Bhattacharyya, S., & Roymahapatra, G. (2025). Cu(I)-Decorated Five-membered Aromatic Heterocyclic Complexes; A Potential H2 Storage System. *ES Chemistry and Sustainability*, *2025*(02), 1311. DOI: 10.30919/escs1311

Bag, A., Giri, S., Dhaiveegan, P., Jhalgam, R. T. T., De, G. C., Ganguly, J., & Roymahapatra, G. (2024). ES. *Energy & Environment*, *2024*(23), 1115. DOI: 10.30919/esee1115

Bag, A., Sinha, S., Das, H. S., Giri, S., De, G. C., Maity, S., Xu, B. B., Guo, Z., & Roymahapatra, G. (2024). Hydrogen Storage Efficiency of the Ag (I)/Au (I) Decorated Five-Member Aromatic Heterocyclic (AH) Compounds: A Theoretical Investigation. *Engineering and Science*, *2024*(28), 1062. DOI: 10.30919/es1062

Bailey, D., & Solomon, G. (2004). Pollution prevention at ports: Clearing the air. *Environmental Impact Assessment Review*, *24*(7), 749–774. DOI: 10.1016/j.eiar.2004.06.005

Bajpai, P. K., Singh, J., & Madaan, A. (2014). Development and applications of bio-composites. *Journal of Reinforced Plastics and Composites*, *33*(20), 1854–1866.

Balakumar, P., Vinopraba, T., & Chandrasekaran, K. (2023). Deep learning based real time Demand Side Management controller for smart building integrated with renewable energy and Energy Storage System. *Journal of Energy Storage*, *58*, 106412. DOI: 10.1016/j.est.2022.106412

Bamoharram, F. F., Heravi, M. M., Roshani, M., & Jahangir, M. (2007). Catalytic Method for Synthesis of Aspirin by a Green, Efficient and Recyclable Solid Acid Catalyst (Preyssler' s Anion) at Room Temperature. *Journal of the Chinese Chemical Society (Taipei)*, *54*(4), 1017–1020. DOI: 10.1002/jccs.200700146

Bandyopadhyay, A., Bandyopadhyay, S., & Biswas, M. N. (2014). Fabrication and operation of an online photoelectric particulate matter monitoring device. *Clean Technologies and Environmental Policy*, *17*(1), 265–271. DOI: 10.1007/s10098-014-0782-z

Bankar, A., Joshi, B., Kumar, A. R., & Zinjarde, S. (2010). Banana Peel Extract Mediated Novel Route for the Synthesis of Silver Nanoparticles. *Colloids and Surfaces. A, Physicochemical and Engineering Aspects*, *368*(1-3), 58–63. DOI: 10.1016/j.colsurfa.2010.07.024

Barwant, M. M., Ogidi, O. I., Yogita, C., & Munje, R. (2025). Study of Consumers Choices on Pesticides Use and Sustainability. In Babaniyi, B. R., & Babaniyi, E. E. (Eds.), *The Interplay of Pesticides and Climate Change* (pp. 469–516). Springer Nature Switzerland., DOI: 10.1007/978-3-031-81669-7_18

Baykara, S. Z. (2018). Hydrogen: A brief overview on its sources, production and environmental impact. *International Journal of Hydrogen Energy*, *43*(23), 10605–10614. https://sdg.iisd.org/news/world-population-to-reach-9-9-billion-by-2050/. DOI: 10.1016/j.ijhydene.2018.02.022

Belliveau, B. H., Starodub, M. E., Cotter, C., & Trevors, J. T. (1987). Metal Resistance and Accumulation in Bacteria. *Biotechnology Advances*, *5*(1), 101–127. DOI: 10.1016/0734-9750(87)90006-1 PMID: 14543146

Betts, K. (2015). How industrial applications in green chemistry are changing our world. *Journal of the American Chemical Society*, •••, 1–28.

Bhalla, B., Saini, M. S., & Jha, M. K. (2013). Effect of age and seasonal variations on leachate characteristics of municipal solid waste landfill. *International Journal of Research in Engineering and Technology*, *2*(8), 223–232. DOI: 10.15623/ijret.2013.0208037

Bhardwaj, L. K., Rath, P., & Tokas, R. (2024). E-Waste Management in Developing Countries: Current Practices, Challenges, Disposal, and Impact on Human Health and Environment. *Sustainable Solutions for E-Waste and Development*, 90-102.

Bhardwaj, L. K., Rath, P., Jain, H., & Choudhury, M. (2025). Exploring the effects of e-waste on soil, water quality and human health. *Discover Civil Engineering*, *2*(1), 12. Advance online publication. DOI: 10.1007/s44290-025-00167-2

Bhardwaj, L. K., Sharma, S., & Jindal, T. (2023). Estimation of physico-chemical and heavy metals in the lakes of Grovnes & Broknes Peninsula, Larsemann Hill, East Antarctica. *Chemistry Africa*, *6*(5), 2677–2694. DOI: 10.1007/s42250-023-00668-6

Bhattacharyya, P., Pathak, H., & Pal, S. (2020). *Climate smart agriculture: Concepts, challenges, and opportunities*.

Bhuiyan, M. M. H., & Siddique, Z. (2025). Hydrogen as an alternative fuel: A comprehensive review of challenges and opportunities in production, storage, and transportation. *International Journal of Hydrogen Energy*, *102*, 1026–1044.

Bhutia, Z. T., Panjikar, P. C., Iyer, S., Chatterjee, A., & Banerjee, M. (2020). Iodine Promoted E ffi cient Synthesis of 2 - Arylimidazo[1,2 - a]pyridines in Aqueous Media: A Comparative Study between Micellar Catalysis and an " On-Water " Platform. *ACS Omega*, *5*(22), 13333–13343. DOI: 10.1021/acsomega.0c01478 PMID: 32548520

Biswas, A. K., Kumar, S., Babu, S. S., Bhattacharyya, J. K., & Chakrabarti, T. (2010). Studies on environmental quality in and around municipal solid waste dumpsite. *Resources, Conservation and Recycling*, *55*(2), 129–134. DOI: 10.1016/j.resconrec.2010.08.003

Blomqvist, A., Araújo, C. M., Srepusharawoot, P., & Ahuja, R. (2007). Li-decorated metal–organic framework 5: A route to achieving a suitable hydrogen storage medium. *Proceedings of the National Academy of Sciences of the United States of America*, *104*(51), 20173–20176. www.pnas.org/cgi/doi/10.1073/pnas.0708603104. DOI: 10.1073/pnas.0708603104 PMID: 18093960

Bogda, D., Pielichowski, J., & Boro, A. (1996). Remarkable fast microwave-assisted N-alkylation of phthalimide in dry media. *Synlett*, *1996*(09), 873–874. DOI: 10.1055/s-1996-5587

Boorboor Ajdari, F., Asghari, P., Molaei Aghdam, A., Abbasi, F., Rao, R. P., Abbasi, A., Ghasemi, F., Ramakrishna, S., & Mikaeili Chahartagh, N. (2024). Silicon Solid State Battery: The Solid-State Compatibility, Particle Size, and Carbon Compositing for High Energy Density. *Advanced Functional Materials*, *34*(30), 2314822. Advance online publication. DOI: 10.1002/adfm.202314822

Bordes, P., Pollet, E., & Avérous, L. (2009). Nano-biocomposites: Biodegradable polyester/nanoclay systems.[[Doi:]. *Progress in Polymer Science*, *34*(2), 125–155. DOI: 10.1016/j.progpolymsci.2008.10.002

Braud, C., Devarieux, R., Atlan, A., Ducos, C., & Vert, M. (1998). Capillary zone electrophoresis in normal or reverse polarity separation modes for the analysis of hydroxy acid oligomers in neutral phosphate buffer. *Journal of Chromatography. B, Biomedical Sciences and Applications*, *706*(1), 73–82. DOI: 10.1016/S0378-4347(97)00468-4 PMID: 9544809

Breynaert, E., Houllebergh s, M., Radhakrishnan, S., Grübel, G., Taulelle, F., & Martens, J. A. (2020). Water as a tuneable solvent: A perspective. *Chemical Society Reviews*, *49*(9), 2557–2569. DOI: 10.1039/C9CS00545E PMID: 32238999

Brizga, J., Hubacek, K., & Feng, K. (2020). The Unintended Side Effects of Bioplastics: Carbon, Land, and Water Footprints. *One Earth*, *3*(1), 45–53. DOI: 10.1016/j.oneear.2020.06.016

Bulucea, C. A., Mastorakis, N. E., Bulucea, C. A., Boteanu, N. I. C. U. L. A. E., & Stinga, A. U. R. E. L. (2010). Systemic approach of hazardous and non-hazardous waste management. In *4th WSEAS International Conference on Energy Planning, Energy Saving, Environmental Education, EPESE* (Vol. 10, pp. 181-189).

Burgess, S. K., Karvan, O., Johnson, J. R., Kriegel, R. M., & Koros, W. J. (2014). Oxygen sorption and transport in amorphous poly (ethylene furanoate). *Polymer*, *55*(18), 4748–4756. DOI: 10.1016/j.polymer.2014.07.041

Burgess, S. K., Mikkilineni, D. S., Daniel, B. Y., Kim, D. J., Mubarak, C. R., Kriegel, R. M., & Koros, W. J. (2014). Water sorption in poly (ethylene furanoate) compared to poly (ethylene terephthalate). Part 1: Equilibrium sorption. *Polymer*, *55*(26), 6861–6869. DOI: 10.1016/j.polymer.2014.10.047

Burgess, S. K., Mikkilineni, D. S., Daniel, B. Y., Kim, D. J., Mubarak, C. R., Kriegel, R. M., & Koros, W. J. (2014). Water sorption in poly (ethylene furanoate) compared to poly (ethylene terephthalate). Part 2: Kinetic sorption. *Polymer*, *55*(26), 6870–6882. DOI: 10.1016/j.polymer.2014.10.065

Burgos, N., Armentano, I., Fortunati, E., Dominici, F., Luzi, F., Fiori, S., Cristofaro, F., Visai, L., Jiménez, A., & Kenny, J. M. (2017). Functional properties of plasticized bio-based poly (lactic acid)/poly(hydroxybutyrate) films for active food packaging. *Food and Bioprocess Technology*, *10*(4), 770–780. DOI: 10.1007/s11947-016-1846-3

Calabrese, M., Portarapillo, M., Di Nardo, A., Venezia, V., Turco, M., Luciani, G., & Di Benedetto, A. (2024). Hydrogen Safety Challenges: A Comprehensive Review on Production, Storage, Transport, Utilization, and CFD-Based Consequence and Risk Assessment. *Energies*, *2024*(17), 1350. DOI: 10.3390/en17061350

Cann, M. C. (1999). Bringing State-of-the-Art, Applied, Novel, Green Chemistry to the Classroom by Employing the Presidential Green Chemistry Challenge Awards. *Journal of Chemical Education*, *76*(12), 1639–1641. DOI: 10.1021/ed076p1639

Cao, S., & Yu, J. (2014). g-C_3N_4-Based Photocatalysts for Hydrogen Generation. *The Journal of Physical Chemistry Letters*, *5*(12), 2101–2107. DOI: 10.1021/jz500546b PMID: 26270499

Cariolet, J. M., Colombert, M., Vuillet, M., & Diab, Y. (2018). Assessing the resilience of urban areas to traffic-related air pollution: Application in greater Paris. *The Science of the Total Environment*, *615*(615), 588–596. DOI: 10.1016/j.scitotenv.2017.09.334 PMID: 28988095

Castro, L., Blázquez, M. L., Muñoz, J. A., González, F., & Ballester, A. (2013). Biological Synthesis of Metallic Nanoparticles Using Algae. *IET Nanobiotechnology / IET*, *7*(3), 109–116. DOI: 10.1049/iet-nbt.2012.0041 PMID: 24028809

Census of India. (2011). *Rural and Urban Distribution of population*. Government of India.

Chae, Y., & Park, J. (2011). Quantifying costs and benefits of integrated environmental strategies of air quality management and greenhouse gas reduction in the Seoul metropolitan area. *Energy Policy*, *39*(9), 5296–5308. DOI: 10.1016/j.enpol.2011.05.034

Chakraborty, T., & Akhtar, N. (2021). Biofertilizers: Characteristic Features and Applications. In Inamuddin, M. I. Ahamed, R. Boddula, & M. Rezakazemi (Eds.), *Biofertilizers* (1st ed., pp. 429–489). Wiley. DOI: 10.1002/9781119724995.ch15

Chakravarty, A., Ahmad, I., Singh, P., Ud Din Sheikh, M., Aalam, G., Sagadevan, S., & Ikram, S. (2022). Green Synthesis of Silver Nanoparticles Using Fruits Extracts of Syzygium Cumini and Their Bioactivity. *Chemical Physics Letters*, *1*, 139493. DOI: 10.1016/j.cplett.2022.139493

Chandna, N., Chandak, N., Kumar, P., Kapoor, J. K., & Sharma, P. K. (2013). Metal- and solvent-free synthesis of N-sulfonylformamidines. *Green Chemistry*, *15*(8), 2294–2301. DOI: 10.1039/c3gc40797g

Chao, K. (2008). A new look at the cross-impact matrix and its application in futures studies. *Journal of Futures Studies*, *12*(4), 45–52.

Chataut, G., Bhatta, B., Joshi, D., Subedi, K., & Kafle, K. (2023). Greenhouse gases emission from agricultural soil: A review. *Journal of Agriculture and Food Research*, *11*, 100533. DOI: 10.1016/j.jafr.2023.100533

Chattaraj, S., Srinivasu, K., Mondal, S., & Ghosh, S. K. (2015). Hydrogen trapping ability of the pyridine–lithium+ (1: 1) complex. *The Journal of Physical Chemistry A, 119*(12), 3056–3063.

Chatterjee, D., Kuotsu, R., James Kikon, Z., Sarkar, D., Ao, M., Ray, S. K., Bera, T., & Deka, B. C. (2016). Characterization of vermicomposts prepared from agricultural solid wastes in north eastern hill region of Nagaland, India. *Proceedings of the National Academy of Sciences. India. Section B, Biological Sciences, 86*(4), 823–833. DOI: 10.1007/s40011-015-0538-5

Chen, B., Deng, Q., Yang, G., Zhou, Y., Chen, W., Cai, Y., & Tu, Z. (2023). Numerical study on heat transfer characteristics and performance evaluation of PEMFC based on multiphase electrochemical model coupled with cooling channel. *Energy, 285*, 128933. DOI: 10.1016/j.energy.2023.128933

Chen, J., Hu, X., Lu, L., & Xiao, W. (2016). Exploration of Visible-Light Photocatalysis in Heterocycle Synthesis and Functionalization : Reaction Design and Beyond. *Accounts of Chemical Research, 49*(9), 1911–1923. DOI: 10.1021/acs.accounts.6b00254 PMID: 27551740

Chen, S., Wang, F., & Haroon, M. (2023). The impact of green economic recovery on economic growth and ecological footprint: A case study in developing countries of Asia. *Resources Policy, 85*, 103955. DOI: 10.1016/j.resourpol.2023.103955

Chen, W., Modi, D., & Picot, A. (2023). Soil and phytomicrobiome for plant disease suppression and management under climate change: A review. *Plants, 12*(14), 2736. DOI: 10.3390/plants12142736 PMID: 37514350

Chen, Y., Kang, Y., Zhao, Y., Wang, L., Liu, J., Li, Y., & Li, B. (2021). A review of lithium-ion battery safety concerns: The issues, strategies, and testing standards. *Journal of Energy Chemistry, 59*, 83–99.

Chen, Y., Lennartz, P., Liu, K. L., Hsieh, Y., Scharf, F., Guerdelli, R., Buchheit, A., Grünebaum, M., Kempe, F., Winter, M., & Brunklaus, G. (2023). Towards All-Solid-State Polymer Batteries: Going Beyond PEO with Hybrid Concepts. *Advanced Functional Materials, 33*(32), 2300501. Advance online publication. DOI: 10.1002/adfm.202300501

Cheung, H. Y., Ho, M. P., Lau, K. T., Cardona, F., & Hui, D. (2009). Natural fibre-reinforced composites for bioengineering and environmental engineering applications. *Composites. Part B, Engineering, 40*(7), 655–663. DOI: 10.1016/j.compositesb.2009.04.014

Chigorimbo-murefu, N. T. L., Riva, S., & Burton, S. G. (2009). *Journal of Molecular Catalysis B : Enzymatic* Lipase-catalysed synthesis of esters of ferulic acid with natural compounds and evaluation of their antioxidant properties. *56*, 4, 277–282. https://doi.org/DOI: 10.1016/j.molcatb.2008.05.017

Chu, C., Wu, K., Luo, B., Cao, Q., & Zhang, H. (2023). Hydrogen storage by liquid organic hydrogen carriers: Catalyst, renewable carrier, and technology–a review. *Carbon Resources Conversion*, *6*(4), 334–351.

Clarke, C. J., Tu, W., Levers, O., Bro, A., & Hallett, J. P. (2018). Green and Sustainable Solvents in Chemical Processes. *Chemical Reviews*, *118*(2), 747–800. DOI: 10.1021/acs.chemrev.7b00571 PMID: 29300087

Cordeiro-Costas, M., Villanueva, D., Eguía-Oller, P., & Granada-Álvarez, E. (2023). Intelligent energy storage management trade-off system applied to Deep Learning predictions. *Journal of Energy Storage*, *61*, 106784. DOI: 10.1016/j.est.2023.106784

CPCB (Central pollution Control Board). (2011). *Status report on Municipal Solid waste management*. Central Pollution Control Board.

CPCB (Central pollution Control Board). 2000 Management of municipal solid waste Delhi. See http://www.cpcb.nic.in/divisionsofheadoffice/pcp/ MSW_Report.pdf (accessed 27 June 2015).

Crespi, S., Fagnoni, M., & Ring, F. (2018). *Photocatalyzed Formation of Heterocycles.Topics in Heterocyclic Chemistry, 54*. Springer., DOI: 10.1007/7081_2017_13

Dahoumane, S. A., Yépremian, C., Djédiat, C., Couté, A., Fiévet, F., Coradin, T., & Brayner, R. (2016). Improvement of Kinetics, Yield, and Colloidal Stability of Biogenic Gold Nanoparticles Using Living Cells of Euglena Gracilis Microalga. *Journal of Nanoparticle Research*, *18*(3), 1. DOI: 10.1007/s11051-016-3378-1

Dai, Y., Sun, Q., Wang, W., Lu, L., Liu, M., Li, J., Yang, S., Sun, Y., Zhang, K., Xu, J., Zheng, W., Hu, Z., Yang, Y., Gao, Y., Chen, Y., Zhang, X., Gao, F., & Zhang, Y. (2018). Utilizations of agricultural waste as adsorbent for the removal of contaminants: A review. *Chemosphere*, *211*, 235–253. DOI: 10.1016/j.chemosphere.2018.06.179 PMID: 30077103

Das, H. S. (2025). Life Cycle Assessment of Emerging Battery Technologies for Electric Vehicles. In Optimized Energy Management Strategies for Electric Vehicles (pp. 87-118). IGI Global Scientific Publishing.

Dasgupta, B., Yadav, V. L., & Mondal, M. K. (2013). Seasonal characterization and present status of municipal solid waste (MSW) management in Varanasi, India. *Advances in Environmental Research*, *2*(1), 51–60. DOI: 10.12989/aer.2013.2.1.051

Das, H. S., Mishra, S., & Roymahapatra, G. (2025). Advanced Nano-Structured Materials for Energy Storage Devices. *Advances in Chemical and Materials Engineering, 2025*, 34. DOI: 10.4018/979-8-3693-5320-2.ch001

Dash, M. K.. (2022). S. Sinha, H. S. Das, G. C. De, S. Giri, G.Roymahapatra. *Sustainable Energy Technologies and Assessments, 2022*(52), 102235. DOI: 10.1016/j.seta.2022.102235

Dash, M. K., Chowdhury, S. D., Chatterjee, R., Maity, S., Roymahapatra, G., Huang, M., Bandyopadhyay, M. N., & Guo, Z. (2022). Can Lithium Fluoride (LiF) Be An Efficient Hydrogen Storage Material? An In-Silico Study. *Engineering and Science, 18*. Advance online publication. DOI: 10.30919/es8d658

Dash, M. K., Das, S., Giri, S., Chandra De, G., & Roymahapatra, G. (2021). Comprehensive in silico study on lithiated Triazine isomers and its H2 storage efficiency. *Journal of the Indian Chemical Society, 2021*(98), 100134. DOI: 10.1016/j.jics.2021.100134

Dash, M. K., Sinha, S., Das, H. S., De, G. C., Giri, S., & Roymahapatra, G. (2022). H2 storage capacity of Li-doped five member aromatic heterocyclic superalkali complexes; an in silico study. *Sustainable Energy Technologies and Assessments, 52*, 102235.

Datta, R., & Henry, M. (2006). Lactic Acid: Recent Advances in Products, Processes and Technologies—a Review. *Journal of Chemical Technology and Biotechnology, 81*(8), 1119–1129. DOI: 10.1002/jctb.1486

Davis Cortina, M., Romero de Terreros Aramburu, M., Neves, A. M., Hurtado, L., Jepsen, J., & Ulmer, U. (2024). The Integration of Thermal Energy Storage Within Metal Hydride Systems: A Comprehensive Review. *Inorganics, 2024*(12), 313. DOI: 10.3390/inorganics12120313

de Jong, E., Dam, M. A., Sipos, L., & Gruter, G.-J. M. (2012). Furandicarboxylic acid (FDCA), A Versatile Building Block for a Very Interesting Class of Polyesters. In Smith, P. B., & Gross, R. A. (Eds.), *Biobased monomers, polymers, and materials* (Vol. 1105, pp. 1–13). American Chemical Society., DOI: 10.1021/bk-2012-1105.ch001

de Jong, E., Visser, H. R. A., Dias, A. S., Harvey, C., & Gruter, G.-J. M. (2022). The Road to Bring FDCA and PEF to the Market. *Polymers, 14*(5), 943. DOI: 10.3390/polym14050943 PMID: 35267764 de Silva, C. J. G.Jr, de Medeiros, A. D. M., de Amorim, J. D. P., do Nascimento, H. A., Converti, A., Costa, A. F. S., & Sarubbo, L. A. (2021). Bacterial Cellulose Biotextiles for the Future of Sustainable Fashion: A Review. *Environmental Chemistry Letters, 19*(4), 2967–2980. DOI: 10.1007/s10311-021-01214-x

Delfani, P., Thuraga, V., Banerjee, B., & Chawade, A. (2024). Integrative approaches in modern agriculture: IoT, ML and AI for disease forecasting amidst climate change. *Precision Agriculture, 25*(5), 2589–2613. DOI: 10.1007/s11119-024-10164-7

Demirbas, A. (2011). Waste management, waste resource facilities and waste conversion processes. *Energy Conversion and Management, 52*(2), 1280–1287. DOI: 10.1016/j.enconman.2010.09.025

Deng, D., Zhang, L., Dong, M., Samuel, R. E., Ofori-Boadu, A., & Lamssali, M. (2020). Radioactive waste: A review. *Water Environment Research, 92*(10), 1818–1825. DOI: 10.1002/wer.1442 PMID: 32860717

Detrembleur, C., Mazza, M., Lou, X., Halleux, O., Lecomte, P., Mecerreyes, D., Hedrick, J. L., & Jérôme, R. (2000). New functional aliphatic polyesters by chemical modification of copolymers of ε-caprolactone with γ-(2-bromo-2-methylpropionate)-ε-caprolactone, γ-bromo-ε-caprolactone, and a mixture of β- and γ-ene-ε-caprolactone. *Macromolecules, 33*(21), 7751–7760. DOI: 10.1021/ma000488o

Dhanaraju, M., Chenniappan, P., Ramalingam, K., Pazhanivelan, S., & Kaliaperumal, R. (2022). Smart farming: Internet of Things (IoT)-based sustainable agriculture. *Agriculture, 12*(10), 1745. DOI: 10.3390/agriculture12101745

Dhania, S., Bernela, M., Rani, R., Parsad, M., Grewal, S., Kumari, S., & Thakur, R. (2022). Scaffolds the backbone of tissue engineering: Advancements in use of polyhydroxyalkanoates (PHA). *International Journal of Biological Macromolecules, 208*, 243–259. DOI: 10.1016/j.ijbiomac.2022.03.030 PMID: 35278518

Dhanke, J. A., Srivastava, D., Menaga, D., Raj, R., Kumar, K. V., Jangir, P., & Mani, P. (2024). Climate-Based AI-Powered Precision Irrigation: Sustainably Smart Agriculture Frameworks for Maximum Crop Yields. *Remote Sensing in Earth Systems Sciences*, •••, 1–12.

Dhuldhaj, U. P., Singh, R., & Singh, V. K. (2023). Pesticide contamination in agro-ecosystems: Toxicity, impacts, and bio-based management strategies. *Environmental Science and Pollution Research International, 30*(4), 9243–9270. DOI: 10.1007/s11356-022-24381-y PMID: 36456675

Dhupia, J., Adnanes, A. K., Lee, K. M., & Kennedy, L. (2011) Electrification of port and port operations. *Proceedings of MTEC2011*, pp. 1–6

Diaz-alvarez, A. E., Francos, J., Crochet, P., Cadierno, V. Recent advances in the use of glycerol as green solvent for synthetic organic chemistry. *Current Green Chem.*, 1, 51-65DOI: 10.2174/1570178620666230418093820

Díez-Pascual, A., & Díez-Vicente, A. (2014). Poly(3-hydroxybutyrate)/ZnO bionanocomposites with improved mechanical, barrier, and antibacterial properties. *International Journal of Molecular Sciences*, *15*(6), 10950–10973. DOI: 10.3390/ijms150610950 PMID: 24941255

Diop, M., Chirinda, N., Beniaich, A., El Gharous, M., & El Mejahed, K. (2022). Soil and water conservation in Africa: State of play and potential role in tackling soil degradation and building soil health in agricultural lands. *Sustainability (Basel)*, *14*(20), 13425. DOI: 10.3390/su142013425

Dissanayake, K., & Kularatna-Abeywardana, D. (2024). A review of supercapacitors: Materials, technology, challenges, and renewable energy applications. *Journal of Energy Storage*, *96*, 112563. DOI: 10.1016/j.est.2024.112563

Do, J. (2017). Mechanochemistry: A Force of Synthesis. *ACS Cent. Sci.* 3, 1, 13–19https://doi.org/DOI: 10.1021/acscentsci.6b00277

Doran, J. W. (1980). Microbial changes associated with residue management with reduced tillage. *Soil Science Society of America Journal*, *44*(3), 518–524. DOI: 10.2136/sssaj1980.03615995004400030016x

Dunn, P. J., Hettenbach, K., Ct, U. K., Pharmaceuticals, I., Plant, R. A. P. I., & Box, P. O. (2004). The development of an environmentally benign synthesis of sildenafil citrate (Viagra ™) and its assessment by Green Chemistry metrics. *Green Chemistry*, *6*(1), 43–48. DOI: 10.1039/B312329D

Dutta, A. K. (2023). Metal Chalcogenides-Based Multifunctional Nanomaterials for Solar-Light-Driven Photo-Catalysis and Electro-Chemical Sensing. In Kumar, A. (Ed.), *Innovative Multifunctional Nanomaterial for Photocatalysis, Sensing, and Imaging* (pp. 1–16). IGI Global Scientific Publishing., DOI: 10.4018/978-1-6684-8743-3.ch001

Eberle, U., Felderhoff, M., & Schüth, F. (2009). Chemical and physical solutions for hydrogen storage. *Angewandte Chemie International Edition*, *48*(36), 6608–6630. DOI: 10.1002/anie.200806293 PMID: 19598190

El Achaby, M., El Miri, N., Aboulkas, A., Zahouily, M., Bilal, E., Barakat, A., & Solhy, A. (2017). Processing and properties of eco-friendly bio-nanocomposite films filled with cellulose nanocrystals from sugarcane bagasse. *International Journal of Biological Macromolecules*, *96*, 340–352. DOI: 10.1016/j.ijbiomac.2016.12.040 PMID: 27988293

Elalfy, D. A., Gouda, E., Kotb, M. F., Bureš, V., & Sedhom, B. E. (2024). Comprehensive review of energy storage systems technologies, objectives, challenges, and future trends. *Energy Strategy Reviews*, *54*, 101482. DOI: 10.1016/j.esr.2024.101482

Elbeshbishy, E., & Okoye, F. (2019). Improper disposal of household hazardous waste: Landfill/municipal wastewater treatment plant. *Municipal Solid Waste Management, 183*.

Elimelech, M., & Phillip, W. A. (2011). The future of seawater desalination: Energy, technology, and the environment. *Science*, *333*(6043), 712–717. DOI: 10.1126/science.1200488 PMID: 21817042

El-Kassas, H. Y., & El-Sheekh, M. M. (2014). Cytotoxic Activity of Biosynthesized Gold Nanoparticles with an Extract of the Red Seaweed Corallina Officinalis on the MCF-7 Human Breast Cancer Cell Line. *Asian Pacific Journal of Cancer Prevention : APJCP*, *15*(10), 4311–4317. DOI: 10.7314/APJCP.2014.15.10.4311 PMID: 24935390

Entezari, A., Aslani, A., Zahedi, R., & Noorollahi, Y. (2023). Artificial intelligence and machine learning in energy systems: A bibliographic perspective. *Energy Strategy Reviews*, *45*, 101017. DOI: 10.1016/j.esr.2022.101017

Erisman, J. W., & Schaap, M. (2004). The need for ammonia abatement with respect to secondary PM reductions in Europe. *Environmental Pollution*, *129*(1), 159–163. DOI: 10.1016/j.envpol.2003.08.042 PMID: 14749079

Ewan, B. C. R., & Allen, R. W. K. (2005). A figure of merit assessment of the routes to hydrogen. *International Journal of Hydrogen Energy*, *30*(8), 809–819. DOI: 10.1016/j.ijhydene.2005.02.003

Fabra, M. J., Pardo, P., Martínez-Sanz, M., Lopez-Rubio, A., & Lagarón, J. M. (2016). Combining polyhydroxyalkanoates with nanokeratin to develop novel biopackaging structures. *Journal of Applied Polymer Science*, *133*(2), app.42695. Advance online publication. DOI: 10.1002/app.42695

Fajersztajn, L., Veras, M., Barrozo, L. V., & Saldiva, P. (2013). Air pollution: A potentially modifiable risk factor for lung cancer. *Nature Reviews. Cancer*, *13*(9), 674–678. DOI: 10.1038/nrc3572 PMID: 23924644

Fann, N., Roman, H. A., Fulcher, C. M., Gentile, M. A., Hubbell, B. J., Wesson, K., & Levy, J. I. (2011). Maximizing health benefits and minimizing inequality: Incorporating local-scale data in the design and evaluation of air quality policies. *Risk Analysis, 31*(6), 908–922. DOI: 10.1111/j.1539-6924.2011.01629.x PMID: 21615761

Fan, P., Yu, H., Xi, B., & Tan, W. (2022). A review on the occurrence and influence of biodegradable microplastics in soil ecosystems: Are biodegradable plastics substitute or threat? *Environment International, 163*, 107244. DOI: 10.1016/j.envint.2022.107244 PMID: 35436719

Farrell, A. E., Keith, D. W., & Corbett, J. J. (2003). A strategy for introducing hydrogen into transportation. *Energy Policy, 31*(13), 1357–1367. DOI: 10.1016/S0301-4215(02)00195-7

Federsel, H. J., Moody, T. S., & Taylor, S. J. (2021). Recent trends in enzyme immobilization—Concepts for expanding the biocatalysis toolbox. *Molecules (Basel, Switzerland), 26*(9), 2822. DOI: 10.3390/molecules26092822 PMID: 34068706

Fei, X., Wang, J., Zhang, X., Jia, Z., Jiang, Y., & Liu, X. (2022). Recent Progress on Bio-Based Polyesters Derived from 2,5-Furandicarbonxylic Acid (FDCA). *Polymers, 14*(3), 625. DOI: 10.3390/polym14030625 PMID: 35160613

Felderhoff, M., Weidenthaler, C., von Helmolt, R., & Eberle, U. (2007). Hydrogen storage: The remaining scientific and technological challenges. *Physical Chemistry Chemical Physics, 9*(21), 2643–2653. DOI: 10.1039/b701563c PMID: 17627309

Feng, X., Jiang, Y., Zhao, J., Miao, M., Cao, S., Fang, J., & Shi, L. (2015). Easy synthesis of photoluminescent N-doped carbon dots from winter melon for bioimaging. *RSC Advances, 5*(40), 31250–31254. DOI: 10.1039/C5RA02271A

Fenibo, E. O., Ijoma, G. N., & Matambo, T. (2022). Biopesticides in Sustainable Agriculture: Current Status and Future Prospects. In Mandal, S. D., Ramkumar, G., Karthi, S., & Jin, F. (Eds.), *New and Future Development in Biopesticide Research: Biotechnological Exploration* (pp. 1–53). Springer Nature Singapore., DOI: 10.1007/978-981-16-3989-0_1

Fernández, J. G., Fernández-Baldo, M. A., Berni, E., Camí, G., Durán, N., Raba, J., & Sanz, M. I. (2016). Production of Silver Nanoparticles Using Yeasts and Evaluation of Their Antifungal Activity against Phytopathogenic Fungi. *Process Biochemistry (Barking, London, England), 51*(9), 1306–1313. DOI: 10.1016/j.procbio.2016.05.021

Fernandez-Lafuente, R., Rosell, C. M., & Guisán, J. (1991). Enzyme reaction engineering: synthesis of antibiotics catalysed by stabilized penicillin G acylase in the presence of organic cosolvents. *Enzyme and microbial technology, 13*(11), 898-905. https://doi.org/ rights and contentDOI: 10.1016/0141-0229(91)90106-KGet

Fesharaki, P. J., Nazari, P., Shakibaie, M., Rezaie, S., Banoee, M., Abdollahi, M., & Shahverdi, A. R. (2010). Biosynthesis of Selenium Nanoparticles Using Klebsiella pneumoniae and Their Recovery by a Simple Sterilization Process. *Brazilian Journal of Microbiology : Publication of the Brazilian Society for Microbiology*, *41*(2), 461–466. DOI: 10.1590/S1517-83822010000200028 PMID: 24031517

Fico, D., Rizzo, D., Casciaro, R., & Esposito Corcione, C. (2022). A Review of Polymer-Based Materials for Fused Filament Fabrication (FFF): Focus on Sustainability and Recycled Materials. *Polymers*, *14*(3), 465. DOI: 10.3390/polym14030465 PMID: 35160455

Finch, C. (1985). Anionic polymerization to cationic polymerization. In Mark, H. F., Bikales, N. M., Overberger, C. G., & Menges, G. (Eds.), *Encyclopedia of Polymer Science and Engineering* (Vol. 2). Wiley-Interscience.,

Foong, S. Y., Chan, Y. H., Yek, P. N. Y., Lock, S. S. M., Chin, B. L. F., Yiin, C. L., Lan, J. C.-W., & Lam, S. S. (2024). Microwave-assisted pyrolysis in biomass and waste valorisation: Insights into the life-cycle assessment (LCA) and techno-economic analysis (TEA). *Chemical Engineering Journal*, *491*, 151942. DOI: 10.1016/j.cej.2024.151942

Fortunati, E., Armentano, I., Iannoni, A., & Kenny, J. (2010). Development and thermal behaviour of ternary PLA matrix composites. *Polymer Degradation & Stability*, *95*(11), 2200–2206. DOI: 10.1016/j.polymdegradstab.2010.02.034

Friel, S., Dangour, A. D., Garnett, T., Lock, K., Chalabi, Z., Roberts, I., Butler, A., Butler, C. D., Waage, J., McMichael, A. J., & Haines, A. (2009). Public health benefits of strategies to reduce greenhouse-gas emissions: Food and agriculture. *Lancet*, *374*(9706), 2016–2025. DOI: 10.1016/S0140-6736(09)61753-0 PMID: 19942280

Gadipelli, S., & Guo, Z. X. (2015). Graphene-based materials: Synthesis and gas sorption, storage and separation. *Progress in Materials Science*, *69*, 1–60.

Gandhi, S., Jain, S., Bhatia, D., & Yadav, P. (2024). *Azadirachta indica:* derived, red-emitting, carbon nanoparticles for cellular bioimaging and potential therapeutics, *bioRxiv*, 02,26, 582094; DOI: 10.1101/2024.02.26.582094

Gandini, A., Silvestre, A. J. D., Neto, C. P., Sousa, A. F., & Gomes, M. (2008). The furan counterpart of poly (ethylene terephthalate): An alternative material based on renewable resources. *Journal of Polymer Science. Part A, Polymer Chemistry*, *47*(1), 295–298. DOI: 10.1002/pola.23130

Garg, D., Sridhar, K., Stephen Inbaraj, B., Chawla, P., Tripathi, M., & Sharma, M. (2023). Nano-biofertilizer formulations for agriculture: A systematic review on recent advances and prospective applications. *Bioengineering (Basel, Switzerland)*, *10*(9), 1010. DOI: 10.3390/bioengineering10091010 PMID: 37760112

Gaumet, M., Vargas, A., Gurny, R., & Delie, F. (2008). Nanoparticles for Drug Delivery: The Need for Precision in Reporting Particle Size Parameters. *European Journal of Pharmaceutics and Biopharmaceutics*, *69*(1), 1–9. DOI: 10.1016/j.ejpb.2007.08.001 PMID: 17826969

Gawande, M. B., Shelke, S. N., Zboril, R., & Varma, R. S. (2014). Microwave-assisted chemistry: Synthetic applications for rapid assembly of nanomaterials and organics. *Accounts of Chemical Research*, *47*(4), 1338–1348. DOI: 10.1021/ar400309b PMID: 24666323

Gedda, G., Sankaranarayanan, S. A., Putta, C. L., Gudimella, K. K., Rengan, A. K., & Girma, W. M. (2023). Green synthesis of multi-functional carbon dots from medicinal plant leaves for antimicrobial, antioxidant, and bioimaging applications. *Scientific Reports*, *13*(1), 6371. DOI: 10.1038/s41598-023-33652-8 PMID: 37076562

Ge, R., Boyce, A. M., Shui Zhang, Y., Shearing, P. R., Cumming, D. J., & Smith, R. M. (2023). Discrete element method and electrochemical modelling of lithium ion cathode structures characterised by X-ray computed tomography. *Chemical Engineering Journal*, *465*, 142749. DOI: 10.1016/j.cej.2023.142749

Getahun, S., Kefale, H., & Gelaye, Y. (2024). Application of Precision Agriculture Technologies for Sustainable Crop Production and Environmental Sustainability: A Systematic Review. *TheScientificWorldJournal*, *2024*(1), 2126734. DOI: 10.1155/2024/2126734 PMID: 39421732

Ge, X., Song, W., Chen, X., Qian, C., Zhou, S., & Liu, X. (2021). Tandem micellar catalysis and cloud point extraction process for CS coupling reaction in water. *Colloids and Surfaces. A, Physicochemical and Engineering Aspects*, *616*, 126263. DOI: 10.1016/j.colsurfa.2021.126263

Ghanbarzadeh, B., Almasi, H., & Entezami, A. A. (2011). Improving the barrier and mechanical properties of corn starch-based edible films: Effect of citric acid and carboxymethyl cellulose. *Industrial Crops and Products*, *33*(1), 229–235. DOI: 10.1016/j.indcrop.2010.10.016

Gharat, N. N., & Rathod, V. K. (2020). Ultrasound-assisted organic synthesis. In *Green sustainable process for chemical and environmental engineering and science* (pp. 1–41). Elsevier., DOI: 10.1016/B978-0-12-819540-6.00001-2

Ghomi, E. R., Khosravi, F., Ardahaei, A. S., Dai, Y., Neisiany, R. E., Foroughi, F., Wu, M., Das, O., & Ramakrishna, S. (2021). The Life Cycle Assessment for Polylactic Acid (PLA) to Make It a Low-Carbon Material. *Polymers, 13*(11), 1854. DOI: 10.3390/polym13111854 PMID: 34199643

Ghorbani, B., Zendehboudi, S., Saady, N. M. C., & Dusseault, M. B. (2023). Hydrogen storage in North America: Status, prospects, and challenges. *Journal of Environmental Chemical Engineering, 11*(3), 109957. DOI: 10.1016/j.jece.2023.109957

Ghose, S., Bhattacharjee, B., Rynjah, D., & Laloo, D. (2025). Pesticides and Allergens. In Odoh, U. E., Gurav, S. S., & Chukwuma, M. O. (Eds.), *Pharmacognosy and Phytochemistry* (1st ed., pp. 315–334). Wiley., DOI: 10.1002/9781394203680.ch16

Gidde, M. R., Todkar, V. V., & Kokate, K. K. (2008, March). Municipal solid waste management in emerging mega cities: a case study of Pune city. In *Proceedings of Indo Italian Conference on Green and Clean Environment,* Pune, India *(*March 20-21*)*.

Gidley, M. J., Lillford, P. J., Rowlands, D. W., Lang, P., Dentini, M., Crescenzi, W., Edwards, M., Fanutti, C., & Reid, J. S. G. (1991). Structure and solution properties of tamarind seed polysaccharide. *Carbohydrate Research, 214*(2), 299–314. DOI: 10.1016/0008-6215(91)80037-N PMID: 1769022

Giglio, E., Luzzani, G., Terranova, V., Trivigno, G., Niccolai, A., & Grimaccia, F. (2023). An Efficient Artificial Intelligence Energy Management System for Urban Building Integrating Photovoltaic and Storage. *IEEE Access : Practical Innovations, Open Solutions, 11*, 18673–18688. DOI: 10.1109/ACCESS.2023.3247636

Giuliano, A., Gioiella, F., Sofia, D., & Lotrecchiano, N. (2018a). A novel methodology and technology to promote the social acceptance of biomass Power plants avoiding Nimby syndrome. *Chemical Engineering Transactions, 67*, 307–312.

Gleick, P. H. (2014). Water, Drought, Climate Change, and Conflict in Syria. *Weather, Climate, and Society, 6*(3), 331–340. DOI: 10.1175/WCAS-D-13-00059.1

Godavitarne, C., Robertson, A., Peters, J., & Rogers, B. (2017). Biodegradable Materials. *Orthopaedics and Trauma, 31*(6), 316–320. DOI: 10.1016/j.mporth.2017.07.011

Gomes, M., Gandini, A., Silvestre, A. J. D., & Reis, B. (2011). Synthesis and characterization of poly (2,5-furan dicarboxylate)s based on a variety of diols. *Journal of Polymer Science. Part A, Polymer Chemistry, 49*(16), 3759–3768. DOI: 10.1002/pola.24812

Gontard, N., Sonesson, U., Birkved, M., Majone, M., Bolzonella, D., Celli, A., Angellier-Coussy, H., Jang, G. W., Verniquet, A., Broeze, J., Schaer, B., Batista, A. P., & Sebok, A. (2018). A research challenge vision regarding management of agricultural waste in a circular bio-based economy. *Critical Reviews in Environmental Science and Technology*, *48*(7–8), 614–654. DOI: 10.1080/10643389.2018.1471957

Green Dot Bioplastics. (2022). A Straightforward Explanation of Biodegradable vs. Compostable vs. Oxo-Degradable Plastics. Retrieved October 10, 2022, from https://www.greendotbioplastics.com/biodegradable-vs-compostable-vs-oxo-1479degradable-plastics-a-straightforward-explanation/

Gruber, J. V. (1999). Polysaccharide-Based Polymers in Cosmetics. In Goddard, E. D., & Gruber, J. V. (Eds.), *Principles of Polymer Science and Technology in Cosmetic and Personal Care* (Vol. 22, pp. 339–403). Marcel Dekker Inc.

Gudimella, K., Gedda, G., Kumar, P. S., Babu, B. K., Yamajala, B., Rao, B. V., Singh, P. P., Kumar, D., & Sharma, A. (2022). Novel synthesis of fluorescent carbon dots from bio-based Carica Papaya Leaves: Optical and structural properties with antioxidant and anti-inflammatory activities. *Environmental Research*, *204*, 111854. DOI: 10.1016/j.envres.2021.111854 PMID: 34437850

Guo, C., & Wang, C. (2020). The theoretical research of hydrogen storage capacities of Cu3Bx (X=1–4) compounds under ambient conditions. *International Journal of Hydrogen Energy*, *45*(46), 24947–24957. DOI: 10.1016/j.ijhydene.2020.06.089

Guo, S., Li, J., Wang, Y., & Wang, Z. (2024). Electrochemical-thermal coupling model of lithium-ion battery at ultra-low temperatures. *Applied Thermal Engineering*, *240*, 122205. DOI: 10.1016/j.applthermaleng.2023.122205

Gupta, P., Sharma, A., & Bhardwaj, L. K. (2024). Solid waste management (SWM) and its effect on environment & human health. Futuristic Trends in Agriculture Engineering & Food Sciences Volume 3; 91-101. IIP Proceedings. https://www.doi.org/10.58532/V3BCAG4P1CH7

Gupta, B. D., Pathak, A., & Semwal, V. (2019). Carbon-Based Nanomaterials for Plasmonic Sensors: A Review. *Sensors (Basel)*, *19*(16), 3536. DOI: 10.3390/s19163536 PMID: 31412590

Gupta, B., Revagade, N., & Hilborn, J. (2007). Poly (lactic acid) fiber: An overview. *Progress in Polymer Science*, *32*(4), 455–482. DOI: 10.1016/j.progpolymsci.2007.01.005

Hadj, T. B. (2021). Nonlinear impact of biomass energy consumption on ecological footprint in a fossil fuel–dependent economy. *Environmental Science and Pollution Research International*, *28*(48), 69329–69342. DOI: 10.1007/s11356-021-15521-x PMID: 34296408

Hagemeyer, A., & Clara, S. (2016). *Materials : Catalysts $. March 2015*, 1–8. https://doi.org/DOI: 10.1016/B978-0-12-803581-8.01121-8

Hahladakis, N. J., Costas, A. V., Weber, R., Iacovidou, E., & Purnell, P. (2018). An overview of chemical additives present in plastics: Migration, release, fate and environmental impact during their use, disposal and recycling. *Journal of Hazardous Materials*, *344*, 179–199. DOI: 10.1016/j.jhazmat.2017.10.014 PMID: 29035713

Hameed, S., Khalil, A. T., Ali, M., Numan, M., Khamlich, S., Shinwari, Z. K., & Maaza, M. (2019). Greener Synthesis of ZnO and Ag-ZnO Nanoparticles Using Silybum marianum for Diverse Biomedical Applications. *Nanomedicine (London)*, *14*(6), 655–673. DOI: 10.2217/nnm-2018-0279 PMID: 30714480

Han, C., Kundu, B. K., Liang, Y., & Sun, Y. (2024). Near-Infrared Light-Driven Photocatalysis with an Emphasis on Two-Photon Excitation: Concepts, Materials, and Applications. *Advanced Materials*, *36*(5), 2307759–2307772. DOI: 10.1002/adma.202307759 PMID: 37703435

Han, H., Liu, Z., Li, J., & Zeng, Z. (2024). Challenges in remote sensing based climate and crop monitoring: Navigating the complexities using AI. *Journal of Cloud Computing (Heidelberg, Germany)*, *13*(1), 34. DOI: 10.1186/s13677-023-00583-8

Han, Z., Fang, R., Chu, D., Wang, D.-W., & Ostrikov, K. (2023). Introduction to Supercapacitors. *Nanoscale Advances*, *5*(16), 4015–4017. DOI: 10.1039/D3NA90074F PMID: 37560421

Haque, S. E. (2021). How effective are existing phosphorus management strategies in mitigating surface water quality problems in the US? *Sustainability (Basel)*, *13*(12), 6565. DOI: 10.3390/su13126565

Hasan, M. M., Haque, R., Jahirul, M. I., Rasul, M. G., Fattah, I. M. R., Hassan, N. M. S., & Mofijur, M. (2025). Advancing energy storage: The future trajectory of lithium-ion battery technologies. *Journal of Energy Storage*, *120*, 116511.

Hassan, M., Mohanty, A. K., & Misra, M. (2024). 3D printing in upcycling plastic and biomass waste to sustainable polymer blends and composites: A review. *Materials & Design*, *237*, 112558. DOI: 10.1016/j.matdes.2023.112558

Hassan, Q., Sameen, A. Z., Olapade, O., Alghoul, M., Salman, H. M., & Jaszczur, M. (2023). Hydrogen fuel as an important element of the energy storage needs for future smart cities. *International Journal of Hydrogen Energy*, *48*(78), 30247–30262. DOI: 10.1016/j.ijhydene.2023.03.413

Hassan, Q., Sameen, A. Z., Salman, H. M., Jaszczur, M., & Al-Jiboory, A. K. (2023). RETRACTED: Hydrogen energy future: Advancements in storage technologies and implications for sustainability. *Journal of Energy Storage*, *72*, 108404. DOI: 10.1016/j.est.2023.108404

Hazarika, R., & Saikia, A. (2020). Landfill site suitability analysis using AHP for solid waste management in the Guwahati Metropolitan Area, India. *Arabian Journal of Geosciences*, *13*(21), 1–14. DOI: 10.1007/s12517-020-06156-2

Helbig, W. A. (1946). Activated carbon. *Journal of Chemical Education*, *23*(2), 98–104. DOI: 10.1021/ed023p98

Heredia-Guerrero, J. A., Heredia, A., Domínguez, E., Cingolani, R., Bayer, I. S., Athanassiou, A., & Benítez, J. J. (2017). Cutin from agro-waste as a raw material for the production of bioplastics. *Journal of Experimental Botany*, *68*(19), 5401–5410. DOI: 10.1093/jxb/erx272 PMID: 28992177

Hidalgo, D., Martín-Marroquín, J. M., & Corona, F. (2019). A multi-waste management concept as a basis towards a circular economy model. *Renewable & Sustainable Energy Reviews*, *111*, 481–489. DOI: 10.1016/j.rser.2019.05.048

Himaja, M., Poppy, D., & Asif, K. (2011). Green technique-solvent free synthesis and its advantages. *International Journal of Research in Ayurveda and Pharmacy*, *2*(4).

Holten, C. H. (1971). *Lactic Acid: Properties and Chemistry of Lactic Acid and Derivatives*. Verlag Chemie. Retrieved from https://www.cabidigitallibrary.org/action/doSearch?do=Lactic+acid.+Properties+and+chemistry+of+lactic+acid+and+derivatives

Hoque, A. (2024). Artificial Intelligence in Post-Harvest Drying Technologies: A Comprehensive Review on Optimization, Quality Enhancement, and Energy Efficiency. *International Journal of Science and Research (Raipur, India)*, *13*(11), 493–502. DOI: 10.21275/SR241107163717

Hoque, A., & Padhiary, M. (2024). Automation and AI in Precision Agriculture: Innovations for Enhanced Crop Management and Sustainability. *Asian Journal of Research in Computer Science*, *17*(10), 95–109. DOI: 10.9734/ajrcos/2024/v17i10512

Hoque, A., Padhiary, M., Prasad, G., & Kumar, K. (2025). Real-Time Data Processing in Agricultural Robotics. In Bora, D. J., & Bania, R. K. (Eds.), *Computer Vision Techniques for Agricultural Advancements* (pp. 431–468). IGI Global., DOI: 10.4018/979-8-3693-8019-2.ch014

Hou, Y., Velthof, G. L., & Oenema, O. (2015). Mitigation of Ammonia, nitrous oxide and methane emissions from manure management chains: A meta-analysis and integrated assessment. *Global Change Biology*, *21*(3), 1293–1312. DOI: 10.1111/gcb.12767 PMID: 25330119

Huq, T., Salmieri, S., Khan, A., Khan, R. A., Le Tien, C., Riedl, B., Fraschini, C., Bouchard, J., Uribe-Calderon, J., Kamal, M. R., & Lacroix, M. (2012). Nanocrystalline cellulose (NCC) reinforced alginate-based biodegradable nanocomposite films. *Carbohydrate Polymers*, *90*(4), 1757–1763. DOI: 10.1016/j.carbpol.2012.07.065 PMID: 22944444

Hussain, I., Lamiel, C., Javed, M. S., Ahmad, M., Sahoo, S., Chen, X., Qin, N., Iqbal, S., Gu, S., Li, Y., Chatzichristodoulou, C., & Zhang, K. (2023). MXene-based heterostructures: Current trend and development in electrochemical energy storage devices. *Progress in Energy and Combustion Science*, *97*, 101097. DOI: 10.1016/j.pecs.2023.101097

Ibáñez-Forés, V., Bovea, M. D., & Azapagic, A. (2013). Assessing the sustainability of best available techniques (BAT): Methodology and application in the ceramic tiles industry. *Journal of Cleaner Production*, *51*, 162–176. DOI: 10.1016/j.jclepro.2013.01.020

Ibrahim, N. I., Shahar, F. S., Sultan, M. T. H., Md Shan, A. U., Safri, S. N. A., & Yazik, M. H. M. (2021). Overview of Bioplastic Introduction and Its Applications in Product Packaging. *Coatings*, *11*(11), 1423. DOI: 10.3390/coatings11111423

Ijaz, F., Shahid, S., Khan, S. A., Ahmad, W., & Zaman, S. (2017). Green Synthesis of Copper Oxide Nanoparticles Using Abutilon Indicum Leaf Extract: Antimicrobial, Antioxidant and Photocatalytic Dye Degradation Activitie. *Tropical Journal of Pharmaceutical Research*, *16*(4), 743. DOI: 10.4314/tjpr.v16i4.2

Ilahi, N., Haleem, A., Iqbal, S., Fatima, N., Sajjad, W., & Sideeq, A.. (2021). Biosynthesis of Silver Nanoparticles Using Endophytic Fusarium Oxysporum Strain NFW16 and Their In Vitro Antibacterial Potential. *Microscopy Research and Technique*, *1*, 1. DOI: 10.1002/jemt.24018 PMID: 34888986

International Solid Waste Association (ISW). (2024). *Global Waste Management Outlook 2024: Beyond an Age of Waste, Turning Rubbish into a Resource*. Association ISW.

Invernizzi, S., Montagnoli, F., & Carpinteri, A. (2019). Fatigue assessment of the collapsed XXth century cable-stayed Polcevera bridge in Genoa. *Procedia Structural Integrity*, *18*, 237–244. DOI: 10.1016/j.prostr.2019.08.159

Irham, A., Roslan, M. F., Jern, K. P., Hannan, M. A., & Mahlia, T. I. (2024). Hydrogen energy storage integrated grid: A bibliometric analysis for sustainable energy production. *International Journal of Hydrogen Energy*, *63*, 1044–1087. DOI: 10.1016/j.ijhydene.2024.03.235

Ishchenko, V., Pohrebennyk, V., Borowik, B., Falat, P., & Shaikhanova, A. (2018). Toxic substances in hazardous household waste.

Islam, A., Islam, T., Mahmud, H., Raihan, O., Islam, M. S., Marwani, H. M., & Awual, M. R. (2024). Accelerating the green hydrogen revolution: A comprehensive analysis of technological advancements and policy interventions. *International Journal of Hydrogen Energy*, *67*, 458–486.

Istiak, M. A., Syeed, M. M., Hossain, M. S., Uddin, M. F., Hasan, M., Khan, R. H., & Azad, N. S. (2023). Adoption of Unmanned Aerial Vehicle (UAV) imagery in agricultural management: A systematic literature review. *Ecological Informatics*, *78*, 102305. DOI: 10.1016/j.ecoinf.2023.102305

Jambeck, J. R., Geyer, R., Wilcox, C., Siegler, T. R., Perryman, M., Andrady, A., Narayan, R., & Lavender Law, K. (2015). Plastic waste inputs from land into the ocean. *Science*, *347*(6223), 768–771. DOI: 10.1126/science.1260352 PMID: 25678662

Jawaid, M., & Abdul Khalil, H. P. S. (2011). Cellulosic/synthetic fibre reinforced polymer hybrid composites: A review. *Carbohydrate Polymers*, *86*(1), 1–18. DOI: 10.1016/j.carbpol.2011.04.043

Jaybhaye, R., Mundhe, N., & Dorik, B. (2014). Site suitability for urban solid waste disposal using geoinformatics: A case study of Pune Municipal Corporation, Maharashtra, India. *Int J Adv Remote Sens GIS*, *3*(1), 769–783.

Jevapatarakul, D., T-Thienprasert, J., Payungporn, S., Chavalit, T., Khamwut, A., & T-Thienprasert, N. P. (2020). Utilization of Cratoxylum Formosum Crude Extract for Synthesis of ZnO Nanosheets: Characterization, Biological Activities and Effects on Gene Expression of Nonmelanoma Skin Cancer Cell. *Biomedicine and Pharmacotherapy*, *130*, 110552. DOI: 10.1016/j.biopha.2020.110552 PMID: 32739741

Jeyarani, S., Vinita, N. M., Puja, P., Senthamilselvi, S., Devan, U., Velangani, A. J., Biruntha, M., Pugazhendhi, A., & Kumar, P. (2020). Biomimetic Gold Nanoparticles for its Cytotoxicity and Biocompatibility Evidenced by Fluorescence-Based Assays in Cancer (MDA-MB-231) and Non-cancerous (HEK-293) Cells. *Journal of Photochemistry and Photobiology. B, Biology*, *202*, 111715. DOI: 10.1016/j.jphotobiol.2019.111715 PMID: 31790882

Jiang, G., Hill, D. J., Kowalczuk, M., Johnston, B., Adamus, G., Irorere, V., & Radecka, I. (2016). Carbon sources for polyhydroxyalkanoates and an integrated biorefinery. *International Journal of Molecular Sciences*, *17*(7), 1157. DOI: 10.3390/ijms17071157 PMID: 27447619

Jiang, M., Liu, Q., Zhang, Q., Ye, C., & Zhou, G. (2012). A series of furan-aromatic polyesters synthesized via direct esterification method based on renewable resources. *Journal of Polymer Science. Part A, Polymer Chemistry*, *50*(5), 1026–1036. DOI: 10.1002/pola.25859

Jiang, T., Wang, M., Zhang, W., Zhu, C., & Wang, F. (2024). A comprehensive analysis of agricultural non-point source pollution in China: Current status, risk assessment and management strategies. *Sustainability (Basel)*, *16*(6), 2515. DOI: 10.3390/su16062515

Jiyuan, L., Mingliang, L., Xiangzheng, D., Dafang, Z., Zengxiang, Z., & Di, L. (2002). The land use and land cover change database and its relative studies in China. *Journal of Geographical Sciences*, *12*(3), 275–282. DOI: 10.1007/BF02837545

Joseph, C. S., Prashanth, K. H., Rastogi, N., Indiramma, A., Reddy, S. Y., & Raghavarao, K. (2011). Optimum blend of chitosan and poly(ε-caprolactone) for fabrication of films for food packaging applications. *Food and Bioprocess Technology*, *4*(7), 1179–1185. DOI: 10.1007/s11947-009-0203-1

Jost, V., & Kopitzky, R. (2015). Blending of polyhydroxybutyrate-co-valerate with polylactic acid for packaging applications: Reflections on miscibility and effects on the mechanical and barrier properties. *Chemical and Biochemical Engineering Quarterly*, *29*(2), 221–246. DOI: 10.15255/CABEQ.2014.2257

Kan, A. (2009). General characteristics of waste management: A review. *Energy Education Science and Technology, Part A. Energy Science and Research*, •••, 23.

Kaniuk, Ł., & Stachewicz, U. (2021). Development and Advantages of Biodegradable PHA Polymers Based on Electrospun PHBV Fibers for Tissue Engineering and Other Biomedical Applications. *ACS Biomaterials Science & Engineering*, *7*(12), 5339–5362. DOI: 10.1021/acsbiomaterials.1c00757 PMID: 34649426

Kannah, R. Y., Kumar, M. D., Kavitha, S., Banu, J. R., Tyagi, V. K., & Rajaguru, P.. (2022). Production and recovery of polyhydroxyalkanoates (PHA) from waste streams: A review. *Bioresource Technology*, *366*, 128203. DOI: 10.1016/j.biortech.2022.128203 PMID: 36330969

Kannan, N., Mukunthan, K. S., & Balaji, S. (2011). A Comparative Study of Morphology, Reactivity and Stability of Synthesized Silver Nanoparticles Using Bacillus Subtilis and Catharanthus Roseus (L.) G. *Colloids and Surfaces. B, Biointerfaces*, *86*(2), 378–383. DOI: 10.1016/j.colsurfb.2011.04.024 PMID: 21592749

Kaoje, I. U., Dankani, I. M., & Ishiaku, I. (2016). Site suitability analysis for municipal solid waste disposal in Birnin Kebbi, Nigeria. *IOSR Journal of Humanities and Social Science (IOSR-JHSS), 21*(7), 01-10.

Karande, R. D., Abitha, V. K., Rane, A. V., & Mishra, R. K. (2015). Preparation of polylactide from synthesized lactic acid and effect of reaction parameters on conversion. *Journal of Materials Science and Engineering: Advanced Technology*, *12*(1), 1–37.

Karka, P., Papadokonstantakis, S., & Kokossis, A. (2017). Cradle-to-gate assessment of environmental impacts for a broad set of biomass-to-product process chains. *The International Journal of Life Cycle Assessment*, *22*(9), 1418–1440. DOI: 10.1007/s11367-017-1262-6

Kasama, K., Aoyama, H., & Akai, S. (2020). Enantiodivergent Synthesis of Axially Chiral Biphenyls from σ-Symmetric 1, 1′-Biphenyl-2, 6-diol Derivatives by Single Lipase-Catalyzed Acylative and Hydrolytic Desymmetrization. *European Journal of Organic Chemistry*, *2020*(6), 654–661. DOI: 10.1002/ejoc.201901583

Kaushal, R. K., Varghese, G. K., & Chabukdhara, M. (2012). Municipal solid waste management in India-current state and future challenges: A review. *International Journal of Engineering Science and Technology*, *4*(4), 1473–1489.

Kaya, C. (2025). Optimizing Crop Production With Plant Phenomics Through High-Throughput Phenotyping and AI in Controlled Environments. *Food and Energy Security*, *14*(1), e70050. DOI: 10.1002/fes3.70050

Kelly, A. (Ed.). (2012). *Concise encyclopedia of composite materials*. Elsevier.

Kelly, F. J., Dunster, C., & Mudway, I. (2003). Air pollution and the elderly: Oxidant/antioxidant issues worth consideration. *The European Respiratory Journal*, *40*(40 suppl), 70s–75s. DOI: 10.1183/09031936.03.00402903 PMID: 12762578

Khajuria, A., Matsui, T., Machimura, T., & Morioka, T. (2010). Assessment of the challenge of sustainable recycling of municipal solid waste management in India. *International Journal of Environmental Technology and Management, 13*(2), 171–187. DOI: 10.1504/IJETM.2010.034300

Khalaj, M., Kamali, M., Costa, M. E. V., & Capela, I. (2020). Green Synthesis of Nanomaterials - A Scientometric Assessment. *Journal of Cleaner Production, 267*, 122036. DOI: 10.1016/j.jclepro.2020.122036

Khan, A. A., Laghari, A. A., Rashid, M., Li, H., Javed, A. R., & Gadekallu, T. R. (2023). Artificial intelligence and blockchain technology for secure smart grid and power distribution Automation: A State-of-the-Art Review. *Sustainable Energy Technologies and Assessments, 57*, 103282. DOI: 10.1016/j.seta.2023.103282

Khan, M. S., Hossain, S., & Jahan, I. (2015). *RS and GIS based site suitability analysis for solid waste disposal: A multi criteria analysis of Khulna City area. Proc. of the WasteSafe*. KUET.

Khunprasert, P., Grisdanurak, N., Thaveesri, J., Danutra, V., & Puttitavorn, W. (2008). Radiographic film waste management in Thailand and cleaner technology for silver leaching. *Journal of Cleaner Production, 16*(1), 28–36. DOI: 10.1016/j.jclepro.2006.06.010

Kilic, A., Oral, B., Eroglu, D., & Yildirim, R. (2023). Machine learning for beyond Li-ion batteries: Powering the research. *Journal of Energy Storage, 73*, 109057. DOI: 10.1016/j.est.2023.109057

Kim, J. S., Lee, B. H., & Kim, B. S. (2005). Production of poly(3-hydroxybutyrate-co-4-hydroxybutyrate) by *Ralstonia eutropha. Biochemical Engineering Journal, 23*(2), 169–174. DOI: 10.1016/j.bej.2005.01.016

Kim, J., & Park, C. (2020). Education, skill training, and lifelong learning in the era of technological revolution: A review. *Asian-Pacific Economic Literature, 34*(2), 3–19. DOI: 10.1111/apel.12299

Kister, G., Cassanas, G., Bergounhon, M., Hoarau, D., & Vert, M. (2000). Structural characterization and hydrolytic degradation of solid copolymers of D,L-lactide-co-ε-caprolactone by Raman Spectroscopy. *Polymer, 41*(3), 925–932. DOI: 10.1016/S0032-3861(99)00223-2

Kobayashi, S. (1999). Enzymatic polymerization: A new method of polymer synthesis. *Journal of Polymer Science. Part A, Polymer Chemistry, 37*(16), 3041–3056. DOI: 10.1002/(SICI)1099-0518(19990815)37:16<3041::AID-POLA1>3.0.CO;2-V

Kobayashi, S., Uyama, H., & Ohmae, M. (2001). Enzymatic polymerization for precision polymer synthesis. *Bulletin of the Chemical Society of Japan*, *74*(4), 613–635. DOI: 10.1246/bcsj.74.613

Koech, A. K., Mwandila, G., Mulolani, F., & Mwaanga, P. (2024). Lithium-ion battery fundamentals and exploration of cathode materials: A review. *South African Journal of Chemical Engineering*.

Kogan, F. (2022). New Remote Sensing Vegetation Health Technology. In F. Kogan, *Remote Sensing Land Surface Changes* (pp. 121–148). Springer International Publishing. DOI: 10.1007/978-3-030-96810-6_5

Krajewski, M. (2017). Magnetic-field-induced synthesis of magnetic wire-like micro-and nanostructures. *Nanoscale*, *9*(43), 16511–16545. DOI: 10.1039/C7NR05823C PMID: 29067381

Krishna, G., Singh, R., Gehlot, A., Shaik, V. A., Twala, B., & Priyadarshi, N. (2024). IoT-based real-time analysis of battery management system with long range communication and FLoRa. *Results in Engineering*, *23*, 102770.

Krishnan, S. R., Nallakaruppan, M. K., Chengoden, R., Koppu, S., Iyapparaja, M., Sadhasivam, J., & Sethuraman, S. (2022). Smart water resource management using Artificial Intelligence—A review. *Sustainability (Basel)*, *14*(20), 13384. DOI: 10.3390/su142013384

Krishna, R. S., Mishra, J., Meher, S., Das, S. K., Mustakim, S. M., & Singh, S. K. (2020). Industrial solid waste management through sustainable green technology: Case study insights from steel and mining industry in Keonjhar, India. *Materials Today: Proceedings*, *33*, 5243–5249. DOI: 10.1016/j.matpr.2020.02.949

Kshetri, N. (2021). Blockchain and sustainable supply chain management in developing countries. *International Journal of Information Management*, *60*, 102376. DOI: 10.1016/j.ijinfomgt.2021.102376

Kulkarni, S. (2016). Sustainable agrochemicals for conservation of agriculture and climate change. *Conservation Agriculture:* An Approach to Combat Climate Change in Indian Himalaya, 135-157. https://doi.org/DOI: 10.1007/978-981-10-2558-7

Kumar, K., Tyagi, U., Maity, S. K., & Kumar, G. (2024). Next-Gen Energy Storage Devices: Carbon-Based Materials for Flexible Supercapacitors. In *Innovative Materials for Industrial Applications: Synthesis, Characterization and Evaluation* (pp. 135–168). IGI Global Scientific Publishing. DOI: 10.4018/979-8-3693-7505-1.ch005

Kumar, K., Tyagi, U., Maity, S. K., Singh, S., Nikita, , & Kumar, G. (2025, April). A strategic approach towards the synthesis of jackfruit peel-based activated carbon: Assessment of different activating agents and operating conditions. *Biomass Conversion and Biorefinery*, *15*(8), 12387–12408. Advance online publication. DOI: 10.1007/s13399-024-06030-1

Kumar, K., Tyagi, U., Sirohi, S., Kumar, R., Kumar Maity, S., Nikita, , Singh, S., & Kumar, G. (2025). A critical review on nanostructure-doped carbonized biomass: A new Era in sustainable supercapacitor technology. *Fuel*, *381*, 133707. DOI: 10.1016/j.fuel.2024.133707

Kumar, P. (2025). Nanofertilizers, Nanopesticides for Individual's (Farmer's) Economy. In Kumar, P., & Dubey, R. C. (Eds.), *Nanofertilizers for Sustainable Agriculture* (pp. 383–406). Springer Nature Switzerland., DOI: 10.1007/978-3-031-78649-5_16

Kumar, P., Barrett, D. M., Delwiche, M. J., & Stroeve, P. (2009). Methods for pretreatment of lignocellulosic biomass for efficient hydrolysis and biofuel production. *Industrial & Engineering Chemistry Research*, *48*(8), 3713–3729. DOI: 10.1021/ie801542g

Kumar, P., Kumar Jain, A., Srivastava, J. P., Kumar, R., Saxena, K. K., Prakash, C., & Buddhi, D. (2024). Multiphysics Simulation of the Shape Prediction and Material Removal Rate in Electrochemical Machining Process. *Advances in Materials and Processing Technologies*, *10*(2), 772–784. DOI: 10.1080/2374068X.2023.2192132

Kumar, R., Verma, A., Shome, A., Sinha, R., Sinha, S., Jha, P. K., Kumar, R., Kumar, P., Shubham, , Das, S., Sharma, P., & Vara Prasad, P. V. (2021). Impacts of plastic pollution on ecosystem services, sustainable development goals, and the need to focus on circular economy and policy interventions. *Sustainability (Basel)*, *13*(17), 9963. DOI: 10.3390/su13179963

Kumar, S., Bhattacharyya, J. K., Vaidya, A. N., Chakrabarti, T., Devotta, S., & Akolkar, A. B. (2009). Assessment of the status of municipal solid waste management in metro cities, state capitals, class I cities, and class II towns in India: An insight. *Waste Management (New York, N.Y.)*, *29*(2), 883–895. DOI: 10.1016/j.wasman.2008.04.011 PMID: 18595684

Kumar, S., Smith, S. R., Fowler, G., Velis, C., Kumar, S. J., Arya, S., Rena, , Kumar, R., & Cheeseman, C. (2017). Challenges and opportunities associated with waste management in India. *Royal Society Open Science*, *4*(3), 160764. DOI: 10.1098/rsos.160764 PMID: 28405362

Kunasundari, B., & Sudesh, K. (2011). Isolation and recovery of microbial polyhydroxyalkanoates. *Express Polymer Letters*, *5*(7), 620–634. DOI: 10.3144/expresspolymlett.2011.60

Kurzweil, P. (2010). Gaston Planté and his invention of the lead–acid battery—The genesis of the first practical rechargeable battery. *Journal of Power Sources*, *195*(14), 4424–4434.

Labet, M., & Thielemans, W. (2009). Synthesis of polycaprolactone: A review. *Chemical Society Reviews*, *38*(12), 3484–3504. DOI: 10.1039/b820162p PMID: 20449064

Lai, K. H., Lun, V. Y. H., Wong, C. W. Y., & Cheng, T. C. E. (2011). Green shipping practices in the shipping industry: Conceptualization, adoption, and implications. *Resources, Conservation and Recycling*, *55*(6), 631–638. DOI: 10.1016/j.resconrec.2010.12.004

Lavanya, M.. (2024). An overview of hydrogen storage technologies–Key challenges and opportunities. *Materials Chemistry and Physics*, •••, 129710.

Laveglia, S., Altieri, G., Genovese, F., Matera, A., & Di Renzo, G. C. (2024). Advances in Sustainable Crop Management: Integrating Precision Agriculture and Proximal Sensing. *AgriEngineering*, *6*(3), 3084–3120. DOI: 10.3390/agriengineering6030177

Lazaroiu, G., Gmal Osman, M., & Strejoiu, C.-V. (2023). Performance Evaluation of Renewable Energy Systems: Photovoltaic, Wind Turbine, Battery Bank, and Hydrogen Storage. *Batteries*, *9*(9), 468. DOI: 10.3390/batteries9090468

Lemaoui, T., Darwish, A. S., Almustafa, G., Boublia, A., Sarika, P. R., Jabbar, N. A., Ibrahim, T., Nancarrow, P., Yadav, K. K., Fallatah, A. M., Abbas, M., Algethami, J. S., Benguerba, Y., Jeon, B.-H., Banat, F., & AlNashef, I. M. (2023). Machine learning approach to map the thermal conductivity of over 2,000 neoteric solvents for green energy storage applications. *Energy Storage Materials*, *59*, 102795. DOI: 10.1016/j.ensm.2023.102795

Lepe, P. T., Heredia, K. V., Namur, E. C., Fabián, G. C. S., & García-Enríquez, S. (2024). Economics and commercialization of bioplastics. In *Bioplastics for Sustainability* (pp. 271–309). Elsevier., DOI: 10.1016/B978-0-323-95199-9.00018-4

Leskovac, A., & Petrović, S. (2023). Pesticide use and degradation strategies: Food safety, challenges and perspectives. *Foods*, *12*(14), 2709. DOI: 10.3390/foods12142709 PMID: 37509801

Leurent, F., & Windisch, E. (2011). Triggering the development of electric mobility: A review of public policies. *European Transport Research Review*, *3*(4), 221–235. DOI: 10.1007/s12544-011-0064-3

Li, C. J., & Chan, T. H. (2007). *Comprehensive organic reactions in aqueous media*. John Wiley & Sons. DOI: 10.1002/9780470131442

Li, H., Cao, X., Liu, Y., Shao, Y., Nan, Z., Teng, L., & Bian, J. (2022). Safety of hydrogen storage and transportation: An overview on mechanisms, techniques, and challenges. *Energy Reports*, *8*, 6258–6269.

Lim, J. M., Jang, Y. S., & Van, T. Nguyen, H., Kim, J. S., Yoon, Y., Park, B. J., Seo, D. H., Lee, K.-K., Han, Z., Ostrikov, K. (Ken), & Doo, S. G. (. (2023). Advances in high-voltage supercapacitors for energy storage systems: Materials and electrolyte tailoring to implementation. *Nanoscale Advances*, *5*(3), 615–626. DOI: 10.1039/D2NA00863G PMID: 36756532

Lin, X., Fan, X., Li, R., Li, Z., Ren, T., Ren, X., & Huang, T. S. (2018). Preparation and characterization of PHB/PBAT-based biodegradable antibacterial hydrophobic nanofibrous membranes. *Polymers for Advanced Technologies*, *29*(1), 481–489. DOI: 10.1002/pat.4137

Liu, C., Lombardo, T., Xu, J., Ngandjong, A. C., & Franco, A. A. (2023). An experimentally-validated 3D electrochemical model revealing electrode manufacturing parameters' effects on battery performance. *Energy Storage Materials*, *54*, 156–163. DOI: 10.1016/j.ensm.2022.10.035

Liu, H. C., You, J. X., Chen, Y. Z., & Fan, X. J. (2014). Site selection in municipal solid waste management with extended VIKOR method under fuzzy environment. *Environmental Earth Sciences*, *72*(10), 4179–4189. DOI: 10.1007/s12665-014-3314-6

Liu, Z., Zhang, S., Wang, L., Wei, T., Qiu, Z., & Fan, Z. (2020). High-efficiency utilization of carbon materials for supercapacitors. *Nano Select*, *1*(2), 244–262. DOI: 10.1002/nano.202000011

Li, Y., Guo, W., Stroe, D. I., Zhao, H., Kristensen, P. K., Jensen, L. R., & Gurevich, L. (2024). Evolution of aging mechanisms and performance degradation of lithium-ion battery from moderate to severe capacity loss scenarios. *Chemical Engineering Journal*, *498*, 155588.

Li, Y., Li, W., Yang, X., Kang, Y., Zhang, H., Liu, Y., & Lei, B. (2021). Salvia Miltiorrhiza-Derived Carbon Dots as Scavengers of Reactive Oxygen Species for Reducing Oxidative Damage of Plants. *ACS Applied Nano Materials*, *4*(1), 113–120. DOI: 10.1021/acsanm.0c02419

Li, Y., Wang, S., Qian, S., Liu, Z., Weng, Y., & Zhang, Y. (2024). Depolymerization and Re/Upcycling of Biodegradable PLA Plastics. *ACS Omega*, *9*(12), 13509–13521. DOI: 10.1021/acsomega.3c08674 PMID: 38559974

Li, Y., Wei, Y., Zhu, F., Du, J., Zhao, Z., & Ouyang, M. (2023). The path enabling storage of renewable energy toward carbon neutralization in China. *ETransportation*, *16*, 100226. DOI: 10.1016/j.etran.2023.100226

Lobato-Peralta, D. R., Okoye, P. U., & Alegre, C. (2024). A review on carbon materials for electrochemical energy storage applications: State of the art, implementation, and synergy with metallic compounds for supercapacitor and battery electrodes. *Journal of Power Sources*, *617*, 235140.

Logan, B. E., Hamelers, B., Rozendal, R., Uwe, S., & Jürg, K. (2006). Stefano, Freguia, Peter A., Willy, V., Korneel, R. Microbial fuel cells: Methodology and technology. *Environmental Science & Technology*, *40*(17), 5181–5192. DOI: 10.1021/es0605016 PMID: 16999087

Lomelí-Ramírez, M. G., Kestur, S. G., Manríquez-González, R., Iwakiri, S., De Muniz, G. B., & Flores-Sahagun, T. S. (2014). Bio-Composites of cassava starch-green coconut fiber: Part II—Structure and properties. *Carbohydrate Polymers*, *102*, 576–583. DOI: 10.1016/j.carbpol.2013.11.020 PMID: 24507321

Loos, K., Zhang, R., Pereira, I., Agostinho, B., Hu, H., Maniar, D., Sbirrazzuoli, N., Silvestre, A. J. D., Guigo, N., & Sousa, A. F. A. (2020). Perspective on PEF Synthesis, Properties, and End-Life. *Frontiers in Chemistry*, *8*, 585. DOI: 10.3389/fchem.2020.00585 PMID: 32850625

Luciani, A., Coccoli, V., Orsi, S., Ambrosio, L., & Netti, P. A. (2008). PCL microspheres-based functional scaffolds by bottom-up approach with predefined microstructural properties and release profiles. *Biomaterials*, *29*(35), 4800–4807. DOI: 10.1016/j.biomaterials.2008.09.007 PMID: 18834628

Ma, X., Li, P., Chen, X., Li, S., Sun, W., & Zhong, Q. (2024). Recyclable ionic liquid-mediated synthesis of quinazolin-4 (3H)-ones under metal-free and solvent-free conditions. *Giant*, 18, 100255.Elsevier.https://doi.org/DOI: 10.1016/j.giant.2024.100255

MacDonald, R. T., Pulapura, S. K., Svirkin, Y. Y., Gross, R. A., Kaplan, D. L., Akkara, J., Swift, G., & Wolk, S. (1995). Enzyme-catalyzed ε-caprolactone ring-opening polymerization. *Macromolecules*, *28*(1), 73–78. DOI: 10.1021/ma00105a008

Machín, A., Cotto, M. C., Díaz, F., Duconge, J., Morant, C., & Márquez, F. (2024). Environmental Aspects and Recycling of Solid-State Batteries: A Comprehensive Review. *Batteries*, *2024*(10), 255. DOI: 10.3390/batteries10070255

Madani, S. S., Shabeer, Y., Allard, F., Fowler, M., Ziebert, C., Wang, Z., Panchal, S., Chaoui, H., Mekhilef, S., Dou, S. X., See, K., & Khalilpour, K. (2025). A Comprehensive Review on Lithium-on Battery Lifetime Prediction and Aging Mechanism Analysis. *Batteries*, *2025*(11), 127. DOI: 10.3390/batteries11040127

Madan, Y. (2020). A mini review study on "the generation of bioactive heterocyclic motifs by using sustainable metal nanocatalysed organic synthetic routes". *Materials Today: Proceedings*, *30*, 197–202. DOI: 10.1016/j.matpr.2020.05.742

Madkour, M. H., Heinrich, D., Alghamdi, M. A., Shabbaj, I. I., & Steinbüchel, A. (2013). PHA Recovery from Biomass. *Biomacromolecules*, *14*(10), 2963–2972. DOI: 10.1021/bm4010244 PMID: 23875914

Madurwar, M. V., Ralegaonkar, R. V., & Mandavgane, S. A. (2013). Application of agro-waste for sustainable construction materials: A review. *Construction & Building Materials*, *38*, 872–878. DOI: 10.1016/j.conbuildmat.2012.09.011

Mahdavi, M., Namvar, F., Ahmad, M., & Mohamad, R. (2013). Green Biosynthesis and Characterization of Magnetic Iron Oxide (Fe3O4) Nanoparticles Using Seaweed (Sargassum Muticum) Aqueous Extract. *Molecules (Basel, Switzerland)*, *18*(5), 5954–5964. DOI: 10.3390/molecules18055954 PMID: 23698048

Mainka, A., Zajusz-Zubek, E., & Kaczmarek, K. (2015). PM10 composition in urban and rural nursery schools in upper Silesia, Poland: Trace elements analysis. *International Journal of Environmental Research and Public Health*, *12*(7), 7990–8008. DOI: 10.3390/ijerph120707990 PMID: 26184269

Maizlish, N., Woodcock, J., Co, S., Ostro, B., Fanai, A., & Fairley, D. (2013). Health cobenefits and transportation-related reductions in greenhouse gas emissions in the San Francisco Bay Area. *American Journal of Public Health*, *103*(4), 703–709. DOI: 10.2105/AJPH.2012.300939 PMID: 23409903

Majeed, M., Hakeem, K. R., & Rehman, R. U. (2022). Synergistic Effect of Plant Extract Coupled Silver Nanoparticles in Various Therapeutic Applications- Present Insights and Bottlenecks. *Chemosphere*, *288*, 132527. DOI: 10.1016/j.chemosphere.2021.132527 PMID: 34637861

Majumder, D., Akhter, J., Mandal, A., Roy, R., Mondal, D., Bhatt, R., & Hossain, A. (2021). Precision Input Management for Minimizing and Recycling of Agricultural Waste. In Bhatt, R., Meena, R. S., & Hossain, A. (Eds.), *Input Use Efficiency for Food and Environmental Security* (pp. 567–603). Springer Nature Singapore., DOI: 10.1007/978-981-16-5199-1_19

Ma, N., Yang, D., Riaz, S., Wang, L., & Wang, K. (2023). Aging Mechanism and Models of Supercapacitors: A Review. *Technologies, 11*(2), 38. DOI: 10.3390/technologies11020038

Ma, N., Zhao, W., Wang, W., Li, X., & Zhou, H. (2024). Large scale of green hydrogen storage: Opportunities and challenges. *International Journal of Hydrogen Energy, 50*, 379–396. DOI: 10.1016/j.ijhydene.2023.09.021

Manabe, K., & Kobayashi, S. (2002). Catalytic asymmetric carbon–carbon bond-forming reactions in aqueous media. *Chemistry (Weinheim an der Bergstrasse, Germany), 8*(18), 4094–4101. DOI: 10.1002/1521-3765(20020916)8:18<4094::AID-CHEM4094>3.0.CO;2-G PMID: 12297999

Mandal, M., Roy, A., Das, S., Rakwal, R., Agrawal, G. K., Singh, P., Awasthi, A., & Sarkar, A. (2024). Food waste-based bio-fertilizers production by bio-based fermenters and their potential impact on the environment. *Chemosphere, 353*, 141539. DOI: 10.1016/j.chemosphere.2024.141539 PMID: 38417498

Manickam, P., Mariappan, S. A., Murugesan, S. M., Hansda, S., Kaushik, A., Shinde, R., & Thipperudraswamy, S. P. (2022). Artificial intelligence (AI) and internet of medical things (IoMT) assisted biomedical systems for intelligent healthcare. *Biosensors (Basel), 12*(8), 562. DOI: 10.3390/bios12080562 PMID: 35892459

Mani, S., & Singh, S. (2016). Sustainable municipal solid waste management in India: A policy agenda. *Procedia Environmental Sciences, 35*, 150–157. DOI: 10.1016/j.proenv.2016.07.064

Maraveas, C. (2022). Incorporating artificial intelligence technology in smart greenhouses: Current State of the Art. *Applied Sciences (Basel, Switzerland), 13*(1), 14. DOI: 10.3390/app13010014

Martins, L. M., Moreno, V. F., Rosario, I. S., de Oliveira Graeff, C. F., & Silva Filho, L. C. (2021). Bronsted Acid Mediated Facile Greener Multicomponent Synthesis of 2, 4-Diaryl-quinoline Derivatives in Water. *Orbital: Electron. J. Chem., 14*(1), 9. DOI: 10.17807/orbital.v14i1.1689

Marvaniya, H. M., Modi, K. N., & Sen, D. J. (2011). Greener reactions under solvent free conditions. *Int. J. Drug Dev. Res, 3*(2), 42–51.

Mason, T. J. (1997). Ultrasound in synthetic organic chemistry. *Chemical Society Reviews, 26*(6), 443–451. DOI: 10.1039/cs9972600443

Massaro, M. C., Biga, R., Kolisnichenko, A., Marocco, P., Monteverde, A. H. A., & Santarelli, M. (2023). Potential and technical challenges of on-board hydrogen storage technologies coupled with fuel cell systems for aircraft electrification. *Journal of Power Sources*, *555*, 232397. DOI: 10.1016/j.jpowsour.2022.232397

Mathur, V. (2012). SCOPE OF RECYCLING MUNICIPAL SOLID WASTE IN DELHI AND NATIONAL CAPITAL REGION (NCR). *Integral Review. Journal of Management*, *5*(2).

Mauritz, K. H., & Moore, R. B. (2004). State of Understanding of Nafion. *Chemical Reviews*, *104*(10), 4535–4546. DOI: 10.1021/cr0207123 PMID: 15669162

Mehmood, A., Raina, N., Phakeenuya, V., Wonganu, B., & Cheenkachorn, K. (2023). The current status and market trend of polylactic acid as biopolymer: Awareness and needs for sustainable development. *Materials Today: Proceedings*, *72*, 3049–3055. DOI: 10.1016/j.matpr.2022.08.387

Mehrenjani, J. R., Gharehghani, A., Ahmadi, S., & Powell, K. M. (2023). Dynamic simulation of a triple-mode multi-generation system assisted by heat recovery and solar energy storage modules: Techno-economic optimization using machine learning approaches. *Applied Energy*, *348*, 121592. DOI: 10.1016/j.apenergy.2023.121592

Mennan, H., Jabran, K., Zandstra, B. H., & Pala, F. (2020). Non-chemical weed management in vegetables by using cover crops: A review. *Agronomy (Basel)*, *10*(2), 257. DOI: 10.3390/agronomy10020257

Mhlanga, D. (2023). Artificial Intelligence and Machine Learning for Energy Consumption and Production in Emerging Markets: A Review. *Energies*, *16*(2), 745. DOI: 10.3390/en16020745

Mikołajczyk, T., Mikołajewski, D., Kłodowski, A., Łukaszewicz, A., Mikołajewska, E., Paczkowski, T., Macko, M., & Skornia, M. (2023). Energy Sources of Mobile Robot Power Systems: A Systematic Review and Comparison of Efficiency. *Applied Sciences (Basel, Switzerland)*, *2023*(13), 7547. DOI: 10.3390/app13137547

Ministry of Environment and Forests (MoEF). 2015 The Gazette of India. Municipal solid waste (Management and Handling) rules, New Delhi, India.

Miranda, D., Gonçalves, R., Wuttke, S., Costa, C. M., & Lanceros-Méndez, S. (2023). Overview on Theoretical Simulations of Lithium-Ion Batteries and Their Application to Battery Separators. *Advanced Energy Materials*, *13*(13), 2203874. Advance online publication. DOI: 10.1002/aenm.202203874

Mishra, H., & Mishra, D. (2024). AI for Data-Driven Decision-Making in Smart Agriculture: From Field to Farm Management. In *Artificial Intelligence Techniques in Smart Agriculture* (pp. 173–193). Springer. https://link.springer.com/chapter/10.1007/978-981-97-5878-4_11

Modak, P., Jiemian, Y., Hongyuan, Y., & Mohanty, C. R. (2010). Municipal solid waste management: turning waste into resources. *Shanghai manual: a guide for sustainable urban development in the 21st century*, 1-36.

Mohammed, A. S., Atnaw, S. M., Salau, A. O., & Eneh, J. N. (2023). Review of optimal sizing and power management strategies for fuel cell/battery/super capacitor hybrid electric vehicles. *Energy Reports*, *9*, 2213–2228. DOI: 10.1016/j.egyr.2023.01.042

Mohammed, A., & Zubair, Z. L. (2022). Land Suitability Analysis for Solid Waste Landfills in Kano Metropolis Using Multi-Criteria Evaluation Techniques. *AFRIGIST JOURNAL OF LAND ADMINISTRATION AND ENVIRONMENTAL MANAGEMENT*, *2*(1), 52–68. DOI: 10.54222/afrigist/ajlaem/v2i1.5

Mohammed, H. I., Majid, Z., & Yamusa, Y. B. (2019, February). GIS based sanitary landfill suitability analysis for sustainable solid waste disposal. [). IOP Publishing.]. *IOP Conference Series. Earth and Environmental Science*, *220*, 012056. DOI: 10.1088/1755-1315/220/1/012056

Mondal, S., Pain, T., Sahu, K., & Kar, S. (2021). Large-scale Green synthesis of porphyrins. *ACS Omega*, *6*(35), 22922–22936. DOI: 10.1021/acsomega.1c03534 PMID: 34514263

Morseletto, P. (2020). Targets for a circular economy. *Resources, Conservation and Recycling*, *153*, 104553. DOI: 10.1016/j.resconrec.2019.104553

Morstein, C. E., & Dienwiebel, M. (2021). Graphite lubrication mechanisms under high mechanical load. *Wear*, *477*, 203794. DOI: 10.1016/j.wear.2021.203794

Mulky, L., Srivastava, S., Lakshmi, T., Sandadi, E. R., Gour, S., Thomas, N. A., Shanmuga Priya, S., & Sudhakar, K. (2024). An overview of hydrogen storage technologies – Key challenges and opportunities. *Materials Chemistry and Physics*, *325*, 129710. DOI: 10.1016/j.matchemphys.2024.129710

Muneer, F., Rasul, I., Qasim, M., Sajid, A., & Nadeem, H. (2022). Optimization, Production and Characterization of Polyhydroxyalkanoate (PHA) from Indigenously Isolated Novel Bacteria. *Journal of Polymers and the Environment*, *30*(8), 3523–3533. DOI: 10.1007/s10924-022-02444-y

Munir, S., Azeem, A., Zaman, M. S., & Haq, M. Z. U. (2024). From field to table: Ensuring food safety by reducing pesticide residues in food. *The Science of the Total Environment*, *922*, 171382. DOI: 10.1016/j.scitotenv.2024.171382 PMID: 38432369

Murray, L. J., Dincă, M., & Long, J. R. (2009). Hydrogen storage in metal–organic frameworks. *Chemical Society Reviews*, *38*(5), 1294–1314. DOI: 10.1039/b802256a PMID: 19384439

Muthusamy, M. S., & Pramasivam, S. (2019). Bioplastics—An Eco-Friendly Alternative to Petrochemical Plastics. *Current World Environment*, *14*(1), 49–59. DOI: 10.12944/CWE.14.1.07

Muzaffar, N., Afzal, A. M., Hegazy, H. H., & Iqbal, M. W. (2023). Recent advances in two-dimensional metal-organic frameworks as an exotic candidate for the evaluation of redox-active sites in energy storage devices. *Journal of Energy Storage*, *64*, 107142. DOI: 10.1016/j.est.2023.107142

Mwaka, O., Mwamahonje, A., Nene, W., Rweyemamu, E., & Maseta, Z. (2024). Pesticides use and its effects on grape production: A review. *Sustainable Environment*, *10*(1), 2366555. DOI: 10.1080/27658511.2024.2366555

Nagalingam, M., Kalpana, V. N., Rajeswari, V. D., & Panneerselvam, A. (2018). Biosynthesis, Characterization, and Evaluation of Bioactivities of Leaf Extract-Mediated Biocompatible Gold Nanoparticles from Alternanthera Bettzickiana. *Biotechnology Reports (Amsterdam, Netherlands)*, *19*, e00268. DOI: 10.1016/j.btre.2018.e00268 PMID: 29992102

Nagendrappa, G. (2002). Organic synthesis under solvent-free condition: An environmentally benign procedure—I. *Reson*, *7*(10), 59–68. DOI: 10.1007/BF02835544

Nag, M., Lahiri, D., Sarkar, T., Ghosh, S., Dey, A., Edinur, H. A., Pati, S., & Ray, R. R. (2021). Microbial Fabrication of Nanomaterial and its Role in Disintegration of Exopolymeric Matrices of Biofilm. *Frontiers in Chemistry*, *9*, 1. DOI: 10.3389/fchem.2021.690590 PMID: 34109159

Nahirniak, S., Ray, A., & Saruhan, B. (2023). Challenges and Future Prospects of the MXene-Based Materials for Energy Storage Applications. *Batteries*, *9*(2), 126. DOI: 10.3390/batteries9020126

Nahiyoon, S. A., Ren, Z., Wei, P., Li, X., Li, X., Xu, J., Yan, X., & Yuan, H. (2024). Recent development trends in plant protection UAVs: A journey from conventional practices to cutting-edge technologies—A comprehensive Review. *Drones (Basel)*, *8*(9), 457. DOI: 10.3390/drones8090457

Nair, L. S., & Laurencin, C. T. (2007). Biodegradable polymers as biomaterials. *Progress in Polymer Science*, *32*(8-9), 762–798. DOI: 10.1016/j.progpolymsci.2007.05.017

Narayana, T. (2009). Municipal solid waste management in India: From waste disposal to recovery of resources? *Waste Management (New York, N.Y.)*, *29*(3), 1163–1166. DOI: 10.1016/j.wasman.2008.06.038 PMID: 18829290

Natesan, U., & Sarkar, C. (2008). GIS-based approach for optimized siting of municipal solid waste landfill. *Waste Management, 28*(11).

Navaid, H., Emadi, H., & Watson, M. (2023). A comprehensive literature review on the challenges associated with underground hydrogen storage. *International Journal of Hydrogen Energy*, *48*(28), 10603–10635. DOI: 10.1016/j.ijhydene.2022.11.225

Nawaz, A., & Ahmad, J. N. (2015). Insect Pest Management in Conservation Agriculture. In Farooq, M., & Siddique, K. H. M. (Eds.), *Conservation Agriculture* (pp. 133–155). Springer International Publishing., DOI: 10.1007/978-3-319-11620-4_6

Nghiem, N. P., Kleff, S., & Schwegmann, S. (2017). Succinic Acid: Technology Development and Commercialization. *Fermentation (Basel, Switzerland)*, *3*(2), 26. DOI: 10.3390/fermentation3020026

Ngoc, L. T. N., Moon, J. Y., & Lee, Y. C. (2023, September 27). Plant Extract-Derived Carbon Dots as Cosmetic Ingredients. *Nanomaterials (Basel, Switzerland)*, *13*(19), 2654. DOI: 10.3390/nano13192654 PMID: 37836295

Nicastro, R., Papale, M., Fusco, G. M., Capone, A., Morrone, B., & Carillo, P. (2024). Legal Barriers in Sustainable Agriculture: Valorization of Agri-Food Waste and Pesticide Use Reduction. *Sustainability (Basel)*, *16*(19), 8677. DOI: 10.3390/su16198677

Nishinari, K., Yamatoya, K., & Shirakawa, M. (2000). Xyloglucan. In Phillips, G. O., & Williams, P. A. (Eds.), *Handbook of hydrocolloids* (pp. 247–267). Woodhead Publishing.

Nita, T., & Aurica, P. C. (2011). Biopolymer-Based Green Composites: Mechanical, Thermal, and Physico-Chemical Characterization. *Journal of Polymers and the Environment*, *19*(3), 546–556.

Nobes, G. A., Kazlauskas, R. J., & Marchessault, R. H. (1996). Lipase-catalyzed ring-opening polymerization of lactones: A novel route to poly(hydroxyalkanoates). *Macromolecules*, *29*(15), 4829–4833. DOI: 10.1021/ma951774g

Nosheen, S., Ajmal, I., & Song, Y. (2021). Microbes as biofertilizers, a potential approach for sustainable crop production. *Sustainability (Basel), 13*(4), 1868. DOI: 10.3390/su13041868

Olabi, A. G., Abdelghafar, A. A., Maghrabie, H. M., Sayed, E. T., Rezk, H., Al Radi, M., Obaideen, K., & Abdelkareem, M. A. (2023). Application of artificial intelligence for prediction, optimization, and control of thermal energy storage systems. *Thermal Science and Engineering Progress, 39*, 101730. DOI: 10.1016/j.tsep.2023.101730

Olanibi, O. A., & Emmanuel, A. A. (2022). Site Suitability Analysis of Solid Waste Disposal in Ilesa, Nigeria. *European Journal of Development Studies, 2*(2), 49–59. DOI: 10.24018/ejdevelop.2022.2.2.60

Omar, H. H., & Bahabri, S., FM. El-Gend, A. (2017). Biopotential Application of Synthesis Nanoparticles as Antimicrobial Agents by Using Laurencia Papillosa. *International Journal of Pharmacology, 13*, 303–312. DOI: 10.3923/ijp.2017.303.312

Öncel, M. S., Bektaş, N., Bayar, S., Engin, G., Çalışkan, Y., Salar, L., & Yetiş, Ü. (2017). Hazardous wastes and waste generation factors for plastic products manufacturing industries in Turkey. *Sustainable Environment Research, 27*(4), 188–194. DOI: 10.1016/j.serj.2017.03.006

Orth, N., Munzke, N., Weniger, J., Messner, C., Schreier, R., Mast, M., & Quaschning, V. (2023). Efficiency characterization of 26 residential photovoltaic battery storage systems. *Journal of Energy Storage, 65*, 107299.

Oyedotun, K. O., & Mamba, B. B. (2024). New trends in supercapacitors applications. *Inorganic Chemistry Communications, 170*, 113154. DOI: 10.1016/j.inoche.2024.113154

Oyewole, O. L., Nwulu, N. I., & Okampo, E. J. (2024). Optimal design of hydrogen-based storage with a hybrid renewable energy system considering economic and environmental uncertainties. *Energy Conversion and Management, 300*, 117991.

Padhiary, M., Kumar, R., & Sethi, L. N. (2024). Navigating the Future of Agriculture: A Comprehensive Review of Automatic All-Terrain Vehicles in Precision Farming. *Journal of The Institution of Engineers (India): Series A, 105*(3), 767–782. DOI: 10.1007/s40030-024-00816-2

Padhiary, M., Hoque, A., Prasad, G., Kumar, K., & Sahu, B. (2025). Precision Agriculture and AI-Driven Resource Optimization for Sustainable Land and Resource Management. In Ruiz-Vanoye, J. A., & Díaz-Parra, O. (Eds.), *Smart Water Technology for Sustainable Management in Modern Cities* (pp. 197–232). IGI Global., DOI: 10.4018/979-8-3693-8074-1.ch009

Padhiary, M., & Kumar, R. (2024a). Assessing the Environmental Impacts of Agriculture, Industrial Operations, and Mining on Agro-Ecosystems. In Azrour, M., Mabrouki, J., Alabdulatif, A., Guezzaz, A., & Amounas, F. (Eds.), *Smart Internet of Things for Environment and Healthcare* (Vol. 1165, pp. 107–126). Springer Nature Switzerland., DOI: 10.1007/978-3-031-70102-3_8

Padhiary, M., & Kumar, R. (2024b). Enhancing Agriculture Through AI Vision and Machine Learning: The Evolution of Smart Farming. In Thangam, D. (Ed.), (pp. 295–324). Advances in Computational Intelligence and Robotics. IGI Global., DOI: 10.4018/979-8-3693-5380-6.ch012

Padhiary, M., Roy, P., Dey, P., & Sahu, B. (2024). Harnessing AI for Automated Decision-Making in Farm Machinery and Operations: Optimizing Agriculture. In Hai-Jew, S. (Ed.), (pp. 249–282). Advances in Computational Intelligence and Robotics. IGI Global., DOI: 10.4018/979-8-3693-6230-3.ch008

Padhiary, M., Saha, D., Kumar, R., Sethi, L. N., & Kumar, A. (2024). Enhancing precision agriculture: A comprehensive review of machine learning and AI vision applications in all-terrain vehicle for farm automation. *Smart Agricultural Technology*, *8*, 100483. DOI: 10.1016/j.atech.2024.100483

Padhiary, M., Tikute, S. V., Saha, D., Barbhuiya, J. A., & Sethi, L. N. (2025). Development of an IOT-Based Semi-Autonomous Vehicle Sprayer. *Agricultural Research*, *14*(1), 229–239. DOI: 10.1007/s40003-024-00760-4

Pandeya, S., Gyawali, B. R., & Upadhaya, S. (2025). Factors influencing precision agriculture technology adoption among small-scale farmers in Kentucky and their implications for policy and practice. *Agriculture*, *15*(2), 177. DOI: 10.3390/agriculture15020177

Pandey, D., Kumar, K. S., & Thomas, J. (2024). Supercapacitor electrode energetics and mechanism of operation: Uncovering the voltage window. *Progress in Materials Science*, *141*, 101219. DOI: 10.1016/j.pmatsci.2023.101219

Pan, S., Giri, S., & Chattaraj, P. K. (2011). A computational study on the hydrogen adsorption capacity of various lithium—Doped boron hydrides. *Journal of Computational Chemistry*, *33*(4), 425–434. DOI: 10.1002/jcc.21985 PMID: 22121031

Panwar, S., Duggirala, K. S., Yadav, P., Debnath, N., Yadav, A. K., & Kumar, A. (2023). Advanced diagnostic methods for identification of bacterial foodborne pathogens: Contemporary and upcoming challenges. *Critical Reviews in Biotechnology, 43*(7), 982–1000. DOI: 10.1080/07388551.2022.2095253 PMID: 35994308

Parida, R., Dash, M. K., Giri, S., & Roymahapatra, G. (2021). Aromatic N-Heterocyclic superalkali M@C4H4N2 complexes (M=Li, Na, K); A promising potential hydrogen storage system. *Journal of the Indian Chemical Society, 2021*(98), 100065. DOI: 10.1016/j.jics.2021.100065

Park, D. H., & Kim, B. S. (2011). Production of poly(3-hydroxybutyrate) and poly(3-hydroxybutyrate-co-4-hydroxybutyrate) by *Ralstonia eutropha* from soybean oil. *New Biotechnology, 28*(6), 719–724. DOI: 10.1016/j.nbt.2011.01.007 PMID: 21333767

Parvizi, P., Jalilian, M., Amidi, A. M., Zangeneh, M. R., & Riba, J. R. (2025). From Present Innovations to Future Potential: The Promising Journey of Lithium-Ion Batteries. *Micromachines, 16*(2), 194.

Patel, R. N. (2018). Biocatalysis for synthesis of pharmaceuticals. *Bioorganic & Medicinal Chemistry, 26*(7), 1252–1274. DOI: 10.1016/j.bmc.2017.05.023 PMID: 28648492

Patel, V., Berthold, D., Puranik, P., & Gantar, M. (2015). Screening of Cyanobacteria and Microalgae for Their Ability to Synthesize Silver Nanoparticles with Antibacterial Activity. *Biotechnology Reports (Amsterdam, Netherlands), 5*, 112–119. DOI: 10.1016/j.btre.2014.12.001 PMID: 28626689

Patil, R. S., Kokate, M. R., & Kolekar, S. S. (2012). Bioinspired Synthesis of Highly Stabilized Silver Nanoparticles Using Ocimum Tenuiflorum Leaf Extract and Their Antibacterial Activity. *Spectrochimica Acta. Part A: Molecular and Biomolecular Spectroscopy, 91*, 234–238. DOI: 10.1016/j.saa.2012.02.009 PMID: 22381796

Patti, A., & Acierno, D. (2022). Towards the Sustainability of the Plastic Industry through Biopolymers: Properties and Potential Applications to the Textiles World. *Polymers, 14*(4), 692. DOI: 10.3390/polym14040692 PMID: 35215604

Pelosi, D., Baldinelli, A., Cinti, G., Ciupageanu, D.-A., Ottaviano, A., Santori, F., Carere, F., & Barelli, L. (2023). Battery-hydrogen vs. flywheel-battery hybrid storage systems for renewable energy integration in mini-grid: A techno-economic comparison. *Journal of Energy Storage, 63*, 106968. DOI: 10.1016/j.est.2023.106968

Perez, L., Trüeb, S., Cowie, H., Keuken, M. P., Mudu, P., Ragettli, M. S., Sarigiannis, D. A., Tobollik, M., Tuomisto, J., Vienneau, D., Sabel, C., & Künzli, N. (2015). Transport-related measures to mitigate climate change in Basel, Switzerland: A health-effectiveness comparison study. *Environment International*, *85*, 111–119. DOI: 10.1016/j.envint.2015.08.002 PMID: 26386464

Planning Commission, Government of India. 2014 Report of the task Force on waste to energy (Volume I) in the context of integrated municipal solid waste management. See http:// planningcommission.nic.in/reports/genrep/rep_ wte1205.pdf (accessed 1 July 2015).

Plastics Europe. (2023). *Plastics—the fast facts 2023*. Retrieved May 29, 2024, from https://plasticseurope.org/knowledge-hub/plastics-the-fast-facts-2023/

Platel, R. H., Hodgson, L. M., & Williams, C. K. (2008). Biocompatible initiators for lactide polymerization. *Polymer Reviews (Philadelphia, Pa.)*, *48*(1), 11–63. DOI: 10.1080/15583720701834166

Platnieks, O., Gaidukovs, S., Thakur, V. K., Barkane, A., & Beluns, S. (2021). Bio-based poly (butylene succinate): Recent progress, challenges, and future opportunities. *European Polymer Journal*, *161*, 110855. DOI: 10.1016/j.eurpolymj.2021.110855

Poornima, P., & Nithya, R. (2022). Site suitability analysis for decentralized sustainable solid waste management system for Coimbatore city using GIS. *Sustainability, Agri, Food and Environmental Research, 10*.

Pordanjani, A. H., Aghakhani, S., Afrand, M., Zhang, P., Tang, R., Mahian, O., Wongwises, S., & Rashidi, M. M. (2023). Thermo-electrochemical simulation of the cooling process in a compact battery pack considering various configurations. *Journal of Power Sources*, *553*, 232112. DOI: 10.1016/j.jpowsour.2022.232112

Porritt, S., Shao, L., Cropper, P., & Goodier, C. (2011). Adapting dwellings for heat waves. *Sustainable Cities and Society*, *1*(2), 81–90. DOI: 10.1016/j.scs.2011.02.004

Poškas, P., Kilda, R., Šimonis, A., Jouhara, H., & Poškas, R. (2019). Disposal of very low-level radioactive waste: Lithuanian case on the approach and long-term safety aspects. *The Science of the Total Environment*, *667*, 464–474. DOI: 10.1016/j.scitotenv.2019.02.373 PMID: 30833245

Prasad, G., Padhiary, M., Hoque, A., & Kumar, K. (2025). AI-Driven Personalized Nutrition Apps and Platforms for Enhanced Diet and Wellness. In Alhussaini Hamad, A. M., & Soni, R. (Eds.), *Food in the Metaverse and Web 3.0 Era* (pp. 125–158). IGI Global., DOI: 10.4018/979-8-3693-9025-2.ch006

Price, S., Kuzhiumparambil, U., Pernice, M., & Ralph, P. (2022). Techno-economic analysis of cyanobacterial PHB bioplastic production. *Journal of Environmental Chemical Engineering*, *10*(3), 107502. DOI: 10.1016/j.jece.2022.107502

Pyo, S.-H., Park, J. H., Srebny, V., & Hatti-Kaul, R. (2020). A sustainable synthetic route for biobased 6-hydroxyhexanoic acid, adipic acid, and ε-caprolactone by integrating bio- and chemical catalysis. *Green Chemistry*, *22*(14), 4450–4455. DOI: 10.1039/D0GC01454K

Qiu, C., Jiang, L., Gao, Y., & Sheng, L. (2023). Effects of oxygen-containing functional groups on carbon materials in supercapacitors: A review. *Materials & Design*, *230*, 111952. DOI: 10.1016/j.matdes.2023.111952

Rahman, M. M., Ferdous, K. S., Ahmed, M., Islam, M. T., Khan, M. R., Perveen, A., Ashraf, G. M., & Uddin, M. S. (2021a). Hutchinson-Gilford Progeria Syndrome: An Overview of the Molecular Mechanism, Pathophysiology and Therapeutic Approach. *Current Gene Therapy*, *21*(3), 216–229. DOI: 10.2174/156652322166621 0303100805 PMID: 33655857

Rahman, M. M., Rahaman, M. S., Islam, M. R., Hossain, M. E., Mannan Mithi, F., Ahmed, M., Saldías, M., Akkol, E. K., & Sobarzo-Sánchez, E. (2021b). Multifunctional Therapeutic Potential of Phytocomplexes and Natural Extracts for Antimicrobial Properties. *Antibiotics (Basel, Switzerland)*, *10*(9), 1076. DOI: 10.3390/ antibiotics10091076 PMID: 34572660

Rahman, T., & Alharbi, T. (2024). Data-Driven Degradation Estimation Techniques, and Sustainable Directions for Energy Storage Systems. *Batteries*, *2024*(10), 220. DOI: 10.3390/batteries10070220

Rajababu, D., Surya, S., Padhiary, M., & Modi, H. (2025). Blockchain for Cybersecurity_ Securing Data Transactions and Enhancing Privacy in Digital Systems. 2025 First International Conference on Advances in Computer Science, Electrical, Electronics, and Communication Technologies (CE2CT), 1426–1430. https://doi .org/DOI: 10.1109/CE2CT64011.2025.10939863

Rajgond, V., Mohite, A., More, N., & More, A. (2023). Biodegradable polyester-polybutylene succinate (PBS): A review. *Polymer Bulletin*, *81*(7), 5703–5752. DOI: 10.1007/s00289-023-04998-w

Rajput, R., Prasad, G. A. K. C., & Chopra, A. K. (2009). Scenario of solid waste management in present Indian context.

Ramachandran, T., Sana, S. S., Kumar, K. D., Kumar, Y. A., Hegazy, H. H., & Kim, S. C. (2023). Asymmetric supercapacitors: Unlocking the energy storage revolution. *Journal of Energy Storage*, *73*, 109096. DOI: 10.1016/j.est.2023.109096

Ramirez-Vidal, P., Sdanghi, G., Celzard, A., & Fierro, V. (2022). High hydrogen release by cryo-adsorption and compression on porous materials. *International Journal of Hydrogen Energy*, *47*(14), 8892–8915.

Ranakoti, L., Gangil, B., Kumar Mishra, S., Singh, T., Sharma, S., Ilyas, R. A., & El-Khatib, S. (2022). Critical review on polylactic acid: Properties, structure, processing, biocomposites, and nanocomposites. *Materials (Basel)*, *15*(12), 4312. DOI: 10.3390/ma15124312 PMID: 35744371

Rana, M. M., Uddin, M., Sarkar, M. R., Meraj, S. T., Shafiullah, G. M., Muyeen, S. M., Islam, M. A., & Jamal, T. (2023). Applications of energy storage systems in power grids with and without renewable energy integration — A comprehensive review. *Journal of Energy Storage*, *68*, 107811. DOI: 10.1016/j.est.2023.107811

Rana, R., Ganguly, R., & Gupta, A. K. (2015). An assessment of solid waste management system in Chandigarh City, India. *The Electronic Journal of Geotechnical Engineering*, *20*(6), 1547–1572.

Rangappa, S. M., Siengchin, S., Parameswaranpillai, J., Jawaid, M., & Ozbakkaloglu, T. (2022). Lignocellulosic fiber reinforced composites: Progress, performance, properties, applications, and future perspectives. *Polymer Composites*, *43*(2), 645–691. DOI: 10.1002/pc.26413

Rao, T. S., Panigrahi, S., & Velraj, P. (2022). Transport and disposal of radioactive wastes in nuclear industry. In *Microbial biodegradation and bioremediation* (pp. 419–440). Elsevier.

Rao, Y. S., Mary, K., Mathew, L., & Potty, S. N. (1999). Tamarind (Tamarindus indica L.) research—A Review. *Indian Journal of Arecanut, Spices and Medicinal Plants*, *1*(3), 127–145. https://www.cabidigitallibrary.org/action/doSearch?do=Indian+Journal+of+Arecanut%2C+Spices+and+Medicinal+Plants

Rashid Khan, H., & Latif Ahmad, A. (2025). Supercapacitors: Overcoming current limitations and charting the course for next-generation energy storage. *Journal of Industrial and Engineering Chemistry*, *141*, 46–66. DOI: 10.1016/j.jiec.2024.07.014

Rashmi, I., Karthika, K. S., Roy, T., Shinoji, K. C., Kumawat, A., Kala, S., & Pal, R. (2022). Soil Erosion and Sediments: A Source of Contamination and Impact on Agriculture Productivity. In Naeem, M., Bremont, J. F. J., Ansari, A. A., & Gill, S. S. (Eds.), *Agrochemicals in Soil and Environment* (pp. 313–345). Springer Nature Singapore., DOI: 10.1007/978-981-16-9310-6_14

Rathi, S. (2006). Alternative approaches for better municipal solid waste management in Mumbai, India. *Waste Management (New York, N.Y.)*, *26*(10), 1192–1200. DOI: 10.1016/j.wasman.2005.09.006 PMID: 16288861

Raut, K., Shendge, A., Chaudhari, J., Lamba, R., & Alshammari, N. F. (2024). Modeling and simulation of photovoltaic powered battery-supercapacitor hybrid energy storage system for electric vehicles. *Journal of Energy Storage*, *82*, 110324. DOI: 10.1016/j.est.2023.110324

Ravindran, R., Hassan, S. S., Williams, G. A., & Jaiswal, A. K. (2018). A review on bioconversion of agro-industrial wastes to industrially important enzymes. *Bioengineering (Basel, Switzerland)*, *5*(4), 93. DOI: 10.3390/bioengineering5040093 PMID: 30373279

Raza, A., Hu, Y., Acharki, S., Buttar, N. A., Zubair, M., Syed, N. R., Vishwakarma, D. K., & Elbeltagi, A. (2024). Role of Artificial Intelligence in Revolutionizing Agricultural Technology. In Pandey, K., Kushwaha, N. L., Pande, C. B., & Singh, K. G. (Eds.), *Artificial Intelligence and Smart Agriculture* (pp. 121–145). Springer Nature Singapore., DOI: 10.1007/978-981-97-0341-8_7

Raza, S., Zhang, J., Ali, I., Li, X., & Liu, C. (2020). Recent trends in the development of biomass-based polymers from renewable resources and their environmental applications. *Journal of the Taiwan Institute of Chemical Engineers*, *115*, 293–303. DOI: 10.1016/j.jtice.2020.10.013

Reddy, V. M., Mauger, A., Julien, C. M., Paolella, A., & Zaghib, K. (2020). Brief History of Early Lithium-Battery Development. *Materials (Basel)*, *2020*(13), 1884. DOI: 10.3390/ma13081884 PMID: 32316390

Redhu, N. S., Thakur, Z., Yashveer, S., & Mor, P. (2022). Artificial intelligence: A way forward for agricultural sciences. In *Bioinformatics in Agriculture* (pp. 641–668). Elsevier. DOI: 10.1016/B978-0-323-89778-5.00007-6

Ren, W. X., Geng, Y., Xue, B., Fujita, T., Ma, Z., & Jiang, P. (2012). Pursuing co-benefits in China's old industrial base: A case of Shenyang. *Urban Climate*, *1*(72), 55–64. DOI: 10.1016/j.uclim.2012.07.001

Reza, M. N., Lee, K.-H., Karim, M. R., Haque, M. A., Bicamumakuba, E., Dey, P. K., Jang, Y. Y., & Chung, S.-O. (2025). Trends of Soil and Solution Nutrient Sensing for Open Field and Hydroponic Cultivation in Facilitated Smart Agriculture. *Sensors (Basel), 25*(2), 453. DOI: 10.3390/s25020453 PMID: 39860823

Ribeiro, I., Monteiro, A., Martins, H., Freitas, S., Borrego, C., & Lopes, M. (2015). How does the use of biodiesel affect urban air quality? *International Journal of Environment and Pollution, 58*(1/2), 79–88. DOI: 10.1504/IJEP.2015.076585

Richards, B. S., Hudry, D., Busko, D., Turshatov, A., & Howard, I. A. (2021). Photon Upconversion for Photovoltaics and Photocatalysis: A Critical Review. *Chemical Reviews, 121*(15), 9165–9195. DOI: 10.1021/acs.chemrev.1c00034 PMID: 34327987

Riva, R., Schmeits, S., Jérôme, C., Jérôme, R., & Lecomte, P. (2007). Combination of ring-opening polymerization and "click chemistry": Toward functionalization and grafting of poly(ε-caprolactone). *Macromolecules, 40*(3), 796–803. DOI: 10.1021/ma0624090

Riva, S., Chopineau, J., Kieboom, A. P. G., & Klibanov, A. M. (1988). Protease-catalyzed regioselective esterification of sugars and related compounds in anhydrous dimethylformamide. *Journal of the American Chemical Society, 110*(2), 584–589. DOI: 10.1021/ja00210a045

Rizos, V., Tuokko, K., & Behrens, A. (2017). The circular economy: A review of definitions, processes, and impacts. *CEPS Research Report, 2017/8.* Retrieved June 29, 2024, from http://aei.pitt.edu/id/eprint/85892

Rodriguez de San Miguel, J. A. (2019). Management of air pollution in Mexico. *Management of Environmental Quality, 30*(3), 578–592. DOI: 10.1108/MEQ-05-2018-0099

Rojek, I., Mikołajewski, D., Mroziński, A., & Macko, M. (2023). Machine Learning- and Artificial Intelligence-Derived Prediction for Home Smart Energy Systems with PV Installation and Battery Energy Storage. *Energies, 16*(18), 6613. DOI: 10.3390/en16186613

Roymahapatra, G., Dash, M. K., Ghosh, A., Bag, A., Mishra, S., & Maity, S. (2022). Hydrogen Storage on Lithium Chloride/Lithium Bromide Surface at Cryogenic Temperature. *ES Energy & Environment, 2022*(18), 90–100. DOI: 10.30919/esee8c756

Roymahapatra, G., Dash, M. K., Sinha, S., De, G. C., & Guo, Z. (2022). Theoretical Investigation of Hydrogen Adsorption Efficiency of [Oxadiazole- xLi+] Complexes (x = 1, 2): In Pursuit of Green Fuel Storage. *Engineering and Science, 2022*(19), 114–124. DOI: 10.30919/es8d671

Roy, P., Padhiary, M., Hoque, A., Sahu, B., Roy, D., & Kumar, K. (2025). Machine Learning for Precision Agriculture and Crop Yield Optimization: Techniques and Applications. In Emara, T., Hassan, E., Trinh, T., Li, G., & Saber, A. (Eds.), *The Role of Artificial Intelligence in Advancing Applied Life Sciences* (pp. 189–234). IGI Global., DOI: 10.4018/979-8-3693-9208-9.ch007

Ruan, Y., Liang, Z., Qian, F., Meng, H., & Gao, Y. (2023). Operation strategy optimization of combined cooling, heating, and power systems with energy storage and renewable energy based on deep reinforcement learning. *Journal of Building Engineering*, *65*, 105682. DOI: 10.1016/j.jobe.2022.105682

Saaty, R. W. (1987). The analytic hierarchy process—What it is and how it is used. *Mathematical Modelling*, *9*(3-5), 161–176. DOI: 10.1016/0270-0255(87)90473-8

Sachdev, A., & Gopinath, P. (2015). Green synthesis of multifunctional carbon dots from coriander leaves and their potential application as antioxidants, sensors and bioimaging agents. *Analyst*, *140*(12), 4260–4269. DOI: 10.1039/C5AN00454C PMID: 25927267

Sadeq, A. M., Homod, R. Z., Hussein, A. K., Togun, H., Mahmoodi, A., Isleem, H. F., & Moghaddam, A. H. (2024). Hydrogen energy systems: Technologies, trends, and future prospects. *The Science of the Total Environment*, *939*, 173622.

Saha, D., Padhiary, M., Hoque, A., & Prasad, G. (2025). 3D printing technology for valorization of food processing wastes and byproducts: A systematic review. *Waste Management Bulletin*, *3*(4), 100192. DOI: 10.1016/j.wmb.2025.100192

Sahoo, R. K., & Sahu, S. (2022, August 23). Reversible hydrogen storage capacity of Li and Sc doped novel C_8N_8 cage: Insights from density functional theory. *International Journal of Energy Research*, *46*(15), 22585–22600. DOI: 10.1002/er.8562

Sahoo, S., Sahoo, R., & Nayak, P. L. (2011). Tamarind Seed Polysaccharide: A Versatile Biopolymer for Mucoadhesive Applications. *Journal of Pharmaceutical and Biomedical Sciences*, *8*(1), 1–12.

Sahoo, S., & Timmann, P. (2023). Energy Storage Technologies for Modern Power Systems: A Detailed Analysis of Functionalities, Potentials, and Impacts. *IEEE Access : Practical Innovations, Open Solutions*, *11*, 49689–49729. DOI: 10.1109/ACCESS.2023.3274504

Saketa, Y., Tamene, N., & Melknew, M. (2023). Municipal solid waste disposal site suitability analysis using multi-criteria evaluation in Assosa, Ethiopia. *International Journal of Environmental Science and Technology*, *20*(4), 3815–3830. DOI: 10.1007/s13762-022-04221-9

Sakkaravarthy, S., Jano, N. A., & Vijayakumar, A. (2024). Overcoming Challenges in Traditional Waste Water Treatment Through AI-Driven Innovation. In Garg, M. C. (Ed.), *The AI Cleanse: Transforming Wastewater Treatment Through Artificial Intelligence* (pp. 53–81). Springer Nature Switzerland., DOI: 10.1007/978-3-031-67237-8_3

Saleem, S. R., Zaman, Q. U., Schumann, A. W., & Naqvi, S. M. Z. A. (2023). Variable rate technologies: Development, adaptation, and opportunities in agriculture. In *Precision Agriculture* (pp. 103–122). Elsevier., https://www.sciencedirect.com/science/article/pii/B9780443189531000106 DOI: 10.1016/B978-0-443-18953-1.00010-6

Salkuti, S. R. (2023). Advanced Technologies for Energy Storage and Electric Vehicles. *Energies, 16*(5), 2312. DOI: 10.3390/en16052312

Salman, M. S., Pratthana, C., Lai, Q., Wang, T., Rambhujun, N., Srivastava, K., & Aguey-Zinsou, K.-F. (2022). Catalysis in Solid Hydrogen Storage: Recent Advances, Challenges, and Perspectives. *Energy Technology (Weinheim), 10*(9), 2200433. Advance online publication. DOI: 10.1002/ente.202200433

Samadi, N., Golkaran, D., Eslamifar, A., Jamalifar, H., Fazeli, M. R., & Mohseni, F. A. (2009). Intra/Extracellular Biosynthesis of Silver Nanoparticles by an Autochthonous Strain of Proteus mirabilis Isolated fromPhotographic Waste. *Journal of Biomedical Nanotechnology, 5*(3), 247–253. DOI: 10.1166/jbn.2009.1029 PMID: 20055006

Samal, I., Bhoi, T. K., Mahanta, D. K., Komal, J., & Pradhan, A. K. (2024). Sustainable Plant Protection Measures in Regenerative Farming. In Kumar, S., Meena, R. S., Sheoran, P., & Jhariya, M. K. (Eds.), *Regenerative Agriculture for Sustainable Food Systems* (pp. 387–421). Springer Nature Singapore., DOI: 10.1007/978-981-97-6691-8_12

Samir, A., Ashour, F. H., Hakim, A. A. A., & Bassyouni, M. (2022). Recent advances in biodegradable polymers for sustainable applications. *NPJ Materials Degradation, 6*(1), 68. DOI: 10.1038/s41529-022-00277-7

Sarah, P., Julien, B. M. G., Melissa, I. G., Paolo, M., Adrienne, G., & Noémie, V. (2024). *Plastic Overshoot Day Report 2024*. EA-Earth Action.

Sarigiannis, D. A., Kontoroupis, P., Nikolaki, S., Gotti, A., Chapizanis, D., & Karakitsios, S. (2017). Benefits on public health from transport-related greenhouse gas mitigation policies in southeastern European cities. *The Science of the Total Environment, 579*, 1427–1438. DOI: 10.1016/j.scitotenv.2016.11.142 PMID: 27919555

Sathe, R. Y., & Dhilip Kumar, T. J. (2020). Reversible hydrogen adsorption in Li functionalized [1,1]paracyclophane. *International Journal of Hydrogen Energy, 2020*(45), 12940–12948. DOI: 10.1016/j.ijhydene.2020.03.009

Satpathy, S., Misra, N. K., Shukla, D., Goyal, V., Bhattacharyya, B. K., & Yadav, C. S. (2023). An in-depth study of the electrical characterization of supercapacitors for recent trends in energy storage system. *Journal of Energy Storage*, *57*, 106198. DOI: 10.1016/j.est.2022.106198

Savage, N., & Diallo, M. S. (2005). Nanomaterials and Water Purification: Opportunities and Challenges. *Journal of Nanoparticle Research*, *7*(4-5), 331–342. DOI: 10.1007/s11051-005-7523-5

Scheepers, C. E., Wendel-Vos, G. C. W., den Broeder, J. M., van Kempen, E. E. M. M., van Wesemael, P. J. V., & Schuit, A. J. (2014). Shifting from car to active transport: A systematic review of the effectiveness of interventions. Transp Res. *Transportation Research Part A, Policy and Practice*, *70*, 264–280. DOI: 10.1016/j.tra.2014.10.015

Schreurs, V. V., & Schaafsma, G. (2010). Lactic acid and lactates: In Health and Wellness. *NUTRAfoods : International Journal of Science and Marketing for Nutraceutical Actives, Raw Materials, Finish Products*, *9*(1), 7–16. DOI: 10.1007/BF03223326

Schüth, F. (2009). Challenges in hydrogen storage. *The European Physical Journal. Special Topics*, *176*(1), 155–166. DOI: 10.1140/epjst/e2009-01155-x

Sellitto, M. A., Borchardt, M., Pereira, G. M., & Bubicz, M. E. (2015). Tool for environmental performance assessment of city bus transit operations: Case studies. *Clean Technologies and Environmental Policy*, *17*(4), 1053–1064. DOI: 10.1007/s10098-014-0892-7

Şener, Ş., Sener, E., & Karagüzel, R. (2011). Solid waste disposal site selection with GIS and AHP methodology: A case study in Senirkent–Uluborlu (Isparta) Basin, Turkey. *Environmental Monitoring and Assessment*, *173*(1-4), 533–554. DOI: 10.1007/s10661-010-1403-x PMID: 20213053

Sengupta, D., Melix, P., Bose, S., Duncan, J., Wang, X., Mian, M. R., Kirlikovali, K. O., Joodaki, F., Islamoglu, T., Yildirim, T., Snurr, R. Q., & Farha, O. K. (2023). Air-Stable Cu(I) Metal–Organic Framework for Hydrogen Storage. *Journal of the American Chemical Society*, *2023*(37), 20492–20502. Advance online publication. DOI: 10.1021/jacs.3c06393 PMID: 37672758

Serge-Correales, Y. E., Ullah, S., Ferreira-Neto, E. P., Rojas-Mantilla, H. D., Hazra, C., & Ribeiro, S. J. L. A. (2022). UV-visible-NIR active smart photocatalytic system based on NaYbF4:Tm3+ upconverting particles and Ag_3PO_4/H_2O_2 for photocatalytic processes under light on/light off conditions. *Materials Advances*, *3*(6), 2706–2715. DOI: 10.1039/D1MA01028J

Shah, F., & Wu, W. (2019). Soil and crop management strategies to ensure higher crop productivity within sustainable environments. *Sustainability (Basel)*, *11*(5), 1485. DOI: 10.3390/su11051485

Shah, T. V., & Vasava, D. V. (2019). A glimpse of biodegradable polymers and their biomedical applications. *E-Polymers*, *19*(1), 385–410. DOI: 10.1515/epoly-2019-0041

Shaikh, T. A., Rasool, T., & Lone, F. R. (2022). Towards leveraging the role of machine learning and artificial intelligence in precision agriculture and smart farming. *Computers and Electronics in Agriculture*, *198*, 107119. DOI: 10.1016/j.compag.2022.107119

Shannon, M. A., Bohn, P. W., Elimelech, M., Georgiadis, J. G., Mariñas, B. J., & Mayes, A. M. (2008). Science and technology for water purification in the coming decades. *Nature*, *452*(7185), 301–310. DOI: 10.1038/nature06599 PMID: 18354474

Sharholy, M., Ahmad, K., Mahmood, G., & Trivedi, R. C. (2008). Municipal solid waste management in Indian cities–A review. *Waste Management (New York, N.Y.)*, *28*(2), 459–467. DOI: 10.1016/j.wasman.2007.02.008 PMID: 17433664

Sharma, P., Gaur, V. K., Kim, S. H., & Pandey, A. (2020). Microbial strategies for bio-transforming food waste into resources. *Bioresource Technology*, *299*, 122580. DOI: 10.1016/j.biortech.2019.122580 PMID: 31877479

Shchurov, I. N., Dedov, S. I., Malozyomov, B. V., Shtang, A. A., Martyushev, N. V., Klyuev, R. V., & Andriashin, S. N. (2021). Degradation of Lithium-Ion Batteries in an Electric Transport Complex. *Energies*, *2021*(14), 8072. DOI: 10.3390/en14238072

Sikka, A. K., Alam, M. F., & Mandave, V. (2022). Agricultural water management practices to improve the climate resilience of irrigated agriculture in India. *Irrigation and Drainage*, *71*(S1), 7–26. DOI: 10.1002/ird.2696

Silva, F. M., Queirós, C., Pereira, M., Pinho, T., Barroso, T., Magalhães, S., Boaventura, J., Santos, F., Cunha, M., & Martins, R. C. (2024). Precision Fertilization: A critical review analysis on sensing technologies for nitrogen, phosphorous and potassium quantification. *Computers and Electronics in Agriculture*, *224*, 109220. DOI: 10.1016/j.compag.2024.109220

Singh, G., Capalash, N., Kaur, K., Puri, S., & Sharma, P. (2016). Enzymes: applications in pulp and paper industry. In Agro-industrial wastes as feedstock for enzyme production (pp. 157-172). *Academic Press*. https://doi.org/DOI: 10.1016/B978-0-12-802392-1.00007-1

Singha, A. S., & Thakur, V. K. (2009). Chemical Resistance, Mechanical, and Physical Properties of Biofibers-Based Polymer Composites. *Polymer-Plastics Technology and Engineering*, *48*(7), 736–742. DOI: 10.1080/03602550902824622

Singha, A. S., & Thakur, V. K. (2009). Physical, Chemical, and Mechanical Properties of *Hibiscus sabdariffa* Fiber/Polymer Composite. *International Journal of Polymeric Materials*, *58*(4), 217–225. DOI: 10.1080/00914030802639999

Singh, R., Altaee, A., & Gautam, S. (2020). Nanomaterials in the advancement of hydrogen energy storage. *Heliyon*, *6*(7), e04487. DOI: 10.1016/j.heliyon.2020.e04487 PMID: 32743097

Singhvi, M., Zinjarde, S., & Gokhale, D. (2019). Polylactic acid: Synthesis and biomedical applications. *Journal of Applied Microbiology*, *127*(5), 1612–1626. DOI: 10.1111/jam.14290 PMID: 31021482

Smdani, G., Islam, M. R., Ahmad Yahaya, A. N., & Bin Safie, S. I. (2023). PERFORMANCE EVALUATION OF ADVANCED ENERGY STORAGE SYSTEMS: A REVIEW. *Energy & Environment*, *34*(4), 1094–1141. DOI: 10.1177/0958305X221074729

Sridevi, V., Modi, M., Lakshmi, M. V. V. C., & Kesavarao, L. (2012). A review on integrated solid waste management. *Int. J. Eng. Sci. Adv. Technol*, *2*, 1491–1499.

Srivastava, M., MR, A. K., & Zaghib, K. (2024). Binders for Li-Ion Battery Technologies and Beyond: A Comprehensive Review. *Batteries*, *10*(8), 268.

Srivastava, R., Krishna, V., & Sonkar, I. (2014). Characterization and management of municipal solid waste: A case study of Varanasi city, India. *International Journal of Current Research and Academic Review*, *2*(8), 10–16.

Stark, N. M., & Matuana, L. M. (2021). Trends in sustainable biobased packaging materials: A mini review. *Materials Today Sustainability*, *15*, 100084. DOI: 10.1016/j.mtsust.2021.100084

Statista Research Department. (2024). Distribution of Global Plastic Materials Production in 2022, by Region. Retrieved May 29, 2024, from https://www.statista.com/statistics/281126/global-plastics-production-share-of-various-countries-and-regions/

Stridsberg, K. M., Ryner, M., & Albertsson, A.-C. (2002). *Controlled ring-opening polymerization: Polymers with designed macromolecular architecture*. Springer.,

Stubbe, J., & Tian, J. (2003). Polyhydroxyalkanoate (PHA) homeostasis: The role of the PHA synthase. *Natural Product Reports, 20*(5), 445–457. DOI: 10.1039/b209687k PMID: 14620841

Sudesh, K., Abe, H., & Doi, Y. (2000). Synthesis, structure and properties of polyhydroxyalkanoates: Biological polyesters. *Progress in Polymer Science, 25*(10), 1503–1555. DOI: 10.1016/S0079-6700(00)00035-6

Suhag, A., Yadav, H., Chaudhary, D., Subramanian, S., Jaiwal, R., & Jaiwal, P. K. (2021). Biotechnological interventions for the sustainable management of a global pest, whitefly (*Bemisia tabaci*). *Insect Science, 28*(5), 1228–1252. DOI: 10.1111/1744-7917.12853 PMID: 32696581

Sun, L., Wang, J., Mahmud, S., Jiang, Y., Zhu, J., & Liu, X. (2019). New insight into the mechanism for the excellent gas properties of poly (ethylene 2,5-furandicarboxylate) (PEF): Role of furan ring's polarity. *European Polymer Journal, 118*, 642–650. DOI: 10.1016/j.eurpolymj.2019.06.033

Sun, Q., Jena, P., Wang, Q., & Marquez, M. (2006). First-Principles Study of Hydrogen Storage on $Li_{12}C_{60}$. *Journal of the American Chemical Society, 128*(30), 9741–9745. DOI: 10.1021/ja058330c PMID: 16866529

Taib, N.-A. A. B., Rahman, R., Huda, D., Kuok, K. K., Hamdan, S., Bin Bakri, M. K., Bin Julaihi, M. R. M., & Khan, A. (2022). A review on poly lactic acid (PLA) as a biodegradable polymer. *Polymer Bulletin, 80*(2), 1179–1213. DOI: 10.1007/s00289-022-04160-y

Tanaka, K., & Toda, F. J. C. R. (2000). Solvent-free organic synthesis. *Chemical Reviews, 100*(3), 1025–1074. DOI: 10.1021/cr940089p PMID: 11749257

Tang, Z., Feng, D., Xu, Y., Chen, L., Zhang, X., & Ma, Q. (2023). Safety Issues of Layered Nickel-Based Cathode Materials for Lithium-Ion Batteries: Origin, Strategies and Prospects. *Batteries, 2023*(9), 156. DOI: 10.3390/batteries9030156

Tasneem, O., Tasneem, H., & Xian, X. (2025). Lithium-ion Battery Technologies for Grid-Scale Renewable Energy Storage. Next Research, 100297.

Taurino, R., Bondioli, F., & Messori, M. (2023). Use of different kinds of waste in the construction of new polymer composites. *Materials Today Sustainability, 21*, 100298. DOI: 10.1016/j.mtsust.2022.100298

Tawalbeh, M., Farooq, A., Martis, R., & Al-Othman, A. (2024). Optimization techniques for electrochemical devices for hydrogen production and energy storage applications. *International Journal of Hydrogen Energy*, *52*, 1058–1092. DOI: 10.1016/j.ijhydene.2023.06.264

Thakur, V. K., Thakur, M. K., & Gupta, R. K. (2014). Review: Raw Natural Fiber–Based Polymer Composites. *IJPAC. International Journal of Polymer Analysis and Characterization*, *19*(3), 256–271. DOI: 10.1080/1023666X.2014.880016

Thomas, A. P., Kasa, V. P., Dubey, B. K., Sen, R., & Sarmah, A. K. (2023). Synthesis and commercialization of bioplastics: Organic waste as a sustainable feedstock. *The Science of the Total Environment*, *904*, 167243. DOI: 10.1016/j.scitotenv.2023.167243 PMID: 37741416

Tindall, J. A., Beverly, R. B., & Radcliffe, D. E. (1991). Mulch effect on soil properties and tomato growth using micro-irrigation. *Agronomy Journal*, *83*(6), 1028–1034. DOI: 10.2134/agronj1991.00021962008300060019x

Tomaszewska, A., Chu, Z., Feng, X., O'kane, S., Liu, X., Chen, J., & Wu, B. (2019). Lithium-ion battery fast charging: A review. *ETransportation*, *1*, 100011.

Tong, Z., Baldauf, R. W., Isakov, V., Deshmukh, P., & Max Zhang, K. (2016). Roadside vegetation barrier designs to mitigate near-road air pollution impacts. *The Science of the Total Environment*, *541*, 920–927. DOI: 10.1016/j.scitotenv.2015.09.067 PMID: 26457737

Tripathi, S., Srivastava, P., Devi, R. S., & Bhadouria, R. (2020). Influence of synthetic fertilizers and pesticides on soil health and soil microbiology. In *Agrochemicals detection, treatment and remediation* (pp. 25–54). Elsevier., https://www.sciencedirect.com/science/article/pii/B9780081030172000027 DOI: 10.1016/B978-0-08-103017-2.00002-7

UNESCO. (2020). *The United Nations World Water Development Report 2020: Water and Climate Change*. UNESCO Publishing.

United Nations Environment Program. (2020). *Single-use plastics: A roadmap for sustainability*. Retrieved June 29, 2024, from https://www.unep.org/resources/report/single-use-plastics-roadmap-sustainability

United Nations. Sustainable Development Goal 6: Ensure availability and sustainable management of water and sanitation for all. https://sdgs.un.org/goals/goal6

Upadhyay, A., Chandel, N. S., Singh, K. P., Chakraborty, S. K., Nandede, B. M., Kumar, M., Subeesh, A., Upendar, K., Salem, A., & Elbeltagi, A. (2025). Deep learning and computer vision in plant disease detection: A comprehensive review of techniques, models, and trends in precision agriculture. *Artificial Intelligence Review, 58*(3), 1–64. DOI: 10.1007/s10462-024-11100-x

Vaddula, B. R., Varma, R. S., & Leazer, J. (2013). Mixing with microwaves: Solvent-free and catalyst-free synthesis of pyrazoles and diazepines. *Tetrahedron Letters, 54*(12), 1538–1541. DOI: 10.1016/j.tetlet.2013.01.029

Vasuki, G., & Kumaravel, K. (2008). Rapid four-component reactions in water: Synthesis of pyranopyrazoles. *Tetrahedron Letters, 49*(39), 5636–5638. DOI: 10.1016/j.tetlet.2008.07.055

Velaoras, K., Menegaki, A. N., Polyzos, S., & Gotzamani, K. (2025). The role of environmental certification in the hospitality industry: Assessing sustainability, consumer preferences, and the economic impact. *Sustainability (Basel), 17*(2), 650. DOI: 10.3390/su17020650

Venkataraman, C., Sagar, A. D., Habib, G., Lam, N., & Smith, K. R. (2010). The Indian National Initiative for advanced biomass cookstoves: The benefits of clean combustion. *Energy for Sustainable Development, 14*(2), 63–72. DOI: 10.1016/j.esd.2010.04.005

Venkateswarlu, S., Vallem, S., Umer, M., Jyothi, N. V. V., Giridhar Babu, A., Govindaraju, S., Son, Y., Jong Kim, M., & Yoon, M. (2023). Recent progress on MOF/MXene nanoarchitectures: A new era in coordination chemistry for energy storage and conversion. *Journal of Energy Chemistry, 86*, 409–436. DOI: 10.1016/j.jechem.2023.07.044

Verma, R., Vinoda, K., Papireddy, M., & Gowda, A. (2016). Toxic Pollutants from Plastic Waste—A Review. *Procedia Environmental Sciences, 35*, 701–708. DOI: 10.1016/j.proenv.2016.07.069

Verma, S. K., Prasad, A., & Katiyar, V. (2024). State-of-the-art review on sustainable biodegradable polymers with a market overview for sustainability packaging. *Materials Today Sustainability, 26*, 100776. DOI: 10.1016/j.mtsust.2024.100776

Vicente, D., Proença, D. N., & Morais, P. V. (2023). The Role of Bacterial Polyhydroalkanoate (PHA) in a Sustainable Future: A review on the biological diversity. *International Journal of Environmental Research and Public Health, 20*(4), 2959. DOI: 10.3390/ijerph20042959 PMID: 36833658

Virendra Kumar Saharan. (2014). *Dipak Pinjari, Parag R Gogate, Aniruddha Bhalchandra Pandit. Advanced Oxidation Technologies for Wastewater Treatment.* Industrial Wastewater Treatment, Recycling and Reuse., DOI: 10.1016/B978-0-08-099968-5.00003-9

Vivek, M., Kumar, P. S., Steffi, S., & Sudha, S. (2011). Biogenic Silver Nanoparticles by Gelidiella Acerosa Extract and Their Antifungal Effects. *Avicenna Journal of Medical Biotechnology*, *3*, 143–148. PMID: 23408653

Vlachokostas, C., Achillas, C., Moussiopoulos, N., & Banias, G. (2011). Multicriteria methodological approach to manage urban air pollution. *Atmospheric Environment*, *45*(25), 4160–4169. DOI: 10.1016/j.atmosenv.2011.05.020

Vlaiculescu, A., & Varrone, C. (2022). Sustainable and eco-friendly alternatives to reduce the use of pesticides. In *Pesticides in the natural environment* (pp. 329–364). Elsevier. DOI: 10.1016/B978-0-323-90489-6.00014-8

Wabnitz, C., & Nichols, W. J. (2010). Plastic Pollution: An Ocean Emergency. *Marine Turtle Newsletter*, *129*, 1–4.

Wahab, A., Muhammad, M., Ullah, S., Abdi, G., Shah, G. M., Zaman, W., & Ayaz, A. (2024). Agriculture and environmental management through nanotechnology: Eco-friendly nanomaterial synthesis for soil-plant systems, food safety, and sustainability. *The Science of the Total Environment*, *926*, 171862. DOI: 10.1016/j.scitotenv.2024.171862 PMID: 38527538

Wali, S. B., Hannan, M. A., Abd Rahman, M. S., Alghamdi, H. A., Mansor, M., Ker, P. J., Tiong, S. K., & Mahlia, T. M. I. (2023). Usage count of hydrogen-based hybrid energy storage systems: An analytical review, challenges and future research potentials. *International Journal of Hydrogen Energy*, *48*(89), 34836–34861. DOI: 10.1016/j.ijhydene.2023.05.298

Wang, D., Liu, N., Chen, F., Wang, Y., & Mao, J. (2024). Progress and prospects of energy storage technology research: Based on multidimensional comparison. *Journal of Energy Storage*, *75*, 109710. DOI: 10.1016/j.est.2023.109710

Wang, D., Xue, B., Wang, L., Zhang, Y., Liu, L., & Zhou, Y. (2021). Fungus-mediated green Synthesis of Nano-Silver Using Aspergillus sydowii and its Antifungal/antiproliferative Activities. *Scientific Reports*, *11*(1), 1. DOI: 10.1038/s41598-021-89854-5 PMID: 33990673

Wang, H., Fu, F., Huang, M., Feng, Y., Han, D., Xi, Y., Xiong, W., Yang, D., & Niu, L. (2023). Lignin-based materials for electrochemical energy storage devices. *Nano Materials Science*, *5*(2), 141–160. DOI: 10.1016/j.nanoms.2022.01.002

Wang, J. F., Lu, L. Y., Zhang, P., Qin, Y. J., & Guo, Z. X. (2022). Selective oxidation of primary benzylic alcohols to aldehydes using NaNO3 under ball milling. *Journal of Chemical Research, 46*(4), 17475198221101983. DOI: 10.1177/17475198221101983

Wang, L., Wang, J., Wang, L., Zhang, M., Wang, R., & Zhan, C. (2022). A critical review on nickel-based cathodes in rechargeable batteries. *International Journal of Minerals Metallurgy and Materials, 29*(5), 925–941. DOI: 10.1007/s12613-022-2446-z

Wang, X., Wang, F., Sang, Y., & Liu, H. (2017). Full-Spectrum Solar-Light-Activated Photocatalysts for Light–Chemical Energy Conversion. *Advanced Energy Materials, 7*(23), 1700473–1700489. DOI: 10.1002/aenm.201700473

Wan, Y., Lu, X., Dalai, S., & Zhang, J. (2009). Thermophysical properties of polycaprolactone/chitosan blend membranes. *Thermochimica Acta, 487*(1-2), 33–38. DOI: 10.1016/j.tca.2009.01.007

Waseem, M., Lakshmi, G. S., Ahmad, M., & Suhaib, M. (2025). Energy storage technology and its impact in electric vehicle: Current progress and future outlook. *Next Energy, 6*, 100202.

Winnacker, M. (2019). Polyhydroxyalkanoates: Recent Advances in their Synthesis and Applications. *European Journal of Lipid Science and Technology, 121*(11), 1900101. DOI: 10.1002/ejlt.201900101

Wood, C. D., Cooper, A. I., & DeSimone, J. M. (2004). Green synthesis of polymers using supercritical carbon dioxide. *Current Opinion in Solid State and Materials Science, 8*(5), 325–331. DOI: 10.1016/j.cossms.2005.02.001

Woodcock, J., Edwards, P., Tonne, C., Armstrong, B. G., Ashiru, O., Banister, D., Beevers, S., Chalabi, Z., Chowdhury, Z., Cohen, A., Franco, O. H., Haines, A., Hickman, R., Lindsay, G., Mittal, I., Mohan, D., Tiwari, G., Woodward, A., & Roberts, I. (2009). Public health benefits of strategies to reduce greenhouse-gas emissions: Urban land transport. *Lancet, 374*(9705), 1930–1943. DOI: 10.1016/S0140-6736(09)61714-1 PMID: 19942277

Woodruff, M. A., & Hutmacher, D. W. (2010). The return of a forgotten polymer—Polycaprolactone in the 21st century. *Progress in Polymer Science, 35*(10), 1217–1256. DOI: 10.1016/j.progpolymsci.2010.04.002

World Bank Group (Ed.). World Development Indicators 2012. World Bank Publications; 2012 Apr 18.

World Health Organization (WHO). (2019). Drinking-water. Retrieved from https://www.who.int/news-room/fact-sheets/detail/drinking-water

World Health Organization. (2018). COP24 special report: health and climate change. https://www.energy.gov/eere/articles/hydrogen-clean-flexible-energy-carrier [Access 05/11/2020].

Wu, Y., Si, W., Yan, S., Wu, L., Zhao, W., Zhang, J., Zhang, F., & Fan, J. (2023). Water consumption, soil nitrate-nitrogen residue and fruit yield of drip-irrigated greenhouse tomato under various irrigation levels and fertilization practices. *Agricultural Water Management*, *277*, 108092. DOI: 10.1016/j.agwat.2022.108092

Wu, Y., Zhang, S., Hao, J., Liu, H., Wu, X., Hu, J., Walsh, M. P., Wallington, T. J., Zhang, K. M., & Stevanovic, S. (2017). On-road vehicle emissions and their control in China: A review and outlook. *The Science of the Total Environment*, *574*, 332–349. DOI: 10.1016/j.scitotenv.2016.09.040 PMID: 27639470

Wysel, M. (2023). Data Sharing Platforms: How Value Is Created from Data Produced by Smart Agriculture. In Zhang, Q. (Ed.), *Encyclopedia of Digital Agricultural Technologies* (pp. 257–267). Springer International Publishing., DOI: 10.1007/978-3-031-24861-0_250

Xia, T., Nitschke, M., Zhang, Y., Shah, P., Crabb, S., & Hansen, A. (2015). Traffic-related air pollution and health co-benefits of alternative transport in Adelaide, South Australia. *Environment International*, *74*, 281–290. DOI: 10.1016/j.envint.2014.10.004 PMID: 25454245

Xie, J., & Lu, Y.-C. (2020, May 19). A retrospective on lithium-ion batteries. *Nature Communications*, *11*(1), 2499. DOI: 10.1038/s41467-020-16259-9 PMID: 32427837

Xu, K., Guo, Y., Lei, G., & Zhu, J. (2023). A Review of Flywheel Energy Storage System Technologies. *Energies*, *16*(18), 6462. DOI: 10.3390/en16186462

Yadav, A., Yadav, K., & Abd-Elsalam, K. A. (2023). Nanofertilizers: Types, delivery and advantages in agricultural sustainability. *Agrochemicals*, *2*(2), 296–336. DOI: 10.3390/agrochemicals2020019

Yalpani, M., Marchessault, R. H., Morin, F. G., & Monasterios, C. J. (1991). Synthesis of poly(3-hydroxyalkanoate) (PHA) conjugates: PHA-carbohydrate and PHA-synthetic polymer conjugates. *Macromolecules*, *24*(22), 6046–6049. DOI: 10.1021/ma00022a024

Yang, C., Zhao, W., Deng, K., Zhou, V., Zhou, X., & Hou, Y. (2017). The association between air pollutants and autism spectrum disorders. *Environmental Science and Pollution Research International*, *24*(19), 15949–15958. DOI: 10.1007/s11356-017-8928-2 PMID: 28540549

Yang, H., He, Z., Zhang, M., Tan, X., Sun, K., Liu, H., Wang, N., Guan, L., Wang, C., Wan, Y., Wang, W., Hu, H., & Wu, M. (2023). Reshaping the material research paradigm of electrochemical energy storage and conversion by machine learning. *EcoMat*, *5*(5), e12330. Advance online publication. DOI: 10.1002/eom2.12330

Yang, J., Sudik, A., Wolverton, C., & Siegel, D. J. (2010). High capacity hydrogen storage materials: Attributes for automotive applications and techniques for materials discovery. *Chemical Society Reviews*, *39*(2), 656–675. Retrieved May 11, 2020, from https://www.energy.gov/eere/articles/hydrogen-clean-flexible-energy-carrier. DOI: 10.1039/B802882F PMID: 20111786

Yang, K., Zhang, L., Wang, W., Long, C., Yang, S., Zhu, T., & Liu, X. (2024). Multiscale modeling for enhanced battery health analysis: Pathways to longevity. *Carbon Neutralization*, *3*(3), 348–385. DOI: 10.1002/cnl2.124

Yang, S.-T., Cao, L., Luo, P. G., Lu, F., Wang, X., Wang, H., Meziani, M. J., Liu, Y., Qi, G., & Sun, Y.-P. (2009). Carbon dots for optical imaging in vivo. *Journal of the American Chemical Society*, *131*(32), 11308–11309. DOI: 10.1021/ja904843x PMID: 19722643

Yang, Y., Wu, Z., Yao, J., Guo, T., Yang, F., Zhang, Z., Ren, J., Jiang, L., & Li, B. (2024). An overview of application-oriented multifunctional large-scale stationary battery and hydrogen hybrid energy storage system. *Energy Reviews*, *3*(2), 100068. DOI: 10.1016/j.enrev.2024.100068

Yasir, M., Zeshan, B., Daud, N. H. A., Shahid, I., & Khalid, H. (2021). Characterization of bacteriocin and chitinase producing bacterial isolates with broad-spectrum antimicrobial activities. *SN Applied Sciences*, *3*(8), 1–12. DOI: 10.1007/s42452-021-04740-z

Yazici, İ., Shayea, I., & Din, J. (2023). A survey of applications of artificial intelligence and machine learning in future mobile networks-enabled systems. *Engineering Science and Technology, an International Journal*, *44*, 101455. DOI: 10.1016/j.jestch.2023.101455

Yedla, S. (2005). Modified landfill design for sustainable waste management. *International Journal of Global Energy Issues*, *23*(1), 93–105. DOI: 10.1504/IJGEI.2005.006412

Yehia, R. S., & Al-Sheikh, H. (2014). Biosynthesis and Characterization of Silver Nanoparticles Produced by Pleurotus Ostreatus and Their Anticandidal and Anticancer Activities. *World Journal of Microbiology & Biotechnology*, *30*(11), 2797–2803. DOI: 10.1007/s11274-014-1703-3 PMID: 25053172

Ye, J., Wang, Y., Kang, J., Chen, Y., Hong, L., Li, M., Jia, Y., Wang, Y., Jia, X., Wu, Z., & Wang, H. (2022). Effects of long-term use of organic fertilizer with different dosages on soil improvement, nitrogen transformation, tea yield and quality in acidified tea plantations. *Plants, 12*(1), 122. DOI: 10.3390/plants12010122 PMID: 36616251

Yin, G. Z., & Yang, X. M. (2020). Biodegradable polymers: A cure for the planet, but a long way to go. *Journal of Polymer Research, 27*(2), 38. DOI: 10.1007/s10965-020-2004-1

Yi, X., Kirk, C., & Robertson, N. (2024). Nickel hydroxide-based energy storage devices: Nickel-metal hydride batteries vs. nickel hydroxide hybrid supercapacitors. *Carb Neutrality, 3*(1), 39. DOI: 10.1007/s43979-024-00114-7

Yi, Z., Chen, Z., Yin, K., Wang, L., & Wang, K. (2023). Sensing as the key to the safety and sustainability of new energy storage devices. *Protection and Control of Modern Power Systems, 8*(1), 27. DOI: 10.1186/s41601-023-00300-2

Yoichi, H., & Yutaka, T. (2010). Mechanical, Thermal, and Physico-Chemical Characterization. *Journal of Polymers and the Environment, 18*(2), 326.

Yoichi, H., & Yutaka, T. (2010). Polymer environment research. *Journal of Polymers and the Environment, 18*(1), 116.

Yong, W. T. L., Thien, V. Y., Misson, M., Chin, G. J. W. L., Hussin, S. N. I. S., Chong, H. L. H., Yusof, N. A., Ma, N. L., & Rodrigues, K. F. (2024). Seaweed: A bioindustrial game-changer for the green revolution. *Biomass and Bioenergy, 183*, 107122. DOI: 10.1016/j.biombioe.2024.107122

Younes, B. (2016). Classification, characterization, and the production processes of biopolymers used in the textiles industry. *Journal of the Textile Institute, 108*(5), 674–682. DOI: 10.1080/00405000.2016.1180731

Younger, M., Morrow-Almeida, H. R., Vindigni, S. M., & Dannenberg, A. L. (2008). The built environment, climate change, and health. Opportunities for co-benefits. *American Journal of Preventive Medicine, 35*(5), 517–526. DOI: 10.1016/j.amepre.2008.08.017 PMID: 18929978

Yu, X., Chen, R., Gan, L., Li, H., & Chen, L. (2023). Battery Safety: From Lithium-Ion to Solid-State Batteries. *Engineering (Beijing), 21*, 9–14. DOI: 10.1016/j.eng.2022.06.022

Zangade, S., & Patil, P. (2019). A review on solvent-free methods in organic synthesis. *Current Organic Chemistry, 23*(21), 2295–2318. DOI: 10.2174/1385272823666191016165532

Zanobetti, A., & Schwartz, J. (2005). The effect of particulate air pollution on emergency admissions for myocardial infarction: A multicity Case-crossover analysis. *Environmental Health Perspectives, 113*(8), 978–982. DOI: 10.1289/ehp.7550 PMID: 16079066

Zayed, M. E., Kabeel, A. E., Shboul, B., Ashraf, W. M., Ghazy, M., Irshad, K., Rehman, S., & Zayed, A. A. A. (2023). Performance augmentation and machine learning-based modeling of wavy corrugated solar air collector embedded with thermal energy storage: Support vector machine combined with Monte Carlo simulation. *Journal of Energy Storage, 74*, 109533. DOI: 10.1016/j.est.2023.109533

Zeng, Y., Verhoef, A., Vereecken, H., Ben-Dor, E., Veldkamp, T., Shaw, L., Van Der Ploeg, M., Wang, Y., & Su, Z. (2025). Monitoring and Modeling the Soil-Plant System Toward Understanding Soil Health. *Reviews of Geophysics, 63*(1), e2024RG000836. DOI: 10.1029/2024RG000836

Zeng, L., Peng, J., Zhang, J., Tan, X., Ji, X., Li, S., & Feng, G. (2023). Molecular dynamics simulations of electrochemical interfaces. *The Journal of Chemical Physics, 159*(9), 091001. Advance online publication. DOI: 10.1063/5.0160729 PMID: 37671960

Zhang, H., Gao, S., & Zhou, P. (2023). Role of digitalization in energy storage technological innovation: Evidence from China. *Renewable & Sustainable Energy Reviews, 171*, 113014. DOI: 10.1016/j.rser.2022.113014

Zhang, M., Ma, W., Tao, R., Fan, Q., Zhang, M., Qin, D., Cao, X., Li, J., Xiong, R., & Huang, C. (2024). Nanomaterials: Recent advances in plant disease diagnosis and treatment. *Nano Today, 57*, 102326. DOI: 10.1016/j.nantod.2024.102326

Zhang, X., Gu, P., & Liu, Y. (2019). Decontamination of radioactive wastewater: State of the art and challenges forward. *Chemosphere, 215*, 543–553. DOI: 10.1016/j.chemosphere.2018.10.029 PMID: 30342399

Zhao, Z. Q., Bai, Z. H., Winiwarter, W., Kiesewetter, G., Heyes, C., & Ma, L. (2017). Mitigating ammonia emission from agriculture reduces PM2.5 pollution in the Hai River basin in China. *The Science of the Total Environment, 609*, 1152–1160. DOI: 10.1016/j.scitotenv.2017.07.240 PMID: 28787789

Zhong-Yu, L., Fu, J. K., Jin-Kun, F., Jian-Ming, W., Yue-Ying, L., & Hu, C. (2001). Preliminary Study on the Mechanism of Nonenzymatic Bioreduction of Precious Metal Ions. *Wuli Huaxue Xuebao, 17*(5), 477–480. DOI: 10.3866/PKU.WHXB20010520

Zhou, Q. L. (2015). Transition-metal catalysis and organocatalysis: Where can progress be expected? *Angewandte Chemie*, *55*(18), 5352–5353. DOI: 10.1002/anie.201509164 PMID: 26662619

Zhou, Y., Ouyang, Y., Zhang, Y., Li, Q., & Wang, J. (2023). Machine Learning Assisted Simulations of Electrochemical Interfaces: Recent Progress and Challenges. *The Journal of Physical Chemistry Letters*, *14*(9), 2308–2316. DOI: 10.1021/acs.jpclett.2c03288 PMID: 36847421

Zhu, X., Zhu, Z. Q., Guo, D., Liu, S., Ji, J. J., Tang, J., Yuan, E., Xie, Z.-B., & Le, Z. G. (2020). Cascade cyclization of glycine derivatives with β-ketoesters for polysubstituted 1, 4-dihydropyridines by visible light photoredox catalysis. *Tetrahedron*, *76*(33), 131353. DOI: 10.1016/j.tet.2020.131353

About The Contributors

Shrikaant Kulkarni, Ph.D., is currently working as an Adjunct Professor, at the Faculty of Business, Victorian Institute of Technology, Melbourne, Australia, and Adjunct Professor at Chitkara University, Punjab, India. Dr. Kulkarni has been a senior academician and researcher for 42 years. He has delivered invited lectures and conducted sessions at national and international conferences and faculty development programs. He has guided many major and minor projects in areas like engineering chemistry, green chemistry, nanotechnology, analytical chemistry, catalysis, chemical engineering materials, industrial organization, and management, Advanced AI, and Educational Technology. He has published over 100+ research papers in national and international journals and conferences of repute. He has edited 30+ books including four books in Advanced AI, and authored 75+ book chapters in books published by CRC, Springer Nature, Apple Academic Press, Elsevier, Wiley, and IET. He authored four textbooks in engineering chemistry too. Dr Kulkarni possesses MSc, MPhil, and PhD degrees in Chemistry and Master's degrees in Economics, Business Management, and Political Science.Two of his Patents have been published and two more are applied for grant. He was invited by UNESCO for an invited talk on " Green Chemistry Education for Sustainable Development " at the International Conference on Green Chemistry, Bangkok, and was well received by international Jury members and dignitaries from across the world. E-mail: shrikaant.kulkarni@vit.edu.au Mob: +91-9970663353

Hemantkumar Akolkar, Ph.D., Working as Assistant Professor, Department of Chemistry, Rayat Shikshan Sanstha's Abasaheb Marathe Arts and New Commerce, Science College, Rajapur, Dist- Ratnagiri, Maharashtra, India Dr. Hemantkumar N. Akolkar is an academician and researcher for last 14 years at both undergraduate and postgraduate level. He has taught subjects like Heterocyclic Chemistry, Green Chemistry, Organic Spectroscopy, Organic Stereochemistry, Chemistry of Natural Products during his career. He possesses Masters in Organic chemistry and Ph.D. in

Chemistry from Savitribai Phule Pune University, Pune. He has 60+ research papers in national and international journals and conferences of repute, 6 patents, 3 books and 9 book chapters. His areas of interest are Heterocyclic Chemistry, Synthetic Organic Chemistry and Green Chemistry. He is reviewer of several journals in Chemistry of international repute.

***

A.Revathi, has more than 10 years of teaching experience in Engineering College. She completed her M.Sc., at Government Arts College, Coimbatore and M.Phil., Chemistry in Erode Arts and Science College, Erode. She has Completed Ph.D. in the field of Environmental Science at Kongu Engineering College under Anna university Chennai. She has published more than 15 articles in various reputed international journals. She has presented many papers in international and national conferences organized by various reputed institutions. She received Best Faculty Award in the Department of Chemistry for the Academic year 2020-21. She has organized various national and international webinars and quiz competition for Student community in the field of Chemistry. Her area of interests Wastewater treatment, Adsorption, Water Quality Analysis, Photocatalyst etc.,

Laxmi Kant Bhardwaj (Ph.D. in Environmental Sciences). I was awarded an appreciation award in 2016 at the conference of the Society of Toxicology (STOX). I went to Antarctica through the Indian Scientific Expedition to the Southern Ocean (ISESO) for 60 days. I was awarded a travel grant award from the Department of Science and Engineering Research Board (SERB) and the North American Chemical Residue Workshop (NACRW), USA. I have 15 years of research experience in Food/Pharma/Environment. I have published 16 research/review articles, 3 book chapters, and 3 books. I reviewed several manuscripts and got peer reviewer certificates. I have been appointed as an expert reviewer with AEIC (Academic Exchange Information Center), ABCD indexing, and the Institute of Scholars (InSC). I was selected as a member of IAO (International Association of Oncology), the Institute of Scholars (InSC), ABCD indexing, and Voice of Indian Concern for the Environment (VOICE). I got a Young Researcher award for 2022 from the Institute of Scholars (InSC). I was selected as an editorial board member for the journals Extensive Reviews and the International Journal of Nature Science (IJNS).

Gauraw R.Chopade received his UG in pharmacy in 2023 from North Maharashtra University.After graduation, he joined the Prin.K.M. Kundnani College Of Pharmacy, Mumbai to pursue his M. Pharm degree in Pharmaceutical Chemistry under the supervision of Prof.Dr.Shrikant M.Ghodse (Assistant professor.Prin.K.M. Kundnani College Of Pharmacy, Mumbai)

Ritu Dahiya is an Associate Professor in the Department of Chemistry at Chhotu Ram Arya College, Sonepat, Haryana, India. With a strong academic background in chemistry, she has contributed significantly to the field through teaching and research. Dr. Dahiya is dedicated to fostering knowledge and understanding of chemistry among her students while engaging in research activities aimed at advancing the discipline. Her expertise and passion for education have made her a respected member of the academic community in Sonepat and beyond.

Satyajit M Deshmukh has Master degree in Chemical Technology from SGBA University, Amravati, Maharashtra India (2012) and PhD in Chemical Technology Technology from SGBA University, Amravati, Maharashtra India (2019). He is currently working as professor and Head of Chemical Engineering Department Datta Meghe College of Engineering, Airoli Navi Mumbai .He has experience in Chemical Engineering with Membrane and Separation, Environmental engineering, Energy and Petrochemical

Amit Kumar Dutta is currently working as an Assistant Professor in the Department of Chemistry, Bangabasi Morning College (Affiliated to University of Calcutta). He has Completed M.Sc. from University of Calcutta in 2007 with Inorganic Chemistry specialization and obtained Ph.D. degree in the area of Nano-catalysis from Bengal Engineering and Science University, Shibpur (presently, IIEST, Shibpur) in 2012 under the supervision of Prof. Bibhutosh Adhikary. His area of research interest is the use of semiconducting nano-materials in photo-catalysis and bio-sensing applications. He has published 36 research papers in different peer-reviewed journals and 6 book chapters. He has completed one UGC project and currently one research scholar is working with him. He has been awarded Junior and Senior Research Fellowship (JRF/SRF) through joint CSIR–UGC (NET) and Research Associate Fellowship (CSIR-RA).

Shajitha G. is an Assistant Professor in the Department of Botany at Providence College (Autonomous), located in Coonoor, Tamil Nadu. With a passion for plant sciences, she is dedicated to research and education in botany. Based in the scenic Nilgiris, she contributes to academic excellence and scientific exploration.

Shrikant M. Ghodse received his UG in Pharmacy in 2011 from Savitribai Phule University. After graduation, he joined the Institute of Chemical Technology, Mumbai, to pursue his M. Pharm and Ph.D. degrees in Pharmaceutical Chemistry under the supervision of Prof. Vikas N Telvekar (Dean, Institute of Chemical

Technology, Mumbai). After Ph.D. awarded in year 2017, he joined as Assistant Professor in Department of Pharmaceutical Chemistry, Principal K. M. Kundnani College of Pharmacy, Mumbai (affiliated to University of Mumbai)

Sakshi Madhukar Gurav recieved UG in Pharmacy in 2023 from University of Mumbai . After graduation, she joined the Prin. K.M Kundnani college of pharmacy, Mumbai, to pursue her M. Pharm degrees in Pharmaceutical Chemistry under the supervision of Assistant professor Dr. Shrikant M.Ghodse (Principal K. M. Kundnani College of Pharmacy, Mumbai)

Milan Hait is currently working as an Associate Professor in the Department of Chemistry, Dr. C.V. Raman University, Bilaspur (C.G.), India. He has more than 16.5 years of teaching and 13 years of research experience. He has published 71 research articles, 20 books, and 25 book chapters in reputed publishers. His research interests are in Phytochemistry, Analysis of Water Quality parameters, Bio-adsorbents, Molecular recognition, Bio and Chemosensor, Synthesis and Analysis of biologically active molecules. He is a Fellow of Indian Chemical Society and member of several professional bodies. He is in the editorial board and reviewer of several reputed journals especially, executive editor of ES Food & Agroforestry and editorial board member of ES Chemistry and Sustainability.

T.C. Manjunath was born on 6th February 1967. He did his BE from RV College of Engg. in 1989, ME in Robotics & Automation from LD College of Engg. in 1992 with First Rank and PhD in Systems & Control Engg. from IIT Bombay in the year 2007. He is currently working as the Dean Research R D and as Professor in the Dept. of Computer Science & Engineering with specialization in IoT, Cyber Security & Blockchain Technology in Rajarajeswari College of Engineering, Bengaluru, Karnataka. He has completed B.E. Degree from R.V. College of Engg., M.E. Govt.'s LD College of Engg., Gujarat Univ., Ahmedabad in First Rank and Ph.D. in Systems and Control Engineering from the Indian Institute of Technology Bombay IIT Bombay He has got a teaching academic, research and administrative experience of more than 35+ years 10 years as Principal, 8 years as HoD rest as faculty and has published more than 1200+ research papers in various national & international journals and conferences, with 175+ patents, 25 book-chapters, 25 awards and also published 25 text books in various field of science, engineering and technology, completed projects worth Rs. 9 Crores till date apart from guiding 10 PhD scholars. He is a member of 30 Professional Society Chapters and is a fellow of IE, IETE, AMIEE and a chartered engineer. He got the best research scholar award in 2006 in IIT Bombay amongst all the research scholars of the institute. He has got 10 awards in academics and research activities.

Santosh Kumar Nathsharma is a dedicated academician in the field of Chemistry. He serves as a faculty member in the Department of Chemistry at Stewart Science College, located in Cuttack, Odisha. With a passion for teaching and scientific inquiry, Dr. Nathsharma contributes significantly to the academic and intellectual growth of his students. The college, situated in the heart of Cuttack with the postal code 753001, benefits from his expertise and commitment to excellence in education.

P. Selvakumar is currently an Associate Professor in the Department of Science and Humanities at Nehru Institute of Technology, Coimbatore, Tamil Nadu, India. Since 2023, he has served as the Department IQAC Coordinator at the institute. Prior to joining Nehru Institute of Technology, he held positions as an Assistant Professor in the Department of Science and Humanities various engineering colleges in Coimbatore, Tamil Nadu, India, spanning from 2011 to 2023. Dr.P. Selvakumar's academic and research interests lie primarily in the field of phytochemistry. He has made significant contributions to his field, with an impressive record that includes 30 journal publications, 12 patents, and 20 conference papers. Additionally, he has organized 5 seminars and conferences, further demonstrating his commitment to academic and scientific exchange. Throughout his career, Dr. P.Selvakumar has been recognized for his dedication and excellence in teaching, having been honored with several awards, including the Best Teacher Awards. He holds a prominent position in the academic community, contributing actively to research.

P. Selvam is an Associate Professor of Chemistry in the Department of Science and Humanities at Karpaga Vinayaga College of Engineering and Technology, Padalam, Chengalpattu, Tamil Nadu. With extensive experience in teaching and research, he has made significant contributions to the field of chemistry. His expertise spans various domains of chemical sciences, and he is committed to guiding students and fostering academic excellence. Dr. Selvam actively engages in research activities, publications, and academic collaborations to enhance knowledge in his field.

Mrutyunjay Padhiary is an Assistant Professor in the Department of Agricultural Engineering at Assam University, Silchar, India, with a deep academic background. His previous positions included significant teaching and research duties at the Faculty of Technology, UBKV in West Bengal, and Tribhuvan University in Nepal, as well as research experience at IIT Kharagpur. He has been awarded multiple awards, including an International NUS Research Fellowship and an excellent score on the 2001 GATE Exam, demonstrating his proficiency in precision agriculture. He has made substantial contributions to biofuel production, artificial intelligence vision, aquaculture, precision agriculture, agricultural automation, and artificial intelligence. He is currently working on several research projects, which range from autonomous surface vessels to precision agriculture all-terrain vehicles.

Dattatraya Pansare is currently working as an Assistant Professor in the Department of Chemistry at Deogiri College, Chhatrapati Sambhajinagar. He has 12 years of teaching experience and 14 years of research expertise in organic synthesis and medicinal chemistry. He is Ph.D. Research Guide in Chemistry subject. He has Granted one and published Three Indian patents on anticancer drugs. His research output includes 92 Research articles in prestigious international journals and publisher such as Elsevier, RSC, Springer-Nature, Wiley, Bentham & many more with more than 70 indexed in Scopus and 20 listed in UGC-approved journals. His Google Scholar profile reflects 700+ citations, an h-index of 16, and an i10-index of 26. Furthermore, He has authored Three books, contributed 20 book chapters to internationally recognized volumes. He is Reviewer of 48+ international reputed journals.

Aamir.M.Patel received his UG in Pharmacy in 2023 from University of Mumbai. After graduation, he joined Principal K.M.Kundnani College of Pharmacy, Mumbai, to pursue his M. Pharm in Pharmaceutical Chemistry under the supervision of Assistant Professor Dr.Shrikant.M.Ghodse.(Principal K.M.Kundnani College of Pharmacy,Mumbai)

Gourisankar Roymahapatra is currently working as an Associate Professor of Chemistry at the School of Applied Science and Humanities, Haldia Institute of Technology (HIT) Haldia, India. He completed his Ph.D. from Jadavpur University, Kolkata (2014) with N-Heterocyclic Carbene Chemistry as a major field under the mentoring of Dr. Joydev Dinda of HIT Haldia, and Prof. Ambikesh Mahapatra of Jadavpur University. Before joining to Global Institute of Science and Technology (GIST Haldia), as a Lecturer in Chemistry in the year 2011, he worked at MCC PTA Chem. Corp. Pvt. Ltd India (MCPI), Haldia, (2003 ~ 2011) as a Senior Chemist. He joined HIT Haldia in 2015 as an Assistant Professor of Chemistry. His area of research interest includes NHC complexes, organometallics, catalysis, antibiotics, anti-carcinogenic, gas adsorption, super alkali, computational chemistry and DFT. For his contribution in research, he got the 'Distinguished Young Scientist Award in Chemistry -2014'; 'Bharat Gaurabh Award – 2018'; and 'Indian Chemical Society Research Excellent Award 2021'. Dr. Roymahapatra is a Life Fellow of the Indian Chemical Society (FICS), and a Life Fellow of the Institutes of Chemistry

Mohit is a research scholar at the Institute of Management Studies and Research, Maharshi Dayanand University, Rohtak, Haryana, India. His research focuses on the dynamic relationship between brand image and purchase intentions, particularly in the context of brand extensions within the FMCG sector. He has co-authored several scholarly works and continues to contribute to the academic community. Mohit can be reached at mohitattri.rs.imsar@mdurohtak.ac.in

Rohini Shelke is currently working as an Assistant Professor in the Department of Chemistry at Radhabai Kale Mahila Mahavidyalaya, Ahmednagar, Maharashtra, India. With a decade of teaching and research experience in organic synthesis and medicinal chemistry, she has made significant contributions to her field. She has filed One Patent and published Two Indian Patent. She has published 39 research articles in reputed international journals indexed by Scopus and UGC-approved lists [Google Scholar Citations =256, h-index=10, i10-index=10], and contributed 13 book chapters to internationally acclaimed volumes.

Nikitha Sreekantha is currently the Principal & Professor of Dayananda Sagar Business Academy. She has also worked as the Principal at The Oxford College of Business Management in Bengaluru, India. She holds a PhD in Management from Bharathiar University and has over 12 years of experience at Dayananda Sagar Institutions, where she served as Head of the Department. Dr. Sreekantha has co-authored seven textbooks and presented numerous research papers at national and international conferences. She has received several awards, including the Best Woman Educator Award from PERAL Foundation in 2021, recognizing her contributions to commerce and management education.

Index

I

K

L

N

P

R

S

W

www.ingramcontent.com/pod-product-compliance
Lightning Source LLC
LaVergne TN
LVHW080347011026
849161LV00009B/2217